Occurrence and Distribution of Selenium

Editor

Milan Ihnat
Land Resource Research Centre
Agriculture Canada
Ottawa, Ontario, Canada

CRC Press, Inc.
Boca Raton, Florida

Library of Congress Cataloging-in-Publication Data

Occurrence and distribution of selenium / editor, Milan Ihnat.
p. cm.
Bibliography: p.
Includes index.
ISBN 0-8493-4932-X
1.Selenium—Analysis. 2. Selenium—Environmental aspects.
I. Ihnat, M.
QD181.S5032 1989
363.1′79—dc19 88-21481
CIP

Direct all inquiries to CRC Press, Inc., 2000 Corporate Blvd., N.W., Boca Raton, Florida, 33431.

International Standard Book Number 0-8493-4932-X

Library of Congress Card Number 88-21481
Printed in the United States

PREFACE

The objective of this volume is to present a reasonably comprehensive and critical treatment of the occurrence and distribution of selenium in various components of the natural environment. It deals largely with a discussion and tabulation of selenium concentrations, with the intent being a presentation of archival value. Researchers in disciplines of biochemistry, nutrition, clinical science, geology, soil science, aquatic, marine, and atmospheric chemistry, and other environmental sciences should find tabulated here pertinent analytical concentration data.

An effort of this magnitude on diversified topics requires input from a number of individuals. The editor records his appreciation to authors of chapters who accepted the invitation to contribute and who took the in-depth critical approach required for this undertaking. Sincere thanks are due to the editor's wife Mrs. Irene Ihnat for assistance with some items on the seemingly endless list of activities constituting the task of editing. Apologies are due her and the two other peoples in the editor's life, Mark and Natasha, for the time taken away from them by this undertaking. Appreciation is due to Agriculture Canada for permission to carry out this work and to the departmental libraries and the library of the Canada Institute for Scientific and Technical Information, Ottawa, and staff therein for use of facilities. The entire manuscript was competently typed and manipulated on a word processor by Mrs. Karen Burns and Mrs. Kim Edmonds whose efforts are deeply appreciated.

At the outset, the goal was to cover comprehensively all natural commodities. As work progressed, it became evident that due to the usual restrictions encumbering an editor and authors of a multiauthored volume, this would not be achievable. As a result some omissions of subject matter are evident. It is felt, however, that what resulted is a reasonably comprehensive coverage of quantitative analytical concentration data within a good number of topics to be a source of archival information to researchers in a wide range of disciplines dealing with the natural environment.

Land Resource Research Centre, Agriculture Canada, contribution No. 88-51.

Milan Ihnat
Ottawa, Ontario
June 1987

THE EDITOR

Milan Ihnat, Ph.D., is a Senior Research Scientist in the Land Resource Research Centre, Agriculture Canada, Ottawa, Ontario.

Dr. Ihnat received a B.Sc. degree with first class honors in chemistry in 1962 and a Ph.D. in 1967 from McGill University, Montreal, Quebec. Following post doctorate work in Biophysical Chemistry at Columbia University, New York, he was appointed as Research Assistant in 1969 in the Faculty of Medicine and as an Assistant Professor in 1970 in the Department of Physiology, McGill University. He joined the Chemistry and Biology Research Institute, Agriculture Canada, Ottawa, in 1971 as a research scientist, becoming Head of the Analytical Chemistry Services in 1981. He transferred to the present center in 1986.

Dr. Ihnat is a member of the Association of Official Analytical Chemists, Spectroscopy Society of Canada, Chemical Institute of Canada, and the American Chemical Society. He has served in various capacities in the Spectroscopy Society of Canada, IUPAC, and the Association of Official Analytical Chemists including the Editorial Board, Interlaboratory Studies Committee, Eastern Ontario-Quebec Regional Section, and an Associate Referee for analytical methods development within the latter organization. In 1973, he received the Associate Referee of the Year Award from the AOAC and was appointed a Fellow in 1984.

His research interests include the development of quantitative analytical methodology for the determination of macro and trace elements in biological materials, analytical atomic spectroscopy, high-reliability analytical measurements, trace elements in soils and crops, and development of reference materials for chemical composition.

CONTRIBUTORS

Jan Aaseth, Ph.D.
Hedmark Central Hospital
Elverum, Norway

Michael L. Berrow, Ph.D.
Department of Spectrochemistry
The Macaulay Land Use Research Institute
Craigiebuckler, Aberdeen
Scotland

Shier S. Berman, Ph.D.
Division of Chemistry
National Research Council of Canada
Ottawa, Ontario

Gregory A. Cutter, Ph.D.
Department of Oceanography
Old Dominion University
Norfolk, Virginia

Robert A. Duce, Ph.D.
Center for Atmospheric Chemistry Studies
Graduate School of Oceanography
University of Rhode Island
Narragansett, Rhode Island

Richard C. Ewan, Ph.D.
Department of Animal Science
Iowa State University
Ames, Iowa

Milan Ihnat, Ph.D.
Land Resource Research Centre
Agriculture Canada
Ottawa, Ontario

Susan A. Lewis, Ph.D.
Hazleton Laboratories America, Inc.
Analytical Chemistry
Vienna, Virginia

Jerome O. Nriagu, Ph.D.
National Water Research Institute
Canada Centre for Inland Waters
Environment Canada
Burlington, Ontario

Byard W. Mosher, Ph.D.
Institute for the Study of Earth, Oceans and Space (EOS)
University of New Hampshire
Durham, New Hampshire

Anita Schubert, M.S.
Nutrient Composition Laboratory
Beltsville Human Nutrition Research Center
U.S. Department of Agriculture
Beltsville, Maryland

Dallas L. Rabenstein, Ph.D.
Department of Chemistry
University of California
Riverside, California

Khoon Sin Tan, Ph.D.
Department of Chemistry
University of California
Riverside, California

K. W. Michael Siu, Ph.D.
Division of Chemistry
National Research Council of Canada
Ottawa, Ontario

Yngvar Thomassen, Ph.D.
Institute of Occupational Health
Oslo, Norway

Allan M. Ure, Ph.D.
Department of Spectrochemistry
The Macaulay Land Use Research Institute
Craigiebuckler, Aberdeen
Scotland

Claude Veillon, Ph.D.
Vitamin and Mineral Nutrition Laboratory
Beltsville Human Nutrition Research Center
U.S. Department of Agriculture
Beltsville, Maryland

Wayne R. Wolf, Ph.D.
Nutrient Composition Laboratory
Beltsville Human Nutrition Research Center
U.S. Department of Agriculture
Beltsville, Maryland

TABLE OF CONTENTS

Chapter 1

INTRODUCTION

Milan Ihnat

The multifacetted physical and chemical characteristics of selenium have made it the target of investigation and application in numerous disciplines of human endeavor. Of particular significance is the toxicological and nutritional duality of the element. Toxic and deficient regions throughout the world and their impact on animal and human health and well being have spurred voluminous research. Virtually every research venture deals with the analytical determination of selenium concentrations in different natural commodities and systems and the drawing of conclusions based on such information. Although many major works in the form of books, reports, and review articles have appeared dealing with various aspects of selenium, there is no published comprehensive, single-source, compilation of selenium concentration data in natural materials. This volume deals with the occurrence and distribution of selenium in natural materials and systems with the thrust being detailed tabulation and discussion of concentration data. It deals with native or endogenous and adventitious, non-deliberately added selenium and does not treat results of experimental investigations in which selenium has been introduced purposely.

The book begins with a review of the analytical chemistry of the inorganic and organic forms of the element and a discussion of the status and reliability of analytical data together with approaches followed by contributors in extracting, managing, and presenting the information. The volume continues with in-depth treatment, with chapters arranged according to groupings of the different materials and systems. Chapters treat separately plants and agricultural materials, foods, animal tissues, human tissues, geological materials and soils, fresh water, the marine environment, and the atmosphere. Each chapter discusses the occurrence of selenium in and distribution among components of the specified system, includes compilation of concentration data based on comprehensive and critical assessment of the literature, and concludes with a comprehensive listing of the pertinent literature. The book concludes with a chapter on the biogeochemistry of selenium in natural systems. Throughout, consideration is given to both total selenium and its compounds as far as these have been reliably and sufficiently studied in the various environments, but emphasizes the selenium moiety most often measured and reported, namely, total selenium.

Chapter 2

INORGANIC ANALYTICAL CHEMISTRY OF SELENIUM

Susan A. Lewis and Claude Veillon

TABLE OF CONTENTS

I. INTRODUCTION

There has been an increasing interest in the determination of selenium, especially at trace concentrations, in a wide variety of matrices because of its toxicological and physiological importance. It is necessary to have methods to determine selenium in various forms over a wide range of concentrations from micrograms per kilogram (or liter) to purely elemental selenium. Rather than to present a comprehensive review of all the analytical methods used to determine selenium, the purpose of this chapter is to give an overview of various techniques with particular emphasis on those techniques used currently for both routine and research applications. The following techniques will be discussed in detail: fluorometry, neutron activation analysis, mass spectrometry, graphite furnace and hydride generation atomic absorption spectrometry, and gas chromatography methods.

Many literature reviews on selenium analytical methodologies in biological and environmental samples have been published. Earlier reviews are those by Watkinson, 1967,[1] Olson et al., 1973,[2] Alcino and Kowald, 1973,[3] Cooper, 1974,[4] Shendrikar, 1974,[5] and Olson, 1976.[6] Reviews since 1976 have tended to be limited to the determination of selenium in specific matrices[7-13] and by specific techniques.[12] These are by Crosby, in foods;[7] Florence[8] and Robberecht and Van Grieken,[9] in environmental waters; by Raptis et al.,[10] Robberecht and Deelstra,[11] and Fishbein,[12] in environmental and biological samples, especially at trace levels, by Tolg[13] in biological materials; and an atomic spectroscopy review by Verlinden et al.[14] Selenium is known to occur in several forms, both inorganic and organic, but until fairly recently only total concentrations of selenium have been reported. Inorganic speciation measurements are necessary, for example, for the study of toxicity and trace element transport in natural water.[8,9] However, there is a paucity of literature concerning the determination of inorganic species of selenium. Methods for the measurement of inorganic selenium will be treated and, wherever appropriate, the applicability of each analytical method for the determination of selenium species will be addressed. Types of materials discussed are soils, rocks, water, biological materials, and air.

II. QUALITY ASSURANCE

For any analytical method, a quality assurance program must be developed to help ensure the accuracy and precision of the results. All aspects of the analytical procedure, from sampling through method development and validation, must be covered. Thiers[15] in 1957 stated, ''unless the complete history of any sample is known with certainty, the analyst is well advised not to spend his time analyzing it.'' However, for many published methods, sample collection and manipulation are not well documented. Care must be exercised in the choice of collection and storage devices as adsorption and desorption can occur.[10] Pyrex® is reported to be a better container for selenium-containing solutions than plastic.[16,17] Sample treatment must be designed to minimize contamination and to avoid loss of the analyte — particularly important for selenium determinations because of the volatility of some selenium compounds. Standards and reagents must be chosen carefully, and judicious use of sample and reagent blanks is generally mandatory; contamination, however, is not generally a widespread problem in selenium determinations. Appropriate reference materials should be used to validate the method. Once a method has been established, the single most effective way of maintaining the quality of the results obtained is by regular use of well-characterized, suitable quality control materials which are treated in an identical manner to the samples. The analytical method must be capable of reproducing the ''correct'' answer with an acceptable level of precision, and this method precision can be obtained from the repeated analysis of quality control materials preferably expressed as ''day-to-day'' precision — a more realistic measure than ''within-run'' precision. Ideally, the quality control materials should have matrix and analyte concentrations similar to the samples, and more than one material with different levels of analyte should be used if available. Many published methods do not report the establishment of a quality assurance program or the use of reference and quality control materials and the monitoring of blanks.

III. SAMPLE PREPARATION

Most methods for the determination of selenium in various matrices require some sample preparation or pretreatment. As previously stated, the most common problem of selenium determinations is losses due to its volatile nature. Most sample preparations for biologicals involve the initial destruction of the organic constituents, followed by the conversion of the element to the tetravalent state and subsequent determination by a variety of techniques. In the 1973 and 1976 reviews, Olson[2,6] covers most of the earlier, commonly used sample preparation procedures. More recently, Raptis et al.[10] have tabulated most sample preparation techniques, together with a comprehensive list of references. Florence[8] and Robberecht and Van Grieken[9] discuss sample handling and preparation for water samples, and Chau et al.[18] and Jiang et al.[19,20] cover sample collection of air and breath for the determination of volatile selenium compounds. Reamer and Veillon[21] and Robberecht et al.[22] discuss losses of selenium from biological materials by the commonly used digestion procedures. May[23] and Yläranta[24] review losses of selenium by dry ashing. Nondestructive sampling techniques may be applicable for neutron activation analysis and X-ray fluorescence analysis. Specific sample preparations will be covered later in this chapter for the six techniques discussed in detail.

IV. METHODOLOGIES

A classification scheme for ranking the quality of a method has been proposed by the International Federation of Clinical Chemistry,[25] and this scheme can be applied to selenium methodologies. Definitive methods are those that represent the ultimate in quality, but are generally so sophisticated that they are not likely to be in common usage. Such methods are neutron activation analysis and mass spectrometry, with the latter employing isotope-dilution techniques, negating the use of an external standard.[26] Reference methods are those that have been tested against the definitive methods and/or represent the best method available for use under well-controlled routine conditions. Under these definitions the Association of Official Analytical Chemists (AOAC) fluorometric methods for plants[27] and foods[28] and the hydride generation atomic absorption spectrophotometric method for foods[29] can be considered as reference methods, as they have undergone the scrutiny of interlaboratory evaluation. These methods have been further modified for use with many sample types, but these modifications do not automatically confer reference status to these methods for all sample types. The next category of methods are those routinely used, but which may have known or unknown analytical biases or systematic errors depending on how well these methods have been studied and how much documentation exists in the literature. This category includes the newer rapid techniques such as electrothermal atomic absorption spectrometry, preferably with Zeeman® background correction, hydride generation/atomic absorption spectrometry, and gas chromatography coupled with various detectors such as atomic absorption spectrometers or electron capture detection systems. It is probable that with sufficient documentation of performance reliability that one or more of these methods could be elevated to the status of a reference method.

Many methods have been used to determine selenium, some of which are now virtually obsolete, and some, on the other hand, represent procedures which are quite new and experimental. Again, for more specific details, the reader is referred to the reviews and references therein cited earlier in this chapter.

Some of the earlier methods are no longer in use because they lack the necessary sensitivity and specificity. Gravimetric methods are only useful for samples with high selenium content. Volumetric methods for selenium determination are based on an oxidation-reduction reaction and are subject to interferences from other oxidants and reductants. Colorimetric and spectrophotometric methods have been largely replaced by the more sensitive fluorometric procedure. However, many spectrophotometric methods are still employed for the analysis of water.[9] Nephelometry and turbidimetry have been used in the past, again primarily for the analysis of water. Nephelometry, as an analytical technique for biological substances, is now being rediscovered, and with the advent of specific immunochemical reagents and rate-reaction, small-angle, light-scattering nephelometers, this technique may have potential for the determination of certain organic selenium species.

Electrochemical methods for selenium determinations have limited applications. They have been applied to the analysis of waters,[9,10] but generally these methods are not sufficiently sensitive and suffer from many interferences. Polarographic techniques, however, have been used for water[9] and biological materials.[30,31] Differential pulse stripping voltammetry has been used to determine selenium in various sample matrices after careful chemical manipulation.[13] Electrochemical techniques may have potential for the determination of selenium species.[8,9]

Several chromatographic techniques have been used to determine selenium[8-10,12] and some of its chemical species. These include gas chromatography, liquid chromatography, thin layer chromatography, ion-exchange chromatography, and high-performance liquid chromatography. These techniques are separatory methods for both total selenium and selenium species and are generally used in conjunction with suitable detectors such as atomic spectrometers[32-34] and mass spectrometers.[35] The full potential of high-performance liquid chromatography for selenium determinations has yet to be completely explored. Schwedt and Schwarz[32] reviewed this technique as a method for inorganic trace analysis. A book, also by Schwedt,[36] covers other chromatographic methods for inorganic analysis.

All types of atomic spectrometric techniques — absorption, emission, and fluorescence — have been used to determine selenium in a wide variety of matrices. Emission techniques, especially those using plasma sources, are generally coupled with some form of sample pretreatment and preconcentration because of comparatively poor detection limits for direct analysis (~60 μg/l in the solution presented to the instrument).[37,38] Hydride generation techniques have also been coupled to plasma atomic emission systems.[11] Plasma atomic fluorescence techniques have now become a viable alternative to emission and absorption measurements since the advent of the first commercially available system.[38] However, the detection limits (~100 μg/l) are not yet much better than in plasma emission. Flame atomic absorption spectrometry has been used to analyze plant material from seleniferous areas;[39] however, the detection limits are poor (in the range 50 to 3000 μg/l for a variety of flames[40]) for plants with normal levels of the element. Verlinden et al.[14] have reviewed atomic absorption techniques for selenium. With the advent of Zeeman®-effect background correction, graphite furnace (electrothermal) atomic absorption techniques are fast becoming a favored analytical method. Hydride generation techniques coupled with atomic absorption spectrometry have also been widely used, but present many analytical problems.

X-ray emission methods are probably not applicable to the determination of selenium at the concentrations found in many environmental and biological specimens. The detection limits (defined as twice the standard deviation of baseline noise) for X-ray fluorescence, X-ray emission, and proton-induced X-ray emission spectrometry are about 200, 200, and 60 μg/l, respectively.[9,11]

Several of the currently more commonly used methods are discussed below.

A. Neutron Activation Analysis

Because of high sensitivity, high specificity, multielement capability, and little sample pretreatment, neutron activation analysis (NAA) has been extensively used for the determination of selenium over a wide concentration range in many sample types, especially biological samples.[2] Several types of activation have been employed[2] with the most widely used being activation with thermal neutrons. NAA can be used with or without destruction of the organic matter in the sample. Nondestructive NAA based on the measurement of both ^{77m}Se and ^{75}Se has been used. Biological samples, however, generally contain high concentrations of sodium and chloride which can cause a relatively large Compton background from ^{38}Cl and ^{24}Na. Cornelis et al.[41] discuss this and other potential problems for the analysis of biological materials by NAA; generally some form of sample preparation and pretreatment is used to improve sensitivity and avoid interferences.

NAA has been used to determine selenium in various matrices such as soft tissue, eye lenses, blood, plasma, urine, feces, dental enamel, feeds, foods, milk, tobacco, fresh- and seawater, nails, hair, bone, and geological materials.[2,10] Absolute detection limits of 4 and 10 ng have been reported for geological and biological materials, respectively. Relative standard deviations (RSD) of less than 10% are routinely reported, which, with careful sample manipulation, can approach 1%.

NAA, with isotope dilution techniques,[26] could be considered a definitive method. There is at present great interest in the fields of nutrition and medicine in developing NAA methods.[42,43] Of particular interest is the use of NAA to monitor stable isotopes in human metabolic studies. Being nonradioactive, stable isotopes can be used as metabolic tracers in all subjects including at-risk groups such as pregnant females and infants where the use of radioisotopes is contraindicated. Extensive work has been carried out by Janghorbani and co-workers using NAA to measure stable isotopes in selenium metabolic studies in humans.[44-47]

NAA, however, has several limitations for these purposes. Access is required to a nuclear reactor and extensive counting facilities, the cost and time for analysis of samples is considerable, and not all available stable isotopes can be used.

B. Mass Spectrometry

Mass spectrometry (MS) using stable isotope dilution techniques (IDMS) can also be a definitive method. Spark-source mass spectrometry (SSMS) has been used by the U.S. National Bureau of Standards to determine selenium in reference materials.[26] SSMS has also been used for the analysis of solid samples for selenium.[2] This element, however, is not an ideal one for SSMS because of its volatility in a high-temperature spark, leading to low sensitivity and relatively poor precision and accuracy. Selenium is also not considered a good candidate for thermal ionization mass spectrometry (TIMS), the most precise and accurate of the current mass spectrometric techniques, because of its high ionization potential. Fast atom bombardment mass spectrometry (FABMS) apparently shows promise for the determination of selenium, but to date no data are available.[48]

Combined gas chromatography/mass spectrometry (GCMS) techniques have the advantage over other mass spectrometric techniques and NAA due to shorter analysis time and less expensive equipment, but still require expert personnel. Generally, the main limitation of GCMS is the availability of suitable chelating agents for the analyte of interest. For several elements of biological interest no suitable chelate has yet been found. However, in our laboratory we have developed a suitable chelate and method for the determination of selenium.[49] This method was further adapted by our laboratory to a double-isotope method for use in the analysis of samples from metabolic tracer studies.[50] This technique has been used to study intrinsically labeled chicken products,[51] for metabolic tracer studies in pregnant and nonpregnant women,[52] to study urinary excretion in New Zealand women,[53] and for a study investigating the pharmacokinetics of selenium.[54,55] This methodology is currently being used to study the pharmacokinetics of selenium (both as selenite and selenomethionine) in

32 subjects matched by sex and age and to study the effect of fasting and nonfasting states on selenium pharmacokinetics. Some preliminary data have been presented.[56]

One big advantage of the GCMS technique is that it can be used to follow the fate of selenium stable-isotope metabolic tags. This method has proven to be very adaptable to many types of sample matrices and has been used under routine conditions to determine total selenium in foods, diets, human breast milk, infant formula,[57] plasma and serum, red blood cells, feces, and urine. Our sample preparation procedure eliminates the use of perchloric acid and its concomitant problems, but is not very good for samples with high fat content.[58] We have also used this methodology to analyze serum reference materials for the U.S. National Bureau of Standards.[59,60] In our experience, this method has the accuracy expected for a definitive method, and in our hands over a $1^1/_2$ year period, we have found an RSD of about 2% at the 100-μg (Selenium per liter) level. The method has an absolute detection limit of about 50 pg.

C. Fluorometry

The AOAC official fluorometric methods (acid digestion fluorometry, ADF) for the determination of selenium in plants[27] and foods[28] are based on researches and interlaboratory studies conducted by Olson[61] and Ihnat,[62,63] respectively, following in-depth research of Watkinson.[64] It must be reiterated that these methods are only official for plants and foods and that application to other sample matrices does not confer reference or official status on them for the determination of selenium in any other matrix. However, fluorometry has been used for a wide variety of sample types. The method utilizes the fluorescent complex produced by the reaction of selenium in the tetravalent state with 2,3-diaminonaphthalene (DAN). Precise instructions for apparatus, reagents, and procedures are given by the AOAC. The digestion procedure is labor intensive and nitric-perchloric or nitric-perchloric-sulfuric acid mixtures are used for the destruction of organic matter. Reamer and Veillon[58] have developed a procedure for the digestion of biological fluids, eliminating the use of perchloric acid, and using instead a nitric-phosphoric-peroxide mixture. The fluorometric assay has also been automated.[65]

The fluorometric method has been used for the determination of selenium in foods, milk, feeds, plants, blood, serum, urine, tissue (hard and soft), feces, fresh- and seawater, sediments, soil, incinerator waste, metals, semiconductor materials, and ores.[2,10,66] The method has reported RSDs ranging from 2 to 100%[62,63,67] and a consensus of opinion that above a concentration of 100 μg/kg the RSD should be better than 8%. Detection limits of 10 to 100 ng have been reported.[10,62,63,68]

In proper hands, with close attention to detail and the use of a good quality assurance program, the fluorometric method for selenium determination is probably the simplest, the least expensive, and the most versatile of all the methods under discussion. However, in many cases this method has been poorly documented and has also been used inappropriately to measure very low levels of selenium, without full realization of its limitations with regard to precision and detection limits.

D. Gas Chromatography

The determination of selenium by gas chromatography (GC) is based usually on the detection of a volatile piazselenol formed by the reaction of Se(IV), after destruction of the organic matrix, with an aromatic diamine such as 4-nitro-*o*-phenylenediamine. Various detectors[10] have been used with this system including those based on flame photometry, thermal conductivity, atomic absorption spectrometry, mass spectrometry (see mass spectrometry method) and electron capture.[68,70] With electron capture detectors (ECD), the absolute detection limit can be as low as 1 pg. GC RSDs are reported to be from 2 to 4%.[70]

This method has been adapted to the determination of selenium in the following matrices:

water, soil, sludge, blood and blood products, grain, fish, milk, hair, urine, and soft tissues.[10,69,70] The method has also been used to detect selenium species in biological materials and in waters.[9,70]

E. Atomic Absorption Spectrometry

The development of atomic absorption spectrometry (AAS) in 1955 by Walsh[71] has brought about a revolution in the field of trace element analysis. As previously stated, flame AAS can only be used for samples containing high concentrations of selenium as the detection limit varies from 50 to 3000 μg/l. However, approaches based on generation of volatile hydrogen selenide prior to introduction to an atomic absorption spectrometer, and electrothermal atomic absorption spectrometry (especially with graphite furnace atomization) can be used to determine selenium at much lower concentrations. The use of these two techniques has increased dramatically in the last 2 decades and accounts for the majority of recent selenium determinations.

Hydride generation atomic absorption spectrometry (HAAS) offers the advantage of good sensitivity and relatively simple instrumentation for the determination of selenium in a wide variety of matrices. In HAAS, sodium tetraborohydride is added to an acidic solution of the sample containing selenium in the tetravalent state forming gaseous H_2Se which is then stripped from solution with a gas and atomized, generally in a heated quartz cell. A method for foods based on acid digestion-HAAS developed by Ihnat and Miller[72] was subjected to an intensive interlaboratory investigation under the auspices of the AOAC.[73,74] Although accuracy averaged over all laboratories was excellent, variation of results among laboratories, reflecting individual laboratory biases, was deemed unacceptable. Determination precision (hydride generation measurement steps) was also observed to be large and unacceptable and related to the type of hydride generator used. The authors concluded (for both arsenic and selenium) that, "on the basis of information from 28 laboratories using 16 different types of hydride generation apparatus, conducting 700 complete analyses and 2400 analyte quantitations on reagent blanks, three model solutions and 13 different reference food samples with levels of arsenic and selenium ranging from 0 to 15,000 and 0 to 8000 ng/g, respectively, . . . the method suffers from both systematic error and imprecision, and hence recommend that the method not be adopted."[74] Subsequent work by Holak[75] on a closed-system nitric acid digestion procedure together with use of a specified simple hydride generator led to an official AOAC method for foods.

There are reports that the accuracy of this method is very dependent on the decomposition technique used. Welz and co-workers[76,77] have investigated the problems of sample digestion and have shown that provided adequate care is taken to ensure complete sample digestion and conversion of the selenium to selenite, this technique can be accurate. Siemer and Koteel[78] have investigated different HAAS techniques and mention methods of optimizing systems to obtain maximum sensitivity. Reamer and co-workers[79] used radiotracers to evaluate losses in different hydride generation systems.

HAAS has been used to determine selenium in waters, soil, ores, blood and blood products, urine, environmental materials, foods, feeds, glass, metallurgical materials, and petroleum.[10-14] It is reported not to be a good technique for the analysis of all biological samples for selenium.[10]

The technique has been shown to have absolute detection limits ranging from 0.1 to 60 ng.[14] Precision and accuracy vary, according to instrumentation used and the care and expertise of the user, but have been reported to be from 1 to 5% RSD and 70 to 110%, respectively.

For most analysts, electrothermal atomic absorption spectrometry (EAAS), especially graphite furnace atomization atomic absorption spectrometry (GFAAS), is the most appropriate technique to analyze samples for small amounts of selenium. Theoretically, this technique can be used for the rapid, direct determination of selenium in large batches of routine samples. However, GFAAS is not simple nor free from interferences — both spectral and matrix — and/or losses due to the volatility of selenium. The use of pyrolytically coated or total pyrolytic graphite tubes, a L'vov® platform, peak area integration, matrix modification and the use of Zeeman® background techniques have enabled some advances to be made. Spectral interferences from iron and/or phosphate are compensated for by Zeeman® background correction. The use of matrix modification — generally nitrates of nickel, copper, palladium, platinum, silver, molybdenum, and/or magnesium — help to thermally stabilize selenium and allow for ashing temperatures of up to 1200°C.[10,14,80-82] However, only a few sample types are suitable for direct analyses, even with the use of matrix modification. Blood, serum, semen, yeast, food supplements, and waters have been analyzed directly.[10,14,80] For the great majority of sample types, some form of sample pretreatment and/or digestion is necessary. Many commonly used acid mixtures are reported to cause matrix interferences in the graphite furnace.[83] Some of the pretreatments involve elaborate sample preparation and the formation of selenium chelates and their subsequent injection into the graphite tubes. This again raises the question of quantitative sample recoveries and sample contamination, but adventitious contamination with selenium is rare.

GFAAS, after some form of sample preparation, has been used for the analysis of water, foods, urine, milk, and geological and metallurgical samples.[10,14,80] GFAAS has been used as a detector for the determination of selenium species in urine and water after sample treatment and chromatographic separation.[8,9,11] For samples that can be successfully analyzed by GFAAS, the technique offers very good detection limits (20 to 50 pg) and imprecisions that range from 2 to 10% RSD. Many types of semiautomated instruments are commercially available.

V. SUMMARY

The greatest activity in selenium determinations at the moment appears to be with regard to biological materials. Definitive methods like NAA and IDMS have been used to verify other working methods, to establish selenium levels in reference materials, and, more importantly, to investigate selenium metabolism in humans using stable isotope tracer methodology. Considerable effort has been devoted to sample preparation methods, particularly with regard to preventing loss and/or nonquantitative recovery of the analyte. The use of radiotracers like ^{75}Se has helped a great deal in this regard.

Some of the newer reference materials now available will go a long way toward ensuring accurate determinations, permitting method validation, and allowing performance evaluation in interlaboratory trials.

Of the more generally available and widely used working methods for selenium determinations, the fluorometric method remains the method of choice for many types of sample matrices. However, important gains in recent years by other methods with potentially greater speed, convenience, and specificity have been made. While this potential has not yet been realized in all cases, much of the current developmental activity is in these areas.

These newer methods include gas chromatography, with both conventional and atomic spectroscopic detection systems, and hydride generation and direct atomization atomic absorption measurements. The GC methods show promise because of the relatively simple instrumentation needed and the fact that the analyte is separated from the matrix. The advent of Zeeman® background correction systems for atomic absorption spectrometers has greatly facilitated selenium determinations, particularly in biological matrices where iron and phosphorus are also present. Matrix modifiers, such as nickel, which stabilize or reduce the volatility of selenium show considerable promise for direct determinations of selenium, with little or no sample preparation. Finally, most of the problems reported in the literature regarding selenium determinations by hydride generation and AAS appear to be associated with sample preparation procedures. Now recognized, these procedures can be standardized, and the use of this relatively rapid and simple method should gain wider acceptance.

Selenium is somewhat unique among trace elements in that several good, independent analytical methods exist, and contamination is much less of a problem than with other, more ubiquitous elements. This, coupled with recent developments and activity, should greatly increase our knowledge regarding this important and interesting element in the years to come.

REFERENCES

1. **Watkinson, J. H.,** Analytical methods for selenium in biological material, in *Selenium in Biomedicine,* Muth, O. H., Oldfield, J. E., and Weswig, P. H., Eds., AVI Publishing, Westport, CT, 1967, 97.
2. **Olson, O. E., Palmer, I. S., and Whitehead, E. I.,** Determination of selenium in biological materials, in *Methods of Biochemical Analysis,* Glick, D., Ed.; Vol. 21, John Wiley & Sons, New York, 1973, 39.
3. **Alcino, J. F. and Kowald, J. A.,** Analytical methods, in *Organic Selenium Compounds: Their Chemistry and Biology,* Klayman, D. L. and Gunther, W. H. H., Eds., John Wiley & Sons, New York, 1973, chap. 17.
4. **Cooper, W. C.,** Analytical chemistry of selenium, in *Selenium,* Zingaro, R. A. and Cooper, W. C., Eds., Van Nostrand Reinhold, New York, 1974, chap. 10.
5. **Shendrikar, A. D.,** Critical evalutation of analytical methods for the determination of selenium in air, water, and biological materials, *Sci. Total Environ.,* 3, 155, 1974.
6. **Olson, O. E.,** Methods of analysis for selenium, a review, in *Proc. of the Symp. on Selenium — Tellurium in the Environment,* University of Notre Dame, Notre Dame, IN, 1976.
7. **Crosby, N. T.,** Determination of metals in foods, a review, *Analyst,* 102, 225, 1977.
8. **Florence, T. M.,** Speciation of trace elements in waters, *Talanta,* 29, 345, 1982.
9. **Robberecht, H. and Van Grieken, R.,** Selenium in environmental waters; determination, speciation and concentration levels, *Talanta,* 29, 823, 1982.
10. **Raptis, S. E., Kaiser, G., and Tölg, G.,** A survey of selenium in the environment and a critical review of its determination at trace levels, *Fresenius Z. Anal. Chem.,* 316, 105, 1983.
11. **Robberecht, H. J. and Deelstra, H. A.,** Selenium in human urine: determination, speciation and concentration levels, *Talanta,* 31, 497, 1984.
12. **Fishbein, L.,** Overview of analysis of carcinogenic and/or mutagenic metals in biological and environmental samples. I. Arsenic, beryllium, cadmium, chromium and selenium, *Int. J. Environ. Anal. Chem.,* 17, 113, 1984.
13. **Tolg, G.,** Selenium-analysis in biological materials, in *Trace Element Analytical Chemistry in Medicine and Biology,* Vol. 3, Brätter, P. and Schramel, P., Eds., Walter de Gruyter, Berlin, 1984, 95.
14. **Verlinden, M., Deelstra, H., and Adriaenssens, E.,** The determination of selenium by atomic absorption spectrometry, a review, *Talanta,* 28, 637, 1981.
15. **Thiers, R. E.,** Contamination in trace element analysis and its control, in *Methods of Biochemical Analysis,* Vol. 5, Glick, D., Ed., Interscience, New York, 1957, 273.
16. **Shendrikar, A. D. and West, P. W.,** The rate of loss of selenium from aqueous solution stored in various containers, *Anal. Chim. Acta,* 74, 189, 1975.
17. **Moody, J. R. and Lindstrom, R. M.,** Selection and cleaning of plastic containers for storage of trace element samples, *Anal. Chem.,* 49, 2264, 1977.
18. **Chau, Y. K., Wong, P. T. S., and Goulden, P. D.,** Gas chromatography-atomic absorption method for the determination of dimethyl selenide and dimethyl diselenide, *Anal. Chem.,* 47, 2279, 1975.
19. **Jiang, S., de Jonghe, W., and Adams, F.,** Determination of alkylselenide compounds in air by gas chromatography-atomic absorption spectrometry, *Anal. Chim. Acta,* 136, 183, 1982.
20. **Jiang, S., Robberecht, H., and Vanden Berghe, D.,** Elimination of selenium compounds by mice through formation of different volatile selenides, *Experientia,* 39, 293, 1983.
21. **Reamer, D. C. and Veillon, C.,** Preparation of biological materials for determination of selenium by hydride generation-atomic absorption spectrometry, *Anal. Chem.,* 53, 1192, 1981.
22. **Robberecht, H. J., Van Grieken, R. E. Van Den Bosch, P. A., Deelstra, H., and Vanden Berghe, D.,** Losses of metabolically incorporated selenium in common digestion procedures for biological materials, *Talanta,* 29, 1025, 1982.

23. **May, T. W.,** Recovery of endogenous selenium from fish tissues by open system dry ashing, *J. Assoc. Off. Anal. Chem.,* 65, 1140, 1982.
24. **Yläranta, T.,** Loss of selenium from plant material during drying, storage and dry ashing, *Ann. Agric. Fenn.,* 21, 84, 1982.
25. **Saris, N.-E., Ed., Buttner, J., Borth, R., Boutwell, J. H., and Broughton, P. M. G.,** Provisional recommendation on quality control in clinical chemistry, *Clin. Chem.,* 22, 532, 1976.
26. **Veillon, C. and Alvarez, R.,** Determination of trace metals in biological materials by stable isotope dilution in *Metal Ions in Biological Systems,* Vol. 16, Sigel, H., Ed., Marcel Dekker, New York, 1983, 103.
27. **Williams, S., Ed.,** Selenium in plants, fluorometric method, final action, in *Official Methods of Analysis of the Association of Official Analytical Chemists,* 14th ed., Association of Official Analytical Chemists, Arlington, VA, 1984, Sect. 3.102 to 3.107.
28. **Williams, S., Ed.,** Selenium in food, fluorometric method, final action, in *Official Methods of Analysis of the Association of Official Analytical Chemists,* 14th ed., Association of Official Analytical Chemists, Arlington, VA, 1984, Sect. 25.154 to 25.157.
29. **Williams, S., Ed.,** Arsenic, cadmium, lead, selenium and zinc in food, multielement method, first action, in *Official Methods of Analysis of the Association of Official Analytical Chemists,* 14th ed., Association of Official Analytical Chemists, Arlington, VA, 1984, Sect. 25.001 to 25.007.
30. **Christian, G. D., Knoblock, E. C., and Purdy, W. C.,** Polarographic determination of selenium in biological materials, *J. Assoc. Off. Anal. Chem.,* 48, 877, 1965.
31. **Holak, W.,** Determination of arsenic and selenium in foods by electroanalytical techniques, *J. Assoc. Off. Anal. Chem.,* 59, 650, 1976.
32. **Schwedt, G. and Schwarz, A.,** Zur Anwendung der hochdruck-flussigkeits-Chromatographie in der anorganischen Analyse. III. Bestimmung von Selen in Trink-Oberflachen-und Abwasser, *J. Chromatogr.,* 160, 309, 1978.
33. **McCarthy, J. P., Caruso, J. A., and Fricke, F. L.,** Speciation of arsenic and selenium via anion-exchange HPLC with sequential plasma emission detection, *J. Chromatogr. Sci.,* 21, 389, 1983.
34. **Chakraborti, D., Hillman, D. C. J., Irgolic, K. J., and Zingaro, R. A.,** Hitachi Zeeman graphite furnace atomic absorption spectrometer as a selenium-specific detector for ion chromatography, separation and determination of selenite and selenate, *J. Chromatogr.,* 249, 81, 1982.
35. **Mawhinney, T. P.,** Separation and analysis of sulfate, phosphate and other oxyanions as their *tert.* butyldimethylsilyl derivatives by gas-liquid chromatography and mass spectrometry, *J. Chromatogr.,* 257, 37, 1983.
36. **Schwedt, G.,** *Chromatographic Methods in Inorganic Analysis,* Dr. Alfred Hüthig Verlag, Heidelberg, 1981.
37. **Parsons, M. L., Major, S., and Forster, A. R.,** Trace element determination by atomic spectroscopic methods — state of the art, *Appl. Spectrosc.,* 37, 411, 1983.
38. Operation Manual, Plasma/AFS, Baird Corporation, Bedford, MA, 1981.
39. **Olson, O. E., Emerick, R. J., and Palmer, I. S.,** Measurement of selenium in plants of high selenium content by flame atomic absorption analysis, *At. Spectrosc.,* 4, 55, 1983.
40. **Ihnat, M.,** Selenium in foods: evaluation of atomic absorption spectrometric techniques involving hydrogen selenide generation and carbon furnace atomization, *J. Assoc. Off. Anal. Chem.,* 59, 911, 1976.
41. **Cornelis, R., Hoste, J., and Versieck, J.,** Potential interferences inherent in neutron activation analysis of trace elements in biological materials, *Talanta,* 29, 1029, 1982.
42. **Morris, J. S., McKnown, D. M., Anderson, H. D., May, M., Primm, P., Cordts, M., Gebhardt, D., Crowson, S., and Spate, V.,** The determination of selenium in samples having medical and nutritional interest using a fast instrumental neutron activation analysis procedure in *Selenium in Biology and Medicine,* Spallholz, J. E., Martin, J. L., and Ganther, H. E., Eds., AVI Publishing, Westport, CT, 1981, 438.
43. **Janghorbani, M., Ting, B. T. G., and Young, V. R.,** Use of stable isotopes of selenium in human metabolic studies: development of analytical methodology, *Am. J. Clin. Nutr.,* 34, 2816, 1981.
44. **Sirichakwal, P. P., Young, V. R., and Janghorbani, M.,** Absorption and retention of selenium from intrinsically labelled egg and selenite as determined by stable isotope studies in humans, *Am. J. Clin. Nutr.,* 41, 264, 1985.
45. **Janghorbani, M., Christensen, M. J., Nahapetian, A., and Young, V. R.,** Selenium metabolism in healthy adults: quantitative aspects using the stable isotope $^{74}Se\ O_3^{2-}$, *Am. J. Clin. Nutr.,* 35, 647, 1982.
46. **Christensen, M. J., Janghorbani, M., Steinke, F. H., Istfan, N., and Young, V. R.,** Simultaneous determination of absorption of selenium from poultry meat and selenite in young men: application of a triple stable isotope method, *Br. J. Nutr.,* 50, 43, 1983.
47. **Kasper, L. J., Young, V. R., and Janghorbani, M.,** Short-term dietary selenium restriction in young adults: quantitative studies with the stable isotope $^{74}SeO_3^{2-}$, *Br. J. Nutr.,* 52, 443, 1984.
48. **Smith, D.,** Personal communication, 1985.
49. **Reamer, D. C. and Veillon, C.,** Determination of selenium in biological materials by stable isotope dilution gas chromatography-mass spectrometry, *Anal. Chem.,* 53, 2166, 1981.

50. **Reamer, D. C. and Veillon, C.,** A double isotope dilution method for using stable selenium isotopes in metabolic trace studies: analysis by gas chromatography/mass spectrometry (GC/MS), *J. Nutr.*, 113, 786, 1983.
51. **Swanson, C. A., Reamer, D. C., Veillon, C., and Levander, O. A.,** Intrinsic labelling of chicken products with a stable isotope of selenium (^{76}Se), *J. Nutr.*, 113, 793, 1983.
52. **Swanson, C. A., Reamer, D. C., Veillon, C., King, J. C., and Levander, O. A.,** Quantitative and qualitative aspects of selenium utilization in pregnant and nonpregnant women: an application of stable isotope methodology, *Am. J. Clin. Nutr.*, 38, 169, 1983.
53. **Veillon, C., Edmonds, L. J., Robinson, M. F., Thomson, C. D., Morris, V. C., and Levander, O. A.,** Urinary excretion of selenium stable isotope tracers by New Zealand women following supplementation, in *Proc. 5th Int. Symp. on Trace Elements in Man and Animals,* TEMA 5, Mills, C. F., Bremner, I., and Chesters, J. K., Eds., Commonweatlh Agricultural Bureaux, Farnham Royal, U.K., 1985, 495.
54. **Veillon, C., Lewis, S. A., Levander, O. A., McAdam, P. A., Hardison, N., and Wood, L.,** Considerations in using stable isotopes of selenium in metabolic tracer studies, Abstr. Pittsburgh Conf. on Analytical Chemistry and Applied Spectroscopy, New Orleans, 1985, Abstr. No. 015.
55. **Lewis, S. A., McAdam, P. A., Veillon, C., Helzlsouer, K., Patterson, B., and Levander, O. A.,** Measurement of urinary natural and stable selenium (Se) isotopes by isotope dilution mass spectrometry, Abstr. Fed. of American Soc. for Experimental Biology, Annaheim, CA, 1985, Abstr. No. 7369.
56. **McAdam, P. A., Lewis, S. A., Helzlsouer, K., Veillon, C., Patterson, B., and Levander, O. A.,** Absorption of selenite and L-selenomethionine in healthy young men using a ^{74}selenium (^{74}Se) tracer, Abstr. Fed. of American Soc. for Experimental Biology, Annaheim, CA, 1985, Abstr. No. 7368.
57. **Lewis, S. A., Patterson, K. Y., Hardison, N. W., Woolson, J., and Veillon, C.,** The analysis of infant formulae for total selenium (Se) by isotope dilution mass spectrometry (IDMS), in *Production, Regulation and Analysis of Infant Formula: A Topical Conference,* Association of Official Analytical Chemists, Arlington, VA, 1985, 227.
58. **Reamer, D. C. and Veillon, C.,** Elimination of perchloric acid in digestion of biological fluids for fluorometric determination of selenium, *Anal. Chem.*, 55, 1605, 1983.
59. **Veillon, C., Lewis, S. A., Patterson, K. Y., Wolf, W. R., Harnly, J. M., Versieck, J., Vanballenberghe, L., Cornelis, R., and O'Haver, T. C.,** Characterization of a bovine serum reference material for major, minor and trace elements, *Anal. Chem.*, 57, 2106, 1985.
60. **Lewis, S. A., Hardison, N. W., Patterson, K. Y., Woolson, J., and Veillon, C.,** Determination of total and stable isotopes of selenium in biological samples by gas chromatography/mass spectrometry, presented at NCI/NBS Workshop, Gaithersburg, MD, 1985, personal communication.
61. **Olson, O. E.,** Fluorometric analysis of selenium in plants, *J. Assoc. Off. Anal. Chem.*, 52, 627, 1969.
62. **Ihnat, M.,** Fluorometric determination of selenium in foods, *J. Assoc. Off. Anal. Chem.*, 57, 368, 1974.
63. **Ihnat, M.,** Collaborative study of a fluorometric method for determining selenium in foods, *J. Assoc. Off. Anal. Chem.*, 57, 373, 1974.
64. **Watkinson, J. H.,** Fluorometric determination of selenium in biological material with 2,3-diaminonaphthalene, *Anal. Chem.*, 38, 92, 1966.
65. **Brown, M. W. and Watkinson, J. H.,** An automated fluorimetric method for the determination of nanogram quantities of selenium, *Anal. Chim. Acta,* 89, 29, 1977.
66. **Analytical Reviews,** GK Turner Associates, Mountain View, CA, 1972.
67. **Ewan, R. C., Baumann, C. A., and Pope, A. L.,** Determination of selenium in biological materials, *J. Agric. Food Chem.*, 76, 212, 1968.
68. Analytical Methods Committee, Determination of small amounts of selenium in organic matter, *Analyst,* 104, 778, 1979.
69. **McCarthy, T. P., Brodie, B., Milner, J. A., and Bevill, R. F.,** Improved method for selenium determination in biological samples by gas chromatography, *J. Chromatogr.*, 225, 9, 1981.
70. **Cappon, C. J. and Smith, J. C.,** Determination of selenium in biological materials by gas chromatography, *J. Anal. Toxicol.*, 2, 114, 1978.
71. **Walsh, A.,** The application of atomic absorption spectra to chemical analysis, *Spectrochim. Acta,* 7, 108, 1955.
72. **Ihnat, M. and Miller, H. J.,** Analysis of foods for arsenic and selenium by acid digestion, hydride evolution atomic absorption spectrophotometry, *J. Assoc. Off. Anal. Chem.*, 60, 813, 1977.
73. **Ihnat, M. and Miller, H. J.,** Acid digestion, hydride evolution atomic absorption spectrophotometric method for determining arsenic and selenium in foods: collaborative study. I, *J. Assoc. Off. Anal. Chem.*, 60, 1414, 1977.
74. **Ihnat, M. and Thompson, B. K.,** Acid digestion, hydride evolution atomic absorption spectrophotometric method for determining arsenic and selenium in foods: collaborative study. II. Assessment of collaborative study, *J. Assoc. Off. Anal. Chem.*, 63, 814, 1980.
75. **Holak, W.,** Analysis of foods for lead, cadmium, copper, zinc, arsenic, and selenium, using closed system sample digestion: collaborative study, *J. Assoc. Off. Anal. Chem.*, 63, 485, 1980.

76. **Welz, B., Melcher, M., and Schlemmer, G.,** Accuracy of the selenium determination in human body fluids using atomic absorption spectrometry, in *Trace Element Analytical Chemistry in Medicine and Biology*, Vol. 3, Brätter, P. and Schramel, P., Eds., Walter de Gruyter, Berlin, 1984, 207.
77. **Welz, B., Melcher, M., and Nève, J.,** Determination of selenium in human body fluids by hydride-generation atomic absorption spectrometry-optimization of sample decomposition, *Anal. Chim. Acta,* 165, 131, 1984.
78. **Siemer, D. D. and Koteel, P.,** Comparisons of methods of hydride generation atomic absorption spectrometric arsenic and selenium determination, *Anal. Chem.*, 49, 1096, 1977.
79. **Reamer, D. C., Veillon, C., and Tokousbalides, P. T.,** Radiotracer techniques for evaluation of selenium hydride generation systems, *Anal. Chem.*, 53, 245, 1981.
80. **Slavin, W., Carnrick, G. R., Manning, D. C., and Pruszkowska, E.,** Recent experiences with the stabilized temperature platform furnace and Zeeman background correction, *At. Spectrosc.*, 4, 69, 1983.
81. **Ediger, R. D.,** Atomic absorption analysis with graphite furnace using matrix modification, *At. Absorpt. Newsl.*, 14, 127, 1975.
82. **Fernandez, F. J., Myers, S. A., and Slavin, W.,** Background correction in atomic absorption utilizing the Zeeman effect, *Anal. Chem.*, 52, 741, 1980.
83. **Slavin, W. and Manning, D. C.,** Graphite furnace interferences, a guide to the literature, *Prog. Anal. At. Spectrosc.*, 5, 243, 1982.

Chapter 3

ANALYTICAL CHEMISTRY OF ORGANIC AND BIOCHEMICAL SELENIUM

Khoon Sin Tan and Dallas L. Rabenstein

TABLE OF CONTENTS

I. INTRODUCTION

The majority of the reported methods for the determination of selenium in organic and biological matrices involve destruction of the organic material by a digestion procedure which at the same time converts the selenium to Se(IV). As described in Chapter 2, Se(IV) is then determined by one of a variety of methods, including fluorometry, atomic absorption, mass spectrometry, and gas chromatography, after conversion to a suitable derivative.[1-9] These methods give the total selenium content of the sample rather than the concentrations of specific selenium-containing compounds and will not be discussed further here since they are considered in the preceding chapter.

The purpose of this chapter is to discuss methods for the determination of specific organic and biochemically important compounds which contain selenium. Despite the widespread distribution of selenium and its importance both as an essential and a toxic element, relatively few methods have been reported for the speciation analysis of organoselenium compounds; it is hoped that this review will provide the impetus for additional research in this area. As methods which have been reported for the determination of organic and biochemical forms of selenium have generally been compound specific, rather than general methods for all organic and biochemical compounds of selenium, the approach taken in this chapter is to review the analytical chemistry of specific selenium-containing organic and biochemical compounds.

II. DETERMINATION OF BIOLOGICAL FORMS OF SELENIUM

The residue of selenocysteine (**I**), which is found in glutathione peroxidase and several other selenium-containing proteins, is the predominant form of selenium in biological tissues.[10-12]

$$H\text{–}Se\text{–}CH_2\text{–}\underset{\displaystyle NH_2}{\underset{|}{CH}}\text{–}COOH$$

(I)

Selenomethionine (**II**) is found in relatively large amounts in bacterial thiolase;[13]

$$CH_3\text{–}Se\text{–}CH_2\text{–}CH_2\text{–}\underset{\displaystyle NH_2}{\underset{|}{CH}}\text{–}COOH$$

(II)

however, its abundance and significance in animal tissues is not known. Low-molecular-weight selenium-containing biological molecules include selenotrisulfides,[14] selenocystine,[15,16] dimethylselenide,[17] trimethylselenonium salts,[18] and acid volatile selenium.[19]

A. Selenoproteins

Although several selenoproteins have been identified, glutathione peroxidase is the only mammalian selenoprotein for which specific assays of activity have been developed. These are described in the following section. Other selenoproteins can be assayed by measuring the selenium content of separated protein fractions[20] or, for selenocysteine-containing proteins, by determining the concentration of selenocysteine in protein hydrolysates or enzymatic digests. At least four selenium-dependent enzymes contain selenium in the form of selenocysteine residues: glutathione peroxidase, which is found in mammals and birds, glycine reductase, certain formate dehydrogenases, and a hydrogenase, which are of bacterial origin. The occurrence of selenocysteine in proteins and methods for its characterization in proteins have been reviewed recently.[20-22]

Selenocysteine in proteins is highly acid labile, and the unprotected selenoamino acid is susceptible to spontaneous oxidation and decomposition, with elimination of selenium. Therefore, conversion of selenocysteine to an *Se*-alkyl derivative is required before hydrolysis or digestion of the protein to its constituent amino acids. One procedure involves borohydride reduction of the selenocysteine residue to the selenol (RSeH) form which is then alkylated by reaction with iodoacetamide or iodoacetic acid.[21] Alternatively, the *Se*-carboxyethyl derivative can be prepared using 3-bromopropionate or the *Se*-aminoethyl derivative using ethylenimine.[23] The derivatized seleno protein is then subjected to acid hydrolysis or enzymatic digestion, followed by determination of the selenocysteine derivative by chromatography using an amino acid analyzer. The selenocysteine alkyl ethers are eluted from the column a few minutes after the corresponding sulfur analogues.

Other techniques for the identification of selenocysteine in proteins are high-performance liquid chromatography (HPLC) and mass spectrometry (MS) of dinitrobenzene derivatives.[22] To illustrate, the procedure described for the identification of selenocysteine in glutathione peroxidase involves derivatization of the borohydride-reduced enzyme by reaction with *1*-fluoro-2,4-dinitrobenzene to form the *Se*-dinitrobenzene derivative. Following dialysis and hydrolysis, **III** was separated from the mixtures by reverse-phase HPLC,

$$\text{DNP–Se–CH}_2\text{–}\underset{\displaystyle\text{NH}_2}{\underset{|}{\text{CH}}}\text{–CO}_2\text{H}$$

(III)

converted to **IV** by a Smiles rearrangment in the presence of methyl iodide,

$$\text{CH}_3\text{–Se–CH}_2\text{–}\underset{\displaystyle\text{NH–DNP}}{\underset{|}{\text{CH}}}\text{–CO}_2\text{H}$$

(IV)

then the free carboxylate group was esterified to give the more volatile derivative **V** for mass spectrometric measurement.

$$\text{CH}_3\text{–Se–CH}_2\text{–}\underset{\displaystyle\text{NH–DNP}}{\underset{|}{\text{CH}}}\text{–CO}_2\text{CH}_3$$

(V)

The reagents and conditions used for the formation of these derivatives are described in the literature.[22,24]

The dinitrobenzene derivative of selenocysteine has a high extinction coefficient (λ_{max} = 335nm, $\xi_{335nm} = 1.42 \times 10^{-4}\ M^{-1}\ \text{cm}^{-1}$) and can therefore be easily identified after thin layer chromatographic (TLC) and HPLC separations using spectrophotometric measurement in the near-UV spectral region.[22] The sensitivity of the technique can be further enhanced by using labeled (e.g., ^{75}Se) compounds.

Tappel et al.[20] have proposed a method for the quantitative determination of selenocysteine-containing proteins, including glutathione peroxidase, which is based on the *in vivo* labeling of selenoproteins with ^{75}Se by the addition of [^{75}Se] selenious acid to drinking water. The ^{75}Se is incorporated biosynthetically, and then the [^{75}Se] carboxymethylselenocysteine derivative is formed by addition of sodium iodoacetate to the ^{75}Se-labeled homogenate or protein fraction, followed by acid hydrolysis. A known amount of [^{3}H] carboxymethylselenocysteine is added as a marker to determine recovery. The hydrolysate is applied to an amino acid analyzer column, and the ^{3}H and ^{75}Se activities are determined in fractions collected. The total ^{75}Se as selenocysteine in the sample is calculated by dividing the measured ^{75}Se radioactivity by the fractional recovery of the marker, as indicated by the ^{3}H radioactivity.

Advantage can also be taken of the absorbance of the selenol (–SeH) group at physiological pH. Because the pK_a of the selenol group of selenocysteine and selenocysteine residues in proteins is 5 to 6, the selenol group will be anionic (RSe^-) at physiological pH and amenable to spectrophotometric assay. This approach was used to determine the selenol content of the selenoprotein component of the glycine reductase complex.[25] The increase in absorbance at 243 nm following reduction of the selenium in the protein to the selenol form by reaction with KBH_4 at pH 7 to 8 was equivalent to the known selenocysteine content of 1 mol/mol of protein.

Selenomethionine-containing proteins can be determined by chromatography of the protein hydrolysate on an amino acid analyzer.[26-28] Another method involves detection of the *Se*-adenosyl-[^{75}Se] selenomethionine fraction from the chromatographic separation.[29]

1. Glutathione Peroxidase

Glutathione peroxidase (glutathione: H_2O_2 oxidoreductase, EC 1.11.1.9) is a selenoenzyme which catalyzes the metabolism of H_2O_2 and a variety of organic hydroperoxides. There is a close correlation between glutathione peroxidase activity and selenium concentration in the blood of humans,[30,31] rats,[32] sheep,[33-35] cattle,[34,36] and chickens,[37] and thus determination of enzyme activity provides a convenient method for the assessment of selenium status.[38]

Several methods have been developed for the determination of glutathione peroxidase activity. The method of choice for biological materials is based on the coupled reaction sequence:[39,40]

$$2GSH + ROOH \text{ Glutathione Peroxidase } GSSG + ROH + H_2O$$
$$GSSG + NADPH + H^+ \text{ Glutathione Reductase } 2GSH + NADP$$

In this system, the glutathione peroxidase-catalyzed oxidation[39-41] of reduced glutathione (γ-L-glutamyl-L-cysteinylglycine, GSH) by hydroperoxide (*t*-butylhydroperoxide, cumene hydroperoxide, or H_2O_2) to oxidized glutathione (GSSG) is coupled to the glutathione reductase-catalyzed oxidation of β-nicotinamide adenine dinucleotide phosphate, reduced form (NADPH), to NADP. Conditions are adjusted so that the rate of oxidation of NADPH is limited by the glutathione peroxidase activity. The disappearance of NADPH can readily be monitored using its spectral properties, e.g., its absorbance[39,40] or fluorescence.[33,42] The assay based on absorbance measurements involves pipetting buffer, glutathione reductase and GSH solutions and the enzyme sample into a semimicrocuvette, and preincubation at 37°C. Then NADPH and *t*-butylhydroperoxide (TBH) solutions are added and the decrease in absorbance is monitored at 340 (or 365) nm for about 5 min. The glutathione peroxidase activity is obtained from the reaction rate after correction for any nonenzymic contribution.

This method has been used to determine glutathione peroxidase activity in human tissue extracts.[43]

For the determination of glutathione peroxidase activity in tissues which also contain glutathione *S*-transferases, for which TBH is a substrate, a modified procedure involving the replacement of TBH by H_2O_2 and the blockage of catalase by the addition of sodium azide should be used.[40]

A simple fluorescent spot test based on the above coupled reaction sequence has been developed for detecting glutathione peroxidase deficiency in sheep.[33,42] The procedure involves the addition of 2 μl of whole blood to 200 μl of screening mixture which contains GSH, glutathione reductase, NADPH, and other reagents. The solution is mixed, the red blood cells are allowed to hemolyze, and then TBH is added. Drops of the mixture are spotted on nonfluorescent absorbent paper at various times after the start of the reaction, the spots are dried immediately under a stream of warm air, and then viewed under long-wavelength UV light (320 to 380 nm). NADPH in the spot causes it to fluoresce. The NADPH is being consumed by the coupled reaction sequence at a rate determined by the glutathione peroxidase activity, and thus the time at which fluorescence ceases provides a measure of the enzyme activity. The spot test method is reported to be simple and rapid and to have sufficient sensitivity and repeatability to allow it to be used for the routine screening of large numbers of blood samples for selenium status.[33]

B. Selenoamino Acids

Selenoamino acids and their sulfur isologues generally occur together in biological samples. Thus, analytical methods for the determination of the selenium-containing compounds involve a separation prior to the measurement step. Low-molecular-weight selenoamino acids have been separated from their isologous sulfur amino acids and determined with an amino acid (ion exchange) analyzer,[26,44-47] and by paper[47] and gas chromatography.[48] *Se*-methylselenocysteine, selenocystine, selenomethionine, selenocystathionine,[26] and the selenic acids selenohypotaurine and selenohomohypotaurine[47] have been separated from their sulfur isologues by ion-exchange chromatography. Quantitation was achieved by measuring the absorbance of the ninhydrin derivatives.[47] The oxidation products of selenoamino acids (seleninic and selenonic acids) can be separated from the corresponding sulfinic and sulfonic acids by paper chromatography.[47] Selenocystine and selenomethionine can be separated from cystine and methionine and quantitatively analyzed by gas chromatography with flame ionization detection.[48] Procedures have been recommended for distinguishing chromatographic peaks of selenium compounds from those of sulfur compounds in chromatograms obtained with a flame photometric detector.[49] The problem arises because the band systems of S_2 and Se_2 overlap extensively, and thus it is not possible to select a wavelength at which only selenium compounds will respond. Optical discrimination with a dual-channel detector is recommended as the best choice; with single-channel detectors, the peaks can be distinguished on the basis of the greater quenching of the sulfur peak by methane doping.

The electrochemistry of selenium (RSeH and RSeSeR) and the corresponding sulfur compounds at the mercury electrode have been studied extensively.[50,51] Although normal polarographic methods have not found widespread use for the determination of selenium-containing compounds, cathodic stripping voltammetry shows promise. Selenocystine can be determined in the presence of cysteine and cystine by cathodic stripping voltammetry at the hanging mercury drop electrode.[52] By careful choice of the potential at which the selenocystine is deposited, the detection limit for selenocystine is 5×10^{-10} *M* in the presence of 100- fold higher concentrations of cysteine and cystine. Although no reports of the use of liquid chromatography with electrochemical detection for the determination of selenols and diselenides have appeared in the literature, the success of the technique for the determination of thiols[53-55] and disulfides[55] and the similarity of the electrochemical behavior of sulfur-containing and selenium-containing compounds suggests that it should be a powerful method for the trace analysis of selenols and diselenides.

C. Metabolites

1. Dimethylselenide and Dimethyldiselenide

Dimethylselenide and dimethyldiselenide are major selenium metabolites in biological systems, e.g., from the metabolism of inorganic selenide salts by fungi,[56,57] plants,[58] and animals.[59] These volatile compounds can be determined at the microgram level by gas chromatography either by direct injection or by trapping the volatile compounds on an adsorbent and desorbing them by elevating the temperature or by solvent elution. A particularly attractive method, developed for the analysis of biologically generated volatile selenium compounds in the atmosphere of lakewater sediment systems, combines the separation power of gas chromatography with the high sensitivity and specificity of silica furnace atomic absorption spectrometry.[60] The procedure involves sucking a known volume of the atmosphere to be analyzed through a glass U-tube trap packed with 3% OV-1 on Chromosorb® W and immersed in a dry ice-methanol bath. After sampling, the trap is heated to about 100°C, and the adsorbed compounds are swept into the gas chromatographic column for separation. Calibration curves, expressed as peak area as a function of absolute weight of selenium, were linear for both CH_3SeCH_3 and $CH_3SeSeCH_3$ up to 50 ng. As little as 0.1 ng Se could be determined with certainty.[60]

2. Trimethylselenonium Ion

Trimethylselenonium ion (TMSe) is an important metabolite, being detectable in urine at doses of selenite insufficient to result in respiratory excretion of dimethylselenide. TMSe in urine can be determined by the use of radiotracer ^{75}Se.[61-64] [^{75}Se] Trimethylselenonium ion is separated from other selenium-containing compounds in urine by ion-exchange chromatography, using both anion- and cation-exchange columns.[64] HCl is removed from the eluate by repeated evaporation and dissolution in water, the residue is redissolved in HNO_3 to convert Se(VI) to Se(IV), which is precipitated with ammonium pyrrolidinecarbodithioate, and the selenium is determined either by neutron activation analysis or by fluorometry.

In an alternative method, the TMSe in urine is also separated from other selenium-containing compounds in urine by anion-exchange/cation-exchange chromatography in tandem; the TMSe is selectively captured on the cation-exchange resin.[65] The resin is dried, irradiated with neutrons, and radioassayed for ^{77m}Se activity. A lower limit of detection of 10 ppb for total selenium and for TMSe was reported. The reported large relative standard deviation of 35% was attributed to the neutron flux of the reactor.

3. Hydrogen Selenide and Other Volatile Selenols

Hydrogen selenide can be produced from sodium selenite through a combination of nonenzymatic and enzymatic reactions involving glutathione and glutathione reductase.[66] Also, subcellular liver organelles have been shown to contain an acid volatile form of selenium,[19] which is assumed to be H_2Se. Several methods have been described for determining H_2Se based on the use of trapping reagents such as lead acetate and silver nitrate, which form insoluble selenides. A disadvantage of such procedures is the inability to distinguish between hydrogen selenide and organic selenols. In an alternative and more general procedure for the determination of H_2Se and volatile organic selenols (e.g., CH_3SeH), *1*-fluoro-2,4-dinitrobenzene is used as the trapping reagent.[67] The selenols react rapidly to form stable dinitrophenyl selenoethers that can be extracted into benzene and unequivocally identified by TLC, HPLC, or MS.

III. SPECTROSCOPIC MEASUREMENTS ON ORGANIC SELENIUM COMPOUNDS

Although spectroscopic methods are not used for the trace determination of organic and biochemical selenium, they have been widely used to study the chemistry and biochemistry of selenium. Low-molecular-weight compounds have received the most attention. Results obtained by infrared, Raman, ultraviolet, chiroptical, X-ray diffraction, and nuclear magnetic resonance (NMR) spectroscopy have been reviewed.[68,69]

Of particular interest is the use of ^{77}Se-NMR spectroscopy, which is potentially an important method for the characterization of selenium in biological systems.[69] The ^{77}Se isotope occurs naturally with an abundance of 7.58%; it has a spin of one half and a NMR sensitivity of 6.93×10^{-3} relative to the sensitivity of ^{1}H. The feasibility of observing selenium incorporated into large biological molecules has been demonstrated.[70,71] The selenium in selenotrisulfides, RSSeSR, which are formed by reaction of the thiol RSH with selenite, can readily be detected at concentrations in the millimolar or higher ranges by ^{77}Se-NMR.[72] Also, ^{77}Se-NMR spectra can be obtained for alkylselenylsulfides (RSSeR′) formed by the reaction of thiols with diselenides and the reaction of selenols with disulfides.[72] ^{77}Se-NMR can be used to probe the chemical properties of selenium-containing compounds, with the first report of the use of ^{77}Se-NMR to investigate protein-selenol binding recently published.[71] Thus, ^{77}Se-NMR appears to be a promising analytical method for the characterization of selenium in organic and biochemical molecules.

ACKNOWLEDGMENT

Support by the National Institutes of Health grant GM 3700 is gratefully acknowledged.

REFERENCES

1. **Olson, O. E., Palmer, I. S., and Whitehead, E. I.,** Determination of selenium in biological materials, in *Methods of Biochemical Analysis,* Vol. 21, Glick, D., Ed., John Wiley & Sons, New York, 1973, 39.
2. **Campbell, A. D.,** Critical evaluation of analytical methods for the determination of trace elements in various matrices. I. Determination of selenium in biological materials and water, *Pure Appl. Chem.,* 56, 645, 1984.
3. **Bem, E. M.,** Determination of selenium in the environment and in biological material, *Environ, Heatlh Perspect.,* 37, 183, 1981.
4. **Dilli, S. and Sutikno, I.,** Analysis of selenium at the ultra-trace level by gas chromatography, *J. Chromatogr.,* 300, 265, 1984.
5. **Robberecht, H. J. and Deelstra, H. A.,** Selenium in human urine. Determination, speciation and concentration levels, *Talanta,* 31, 497, 1984.
6. **Shendrikar, A. D.,** Critical evaluation of analytical methods for the determination of selenium in air, water, and biological materials, *Sci. Total Environ.,* 3, 155, 1974.
7. **Masson, M. R.,** Determination of selenium and tellurium in organic compounds and organic materials — a review, *Mikrochim. Acta,.* 419, I, 1976.
8. **Robberecht, H. and Van Grieken, R.,** Selenium in environmental waters: determination, speciation and concentration levels, *Talanta,* 29, 823, 1982.
9. **Raptis, S. E., Kaiser, G., and Tölg, G.,** A survey of selenium in the environment and a critical review of its determination at trace levels, *Fresenius Z. Anal. Chem.,* 316, 105, 1983.
10. **Shamberger, R. J.,** *Biochemistry of Selenium,* Plenum Press, New York, 1983.
11. **Beilstein, M. A., Tripp, M. J., and Whanger, P. D.,** Evidence for selenocysteine in ovine tissue organelles, *J. Inorg. Biochem.,* 15, 339, 1981.
12. **Hawkes, W. C., Wilhelmsen, E. C., and Tappel, A. L.,** Abundance and tissue distribution of selenocysteine-containing proteins in the rat, *J. Inorg. Biochem.,* 23, 77, 1985.

13. **Hartmanis, M. G. N. and Stadtman, T. C.,** Isolation of a selenium-containing thiolase from *Clostridium kluyveri:* identification of the selenium moiety of selenomethionine, *Proc. Natl. Acad. Sci. U.S.A.*, 79, 4912, 1982.
14. **Jenkins, K. J.,** Evidence for the absence of selenocystine and selenomethionine in the serum proteins of chicks administered selenite, *Can. J. Biochem.*, 46, 1417, 1968.
15. **Godwin, K. O. and Fuss, C. N.,** The entry of selenium into rabbit protein following the administration of $Na_2^{75}SeO_3$, *Aust. J. Biol. Sci.*, 25, 865, 1972.
16. **Olson, O. E. and Palmer, I. S.,** Selenoamino acids in tissues of rats administered inorganic selenium, *Metabolism,* 25, 299, 1976.
17. **Ganther, H. E.,** Enzymic synthesis of dimethyl selenide from sodium selenite in mouse liver extracts, *Biochemistry,* 5, 1089, 1966.
18. **Byard, J. L.,** Trimethyl selenide. A urinary metabolite of selenite, *Arch. Biochem. Biophys.*, 130, 556, 1969.
19. **Diplock, A. T., Caygill, C. P. J., Jeffrey, E. H., and Thomas, C.,** The nature of the acid-volatile selenium in the liver of the male rat, *Biochem. J.*, 134, 283, 1973.
20. **Tappel, A. L., Hawkes, W. C., Wilhelmsen, E. C., and Motsenbocker, M. A.,** Selenocysteine-containing proteins and glutathione peroxidase, *Methods Enzymol.*, 107, 602, 1984.
21. **Stadtman, T. C.,** Occurrence and characterization of selenocysteine in proteins, *Methods Enzymol.*, 107, 576, 1984.
22. **Ganther, H. E., Kraus, R. J., and Foster, S. J.,** Identification of selenocysteine by high-performance liquid chromatography and mass spectrometry, *Methods Enzymol.*, 107, 582, 1984.
23. **Cone, J. E., Martín del Río, R., Davis, J. N., and Stadtman, T. C.,** Chemical characterization of the selenoprotein component of clostridial glycine reductase: identification of selenocysteine as the organoselenium moiety, *Proc. Natl. Acad. Sci. U.S.A.*, 73, 2659, 1976.
24. **Kraus, R. J., Foster, S. J., and Ganther, H. E.,** Identification of selenocysteine in glutathione peroxidase by mass spectroscopy, *Biochemistry,* 22, 5853, 1983.
25. **Cone, J. E., Martín del Río, R., and Stadtman, T. C.,** Clostridial glycine reductase complex. Purification and charcterization of the selenoprotein component, *J. Biol. Chem.*, 252, 5337, 1977.
26. **Walter, R., Schlesinger, D. H., and Schwartz, I. L.,** Chromatographic separation of isologous sulfur- and selenium-containing amino acids: reductive scission of the selenium-selenium bond by mercaptans and selenols, *Anal. Biochem.*, 27, 231, 1969.
27. **Benson, J. V., Jr. and Patterson, J. A.,** Accelerated chromatographic analysis of selenocysteine and selenomethionine, *Anal. Biochem.*, 29, 130, 1969.
28. **Martin, J. L. and Gerlach, M. L.,** Separate elution by ion-exchange chromatography of some biologically important selenoamino acids, *Anal. Biochem.*, 29, 257, 1969.
29. **Sliwkowski, M. X.,** Characterization of selenomethionine in proteins, *Methods Enzymol.*, 107, 620, 1984.
30. **Thomson, C. D., Rea, H. M., Doesburg, V. M., and Robinson, M. F.,** Selenium concentrations and glutathione peroxidase activities in whole blood of New Zealand residents, *Br. J. Nutr.*, 37, 457, 1977.
31. **Rea, H. M., Thomson, C. D., Campbell, D. R., and Robinson, M. F.,** Relation between erythrocyte selenium concentrations and glutathione peroxidase (EC 1.11.1.9) activities of New Zealand residents and visitors to New Zealand, *Br. J. Nutr.*, 42, 201, 1979.
32. **Hafeman, D. G., Sunde, R. A., and Hoekstra, W. G.,** Effect of dietary selenium on erythrocyte and liver glutathione peroxidase in the rat, *J. Nutr.*, 104, 580, 1974.
33. **Peter, D. W.,** Modified fluorescent spot test for glutathione peroxidase and selenium concentration in sheep blood, *Vet. Rec.*, 107, 193, 1980.
34. **Wilson, P. S. and Judson, G. J.,** Glutathione peroxidase activity in bovine and ovine erythrocytes in relation to blood selenium concentration, *Br. Vet. J.*, 132, 428, 1976.
35. **Oh, S. H., Pope, A. L., and Hoekstra, W. G.,** Dietary selenium requirement of sheep fed a practical-type diet as assessed by tissue glutathione peroxidase and other criteria, *J. Anim. Sci.*, 42, 984, 1976.
36. **Smith, P. J., Tappel, A. L., and Chow, C. K.,** Glutathione peroxidase activity as a function of dietary selenomethionine, *Nature (London),* 247, 392, 1974.
37. **Noguchi, T., Cantor, A. H., and Scott, M. L.,** Mode of action of selenium and vitamin E in prevention of exudative diathesis in chicks, *J. Nutr.*, 130, 1502, 1973.
38. **Hoekstra, W. G.,** Glutathione peroxidase activity of animal tissues as an index of selenium status, in *Trace Substances in Environmental Health,* Vol. 9, Hemphill, D. D. Ed., University of Missouri, Columbia, 1975, 331.
39. **Paglia, D. E. and Valentine, W. N.,** Studies on the quantitative and qualitative characterization of erythrocyte glutathione peroxidase, *J. Lab. Clin. Med.*, 70, 158, 1967.
40. **Flohé, L. and Günzler, W. A.,** Assays of glutathione peroxidase, *Methods Enzymol.*, 105, 114, 1984.
41. **Tappel, A. L.,** Glutathione peroxidase and hydroperoxides, *Methods Enzymol.*, 52, 506, 1978.
42. **Board, P. G. and Peter, D. W.,** A simple test for glutathione peroxidase and selenium deficiency, *Vet. Rec.*, 99, 144, 1976.

43. **Carmagnol, F., Sinet, P. M., and Jerome, H.,** Selenium-dependent and non-selenium-dependent glutathione peroxidases in human tissue extracts, *Biochim. Biophys. Acta,* 759, 49, 1983.
44. **Walter, R. and du Vigneaud, V.,** 6-Hemi-L-selenocystine-oxytocin and *1*-deamino-6-hemi-L-selenocystine-oxytocin, highly potent isologs of oxytocin and *1*-deamino-oxytocin, *J. Am. Chem. Soc.,* 87, 4192, 1965.
45. **McConnell, K. P. and Wabnitz, C. H.,** Elution of selenocystine and selenomethionine from ion-exchange resins, *Biochim. Biophys. Acta,* 86, 182, 1964.
46. **Martin, J. L. and Cummins, L. M.,** Separate elution of selenocystine and selenomethionine by ion-exchange technique, *Anal. Biochem.,* 15, 530, 1966.
47. **De Marco, C., Cossu, P., Dernini, S., and Rinaldi, A.,** Chromatographic separation of selenohypotaurine, selenotaurine, selenohomohypotaurine and selenohomotaurine, *J. Chromatogr.,* 129, 369, 1976.
48. **Caldwell, K. A. and Tappel, A. L.,** Separation by gas-liquid chromatography of silylated derivatives of some sulfo- and selenoamino acids and their oxidation products, *J. Chromatogr.,* 32, 635, 1968.
49. **Hancock, J. R., Flinn, C. G., and Aue, W. A.,** Means of distinguishing selenium peaks from sulfur peaks in gas chromatography with a flame photometric detector, *Anal. Chim. Acta,* 116, 195, 1980.
50. **Nygård, B.,** Polarographic investigations of organic selenium compounds. III. Polarography of selenocystine-selenocysteine, *Ark. Kemi,* 27, 341, 1967.
51. **Danielsson, R., Johansson, B.-L., Nygård, B., and Persson, B.,** Polarographic reductive behaviour of organic selenium and sulfur compounds characterized by mercury assisted chemical pre-reactions, *Chem. Scr.,* 20, 19, 1982.
52. **Grier, R. A. and Andrews, R. W.,** Cathodic stripping voltammetry of selenocystine, cystine, and cysteine in dilute aqueous acid, *Anal. Chim. Acta,* 124, 333, 1981.
53. **Saetre, R. and Rabenstein, D. L.,** Determination of penicillamine in blood and urine by high performance liquid chromatography, *Anal. Chem.,* 50, 276, 1978.
54. **Rabenstein, D. L. and Saetre, R.,** Analysis for glutathione in blood by high-performance liquid chromatography, *Clin. Chem.,* 24, 1140, 1978.
55. **Allison, L. A. and Shoup, R. E.,** Dual electrode liquid chromatography detector for thiols and disulfides, *Anal. Chem.,* 55, 8, 1983.
56. **Fleming, R. W. and Alexander, M.,** Dimethylselenide and dimethyltelluride formation by a strain of *Penicillium, Appl. Microbiol.,* 24, 424, 1972.
57. **Barkes, L. and Fleming, R. W.,** Production of dimethylselenide gas from inorganic selenium by eleven soil fungi, *Bull. Environ. Contam. Toxicol.,* 12, 308, 1974.
58. **Evans, C. S., Asher, C. J., and Johnson, C. M.,** Isolation of dimethyl diselenide and other volatile selenium compounds from *Astragalus racemosus* (Pursh.), *Aust. J. Biol. Sci.,* 21, 13, 1968.
59. **Vlasáková, V., Beneš, J., and Pařizek, J.,** Application of gas chromatography for the analysis of trace amounts of volatile ^{75}Se metabolites in expired air, *Radiochem. Radioanal. Lett.,* 10, 251, 1972.
60. **Chau, Y. K., Wong, P. T. S., and Goulden, P. D.,** Gas chromatography-atomic absorption method for the determination of dimethyl selenide and dimethyl diselenide, *Anal. Chem.,* 47, 2279, 1975 (and references therein).
61. **Janghorbani, M., Ting, B. T. G., and Young, V. R.,** Use of stable isotopes of selenium in human metabolic studies: development of analytical methodology, *Am. J. Clin. Nutr.,* 34, 2816, 1981.
62. **Foster, S. J. and Ganther, H. E.,** Synthesis of [^{75}Se] trimethylselenonium iodide from [^{75}Se] selenocystine, *Anal. Biochem.,* 137, 205, 1984.
63. **Palmer, I. S., Fischer, D. D., Halverson, A. W., and Olson, O. E.,** Identification of a major selenium excretory product in rat urine, *Biochim. Biophys. Acta,* 177, 336, 1969.
64. **Nahapetian, A. T., Young, V. R., and Janghorbani, M.,** Measurement of trimethylselenonium ion in human urine, *Anal. Biochem.,* 140, 56, 1984.
65. **Blotcky, A. J., Hansen, G. T., Opelanio-Buencamino, L. R., and Rack, E. P.,** Determination of trimethylselenonium ion in urine by ion-exchange chromatography and molecular neutron activation analysis, *Anal. Chem.,* 57, 1937, 1985.
66. **Ganther, H. E.,** Metabolism of hydrogen selenide and methylated selenides, in *Advances in Nutritional Research,* Vol. 2, Draper, H. H., Ed., Plenum Press, New York, 1979, 107.
67. **Ganther, H. E. and Kraus, R. J.,** Identification of hydrogen selenide and other volatile selenols by derivatization with *l*-fluoro-2,4-dinitrobenzene, *Anal. Biochem.,* 138, 396, 1984.
68. **Klayman, D. L. and Günther, W. H. H., Eds.,** *Organic Selenium Compounds: Their Chemistry and Biology,* Wiley-Interscience, New York, 1973.
69. **Odom, J. D.,** Selenium biochemistry-chemical and physical studies, *Struct. Bonding (Berlin),* 54, 1, 1983.
70. **Luthra, N. P., Costello, R. C., Odom, J. D., and Dunlap, R. B.,** Demonstration of the feasibility of observing nuclear magnetic resonance signals of ^{77}Se covalently attached to proteins, *J. Biol. Chem.,* 257, 1142, 1982.

71. **Mullen, G. P., Dunlap, R. B., and Odom, J. D.,** Selenium-77 nuclear magnetic resonance investigation of a protein-selenoligand complex: interaction of α-chymotrypsin with (phenylselenyl) acetate, *J. Am. Chem. Soc.,* 107, 7187, 1985.
72. **Tan, K. S. and Rabenstein, D. L.,** Unpublished results.

Chapter 4

THE LITERATURE OF SELENIUM AND THE STATUS AND TREATMENT OF ANALYTICAL DATA

Milan Ihnat and Wayne R. Wolf

TABLE OF CONTENTS

I. INTRODUCTION

This chapter covers a brief summary of the historical and current literature on selenium and includes a listing of some of the major works on various facets of the element. It also addresses the status of the analytical information in the current and recent past literature and summarizes the approaches taken by authors of this volume in assessing and selecting information for inclusion in their contributions.

II. THE LITERATURE OF SELENIUM

Beginning with the discovery of selenium by Berzelius and Gahn in 1817 and the first reference dealing with the discovery of this new element,[1] the literature on this interesting element of industrial, toxicological, nutritional, and medical significance has mushroomed. Early reports dealt with its mineralogy, crystallography, physical and chemical constants, electrical properties, optical properties, inorganic chemistry, organic chemistry, physical chemistry, analytical chemistry, electrical cells, and industrial applications. A small number of publications appeared during the century following the discovery of the element dealing with various biochemical aspects, with the first paper on this subject apparently by Cameron[2] on the absorption of selenium by plants. With the suggestion and establishment of selenium as a toxic agent in food and feed crops in the 3rd and 4th decades of the 20th century, research on selenium toxicity was a major aspect added to the above list. Much work was carried out on the occurrence of selenium in crops and selenium accumulator plants growing on seleniferous formations in the U.S., with in-depth reports by researchers in South Dakota and Wyoming. The discovery by Schwarz and Foltz[3] of the nutritional essentiality of selenium added another significant dimension to this intriguing element, with much current reporting on the relation of selenium deficiency to animal and human disorders and the possible protective role of the element in cancer.

The literature on selenium is overwhelming. A listing by Doty[4] of titles of work relating to selenium, published between 1817 and 1925, owned by the New York Public Library in 1926 records 1526 publications. *Selenium and Tellurium Abstracts,* begun in 1959, published its last issue 25 years later with the 74,874th abstract on selenium and tellurium. Scientific publications on selenium continue to be abstracted by several abstracting services with thorough coverage by *Chemical Abstracts, Biological Abstracts, Science Citation Index, USDA Bibliography of Agriculture*, Abstract series of the Commonwealth Agricultural Bureaux, Index Medicus and Excerpta Medica. Some major works on selenium in the form of books, reports, review articles, and proceedings of conferences dealing with topics covered in this volume are given in the list of references as a guide for the reader.[5-63] Several articles of historical interest and of non-English origin are included for information. Articles in journals and other reports covering specific topics may be found in the rather comprehensive listings of references included in each chapter in this volume.

III. THE STATUS AND TREATMENT OF ANALYTICAL DATA

Analytical procedures based on acid decomposition/distillation/titrimetry or photometry applied in the 1930s and 1940s to estimating high concentrations of selenium in biological materials were satisfactory for that purpose. One concern noted was the possible loss of the element as volatile compounds during drying of samples prior to decomposition. Development of methods centering on light absorption spectrometry and fluorometry of selenium complexes with benzidine and naphthalene derivatives placed in the hands of analysts more detective (higher power of detection, colloquial term — more sensitive) techniques for measuring endogenous levels in ordinary crops and animal tissues. Unless automated, the fluorometric method is relatively slow, but an excellent performer in respect of precision and accuracy, and is in widespread use. In the hands of appropriately trained, experienced, and competent analysts, this technique and others, such as neutron activation analysis and hydride generation atomic absorption spectrometry, are yielding sufficiently reliable analytical information. Selenium is one of the few trace elements for which contamination during sampling and chemical analysis is usually of little concern. Data reported in the literature were thus generally expected to be of reasonable quality, especially when good analytical approaches are taken.

The charge to contributing authors was to undertake a comprehensive review and critical assessment of information in the literature in order to extract and present reliable analytical data. Responses to a number of specific points regarding criteria for data selection and assessment were solicited. These are in summary as follows: all available published data were sought, largely contained in refereed scientific journals. In instances when a sufficient body of data was deemed not to exist in such literature or for the sake of completeness in coverage, information was acquired from reports of various kinds. In general, only data determined using established quantitiative analytical methods applicable to the matrices analyzed were selected, being evaluated as objectively as possible, but with an inevitable element of subjective judgement. Criteria for data selection were based on documentation of sampling and analytical methodology information, and all reasonable data, regardless of method used, were accepted. Some authors did not include data unless sample type was clearly identified. Others reported that it was difficult to evaluate existing data quality based on the type of sampling system employed. Attempts were made to exclude results which did not make sense within the discipline covered or where there appeared to be problems with sample contamination. Although most often published accounts contain an insufficient presentation of quality control information regarding method performance characteristics at the time of application, efforts were made by contributing authors to use whatever information was available in judging data reliability.

A systematic quantitative approach to the critical evaluation of published analytical data was undertaken by Wolf and Schubert in extracting information on the selenium content of a core group of U.S. foods. This approach, developed at one of the author's (WRW) institutions for ongoing research[64,65] is summarized below.

To evaluate each published report of nutrient data, a set of criteria for five general categories was developed: number of samples, analytical method, sample handling, sampling plan, and analytical quality control. A literature search yielded approximately 70 articles published after 1960 that report analytical selenium data for foods grown and processed in the U.S. and Canada, as well as related methodology papers. Recent Food and Drug Administration Total Diet Study analyses,[66] unpublished at the time of evaluation, were also included because they provide extensive data on cooked foods and mixtures, unlike most other studies, which have analyzed primarily raw and uncooked foods.

The basis for defining selenium-specific criteria within each of the five general categories, summarized in Table 1, was provided by examination of the published reports of determinations of selenium in foods together with a knowledge of accepted analytical methodology, sample handling procedures, and quality control measures for this nutrient, as well as knowledge of statistical methods. The data from each study were rated by food item within each category on a scale from 0 (unacceptable) to 3 (most desirable). The ratings and selenium values for each study were noted on the corresponding worksheet for each food item. Table 2 reproduces, as an example, the worksheet for egg white, which illustrates most of the possible combinations of ratings.

For each set of ratings from each study, a quality index (QI) was assigned. When three or more individual ratings were 0 or when the analytical method was rated 0, the QI was 0; otherwise, the five ratings were averaged. Selenium values from individual studies that

Table 1
DATA QUALITY CRITERIA REQUIREMENTS

Criteria categories	Ratings: 3	2	1	0
Number of samples	>10; standard deviation, standard error, or raw data reported	3—10	1—2; explicitly stated or not specified	—
Analytical method	Official fluorometric (reference given or other method documented by a complete published write-up with validation studies for foods analyzed, including use of an appropriate reference material [RM] where available), 95—105% recoveries on food similar to sample analyzed in same or other paper; Se concentration above quantitation limit of the method	Modified fluorometric or other method, some documentation, incomplete validation studies for foods analyzed; must include 90—110% recoveries on food similar to sample analyzed (or good recovery but no statistics given), and/or use of other method (official fluorometric, isotope dilution, or NAA) on same sample with good agreement (within 10%)	Nonfluorometric method, partially described; 80—90% or >110% recoveries on food similar to sample; or use of comparison method or recoveries on food only somewhat related to sample (animal/plant)	No documentation of method, no reference or inaccessible reference given, no validation studies or poor agreement (>10%) of test method with comparison method on same sample
Sample handling	Complete documentation of procedures including validation of homogenization method, details of food preparation and storage, and moisture changes monitored	Pertinent procedures documented, seem reasonable, but some details not reported	Only edible portion analyzed	Totally inappropriate procedures or no documentation of criteria pertinent to food analyzed
Sampling plan	Multiple geographical sampling with complete description; sample is representative of brands/varieties commonly consumed or commercially used	1 or 2 geographic areas sampled; sample is representative	Sample representative of small percent of U.S. and/or origin not clear	Not described or sample not representative
Analytical quality control	Optimum accuracy and precision of method monitored and indicated explicitly by data	Documentation of assessment of both accuracy and precision of method; acceptable accuracy and precision	Some description of minimally acceptable accuracy and precision of method	No documentation of accuracy and/or precision

Table 2
WORKSHEET FOR RAW EGG WHITE

Description[a]	Number of samples (actual)	Analytical method	Sample handling	Sampling plan	Analytical quality control	Quality index[b]	Mean selenium concentration ± SD (μg/kg)	Comments
	Data quality criteria ratings							
White	1 (2)	2	1	2	0	1.2	51	Duplicates, no quality control documentation
Albumen	1 (?)	1	1	0	0	0.6	80	No sampling plan, quality control documentation
White	2 (4)	2	2	2	0	1.6	350 ± 120	No sampling plan documentation
White	1 (1)	0	1	2	0	0	53	No method validation
White	3 (18)	2	1	0	0	1.2	150	Canadian
White part	1 (1)	2	2	0	0	1.0	24	Triplicates, Canadian
White	2 (9)	0	0	1	0	0	51 ± 29	No method validation
White	1 (1)	1	0	0	0	0	172	Hg-contaminated feed
						Sum = 5.0		
							Grand mean = 140 Range = 20—350	Confidence code = b

[a] Each line of entry is from a different reference.

[b] A quality index ≥1.0 is required for a datum to be considered acceptable. The grand mean is calculated from the acceptable means; the confidence code is derived from the sum of the quality indices of the acceptable studies.

Table 3
ASSIGNMENT AND MEANING OF CONFIDENCE CODES

Sum of quality indices	Confidence code	Meaning of confidence code
>6.0	a	The user can have considerable confidence in this value
3.4 to 6.0	b	The user can have confidence in this value; however, some problems exist regarding the data on which the value is based
1.0 to <3.4	c	The user can have less confidence in this value due to limited quantity and/or quality of data

expressed selenium content on a fresh weight basis and that had a corresponding QI of 1.0 or greater were averaged to yield a grand mean selenium concentration value. Several studies were eliminated solely because selenium content was expressed only on a dry weight basis, with no moisture levels reported. A confidence code (CC), indicating the confidence a user can have in the grand mean selenium value for each food, was assigned as illustrated in Table 3. This CC is based on the sum of the QIs equal to or greater than 1.0. This critical approach to extracting reliable analytical information was applied to foods reported in Chapter 6.

ACKNOWLEDGMENT

Land Resource Centre, Agriculture Canada, contribution No. 87-68.

REFERENCES

1. **Berzelius, J. J.,** Lettre de M. Berzelius à M. Berthollet sur deux métaux nouveaux, *Ann. Chim. Phys.*, Ser. 2, 7, 199, 1817.
2. **Cameron, C. A.,** Preliminary note on the absorption of selenium by plants, *Proc. R. Dublin Soc.*, 2, 231, 1879.
3. **Schwarz, K. and Foltz, C. M.,** Selenium as an integral part of factor 3 against dietary necrotic liver degeneration, *J. Am. Chem. Soc.*, 79, 3292, 1957.
4. **Doty, M. F.,** *Selenium, a List of References,* New York Public Library, 1927.
5. **Beath, O. A., Eppson, H. F., and Gilbert, C. S.,** Selenium and other toxic minerals in soils and vegetation, *Wyo. Agric. Exp. Stn. Bull.*, 206, 1935.
6. **Byers, H. G.,** Selenium occurrence in certain soils in the United States with a discussion of related topics, *U.S. Dep. Agric. Tech. Bull.*, 482, 1935.
7. **Byers, H. G.,** Selenium occurrence in certain soils in the United States with a discussion of related topics, second report, *U.S. Dep. Agric. Tech. Bull.*, 530, 1936.
8. **Knight, S. H. and Beath, O. A.,** The occurrence of selenium and seleniferous vegetation in Wyoming. I. The rocks and soils of Wyoming and their relations to the selenium problem. II. Seleniferous vegetation of Wyoming, *Wyo. Agric. Exp. Stn. Bull.*, 221, 1937.
9. **Moxon, A. L.,** Alkali disease or selenium poisoning, *S. D. Agric. Exp. Stn. Bull.*, 311, 1937.
10. **Moxon, A. L., Olson, O. E., and Searight, W. V.,** Selenium in rocks, soils and plants, *S. D. Agric. Exp. Stn. Tech. Bull.*, 2, 1939.
11. **Byers, H. G., Miller, J. T., Williams, K. T., and Lakin, H. W.,** Selenium occurrence in certain soils in the United States with a discussion of related topics: third report, *U.S. Dep. Agric. Tech. Bull.*, 601, 1938.
12. **Williams, K. T., Lakin, H. W., and Byers, H. G.,** Selenium occurrence in certain soils in the United States, with a discussion of related topics; fourth report, *U.S. Dep. Agric. Tech. Bull.*, 702, 1940.
13. **Williams, K. T., Lakin, H. W., and Byers, H. G.,** Selenium occurrence in certain soils in the United States, with a discussion of related topics: fifth report, *U.S. Dep. Agric. Tech. Bull.*, 758, 1941.
14. **Lakin, H. W. and Byers, H. G.,** Selenium occurrence in certain soils in the United States, with a discussion of related topics: sixth report, *U.S. Dep. Agric. Tech. Bull.*, 783, 1941.

15. **Searight, W. V. and Moxon, A. L.,** Selenium in glacial and associated deposits, *S. D. Agric. Exp. Stn. Tech. Bull.*, 5, 1945.
16. **Beath, O. A., Hagner, A. F., and Gilbert, C. S.,** Some rocks and soils of high selenium content, *Wyo. Geol. Surv. Bull.*, 36, 1946.
17. **Lakin, H. W. and Byers, H. G.,** Selenium occurrence in certain soils in the United States, with a discussion of related topics: seventh report, *U.S. Dep. Agric. Tech. Bull.*, 950, 1948.
18. **Moxon, A. L.,** Selenium: its occurrence in rocks and soils, absorption by plants, toxic action in animals, and possible essential role in animal nutrition, in *Trace Elements,* Lamb, C. A., Bentley, O. G., and Beattie, J. M., Eds., Academic Press, New York, 1958, 175.
19. **Anderson, M. S., Lakin, H. W., Beeson, K. C., Smith, F. F., and Thacker, E.,** Selenium in agriculture, *U.S. Dep. Agric. Handbook,* 200, 1961.
20. **Rosenfeld, I. and Beath, O. A.,** *Selenium, Geobotany, Biochemistry, Toxicity, and Nutrition,* Academic Press, New York, 1964.
21. **Muth, O. H., Oldfield, J. E., and Weswig, P. H., Eds.,** *Selenium in Biomedicine,* AVI Publishing, Westport, CT, 1967.
22. **Klenha, J.,** Biochemistry of selenium (in Czech), *Chem. Listy,* 60, 1656, 1966.
23. **Mills, C. F., Bremner, I., Chesters, J. K., and Quarterman, J., Eds.,** *Trace Element Metabolism in Animals,* Churchill Livingstone, Edinburgh, 1970.
24. National Research Council (U.S.), *Selenium in Nutrition,* National Academy of Sciences, Washington, D.C., 1971.
25. **Bisbjerg, B.,** *Studies on Selenium in Plants and Soils,* Riso Rep. No. 200, Danish Atomic Energy Commission, Roskilde, 1972.
26. **Lisk, D. J.,** Trace metals in soils, plants and animals, *Adv. Agron.*, 24, 267, 1972.
27. **Jenkins, K. J. and Hidiroglou, M.,** Review of selenium/vitamin E responsive problems in livestock. Case for selenium as a feed additive in Canada, *Can. J. Anim. Sci.*, 52, 591, 1972.
28. **Darby, W.,** Trace Elements in human nutrition, *WHO Tech. Rep. Ser.*, 532, 1973.
29. **Underwood, E. J.,** Trace elements, in *Toxicants Occurring Naturally in Foods,* National Academy of Sciences, Washington, D.C., 1973, 43.
30. **Allaway, W. H.,** Selenium in the food chain, *Cornell Vet.*, 63, 151, 1973.
31. **Klayman, D. L. and Gunther, W. H. H., Eds.,** *Organic Selenium Compounds: Their Chemistry and Biology,* Wiley-Interscience, Toronto, 1973.
32. **Zingaro, R. A. and Cooper, W. C., Eds.,** *Selenium,* Van Nostrand Reinhold, New York, 1974.
33. **Hoekstra, W. G., Suttie, J. W., Ganther, H. E., and Mertz, W., Eds.,** *Trace Element Metabolism in Animals — 2,* TEMA 2, University Park Press, Baltimore, 1974.
34. **Ermakov, V. V. and Koval'skii, V. V.,** *Biological Significance of Selenium* (in Russian) Nauka, Moscow, 1974; as cited in *Chem. Abst.*, 83-39162t.
35. **Stadtman, T. C.,** Selenium biochemistry, *Science,* 183, 915, 1974.
36. **Gasanov, G. G., Ed.,** *Selenium in Biology, Proceedings of a Scientific Conference* (in Russian), Elm, Baku, Azerbaijan S.S.R., 1974; as cited in *Chem. Abst.*, 82-166296u.
37. **Nicholas, D. J. D. and Egan, A. R., Eds.,** *Trace Elements in Soil-Plant-Animal Systems,* Academic Press, New York, 1975.
38. **Frost, D. V.,** Selenium in biology, *Annu. Rev. Pharmacol.*, 15, 259, 1975.
39. National Research Council (U.S.), *Medical and Biological Effects of Environmental Pollutants: Selenium,* National Academy of Sciences, Washington, D.C., 1976.
40. **Industrial Health Foundation,** *Proc. of the Symp. on Selenium-Tellurium in the Environment,* Industrial Health Foundation, Pittsburgh, 1976.
41. **Ganther, H. E., Hafeman, D. G., Lawrence, R. A., Serfass, R. E., and Hoekstra, W. G.,** Selenium and glutathione peroxidase in health and disease — a review, in *Trace Elements in Human Health and Disease,* Vol. 2, Prasad, A. S. and Oberleas, D., Eds., Academic Press, New York, 1976, 165.
42. **Burk, R. F.,** Selenium in man, in *Trace Elements in Human Health and Disease,* Vol. 2, Prasad, A. S. and Oberleas, D., Eds., Academic Press, New York, 1976, 105.
43. **Levander, O. A.,** Selected aspects of the comparative metabolism and biochemistry of selenium and sulfur, in *Trace Elements in Human Health and Disease,* Vol. 2, Prasad, A. S. and Oberleas, D., Eds., Academic Press, New York, 1976, 135.
44. **Johnson, C. M.,** Selenium in the evironment, *Residue Rev.*, 62, 101, 1976.
45. **Gissel-Nielsen, G.,** *Control of Selenium in Plants,* Riso Rep. No. 370, Denmark Research Establishment, Roskilde, 1977.
46. **Harr, J. R.,** Biological effects of selenium, in *Toxicity of Heavy Metals in the Environment, Part 1,* Oehme, F. W., Ed., Marcel Dekker, New York, 1978, 393.
47. **Duchaigne, A. and Arvy, M. P.,** Selenium in biology (in French), *Annee Biol.*, 17, 529, 1978.

48. **Kirchgessner, M., Roth-Maier, D. A., Roth, H.-P., Schwarz, F. J., and Weigand, E., Eds.,** *Trace Element Metabolism in Man and Animals — 3,* TEMA 3, Arbeitskreis fur Tierernahrungsforschung, Weihenstephen, W. Germany, 1978.
49. **Stadtman, T. C.,** Some selenium-dependent biochemical processes, *Adv. Enzymol. Relat. Areas Mol. Biol.,* 48, 1, 1979.
50. **Dunckley, J. V., Ed.,** *Proc. of the New Zealand Workshop on Trace Elements in New Zealand,* University of Otago, Dunedin, 1981.
51. **Spallholz, J. E., Martin, J. L., and Ganther, H. E., Eds.,** *Selenium in Biology and Medicine,* AVI Publishing, Westport, CT, 1981.
52. **Brown, T. A. and Shrift, A.,** Selenium: toxicity and tolerance in higher plants, *Biol. Rev. Cambridge Philos. Soc.,* 57, 59, 1982.
53. **Levander, O. A.,** Selenium: biochemical actions, interactions, and some human health implications, *Curr. Top. Nutr. Dis.,* 6, 345, 1982.
54. **Robinson, M. F.,** Clinical effects of selenium deficiency and excess, *Curr. Top. Nutr. Dis.,* 6, 325, 1982.
55. **Gawthorne, J. M., Howell, J. McC., and White, C. L., Eds.,** *Trace Element Metabolism in Man and Animals,* TEMA 4, Springer-Verlag, Berlin, 1982.
56. **Shamberger, R. J.,** *Biochemistry of Selenium,* Plenum Press, New York, 1983.
57. National Research Council (U.S.), *Selenium in Nutrition,* Revised ed., National Academy Press, Washington, D.C., 1983.
58. **Sharma, S. and Singh, R.,** Selenium in soil, plant, and animal systems, *CRC Crit. Rev. Environ. Control,* 13, 23, 1983.
59. **Barbezat, G. O., Casey, C. E., Reasbeck, P. G., Robinson, M. F., and Thompson, C. D.,** Selenium, *Curr. Top. Nutr. Dis.,* 12, 231, 1984.
60. **Combs, G. F., Jr. and Combs, S. B.,** The nutritional biochemistry of selenium, *Annu. Rev. Nutr.,* 4, 257, 1984.
61. **Gissel-Nielsen, G., Gupta, U. C., Lamand, M., and Westermarck, T.,** Selenium in soils and plants and its importance in livestock and human nutrition, *Adv. Agron.,* 37, 397, 1984.
62. **Girling, C. A.,** Selenium in agriculture and the environment, *Agric. Ecosyst. Environ.,* 11, 37, 1984.
63. **Mills, C. F., Bremner, I., and Chesters, J. K., Eds.,** *Trace Elements in Man and Animals,* TEMA 5, Commonwealth Agricultural Bureaux, Farnham Common, UK, 1985.
64. **Holden, J. M., Schubert, A., Wolf, W. R., and Beécher, G. R.,** A system for evaluating the quality of published nutrient data: selenium, a test case, in *Food and Nutrition Bulletin,* Suppl. 12, Food Composition Data: A User's Perspective, Rand, W. T., Windham, C. T., Wyse, B., and Young, V. R., Eds., United Nations University, Tokyo, 1987, 177.
65. **Schubert, A., Holden, J. M., and Wolf, W. R.,** Selenium content of a core group of foods based on a critical evaluation of published analytical data, *J. Am. Diet. Assoc.,* 87, 285, 1987.
66. **Pennington, J. A. T., Young, B. E., Wilson, D. B., Johnson, R. D., and Vanderveen, J. E.,** Mineral content of foods and total diets: the selected minerals in foods survey, 1982—1984, *J. Am. Diet. Assoc.,* 86, 876, 1986.

Chapter 5

PLANTS AND AGRICULTURAL MATERIALS

Milan Ihnat

TABLE OF CONTENTS

I. INTRODUCTION

Selenium plays a significant role in animal and human well being and health, acting either as a toxic constituent or an essential nutrient.[1-7] Plants and products derived therefrom constitute one of the important vehicles effecting the transfer of selenium from soil to humans directly or indirectly via the consumption of products of animal origin. The impact of this route is considerable as the bulk of selenium acquired by humans is by ingestion of food.

The occurrence and significance of selenium in plants and agriculture has been the subject of numerous publications ever since the detection[8] and the first quantitative measurement[9] of the element in plants. Extensive work beginning in the 1930s by many workers, particularly those in the U.S. Department of Agriculture, Agricultural Experiment Station of the South Dakota State College of Agriculture and Mechanic Arts, and the Agricultural Experiment Station of the University of Wyoming, has been conducted on the occurrence of selenium in soils and plants of the western U.S. Much of the research has been directed toward elucidation of the toxicity to animals of selenium accumulator plants growing on seleniferous soils discussed in Section V. Concomitantly, analyses were also carried out on food and feed crops with early reports on composition by Robinson,[9] Byers,[10] Byers and Lakin,[11] Thorvaldson and Johnson,[12] Lakin and Byers,[13] and Williams et al.[14] Early reports on toxic effects of high selenium content feed crops and selenium accumulator plants on livestock are by Franke, Tully, Moxon, Olson, Palmer, and co-workers following studies begun in 1928 by Franke and co-workers in South Dakota.[15-17] More recent and other pertinent references regarding the selenium content of a variety of feed and other crops, derived products, and related agricultural materials are contained in the tables in this chapter. Good detailed expositions on selenium in the agricultural context may be found in reports by Anderson et al.,[18] Underwood,[19] Moxon,[20,21] and Rosenfeld and Beath[22] as well as in proceedings of conferences on selenium or trace elements.[1,2,23-28]

Plants, raw agricultural crop commodities, and other products relevant to animal feeds and rations and their selenium concentrations are the central themes of this chapter. Other agriculturally related commodities and resources are treated in other chapters of this volume (foods, Chapter 6; animal tissues, Chapter 7; soils, Chapter 9). Invariably, total selenium concentration data are presented. The format is tabular with subject divisions: feed crops, feed ingredients and rations, other plants and products, and selenium accumulator plants. Data treatment and presentation is moderately comprehensive covering a fair sampling of the (mainly English) literature, giving some ideas of selenium levels in a variety of plants and agricultural commodities.

II. FEED CROPS

The first group of plants considered contains those constituting animal feeds grazed directly in the pasture (unprocessed) or offered to animals as freshly cut forage, hay, or ensiled material (partially processed). Crops which, due to economics or local convention, serve as both feed and food sources are included. Selenium concentration data and other information are presented in Table 1. Crops are listed alphabetically by common name with the taxonomic name included; for each crop, information is given by country (with indication of regions, states, provinces, or cities if that information is available) and arranged in alphabetical order for each continent in the following sequence: North America, South America, Europe, Africa, Asia, and Oceania. The Remarks column contains information on plant components analyzed, soil type, observed relationships between crops and soil selenium levels, crop notation used by the authors, and other details of treatment in experimental studies. Additional details are to be found in the footnotes to the table.

Crop components are typically those customarily offered to or consumed by the animal,

Table 1
CONCENTRATION OF SELENIUM IN ANIMAL FEED CROPS[a]

Geographical location[b]	Selenium concentration (mg/kg)[c] Range	Mean	N[d]	Date[e]	Remarks[f]	Ref.
Alfalfa (Lucerne), *Medicago sativa* L.[g]						
Creston, BC, Canada (6)	—	0.11 ± 0.16	—	1973	1—4-year stands, Lister (orthic gray luvisol) with 0.27 mg Se/kg	59
Creston, BC, Canada (6)	—	0.53 ± 0.72	—	1973	1—4-year stands, Kuskanook (carbonated orthic gleysol) with 0.27 mg Se/kg	59
Ontario and Quebec, Canada	0.020—0.167	0.072	16	1974	W and w/o phosphate treatment, 8 different clay and loam soils, greenhouse pot experiments	60
Denmark (6)	0.084—0.195	0.13	7	1965		61
Jutland, Denmark (30)	—	0.035 ± 0.004[h]	30	1973	30 fields in Jutland	62
de la Roche-Posay, Vienne, France	0.07—0.19	0.10 m	192	1977	Spring vegetative cycle, calcareous soil	63
de la Roche-Posay, Vienne, France	0.09—0.28	0.15 m	192	1977	Leaves, spring vegetative cycle, calcareous soil	63
de la Roche-Posay, Vienne, France	0.04—0.09	0.055 m	192	1977	Stems, spring vegetative cycle, calcareous soil	63
Limoges, Haute-Vienne, France	0.020—0.073	0.034	96	1975	Spring cycle, granitic soil	63
Limoges, Haute-Vienne, France	0.03—0.09	0.035 m	80	1975	Fall 2nd growth, granitic soil	63
Gujarat State, India	—	0.82	—	1970	ADPLAS method	64
Naot-Mordechai, Israel	1.4—44.0	11.1	14	1957	ADDT method	65
Other locations, Israel	0.1—6.0	1.2 m	14	1957	ADDT method	65
New Zealand	—	0.041	—	1983	Stems/stolons, Takapau stony loam	66
New Zealand	—	0.089	—	1983	Leaves, Takapau stony loam	66
New Zealand (13)	0.006—0.332	0.012 m	13	1983	13 different soils	66
Ultana region, Sweden (1)	—	0.040	1	1968		67
Kimberly, ID, U.S.	—	0.12	15	1966—67	5 cuttings from 3 control field plots	68
Kimberly, ID, U.S.	—	0.043	15	1972	1/10 bloom, Portneuf silt loam, controls from greenhouse pot studies	69
Kimberly, ID, U.S.	—	0.070	3	1972	Field study control	69
	0.107—0.149	—	6	1972	Added phosphate	
Kimberly, ID, U.S.	0.048—0.190	0.084 m	14	1972	Greenhouse pot studies using 14 different soils from northwest U.S. (ID, OR, MT, WA), w/phosphate in 2nd datum	69
	0.049—0.214	0.096 m	14			

Table 1 (continued)
CONCENTRATION OF SELENIUM IN ANIMAL FEED CROPS[a]

Geographical location[b]	Selenium concentration (mg/kg)[c] Range	Mean	N[d]	Date[e]	Remarks[f]	Ref.
New Mexico, U.S.	1.3—1.8	—	—	1974		70
New Mexico, U.S.	—	<0.05	10	1974	Alfalfa hay	70
Ithaca, NY, U.S.	0.08—0.11	—	—	1965	W and w/o sulfur treatment, Hudson silty clay loam, greenhouse plot experiments	30
Jefferson County, OR, U.S.	0.01—0.09	—	5	1962—63	2-year-old stand, Madras sandy loam	71
Tennessee, U.S.	0.064—0.110	0.086	4	1976	Alfalfa hay, from state experiment stations and farms	72
Tennessee, U.S.	0.078—0.125	0.104	3	1976	Denoted silage, from state experiment stations and farms	72
Washington, U.S.; British Columbia, Canada	0.003—0.171	0.047 ± 0.043	33	1976—77	Alfalfa hay	73
Barley, *Hordeum distichum* L.						
Roskilde, Denmark	0.009—0.021	0.020	—	1972—76	Grain, field experiments at Riso Natl. Laboratory	74
Roskilde, Denmark	0.011—0.032	0.021	—	1972—76	Straw, field experiments at Riso Natl. Laboratory	74
Barley grain, *Hordeum vulgare* L.						
Alberta, Canada	0.005—2.213	0.211 ± 0.012	428	1971—74	Submitted by farmers and ranchers to Agricultural Soil and Feed Testing Laboratory, broad representation of feed in Alberta	75
Creston region, BC, Canada	0.06—0.58	0.25 ± 0.14	17	1966—73		76
Peace region, BC, Canada	0.04—0.80	0.16 ± 0.14	56	1966—73		76
Thompson region, BC, Canada	0.04—0.99	0.32 ± 0.23	37	1966—73		76
Ontario and Quebec, Canada	0.20—0.10	0.04 ± 0.008[h]	9	1971		77
Prince Edward Island, Canada (13)	0.006—0.038	0.016 ± 0.010	13	1971	13 different sampling sites covering major soil types	78
Prince Edward Island, Canada	0.010—0.029	0.018 ± 0.005	22	1971	Several locations from research station plots	78
Prince Edward Island, Canada (16)	0.006—0.040	0.020	16	1971—72	4 soil series represented	79
Saskatchewan, Canada (16)	—	0.23 ± 0.18	16	1964—68	From test plots at 16 locations representing 2 replicates of each of the major associations (loam and clay) in each of four main soil zones	80
Western Canada	0.20—0.78	0.35 ± 0.020[h]	24	1971		77
Denmark	<0.005—0.10	0.015 m	160	1972	Samples from all over Denmark	81
Denmark (21)	0.005—0.025	0.017	21	1972	1st year of 4-year field experiment	81

Denmark (21)	0.005—0.040	0.019	100	1972—75	4-year field experiment	74
Denmark	0.002—0.110	0.018	318	1972—73	From farms well distributed over Denmark, higher conc. in crops from higher clay content soils	62
Limerick County, Ireland	—	7.2	—	1953	High organic matter dark soil with 40—88 mg Se/kg, ADDT method	79
Finland (10)	0.002—0.013[i]	0.005[i]	23	1968—69	From 10 field stations of the Agricultural Research Centre	82
Sweden (8)	0.006—0.022	0.011	8	1968	Soil Se range 0.221—0.945 mg/kg, no correlation found between plant and soil Se levels	67
Barley, *Hordeum vulgare* L. var. Tern						
Denmark	—	0.016	4	1973	Seed — Grown on mineral soil in outdoor pot experiments	83
	—	0.015	4	1973	Grain	
	—	0.019	4	1973	Straw	
Barley and oat grain						
British Columbia, Canada	0.03—1.26	0.20 ± 0.18	178	1966—72		76
Beet (sugar, fodder, forage), *Beta vulgaris* L.						
Roskilde, Denmark	0.037—0.174	0.086	5	1972—76	Top — Control data from field experiments at Riso Natl. Laboratory	74
Roskilde, Denmark	0.015—0.035	0.026	5	1972—76	Root	
Lolland, Denmark	—	0.034 ± 0.001[h]	4	1973	Root — Collections from state experimental farms in Lolland, Sealand, and Bornholm and from a few private farms in North Sealand	62
Sealand and Bornholm, Denmark	—	0.007	17	1973	Root	
North Sealand, Denmark	—	0.016 ± 0.002[h]	6	1973	Root	
Lolland, Denmark	—	0.132 ± 0.020[h]	3	1973	Top	
Sealand and Bornholm, Denmark	—	0.052	13	1973	Top	
North Sealand, Denmark	—	0.049 ± 0.013[h]	4	1973	Top	
Finistère region, Brittany, France	0.025—0.062	0.039	6	1975	Roots	84
Brittany, France	0.025—0.062	0.039	6	1970—78	Roots, appears to be same information as from Reference 84	85
Bahiagrass, *Paspalum notatum*						
Southeastern U.S.	0.031—0.148	0.086 ± 0.031	41	1977—79	Providence silt loam, data averaged over different varieties, significant seasonal trends	86
Bermudagrass, *Cynodon dactylon* L.						
Gujarat State, India	—	0.20	—	1970	ADPLAS method	64
Punjab State, India	0.04—0.29	0.19	4	1977	ADPLAS method	87
Naot-Mordechai, Israel	0.2—8.9	4.6	10	1957	ADDT method	65
Other locations, Israel	0.2—1.2	0.4 m	9	1957	ADDT method	65
Kenya	0.084—0.098	0.091 ± 0.009	2	1973	Denoted stargrass	88
Tennessee, U.S.	0.043—0.157	0.101	44	1976	Bermuda green chop — From state experimental stations and farms	72
Tennessee, U.S.	0.050—0.185	0.118	2	1976	Bermuda hay	72
Southeastern U.S.	0.062—0.448	0.164 ± 0.098	27	1977—79	Providence silt loam, data averaged over different varieties	86

Table 1 (continued)
CONCENTRATION OF SELENIUM IN ANIMAL FEED CROPS[a]

Geographical location[b]	Selenium concentration (mg/kg)[c]					
	Range	Mean	N[d]	Date[e]	Remarks[f]	Ref.
Birdsfoot trefoil, *Lotus corniculatus* L.						
Kapuskasing, Ontario, Canada	—	0.031	80	1966	Swards established in 1963 on gray wooded clay soils, 2 harvests with 10 quadruplicate weekly samplings	89
Bromegrass, *Bromus inermis* Leyss.						
Kapuskasing, Ontario, Canada	—	0.038	80	1966	Swards established in 1963 on gray wooded clay soils, 2 harvests with 10 quadruplicate weekly samplings	89
Browntop, *Agrostis tenuis* Sibth.						
New Zealand	0.007—0.035	0.018	16	1959—60	3—4 cuts over 559 d, Atiamuri sand, control data from plot trials on existing pasture	90
New Zealand	—	0.027	—	1966	Ruakura Hill Station Farm at Whatawhata	90
New Zealand	0.009—0.014	0.012	24	1967	Atiamuri sand cores from browntop pasture for pot trials, 4 cuts	91
Cabbage forage						
Finistère region, Brittany, France	0.021—0.035	0.026	18	1975		84
Finistère region, Brittany, France	0.032—0.051	0.046	5	1975	Leaves and stems	84
Brittany, France	0.021—0.051	0.036	23	1970—78	Appears to be same information as from Reference 84	85
Clover, Alexandrian or Egyptian, *Trifolium alexandrium*						
Gujarat State, India	—	0.87	—	1970	ADPLAS method	64
Punjab State, India	0.04—0.17	0.10	7	1977	ADPLAS method	87
Naot-Mordechai, Israel	2.4—26.9	16.1	7	1957	ADDT method	65
Other locations, Israel	0.2—1.3	0.4 m	7	1957	ADDT method	65
Clover, subterranean, *Trifolium subterraneum* L.						
Australia	0.05—0.08	0.06	10	1982	Yellow podzolic soil derived from dactite, glasshouse trials, tops removed before flowering and analyzed, 2nd datum w added S	92
Australia	0.05—0.14	0.09	37	1982		
Near Canberra, Australia	—	0.01	2	1982	Field study on yellow podzolic soil, w and w/o $CaSO_4$ treatment, sampled 4—17 months after treatment, no S effect	92
Near Cooma, NSW, Australia	—	0.17	2	1982	Field study on black earth derived from basalt, w and w/o $CaSO_4$ treatment, sampled 20—21 months after treatment, no S effect	92

Esperance distinct, Western Australia	0.01—0.02	—	—	1963,65		93
Clover, red, *Trifolium pratense* L.						
Prince Edward Island, Canada (41)	0.001—0.029	0.012 ± 0.005	55	1971	41 sampling sites covering major soil types	78
Prince Edward Island, Canada (66)	0.005—0.031	0.015	66	1971—72	8 soil series with 0.09—0.60 mg Se/kg	94
Roskilde, Denmark	0.026—0.060	0.040	—	1972—76	Field experiments at Riso Natl. Laboratory	74
Ireland	—	7.5	2	1972	Glasshouse pot experiments using peat from Limerick county with 93 mg Se/kg, 1st and 2nd data from controls and treatments with non-Se additives, ADPLAS method with poor detection limit estimated at 3—10 mg/kg	95
	2.5—14.8	7.0	38			
Ultana region, Sweden	—	0.18	1	1968	Soil Se 0.199 mg/kg	67
Tennessee, U.S.	0.061—0.088	0.075	2	1976	Denoted red clover and straw from state experiment stations and farms	72
Clover, white, *Trifolium repens* L.						
Denmark (7)	0.108—0.207	0.17	11	1965		61
Naot-Mordechai, Israel	—	1.7	1	1957	ADDT method	65
New Zealand	0.007—0.027	0.014	36	1964—65	Oruanui sand (a yellow-brown pumice), control data from plot experiments, 9 cuts from existing sward containing browntop, cocksfoot, and clover	96
New Zealand	—	0.010	—	1983	Stems/stolons, pastures on Takapau stony loam	66
New Zealand	—	0.015	—	1983	Leaves, pastures on Takapau stony loam	66
New Zealand	0.005—0.017	0.010	12	1959—60	3—4 cuts over 397 d, Atiamuri sand, control data from plot trials on existing pasture	90
New Zealand	—	0.013	—	1966	Ruakura Hill Station Farm at Whatawhata	90
New Zealand	—	0.015	—	1983	Flowers/seedheads } Plants from Pasture on Heretaunga silt loam	66
	—	0.022	—	1983	Terminal bud tissues	
	0.016—0.031	0.018 m	—	1983	Leaves	
	0.010—0.017	0.011 m	—	1983	Stolon tissue	
Clover, unspecified						
Finland (9)	0.004—0.022	0.012	17	1968—69	From 9 field stations of the Agricultural Research Centre	82
Clover, (white + subterranean)						
New Zealand (13)	0.006—0.033	0.014	116	1966	13 localities, mineral soils, 5 cuts from control plot trials with added phosphate	97
Clover (white) + perennial ryegrass						
New Zealand	0.009—0.015	0.012 ± 0.002	8	1971	Rerewhakaaitu soil, Se-deficient area	98

Table 1 (continued)
CONCENTRATION OF SELENIUM IN ANIMAL FEED CROPS[a]

Geographical location[b]	Selenium concentration (mg/kg)[c] Range	Mean	N[d]	Date[e]	Remarks[f]	Ref.
New Zealand	0.009—0.024	0.017 ± 0.005	6	1971	Wallaceville soil, marginally Se-deficient area	98
New Zealand	0.023—0.600	0.208 ± 0.178	8	1971	Whangaehu soil, area known to have high Se conc.	98
New Zealand	0.030—0.038	0.035	16	1963—65	Control data from 4 cuts of trial on existing swards on Horotiu sandy loam (a yellow-brown loam)	96
Clover (white) + perennial ryegrass + Yorkshire fog, *Holcus lanatus* L.						
Ruakura, New Zealand	0.01—0.011	0.010	16	1962	Control data from 4 cuts of trials on 4-year-old pastures on Rukuhia peat at the Moanatuatua area of the Ruakura Agricultural Research Centre	97
Clover, (white) + grass						
Roskilde, Denmark	0.043—0.205	0.074 m	28	1965—66		61
Clover + grass/clover mixtures						
Jutland, Denmark (98)	—	0.044 ± 0.002[h]	98	1973	Collections from 98 fields	62
Corn, grain, *Zea mays*						
Ontario, Quebec, Canada	0.01—0.10	0.04 ± 0.003	45	1971		77
Finistère region, Brittany, France	0.018—0.028	0.024	15	1974		84
Brittany, France	0.018—0.028	0.024	15	1970—78	Appears to be same information as from Reference 84	85
Illinois, U.S.	0.02—0.15	0.05	31	1969	Directly from farms	56
Indiana, U.S.	0.01—0.15	0.04	10	1969	Directly from farms	56
Iowa, U.S.	0.02—0.16	0.05	25	1969	Directly from farms	56
Kansas, U.S.	—	0.99	1	1969	Directly from farms	56
Mississippi, U.S.	0.03—0.04	0.03	5	1969	Directly from farms	56
Missouri, U.S.	0.02—0.09	0.05	4	1969	Directly from farms	56
Montana, U.S.	0.02—0.29	0.09	22	1969	Directly from farms	56
Nebraska, U.S.	0.04—0.81	0.35	6	1969	Directly from farms	56
North Dakota, U.S.	0.09—0.26	0.19	6	1969	Directly from farms	56
Ohio, U.S.	—	0.020	62	1981	Data from a 1973 report of the Ohio Agriculture Research and Development Center	99
South Dakota, U.S.	0.11—2.03	0.40	10	1969	Directly from farms	56
Wisconsin, U.S.	0.02—0.13	0.05	5	1969	Directly from farms	56

Corn, whole plant (forage, silage), *Zea mays*						
British Columbia, Canada	0—0.26	0.08 ± 0.05	235	1966—72	Forage	76
Boundary region, BC, Canada	—	0.07 ± 0.02	3	1966—73	Corn silage	76
Cariboo region, BC, Canada	—	0.15 ± 0.09	10	1966—73	Corn silage	76
Coastal region, BC, Canada	—	0.07 ± 0.04	124	1966—73	Corn silage	76
Fraser Valley, BC, Canada	0.001—0.076	0.016 ± 0.016	32	1976—77	Corn silage described as mainly fodder corn with some cannery residue, *Zea mays saccharata*	73
South Okanagan region, BC, Canada	—	0.06 ± 0.02	5	1966—73	Corn silage	76
Thompson region, BC, Canada	—	0.11 ± 0.06	79	1966—73	Corn silage	76
Ontario, Canada	0.012—0.044	0.027	8	1974	W or w/o phosphate treatment, greenhouse pot experiments on 4 loam soils from eastern Ont.	60
Finistère region, Brittany, France	0.018—0.058	0.032	88	1970—71	Ensilage stage	84
Quimper (Brittany), France	0.013—0.040	—	2	1974—75	Control data from pot experiments using soils rich in organic matter from W. Brittany, 2 crops before efflorescence stage	84
Brittany, France	0.018—0.066	0.034	115	1970—78	Refer to Reference 84	85
Sweden	0.010—0.033	0.017	5	1966—67		100
Southeastern U.S.	0.020—0.081	0.050 ± 0.018	19	1977—79	Silage grown on Providence silt loam	86
Tennessee, U.S.	0.020—0.128	0.053	101	1976	Corn silage, state experiment station and farms	72
Tennessee, U.S.	0.047—0.070	0.059	2	1976	Corn green chop, state experiment stations and farms	72
Fescue, creeping red, *Festuca rubra* L.						
Kapuskasing, Ontario, Canada	—	0.047	80	1966	Swards established in 1963 on gray wooded clay soil, 2 harvests with 10 quadruplicate weekly samplings	89
Fescue, meadow, *Festuca pratensis*						
Limerick County, Ireland (1)	—	117	—	1953	High organic matter dark soil with 40—88 mg Se/kg, ADDT method	101
Orebo, Sweden (1)	—	0.064	1	1968	Soil Se content 0.976 mg/kg	67
Fescue, unspecified						
Brittany, France	0.028—0.117	0.048	64	1970—78		85
Ozark region, MT, U.S.	0.020—0.137	0.046	24	1974	Missouri Soil Region 7	102
Tennessee, U.S.	0.069—0.129	0.097	4	1976	Fescue hay, state experiment stations and farms	72
Flax, *Linum usitatissimum* L.						
Ultana region, Sweden (1)	—	0.006	1	1968	Soil Se content 0.165 mg/kg	67
Fodder						
Denmark	0.014—0.320	0.142	28	1970	Fodder from 19 dairy farms	103

Table 1 (continued)
CONCENTRATION OF SELENIUM IN ANIMAL FEED CROPS[a]

Geographical location[b]	Selenium concentration (mg/kg)[c] Range	Mean	N[d]	Date[e]	Remarks[f]	Ref.
Grasses, forage, silage, hay (mixed or unspecified)						
Near Canberra, Australia	0.01—0.02	0.015	2	1982	Field study on yellow podzolic soil, w and w/o $CaSO_4$ treatment, sampled 4—17 months after treatment, no S effect	92
Near Cooma, NSW, Australia	0.63—0.93	—	2	1982	Field study on black earth derived from basalt, w and w/o $CaSO_4$ treatment, sampled 20—21 months after treatment, effect of S noted	92
British Columbia, Canada	0—1.26	0.21 ± 0.15	422	1966—72	Grass forage	76
Fraser Valley, BC, Canada	0.001—0.069	0.023 ± 0.017	31	1976—77	Grass silage, mainly ensiled orchardgrass	73
Fraser Valley, BC, Canada	0.002—0.056	0.024 ± 0.014	32	1976—77	Grass hay, mainly sun-cured orchardgrass	73
Fraser Valley, BC, Canada	0.001—0.066	0.017—0.018	26	1976—77	Pasture samples, mainly orchardgrass	73
Charlottetown, PEI, Canada	0.008—0.022	0.012	4	1971	Several locations from research station plots	78
Jutland, Denmark (15)	—	0.032 ± 0.004[h]	15	1973	15 fields in Jutland	62
Denmark (4)	0.173—0.251	0.21	4	1965		61
Finland	—	0.014	21	1971—72	Grass } Timothy dominant with orchard-grass	104
Finland	—	0.015	42	1971—72	Silage } and Italian ryegrass at 5 experimental sites	
Finland (7)	0.002—0.048[i]	0.014[i]	30	1968—69	Pasture grasses from 7 field stations of the Agricultural Research Centre	82
France (23)	0.017—0.18	0.037 m	23	1970	Pasture grass, unpublished data of Lamand 1968—69	105
New Zealand (13)	0.005—0.050	0.012 m	13	1983	13 different soils	66
Tennessee, U.S.	0.096—0.279	0.168	4	1976	Mixed grass hay (mostly orchardgrass) from state experiment stations and farms	72
Grasses + legumes (including clovers)						
Alberta, Canada	0.002—2.000	0.176 ± 0.024	143	1971—74	Denoted grass-legume roughage (mainly brome-alfalfa and timothy-clover mixes), submitted to Agricultural Soil and Feed Testing Laboratory, broad representation of feeds in Alberta	75
Cariboo region, BC, Canada	—	0.22 ± 0.13	81 }	1966—73	Authors of Reference 76 omitted samples for which results exceeded the mean by more than 2 SD	76
Chilcotin region, BC, Canada	—	0.19 ± 0.07	39 }			
Coastal region, BC, Canada	—	0.16 ± 0.09	296 }			

Creston region, BC, Canada	—	0.21 ± 0.15	183				
Boundary region, BC, Canada	—	0.14 ± 0.09	25				
Bulkley region, BC, Canada	—	0.14 ± 0.11	27				
East Kootenay region, BC, Canada	—	0.21 ± 0.12	23				
Nicola region, BC, Canada	—	0.27 ± 0.14	36				
Peace region, BC, Canada	—	0.19 ± 0.10	29				
South Okanagan region, BC, Canada	—	0.20 ± 0.18	60				
Thompson region, BC, Canada	—	0.34 ± 0.20	176				
Roskilde, Denmark	—	0.034 ± 0.004	5	1980—81	Ryegrass and white clover, control data from subplots of field trials on 1y old field		106
	—	0.044 ± 0.006	5				
	—	0.059 ± 0.004	6				
	—	0.024 ± 0.004	9				
Hamilton, New Zealand	—	0.030	150	1984	Te Kowhai silt loam	25 sites sampled on each of 2—3 paddocks twice in early and late spring, 2nd datum in each pair is from dung/urine affected areas	107
		0.018	150				
Hamilton, New Zealand	—	0.064	75	1984	Kaipaki peaty loam		
	—	0.038	75				
Taupo, New Zealand	—	0.011	100	1984	Oruanui hill soil		
	—	0.012	100				
Hamilton, New Zealand	—	0.118	125	1984	Waingaro hill soil		
	—	0.070	125				
New Zealand (5)	0.005—0.008	0.0063	6	1983	Severely deficient	2—8 of a total of 17 species of grasses and clover (excluding alfalfa) in 5 pastures with the indicated adequacy for grazing lambs[j]	66
	0.009—0.013	0.011	7	1983	Moderately deficient		
	0.019—0.028	0.022	8	1983	Marginally deficient		
	0.042—0.059	0.051	8	1983	Adequate		
	0.530—0.590	0.560	2	1983	Adequate		
Tennessee, U.S.	0.038—0.160	0.092	11	1976	Hay, from state experiment stations and farms		72
Legumes (legume forages)							
British Columbia, Canada	0—1.26	0.22 ± 0.15	570	1966—72	Denoted forage, authors of Reference 76 omitted results exceeding the mean by more than 2 SD		76
Charlottetown, PEI, Canada	0.016—0.025	0.018	4	1971	Several locations from research station plots		78
Oats, grain, *Avena sativa*							
Creston region, BC, Canada	0.11—0.48	0.25 ± 0.12	19	1966—73			76
Peace region, BC, Canada	0.04—0.35	0.13 ± 0.07	27	1966—73			76

Table 1 (continued)
CONCENTRATION OF SELENIUM IN ANIMAL FEED CROPS[a]

Geographical location[b]	Selenium concentration (mg/kg)[c] Range	Mean	N[d]	Date[e]	Remarks[f]	Ref.
Thompson region, BC, Canada	0.07—1.13	0.28 ± 0.12	27	1966—73		76
Ontario and Quebec, Canada	0.01—0.09	0.06 ± 0.010[h]	18	1971		77
Charlottetown, PEI, Canada	0.011—0.027	0.018 ± 0.006	10	1971	Several locations from research station plots	78
Prince Edward Island, Canada (~24)	0.004—0.043	0.021	24	1971—72	5 soil series	79
Prince Edward Island, Canada (21)	0.004—0.043	0.021 ± 0.011	21	1971	21 sampling sites covering major soil types	78
Saskatchewan, Canada (16)	—	0.26 ± 0.20	16	1964—68	From test plots at 16 locations representing 2 replicates of each of the major associations (loam and clay) in each of four main soil zones	80
Western Canada	0.23—0.43	0.30 ± 0.017[h]	11	1971		77
Denmark	0.003—0.054	0.016	44	1973	From farms well distributed over Denmark, higher conc. in crops from higher clay content soils	62
Finland (10)	0.003—0.018[i]	0.009[i]	14	1968—69	From 10 field stations of the Agricultural Research Centre	82
Sweden (7)	0.009—0.033	0.014	7	1968	Soil Se 0.221—0.945 mg/kg, no correlation found between plant and soil Se levels	67
Oats, whole plant (forage, hay), *Avena sativa*						
British Columbia, Canada	0—0.44	0.15 ± 0.08	113	1966—72	Oat forage	76
Boundary region, BC, Canada	—	0.10 ± 0.08	3	1966—73	Oat forage	76
Bulkley region, BC, Canada	—	0.14 ± 0.05	21	1966—73	Oat forage	76
Cariboo region, BC, Canada	—	0.16 ± 0.08	31	1966—73	Oat forage	76
Chilcotin region, BC, Canada	—	0.21 ± 0.14	3	1966—73	Oat forage	76
Coastal region, BC, Canada	—	0.13 ± 0.06	16	1966—73	Oat forage	76
Creston region, BC, Canada	—	0.18 ± 0.02	6	1966—73	Oat forage	76
East Kootenay region, BC, Canada	—	0.10 ± 0.05	7	1966—73	Oat forage	76
Nicola region, BC, Canada	—	0.05 ± 0.01	2	1966—73	Oat forage	76
Peace region, BC, Canada	—	0.15 ± 0.10	9	1966—73	Oat forage	76

South Okanagan region, BC, Canada	—	0.15 ± 0.08	3	1966—73	Oat forage	76
Thompson region, BC, Canada	—	0.21 ± 0.15	16	1966—73	Oat forage	76
Punjab State, India	Trace—0.155	0.07	3	1977	ADPLAS method	87
Naot-Mordechai, Israel	11.2—22.2	14.5	4	1957	ADDT method	65
Tennessee, U.S.	—	0.059	1	1976	Oat hay from state experiment stations and farms	72
Oats + ryegrass forage						
Southeastern U.S.	0.045—0.151	0.082 ± 0.027	31	1977—79	Providence silt loam	86
Orchardgrass (cocksfoot),[k] *Dactylis glomerata* L.						
Kapuskasing, Ontario, Canada	—	0.041	80	1966	Swards established in 1963 on gray wooded clay soil, 2 harvests with 10 quadruplicate weekly samplings	89
New Zealand	0.005—0.020	0.012	16	1959—60	3—4 cuts over 559 d, Atiamuri sand, control data from plot trials on existing pasture	90
New Zealand	—	0.007	—	1983	Stems/stolons } Numerous plants from pastures on Takapau stony loam	66
	—	0.018	—	1983	Leaves	
	—	0.013	—	1983	Flowers/seedheads	
New Zealand	—	0.015	20	1983	Spikelets } 20 plants from wasteland on Heretaunga mottled silt loam	66
	0.003—0.008	0.006 m	20	1983	Stems	
	0.015—0.017	0.015 m	20	1983	Leaves	
New Zealand (2)	0.013—0.058	0.018 m	9	1983	2 different soils	66
Ultana region, Sweden (1)	—	0.061	1	1968	Soil Se content 0.188 mg/kg	67
Pasture samples (pasture crops, pasture species, herbage, mixed herbage)[l]						
Near Canberra, Australia	0.01—0.02	0.02	6	1982	Field study on yellow podzolic soil, w and w/o $CaSO_4$ treatment, sampled 4—17 months after treatment, no S effect	92
Near Cooma, NSW, Australia	0.51—0.82	—	4	1982	Field study on black earth derived from basalt, w and w/o $CaSO_4$ treatment, sampled 20—21 months after treatment, effect of S noted	92
Jutland, Denmark (28)	0.023—0.225	—	320	1973	28 farms in Jutland, range of means estimated from Figure 2 in Reference 62 varying with season from 0.023 in July/August to 0.225 in March	62
Limerick County, Ireland	150—500	—	—	1953	High organic matter dark soil with 30—324 mg Se/kg, ADDT method	101
Meath County, Ireland	—	37	—	1975	Field with soil Se 42 mg/kg and diagnosed Se toxicity in cattle, no evidence of unusual plants, grass species similar to adjoining fields	108

Table 1 (continued)
CONCENTRATION OF SELENIUM IN ANIMAL FEED CROPS[a]

Geographical location[b]	Selenium concentration (mg/kg)[c]		N[d]	Date[e]	Remarks[f]	Ref.
	Range	Mean				
Chihuahua, Mexico	<0.01—>2.0	0.607 ± 0.930	23	1974	11 specimens <0.01 mg/kg, 4 specimens >2.0 mg/kg	70
Wallaceville, New Zealand	0.013—0.030	0.022	34	1963—64	From 7 paddocks with soil Se 0.41—0.71 (mean 0.56) mg/kg	109
Wallaceville, New Zealand	0.010—0.03	0.018	12	1968—69		109
New Zealand (5)	0.005—0.010	0.006 m	7	1983	Severely deficient (5—15 of a total of 27 different species in 5 pastures with the indicated adequacy for grazing lambs[m])	66
	0.009—0.013	0.010 m	7	1983	Moderately deficient	
	0.019—0.049	0.028 m	15	1983	Marginally deficient	
	0.042—0.332	0.058 m	10	1983	Adequate	
	0.300—1.800	0.530 m	5	1983	Adequate	
Buganda/Busoga areas, Uganda (22)	0.008—0.326	0.107 ± 0.074	63	1965—66	Total aerial parts of 11 pasture species from 22 farms, one high value omitted[n]	110
Ankole, Uganda (45)	0.019—0.409	0.081 ± 0.075	45	1968	Total aerial parts of 4 pasture species from 45 sites[o]	110
Nyakyesasa, Uganda	0.049—0.174	0.093	31	1969—70	Total aerial parts from 4 paddocks	110
Eastern Uganda (13)	0.035—0.930	0.233 ± 0.218	41	1971	Total aerial parts of 11 pasture species from 13 farms[p]	110
Northern Uganda (22)	0.029—0.503	0.088 ± 0.109	67	1973	Total aerial parts of 16 pasture species from 22 sites[q]	110
England, United Kingdom	0.010—0.610	0.12	19	1975	Herbage usually consisting of ryegrass, timothy, or cocksfoot (orchardgrass), but sometimes of indigenous species	111
Reed canarygrass, *Phalaris arundinacea* L.						
Kapuskasing, Ontario, Canada	—	0.049	80	1966	Swards established in 1963 on gray wooded clay soil, 2 harvests with 10 quadruplicate weekly samplings	89
Rye, grain, *Secale cereale*						
Charlottetown, PEI, Canada	—	0.020	1	1971	Research station plot	78
Denmark	0.006—0.072	0.016	25	1973	From farms well distributed over Denmark, sandy soils	62
Finland (8)	0.002—0.007[i]	0.004[i]	9	1968—69	From 8 field stations of the Agricultural Research Centre	82
Ryegrass, Italian, *Lolium multiflorum* Lam.						
Belgium	0.05—0.11	0.075	10	1982	Pot trials with 10 different soil types (0.04—0.27 mg Se/kg) representing 65% of agricultural soils in Belgium, high correlation between Se levels in plants and soils	112

Finland	0.020—0.039	0.027	64	1979	Clay soil	3—4 crops, control data from pot experiments	113
	0.017—0.040	0.027	64	1979	Fine sand		
	0.017—0.020	0.019	48	1980	Clay soil		
	0.014—0.022	0.019	48	1980	Carex peat		
Finland	0.008—0.024	0.014	882	1983	49 mineral soil samples	Control data from pot experiments using spoils from plough layers from different locations in Finland, 3 crops taken, 1st at silage stage, 3rd 2.5 months after start of experiment, both soils had 0.01—0.02 mg Se/kg	114
	0.008—0.018	0.012	306	1983	17 organogenic soil samples		
Finistère region, Brittany, France	0.038—0.134	0.066	24	1974	Fall, young pasture stage		84
	0.028—0.073	0.047	30	1975	Spring, young pasture stage		84
	0.020—0.066	0.043	22	1975	Normal pasture stage		84
	0.019—0.066	0.035	48	1975	Hay stage		84
Finistère region, Brittany, France	0.050—0.085	0.068 m	4	1973—74	Control data from pot experiments using soils rich in organic matter from W. Brittany, 3—4 cuts of 1st and 2nd crop at silage stage		84
	0.030—0.035	0.030 m	3	1974—75			
Brittany, France	0.019—0.134	0.050	422	1970—78	Refer to Reference 84		85
Naot-Mordechai, Israel	—	7.5	1	1957	ADDT method		65
Ryegrass, perennial, *Lolium perenne* L.							
Roskilde, Denmark	0.030—0.072	0.040	11	1972—76	Control data from field experiments at Riso Natl. Laboratory		74
Denmark (8)	0.082—0.152	0.12	11	1965			61
Ireland	9.5—12.8	10.5	12	1972	Glasshouse pot experiments using peat from Limerick county with 93 and 230 mg Se/kg for the 1st and 2nd pairs of data, respectively; within pairs, 1st and 2nd data are from controls and treatments with non Se additives, respectively, ADLAS method with poor detection limit estimated at 3—10 mg/kg		95
	2.5—21.3	6.4	70				
	—	11.2	2				
	<5—39.2	17.3	40				
New Zealand	—	0.021 ± 0.006	17	1976	Spring and fall collections from urine patch		115
	—	0.033 ± 0.007	15	1977	Fall collections from interexcreta locations		115
New Zealand	0.006—0.030	0.016	12	1959—60	3—4 cuts over 559 d, Atiamuri sand, control data from plot trials on existing pasture		90
New Zealand	—	0.020	—	1966	Ruakura Hill Station Farm at Whatawhata		90
New Zealand	—	0.005	—	1983	Stems/stolons	Numerous plants from pastures on Takapau stony loam	66
	—	0.018	—	1983	Leaves		
	—	0.012	—	1983	Flowers/seedheads		

Table 1 (continued)
CONCENTRATION OF SELENIUM IN ANIMAL FEED CROPS[a]

Geographical location[b]	Selenium concentration (mg/kg)[c] Range	Mean	N[d]	Date[e]	Remarks[f]	Ref.
New Zealand	0.016—0.029	0.023 m	11	1983	From urine fertilized areas, Heretaunga silt loam	66
	0.030—0.050	0.037 m	11	1983	From unfertilized areas, Heretaunga silt loam	
Sedge						
British Columbia, Canada	0.03—0.26	0.13 ± 0.06	46	1966—72	Forage	76
Cariboo region, BC, Canada	—	0.12 ± 0.06	24	1966—73	Hay	76
Chilcotin region, BC, Canada	—	0.15 ± 0.06	18	1966—73	Hay	76
Nicola region, BC, Canada	—	0.13 ± 0.07	4	1966—73	Hay	76
Sorghum, *Sorghum vulgare* Pers.						
Gujarat State, India	—	0.68	—	1970	Straw, ADPLAS method	64
Gujarat State, India	—	0.73	—	1970	Green fodder, ADPLAS method	64
Punjab State, India	Trace—0.28	0.13	4	1977	ADPLAS method	87
Naot-Mordechai, Israel	—	0.4	—	1957	ADDT method	65
Southeastern U.S.	0.039—0.112	0.061 ± 0.018	21	1977—79	Silage, Providence silt loam	86
Sorghum + corn silage						
Tennessee, U.S.	0.027—0.126	0.067	23	1976	From state experiment stations and farms	72
Swede, root, *Brassica napus*						
N. Sealand, Denmark	—	0.034 ± 0.005[h]	4	1973	From several private farms	62
Sudangrass, *Sorghum vulgare sudanense* Hitchc.						
Tennessee, U.S.	—	0.056	1	1976	Hay	72
Timothy, *Phleum pratense* L.						
Kapuskasing, Ontario, Canada	—	0.0326	80	1966	Swards established in 1963 on gray wooded clay soil, 2 harvests with 10 quadruplicate weekly samplings	89
Prince Edward Island, Canada (41)	0.001—0.024	0.009 ± 0.006	49	1971	1st and 2nd cuts from 41 sampling sites covering major soil types	78
Prince Edward Island, Canada (66)	0.005—0.023	0.013	66	1971—72	8 soil series with 0.09—0.60 mg Se/kg	79
Denmark (7)	0.100—0.168	0.14	10	1965		61
Finland (13)	0.005—0.039	0.012	44	1968—69	From 13 field stations of the Agricultural Research Centre	82

Finland	—	0.020	4	1981	1 cut in 1981, 3 cuts in 1982—83, control data from field plots on high humus silt at the N. Saro Research Station of the Agricultural Research Center of Finland	116
	—	0.009	24	1982,83		
Finland	—	0.023	56	1982	2—3 cuts at each site, control data from field plots on silt, fine sand and mould containing 21% organic C at 3 research stations	116
Sweden (4)	0.008—0.029	0.015	4	1968	Soil Se range 0.198—0.976 mg/kg, no correlation found between plant and soil Se levels	67
Wheat grain (spring wheat), *Triticum aestivum* L.						
Charlottetown, PEI, Canada	0.009—0.024	0.015 ± 0.005	11	1971	Several locations from research station plots	78
Denmark	0.004—0.067	0.020	41	1973	From farms well distributed over Denmark, higher conc. in crops from higher clay content soils	62
Wheat grain (winter wheat), *Triticum aestivum* L.						
Charlottetown, PEI, Canada	0.013—0.021	0.017 ± 0.003	5	1971	Several locations from research station plots	78
Denmark	0.007—0.087	0.021	40	1973	From farms well distributed over Denmark, higher conc. in crops from higher clay content soils	62
Wheat grain (durum wheat), *Triticum durum* Desf.						
Charlottetown, PEI, Canada	0.024—0.029	0.026	3	1971	Several locations from research station plots	78
Wheat grain (variety unspecified), *Triticum aestivum* L.						
British Columbia, Canada	0.03—1.26	0.32 ± 0.20	120	1966—72		76
Creston region, BC, Canada	0.09—0.75	0.25 ± 0.17	39	1966—73		76
Peace region, BC, Canada	0.06—0.48	0.23 ± 0.12	22	1966—73		76
Thompson region, BC, Canada	0.05—1.26	0.45 ± 0.17	58	1966—73		76
Ontario and Quebec, Canada	0.02—0.14	0.06 ± 0.010	13	1971		77
Saskatchewan, Canada	—	0.28 ± 0.23	16	1964—68	From test plots at 16 locations representing 2 replicates of each of the major associations (loam and clay) in each of four main soil zones	80
Saskatchewan, Canada	0.0—4.0	0.44	2230	1940	Analyses on 230 composite and also on individual samples representing 2230 individual samples, 98.9% of data 0—1.5 mg/kg, method detection limit ca. 0.25 mg/kg	10
Western Canada	0.37—0.82	0.51 ± 0.030	22	1971		77
Roskilde, Denmark	0.016—0.030	0.022	5	1972—76	Control data from field experiments at Riso Natl. Laboratory	74
Finland (3)	0.004—0.007[i]	0.006[i]	5	1968—69	From 3 field stations of the Agricultural Research Centre, one value of 0.085 mg/kg omitted by author of this chapter	82

Table 1 (continued)
CONCENTRATION OF SELENIUM IN ANIMAL FEED CROPS[a]

Geographical location[b]	Selenium concentration (mg/kg)[c] Range	Mean	N[d]	Date[e]	Remarks[f]	Ref.
Montana, U.S.	3—80	7 m	10	1936	Whole plant, soil Se 0.1—2.5 (m = 0.6) mg/kg (n = 68)	117
South Dakota, U.S.	0—60	6 m	63	1936	Heads, plant, stubble, soil Se 0.3—20 (m = 3) mg/kg (n = 348)	117
Colorado, U.S.	0.5—3	2.5 m	4	1941	Wheat and wheat products, grain from elevators, wheat heads from fields	14
Kansas, U.S.	<0.1—5	1 m	44	1941		
Montana, U.S.	<0.1—3	0.5 m	403	1941		
Nebraska, U.S.	<0.1—10	0.5 m	79	1941		
North Dakota, U.S.	<0.1—3	0.5 m	72	1941		
South Dakota, U.S.	<0.1—25	2 m	149	1941		
Wyoming, U.S.	<0.1—3	0.5 m	72	1941		
Wheat straw						
Roskilde, Denmark	0.016—0.028	0.021	5	1972—76	Control data from field experiments at Riso Natl. Laboratory	74
Gujarat State, India	—	0.59	—	1970	ADPLAS method	64
Wheat, *Triticum* spp.						
Naot-Mordechai, Israel	6.1—22.6	12.9	5	1957	ADDT method, tissue not specified	65

[a] Crop components are typically those customarily offered to or consumed by the animal, e.g., aerial parts or standard mixtures of plant components unless otherwise specified.

[b] Crops are typically from normal cultivated or uncultivated fields or experimental fields within the general geographical location (region, state, province, country) specified. Crops from greenhouse plot or pot studies are noted in the remarks column. All data refer to situations where no selenium has purposely been added. Numbers in parenthesis refer to the number of different sites selected for collection of crop material.

[c] Selenium concentrations refer to total selenium reported on a dry matter basis unless otherwise noted. Drying conditions reported: no details, air dried or oven dried with no details, freeze dried, dried at <40°, <60°, and at various temperatures in the range 30—100°C for various times as well as analysis of dry material and conversion of data to analytical dry matter determined by drying at 105°C. Concentration ranges are as reported in the literature or as estimated/calculated from available information by the author of this chapter. Mean concentrations ± standard deviations are as reported or estimated by this author. Median values are noted (m) and were generally estimated by this author. Occasionally ranges are ranges of mean values and means are means of mean values. Values in units other than mg/kg have been converted to mg/kg. The most frequently used analytical method (80%) was acid decomposition-fluorometry followed by acid decomposition-hydride generation atomic absorption spectrometry (10%) with procedures based on neutron activation analysis, light absorption spectrometry, and titrimetry making up the rest. Occasionally, the method of analysis was not reported.

[d] Number of samples analyzed as reported in the literature or calculated or estimated by the present author from information in the literature; a blank indicates data not reported. Some calculations or estimates may be lower limits.

[e] Date of sample collection if available, otherwise date of publication.

[f] This column contains additional information such as tissue components analyzed, soil type and relation of crop selenium to soil selenium levels, age of crop or stand, greenhouse plot or pot experiments, and crop notation used by authors of the reference. Mention of analytical method is because of suspected insufficiency of method and data: ADDT — acid decomposition, distillation, titrimetry; ADPLAS — acid decomposition, precipitation, light absorption spectrometry.

[g] Information has been grouped by crops assigned common and taxonomic names from the literature surveyed. Taxonomic names have been verified with standard botanical literature sources or have been obtained from same. Due to occasionally incomplete crop nomenclature specification in the literature and analyses of crop mixtures, authors' descriptive names of the materials have been included. With the exception of cereal crops, varieties have been omitted.

[h] Mean ± standard error.

[i] Range of mean values, mean is mean of means.

[j] First 18 species listed in footnote m with the exception of alfalfa (lucerne, *Medicago sativa*).

[k] Other entries for this species appear under the heading "Grasses" which includes mixtures of species.

[l] Entries under the general heading "Pasture samples" include those for which the terms pasture samples, pasture species, pasture crops, or mixed herbage are used in the literature. Typically the composition and species are unspecified, and some data are ranges or averages over many species listed in remarks or footnotes. Information for other pasture species defined in the literature or otherwise denoted as "Grasses" or "Grasses + legumes" is under these headings.

[m] Italian ryegrass (*Lolium multiflorum*), *Poa trivialis, Bromus mollis,* barley grass (*Hordeum murinum*), sweet vernal (*Anthoxanthum orderatum*), Chewings fescue (*Festuca rubra*), Timothy (*Phleum pratense*), tall fescue (*Festuca arundinacea*), cocksfoot (*Dactylis glomerata*), meadow fescue (*Festuca pratensis*), browntop (*Agrostis tenuis*), Yorkshire fog (*Holcus lanatus*), perennial ryegrass (*Lolium perenne*), Lucerne (*Medicago sativa*), white clover (*Trifolium repens*), suckling clover (*T. dubium*), subterranean clover (*T. subterraneum*), red clover (*T. pratense*), yellow sour weed (*Oxalis spp.*), yarrow (*Achillea millefolium*), mouse-ear (*Cerastium holosteoides*), dandelion (*Taraxacum officinale*), plantain (*Plantago lanceolata*), daisy (*Bellis perennis*), hawkbit (*Leontodon taraxacoides*), dovesfoot (*Geranium molle*), dock (*Rumex obtusifolius*).

[n] *Branchiaria* spp. *Brachiaria ruziziensis, Brachiaria platynota, Chloris gayana, Cynodon dactylon, Digitaria scalarum, Hyparrhenia rufa, Panicum maximum, Paspalum commersonii, Pennisetum purpureum, Setaria sphacelata.*

[o] *Brachiaria* spp., *Brachiaria jubata, Brachiaria platynata, Setaria sphacelata.*

[p] *Centrosema pubescens, Chloris gayana, Cynodon dactylon, Hyparrhenia filipendula, Hyparrhenia rufa, Panicum maximum, Pennisetum clandestinum, Paspalum commersonii, Pennisetum purpureum, Stylosanthes gracilis, Setaria sphacelata.*

[q] *Brachiaria* spp. *Brachiaria brizantha, Brachiaria decumbens, Centosema pubescens, Chloris gayana, Hyparrhenia* spp., *Hyparrhenia filipendula, Hyparrhenia rufa, Panicum maximum, Paspalum commersonii, Pennisetum purpureum, Setaria* spp., *Setaria trinerva, Sporobolis pyramidalis, Stylosanthes gracilis.*

e.g., above ground parts or standard mixtures of plant components. Crops are typically from normal cultivated or uncultivated fields or experimental fields located within the geographical location specified. Greenhouse plot or pot studies are noted. All concentration data in this table and chapter refer to situations where *no selenium has been added purposely* in any form (selenium salts or soil amendments) and, hence, is taken to reflect natural conditions.

The relatively low, sub-milligram-per-kilogram levels of selenium generally occurring in feed crops dictates the use of a sufficiently detective analytical method for determining concentrations of the element. For this reason concentration information selected in this table and chapter is from publications after the early 1960s, when the detective fluorometric method was developed. Up to then, the low natural levels of selenium could not be reliably or conveniently measured with the old distillation-titrimetric, nephelometric, or photometric procedures. These techniques were, however, suitable to determine the high concentrations of selenium in selenium accumulator plants and other plants growing on seleniferous soils, and such data is included in this chapter.

III. FEED INGREDIENTS AND RATIONS

Although feed crops discussed in Section II are primary ingredients of animal feedstuffs, the term ''feed ingredient'' is used here as a distinct term applying to processed crops, by-products of food processing, and processed fish, meat, and other materials suitable for feeding animals. For commodities subject to little processing, it is conceivable to question placement here or in Section II as a feed crop. Such feed ingredients, together with concentrates (generally high-energy, grain-based protein mixtures with or without vitamin, mineral, and antibiotic mixtures), mineral and vitamin premixes, and feed crops, constitute the final mixed diet or ration. Concentration data for selenium are presented in Table 2 with products divided among the three categories: feed ingredients, concentrates/premixes, and diets/rations.

Selenium concentrations in feed ingredients vary as a function of commodity and geographic origin as well as processing. Levels range from ca. 0.001 mg/kg for products such as lard, cellulose powder, salt, and vitamin mixtures, through ca. 0.01 mg/kg for torula yeast, alfalfa, and grass meals, ca. several milligrams per kilogram for fish meals, to 10 to 30 mg/kg in ground corn, wheat, and barley grain from South Dakota.

Rations include both normal mixed diets used for feeding livestock such as cattle and swine and experimental diets formulated for controlled investigations on poultry, rats, and larger animals. A cooperative study of dietary selenium concentrations in diets that were fed to growing swine in 13 U.S. states was conducted by Ku et al.[29] Corn was used as the main energy source in the majority of the diets, with soybean meal as a primary supplemental protein source in all diets. Selenium concentrations ranged from 0.027 to 0.493 mg/kg for the diets from Virginia and South Dakota, respectively. The authors state that while the variation in selenium content may have been influenced by variations in ingredients used and method of processing, geographical factors associated with selenium content and availability were prime reasons for the wide variation of selenium contents of the rations. Experimental diets listed in the table are of controlled selenium content used to investigate selenium-responsive diseases. A low-selenium diet for chick and rat experiments based on torula yeast, glucose or sucrose, cellulose, and lard with low- and high-selenium ground alfalfa was used to study the value of alfalfa in preventing selenium deficiencies.[30] Basal rations formulated with low-selenium alfalfa contained a total of 0.0004 to 0.0013 mg Se per kilogram and resulted in a high incidence of exudative diathesis in chicks and liver necrosis in rats. In a study of nutritional pancreatic atrophy, Combs et al.[31] were concerned about the use of an amino acid-based, low-selenium diet to elicit deficiency symptoms. In their view, the use of such an artificial diet raises questions about the role of the type of

Table 2
CONCENTRATION OF SELENIUM IN ANIMAL FEED INGREDIENTS AND RATIONS[a]

Geographical location[b]	Selenium concentration (mg/kg)[c]		N[d]	Date[e]	Remarks[f]	Ref.
	Range	Mean				
Feed Ingredients						
Alfalfa meal						
Queensland, Australia	—	0.25	—	1982		118
Canada	0.02—0.14	0.06 ± 0.007	16	1971		77
Indiana, U.S.	0.047—0.0080	0.065	3	1976		119
New York, U.S.	—	0.21	—	1975		120
Ohio, U.S.	0.15—2.22	—	—	1982	For Se deficiency studies	121
Vancouver, BC, Canada	0.03—0.33	0.18 ± 0.09	13	1971—72	Processed and/or imported ingredients	76
Anchovy fish meal						
Chile	0.84—2.60	1.35 ± 0.45	12	1969		122
Peru	—	1.0	1	1973	Factory fish meal, lipid extracted in laboratory	123
Peru	—	1.39 ± 0.20	—	1969		122
South Africa	—	1.2	1	1973	Factory fish meal, lipid extracted in laboratory	123
Aquatic plants (7 species)						
New York, U.S.	0.2—0.4	0.3 m	7	1977	Identified species: clasping leaf pondweed (*Potamogeton richardsonii*), coontail (*Ceretophyllum demersum*), curly leafed pondweed (*Potamogeton crispus*), sago pondweed (*Potamogeton pectinatus*), water milfoil (*Myriophyllum spicatum*), water stargrass (*Heteranthera dubia*), and wild celery (*Vallisneria americana*), used as feeds in rations for lambs	124
Barley						
South Dakota, U.S.	—	15.4	—	1938	Ground toxic barley no. 587	125
Beet pulp						
Sweden	0.044—0.310	0.171	4	1966—67		100
Blood meal						
Queensland, Australia	1.00—1.26	1.13	2	1982		118
Vancouver, BC, Canada	0.61—1.03	0.80 ± 0.16	6	1971—72	Processed and/or imported ingredients	76
Canada	0.51—1.16	0.69 ± 0.060[g]	13	1971		77

Table 2 (continued)
CONCENTRATION OF SELENIUM IN ANIMAL FEED INGREDIENTS AND RATIONS[a]

Geographical location[b]	Selenium concentration (mg/kg)[c]		N[d]	Date[e]	Remarks[f]	Ref.
	Range	Mean				
Blue whiting meal						
Denmark	—	1.57	1	1985	Selected whole meal for test diets	126
Bone meal						
Kenya	—	0.075		1973	Department of Animal Production, Kabete, Kenya	88
Bran						
Vancouver, BC, Canada	0.71—1.26	0.99 ± 0.18	14	1971—72	Processed and/or imported ingredients	76
Brewer's grains						
Canada	0.81—1.07	0.94 ± 0.060[g]	4	1971		77
New York, U.S.	—	0.70	—	1975		120
Brewer's yeast						
Vancouver, BC, Canada	0.29—0.71	0.44 ± 0.15	13	1971—72	Processed and/or imported ingredients	76
Brewer's yeast						
New York, U.S.	—	1.5	—	1975		120
U.S.	—	0.965	2	1961	Debittered	127
U.S.	0.110—0.123	0.116	3	1961		127
Buttermilk						
Canada	0.11—0.13	0.12 ± 0.006[g]	3	1971		77
Calcium carbonate						
Canada	0.04—0.05	0.04 ± 0.003[g]	3	1971		77
Calcium phosphate						
Canada	0.39—1.02	0.65 ± 0.080[g]	11	1971		77
Kenya	—	0.173	—	1973	Department of Animal Production, Kabete, Kenya	88
Ohio, U.S.	—	0.440	—	1973	Ingredient used in conventional swine diets	128
Capelin fish meal						
Finland	0.47—0.68	0.58	2	1973	Laboratory-produced products	123
Norway	0.6—1.6	1.12	7	1973	Factory fish meal, lipid extracted in laboratory	123
Norway	1.04—1.88	1.38	7	1980	Commercial factory fish meal	129

Casein						
Canada	0.17—0.17	0.17	2	1971		77
The Netherlands	—	0.290 ± 0.034[h]	5	1976	Grade A	130
The Netherlands	—	0.121 ± 0.012[h]	5	1976	Grade B	130
The Netherlands	—	0.325 ± 0.036[h]	5	1976	Grade C	130
Cat food						
New York, U.S.	—	0.3 ± 0.0	6	1978	Commercial, dry	131
New York, U.S.	—	4.7 ± 1.2	18	1978	Red meat tuna	131
Cellulose powder						
The Netherlands	—	0.0018 ± 0.0018[h]	3	1976		130
Citharinous fish meal						
Kenya	—	0.390	—	1973	Department of Animal Production, Kabete, Kenya	88
Corn						
Vancouver, BC, Canada	0.22—0.40	0.32 ± 0.06	13	1971—72	Processed and/or imported ingredients	76
Norway	0.03—0.32	0.08	20	1980	Corn/corn grits, imported ingredient	132
New York, U.S.	—	1.0	—	1975		120
Ohio, U.S.	—	0.019	—	1973	Ingredient used in conventional swine diets	128
South Dakota, U.S.	—	29	—	1938	Ground toxic corn no. 570	125
Corn bran						
Kenya	—	0.493	—	1973	Department of Animal Production, Kabete, Kenya	88
Corn gluten feed						
Canada	0.33—0.46	0.38 ± 0.040[g]	3	1971	With distiller's grains	77
Canada	0.12—0.25	0.17 ± 0.010[g]	12	1971		77
Corn gluten meal						
Canada	0.20—0.57	0.31 ± 0.030[g]	11	1971		77
Norway		0.54	1	1980	Commercial material	129
Corn meal						
Kenya	0.041—0.115	—	—	1973	Department of Animal Production, Kabete, Kenya	88
Cottonseed cake						
Kenya	—	0.424	—	1973	Department of Animal Production, Kabete, Kenya	88
Cottonseed meal						
Queensland, Australia	0.34—0.50	0.42	2	1982		118
Norway	0.07—1.1	0.37	14	1980	Imported ingredients	132
New York, U.S.	—	0.96	—	1975		120
Crab meal	0.15—0.34	0.25	2	1969	King crab meal	122
	0.76—6.70	3.73	2	1969	Blue crab meal	122

Table 2 (continued)
CONCENTRATION OF SELENIUM IN ANIMAL FEED INGREDIENTS AND RATIONS[a]

Geographical location[b]	Selenium concentration (mg/kg)[c]		N[d]	Date[e]	Remarks[f]	Ref.
	Range	Mean				
Dextrose						
Indiana, U.S.	—	0.014	1	1976		119
Distiller's corn grain						
Vancouver, BC, Canada	0.25—0.64	0.48 ± 0.15	13	1971—72	Processed and/or imported ingredients	76
Distiller's grains						
Canada	0.16—0.56	0.35 ± 0.040[g]	12	1971		77
Indiana, U.S.	0.479—0.506	0.491	3	1976	Dried	119
Distiller's grains & solubles						
Canada	—	0.43	1	1971		77
New York, U.S.	—	0.32	1	1975	Dried	120
Distiller's solubles						
Canada	0.15—0.59	0.33 ± 0.060[g]	8	1971		77
Vancouver, BC, Canada	0.34—0.92	0.67 ± 0.20	12	1971—72	Processed and/or imported ingredients	76
Egg albumen						
Indiana, U.S.	—	2.7	1	1976	Dried	119
Feather meal						
Vancouver BC, Canada	0.72—1.15	0.90 ± 0.15	13	1971—72	Processed and/or imported ingredients	76
Canada	0.60—0.88	0.72 ± 0.030[g]	8	1971		77
Fish meal	1.25—2.94	1.96 ± 0.110[g]	14	1971		77
Fish solubles						
New York, U.S.	—	2.1	—	1975		120
Grass meal						
Vancouver, BC, Canada	0.03—0.27	0.14 ± 0.08	13	1971—72	Processed and/or imported ingredients	76
Groundnut						
Kenya	—	0.392	—	1973	Department of Animal Production, Kabete, Kenya	88
Herring fish meal						
Canada	2.21—3.36	2.84 ± 0.140[g]	10	1971		77
Vancouver, BC, Canada	1.32—3.02	1.90 ± 0.48	13	1971—72	Processed and/or imported ingredients	76

East coast, Canada	1.30—2.60	1.95 ± 0.37	12	1969	Atlantic sea herring	122
North Sea	1.5—4.2	2.4	4	1973	Factory fish meal (summer), lipid extracted in laboratory	123
Norway	—	2.78 ± 0.67	—	1969		122
Norway	1.6—2.1	1.9	4	1973	Factory fish meal (summer), lipid extracted in laboratory	123
Norway	1.8—2.1	2.0	4	1973	Factory fish meal (winter), lipid extracted in laboratory	123
Shetland, U.K.	1.37, 1.37	1.37	2	1973	Laboratory-produced products	123
New York, U.S.	—	3.6	—	1975		120
Hog kidney						
Ohio, U.S.	—	12.8	—	1982	For Se deficiency studies	121
Lard						
The Netherlands	—	0.0013 ± 0.0010[h]	3	1976		130
Limestone	0.02—0.17	0.07 ± 0.020[g]	11	1971		77
Kenya	—	0.032	—	1973	Department of Animal Production, Kabete, Kenya	88
Ohio, U.S.	—	0.088	—	1973	Ingredients used in conventional swine diets	128
Ohio, U.S.	—	0.01	1	1974		133
Linseed meal						
Canada	0.65—1.51	1.05 ± 0.070[g]	10	1971		77
Indiana, U.S.	0.091—0.159	0.125	2	1976		119
Mackerel fish meal						
Shetland, U.K.	1.46—2.38	1.92	2	1973	Laboratory-produced products	123
North Sea	2.5—4.2	3.4	5	1973	Factory fish meal, lipid extracted in laboratory	123
Norway	5.87—6.46	6.17	2	1980	Commercial factory fish meal	129
Maizena						
The Netherlands	—	0.0022 ± 0.0017[h]	3	1976		130
Malt sprouts						
Canada	0.29—0.97	0.60 ± 0.100[g]	6	1971		77
Vancouver, BC, Canada	0.22-0.88	0.48 ± 0.18	12	1971—72	Processed and/or imported ingredients	76
Meat and bone meal						
Queensland, Australia	0.22-0.49	0.33 ± 0.08	18	1982		118
Kenya	—	0.642	—	1973	Department of Animal Production, Kabete, Kenya	88
New York, U.S.	—	0.69	—	1975		120
Ohio, U.S.	—	0.34	—	1982	For Se deficiency studies	121
Meat meal						
Queensland, Australia	0.36—0.65	0.44	4	1982		118
Vancouver, BC, Canada	0.27—0.82	0.48 ± 0.17	13	1971—72	Processed and/or imported ingredients	76
Canada	0.20—0.66	0.40 ± 0.020[g]	34	1971		77
Ohio, U.S.	0.13—0.24	0.18	2	1974		133
Iowa, U.S.	—	0.84	1	1974		133

Table 2 (continued)
CONCENTRATION OF SELENIUM IN ANIMAL FEED INGREDIENTS AND RATIONS[a]

Geographical location[b]	Selenium concentration (mg/kg)[c]		N[d]	Date[e]	Remarks[f]	Ref.
	Range	Mean				
Menhaden fish meal	—	2.04	—	1982	Lyophilized	134
	—	2.33	—	1982	Commercial	134
	—	2.22 ± 0.72	—	1969		122
Atlantic	0.75—4.20	2.09 ± 0.99	12	1969	Heat transfer process, plus solubles	122
Atlantic & Gulf of Mexico	1.22—3.98	2.22 ± 0.72	12	1969	Wet reduction process, minus solubles	122
Mississippi Delta, U.S.	1.10—3.70	1.93 ± 0.80	12	1969	Wet reduction process, minus solubles	122
New York, U.S.	—	2.6	—	1975		120
Ohio, U.S.	—	4.6	—	1982	For Se deficiency studies	121
Menhaden fish solubles						
Atlantic coast, U.S.	1.4—3.6	2.22 ± 0.443	20	1970	From plants along Atlantic coast	135
Gulf of Mexico	2.2—3.3	2.59 ± 0.335	18	1970	From plants along Gulf coast	135
Milk						
Indiana, U.S.	0.055—0.110	0.066	10	1976		119
Milk replacer (commercial)						
Indiana, U.S.	—	0.115	1	1972		119
Molasses						
Indiana, U.S.	—	0.329	1	1972		119
Norway pout fish meal						
North Sea	1.5—2.3	1.9	3	1973	Factory fish meal, lipid extracted in laboratory	123
Offal meal						
Queensland, Australia	0.66—0.98	0.82	2	1982		118
Pilchard fish meal						
Denmark	—	2.02	—	1985	Selected whole meal for test diets	126
Pollard						
Kenya	—	0.064	—	1973	Department of Animal Production, Kabete, Kenya	88
Poultry by-product meal						
New York, U.S.	—	17	—	1975		120
Protopterous fish meal						
Kenya	—	0.640	—	1973	Department of Animal Production, Kabete, Kenya	88

Rapeseed meal						
Canada	0.46—1.19	0.92	10	1968—69	Grown in Manitoba, Saskatchewan, and Alberta	136
Canada	0.46—1.90	0.98 ± 0.080[g]	19	1971		77
(RSM A), Canada	—	0.82	—	1974	Name indicates location where processed or rapeseed type, number indicates year produced, letters indicate cooperative samples supply by the Rapeseed Association used to test biological activity of Se equivalent to sodium selenite in the prevention of exudative diathesis in chicks; range 0.37—1.76 (mean = 0.95) mg/kg (n = 11)	137
(RSM D), Canada	—	0.83	—			
(RSM E), Canada	—	0.93	—			
(RSM H), Canada	—	0.44	—			
(RSM L), Canada	—	1.63	—			
(RSM P), Canada	—	0.54	—			
(Span-71), Canada	—	0.37	—			
(Span-72), Canada	—	1.76	—			
(Alberta 72), Canada	—	1.65	—			
(Man. 72), Canada	—	0.80	—			
(Sask. 72), Canada	—	0.73	—			
Norway	0.05—1.9	0.55	13	1980	Imported ingredients	132
Rock phosphate						
Canada	—	0.64	1	1971		77
Salmon fish meal						
Vancouver, BC, Canada	1.27—2.57	1.91 ± 0.39	11	1971—76	Processed and/or imported ingredients	76
Screenings, No. 1						
Vancouver, BC, Canada	0.42—1.09	0.68 ± 0.21	13	1971—72	Processed and/or imported ingredients	76
Screenings, Refuse						
Vancouver, BC, Canada	0.41—0.85	0.60 ± 0.12	12	1971—72	Processed and/or imported ingredients	76
Sisal waste						
Kenya	0.036—0.082	0.059 ± 0.015	7	1973	Department of Animal Production, Kabete, Kenya	88
Skim milk						
Canada	0.10—0.17	0.12 ± 0.020[g]	4	1971		77
Indiana, U.S.	—	0.076	1	1976	Dried	119
Smelt fish meal	0.49—1.23	0.95 ± 0.29	6	1969	Smelt and perch	122
Sorghum						
Norway	0.04—0.19	0.11	11	1980	Imported ingredients	132
Soybean cake						
Kenya	—	0.433	—	1973	Department of Animal Production, Kabete, Kenya	88

Table 2 (continued)
CONCENTRATION OF SELENIUM IN ANIMAL FEED INGREDIENTS AND RATIONS[a]

Geographical location[b]	Selenium concentration (mg/kg)[c]		N[d]	Date[e]	Remarks[f]	Ref.
	Range	Mean				
Soybean meal						
Queensland, Australia	—	0.34	—	1982		118
Vancouver, BC, Canada	0.29—0.78	0.56 ± 0.16	14	1971—72	Processed and/or imported ingredients	76
Canada	0.04—0.20	0.12 ± 0.006[g]	32	1971		77
Norway	—	0.42	1	1980	Commercial, solvent extracted	129
Norway	0.07—1.0	0.49	17	1980	Imported ingredients	132
Illinois, U.S.	0.20—0.21	0.20	2	1974		133
Indiana, U.S.	0.088—0.124	0.104	5	1972		119
Indiana, U.S.	—	0.09	1	1974		133
Iowa, U.S.	—	1.04	1	1974		133
New York, U.S.	—	0.54	—	1975		120
Ohio, U.S.	0.08—0.85	—	—	1982	For Se deficiency studies	121
Ohio, U.S.	0.05—0.13	0.10	4	1974		133
Ohio, U.S.	—	0.075	—	1973	Ingredient used in conventional swine diets	128
Sprat & herring meal						
Denmark	1.25, 1.79, 2.00	—	3	1985	Selected whole meals for test diets	126
Sunflower meal						
Queensland, Australia	0.06—0.43	0.25	2	1982		118
Tilapia fish meal						
Kenya	—	0.750	—	1973	Department of Animal Production, Kabete, Kenya	88
Torula yeast						
Indiana, U.S.	0.011—0.022	0.016	2	1972		119
Ohio, U.S.	—	0.04	—	1982	For Se deficiency studies	121
U.S.	0.0045—0.0076	0.0060	7	1961	USP & feed grades from U.S. sources	127
Tuna	—	9.68	—	1975		138
Oregon, U.S.	—	0.9	—	1983	Precooked	139
Oregon, U.S.	—	1.6	—	1983	Canned, from Bumble Bee ® Seafoods	139
Oregon, U.S.	—	1.7	—	1983	Raw	139

Tuna fish meal	3.40—6.20	4.63 ± 0.98	9	1969		122
New York, U.S.	—	5.0	—	1975		120
Urea, feed grade						
Indiana, U.S.	—	0.60	1	1972		119
Wheat	—	2.00	—	1975		138
New York, U.S.	—	2.0	—	1975		120
Kenya	—	0.096	—	1973	Department of Animal Production, Kabete, Kenya	88
South Dakota, U.S.	—	30	—	1938	Ground toxic wheat no. 607	125
Wheat bran	—	1.66	1	1971	Durum, wheat milling by-product	77
Ontario and Quebec, Canada	0.24—0.40	0.32 ± 0.080[g]	2	1971	Wheat milling by-product	77
Kenya	—	0.098	—	1973	Department of Animal Production, Kabete, Kenya	88
Western Canada	0.53—1.28	0.95 ± 0.070[g]	9	1971	Wheat milling by-product	77
Dallas, SD, U.S.	—	5.4	—	1983	From ND, hard red spring wheat	139
Wheat bread						
Dallas, SD, U.S.	—	3.5	—	1983	Whole, from ND, hard red spring wheat	139
Wheat flour						
Dallas, SD, U.S.	—	3.5	—	1983	Whole, from ND, hard red spring wheat	139
Wheat germ						
Queensland, Australia	—	0.23	—	1982		118
Wheat middlings	—	1.32	1	1971	Durum, wheat milling by-product	77
Ontario and Quebec, Canada	—	0.22	1	1971	Wheat milling by-product	77
Western Canada	0.41—0.89	0.70 ± 0.060[g]	8	1971	Wheat milling by-product	77
Wheat mill run						
Queensland, Australia	—	0.22	—	1982		118
Wheat shorts	—	1.70	1	1971	Durum, wheat milling by-product	77
Vancouver, BC, Canada	0.60—1.51	0.86 ± 0.25	14	1971—72	Processed and/or imported ingredients	76
Ontario and Quebec, Canada	0.10—0.28	0.20 ± 0.080[g]	2	1971	Wheat milling by-product	77
Western Canada	0.32—1.36	0.81 ± 0.090[g]	16	1971	Wheat milling by-product	77
Whey						
Canada	0.04—0.11	0.06 ± 0.006[g]	11	1971		77
White fish meal						
Canada	1.65—2.18	1.83 ± 0.080[g]	6	1971		77
Whiting & herring meal						
Denmark	1.52, 1.82, 2.54	—	3	1985	Selected whole meals for test diets	126
Yeast						
Canada	1.20—1.28	1.23 ± 0.020[g]	4	1971		77

Table 2 (continued)
CONCENTRATION OF SELENIUM IN ANIMAL FEED INGREDIENTS AND RATIONS[a]

Geographical location[b]	Selenium concentration (mg/kg)[c] Range	Mean	N[d]	Date[e]	Remarks[f]	Ref.
Concentrates / Premixes						
Broiler finisher concentrate, Queensland, Australia	—	0.41	—	1982		118
Chick starter premix, Queensland, Australia	—	4.8	—	1982		118
Chicken broiler concentrate, Queensland, Australia	—	0.86	—	1982		118
Concentrate (feed, mixture)						
Eastern Fraser Valley, BC, Canada	0.059—0.406	0.156 ± 0.063	69	1976—77	Commercially prepared concentrate feeds based on barley and corn with protein supplements such as soybean meal and rapeseed meal, some also contain urea, molasses, plant by-products or minerals	73
Finland	—	0.7	—	1980	Composition: fat-free milk powder, fish meal, soybean grist, turnip rape grist, meat-bone meal, fodder yeast, molasses, alfalfa meal, vitamin mixture, calcium phosphate, hostaphos, salt, magnesium oxide, copper oxide, iron sulfate, copper iodide, zinc oxide, manganese oxide, cobalt sulfate	140
Norway	—	0.37 ± 0.07	35	1985	Kufor® A 12.5%, low protein concentrate for cattle	141
Norway	—	0.34 ± 0.09	45	1985	Purkefor®, Svinefor®, and Svinefor® 1, compound pig feeds	141
Sweden	0.102—0.188	0.159	3	1966—67		100
Layer premix						
Queensland, Australia	—	0.07	—	1982		118
Mixed concentrates for swine and poultry						
Norway	0.09—0.34	0.22	45	1980	Standard mixtures of concentrates for swine and poultry from different manufacturers (soybean meal, herring meal, blood meal, meat-bone meal)	132

Salt mixture						
The Netherlands	—	0.0022 ± 0.0010[h]	3	1976		130
Trace mineral salt						
Ohio, U.S.	—	0.022	—	1973	Ingredient used in conventional swine diets	128
Vitamin premix						
Ohio, U.S.	—	0.070	—	1973	Ingredient used in conventional swine diets	128
Vitamin premix						
The Netherlands	—	0.0013 ± 0.0010[h]	6	1976		130
Rations / Diets						
Amino acid basal diet	—	0.012	—	1975	Composition of basal diet for production of pancreatic fibrosis in chicks: sucrose, amino acid mix, soybean oil, vitamin mix, mineral mix, sodium taurocholate, oleic acid, and choline chloride; selenium levels augmented with tuna (9.68 mg Se/kg), wheat (2.00 mg Se/kg), organoselenium, and selenite	138
Aquatic ration for lambs						
New York, U.S.	—	0.2	—	1977	Composition: aquatic plants, alfalfa meal, oats (crimped), corn (cracked), wheat bran, soybean meal, molasses, salt, and vitamin supplements	124
Basal diet for chick experiments						
New York, U.S.	0.015—0.040	—	5	1975	Composition: casein, isolated soy protein, torula yeast, stripped corn oil, stripped lard, glucose, choline chloride, vitamin mix, mineral mix, amino acid mix, and cellulose	120
Basal chick/poultry starter diets (low selenium)						
New York, U.S.	—	0.07	—	1971	Composition: yellow corn (low selenium), stabilized grease, soybean meal (50% protein, low selenium), dicalcium phosphate, limestone, managanese sulfate, zinc oxide, iodized salt, vitamin premix, DL-methionine, choline chloride	142
Basal diets for dairy cows						
Missouri, U.S.	—	0.334 ± 0.048	9	1980	Composition: ground corn, dried brewer's grains, soybean meal, linseed meal, molasses, calcium carbonate, and trace mineral salt	143
Basal diets for growing/finishing swine						

Table 2 (continued)
CONCENTRATION OF SELENIUM IN ANIMAL FEED INGREDIENTS AND RATIONS[a]

Geographical location[b]	Selenium concentration (mg/kg)[c] Range	Mean	N[d]	Date[e]	Remarks[f]	Ref.
Michigan, U.S.	0.041, 0.042	0.041	2	1971	Composition: corn, soybean meal (49% crude protein), dicalcium phosphate, limestone, salt, vitamin trace mineral premix, arsanilic acid, and bifuran	144
Michigan, U.S.	0.036, 0.040	—	2	1978	Composition: ground yellow dent corn, ground, solvent-extracted soybean seed, dicalcium phosphate, ground limestone, salt, vitamin-trace mineral premix	145
Basal diets for swine,						
Ontario, Canada	—	0.03	—	1977	Composition: soft dent corn grain, ground, solvent-extracted soybean seed, salt, calcium phosphate, supplement, limestone, trace mineral premix, and vitamin premix; for research to determine effect of supplemental vitamin E and Se on swine reproduction	146
Michigan, U.S.	0.041, 0.042	—	—	1973	Composition: corn, soybean meal, dicalcium phosphate, limestone, salt, trace mineral salt (high Zn), vitamin-trace mineral premix and vitamin-antibiotic premix; compounded at East Lansing, MI, and Brookings, SD, from locally purchased ingredients	147
South Dakota, U.S.	0.236, 0.296	—	—			
South Dakota, U.S.	0.438, 0.448	—	—			
Basal starter diet for chicks and poults (high selenium)						
New York and northeastern U.S.	—	0.68	—	1971	Composition: yellow corn (low selenium), stabilized grease, soybean meal (high selenium), wheat (high selenium), fishmeal, dicalcium phosphate, limestone, manganese sulfate, zinc oxide, iodized salt, vitamin premix, DL-methionine, and choline chloride	142
Basal starter diet for chicks and poults (low selenium)						

New York and northeastern U.S.	—	0.07	—	1971	Composition: yellow corn (low selenium) stabilized grease, soybean meal (low selenium), dicalcium phosphate, limestone, manganese sulfate, zinc oxide, iodized salt, vitamin premix, DL-methionine, and choline chloride	142
Basic diet — beef cow production farm						
Sweden	—	0.07	—	1985		148
Cattle diet (finishing)						
Virginia & Florida, U.S.	0.043—0.121	—	3	1981	Whole shelled corn, 7% corn silage with basal or added protein as soybean meal or linseed meal	99
Cattle diet (growing)						
Virginia & Florida, U.S.	0.048—0.049	—	3	1981	55% corn silage, shelled corn, and soybean meal to give 11.6—16.5% protein (3 levels)	99
Chick starter feed						
Queensland, Australia	0.10—0.33	0.21	4	1982		118
Control ration for lambs						
New York, U.S.	—	0.2	1	1977	Composition: alfalfa meal, oats (crimped), corn (cracked), wheat bran, soybean meal, molasses, salt, and vitamin supplements	124
Experimental all-mash laying rations (control, Nos. 2,3,4)						
South Dakota, U.S.	0, 2.5, 5, 10	—	—	1938	Composition of 4 rations: ground corn (toxic #570, 29 mg Se/kg), ground barley (toxic, #587, 15.4 mg Se/kg), ground wheat (toxic, #607, 30 mg Se/kg), ground normal corn, ground normal barley, ground normal wheat, wheat bran, wheat middlings, meat and bone scraps, alfalfa leaf meal, dried buttermilk, salt and cod liver oil concentrate; high-level Se corn, barley, and wheat absent from controls	125
Experimental chick diet						
Ithaca, NY, U.S.	0.0009, 0.0013	—	2	1965	Composition: torula yeast, glucose, stripped lard, cellulose, L-arginine-HCl, glycine, DL-methionine, DL-phenylalanine, vitamin mix, mineral mix, ground alfalfa; this ration resulted in high incidence of exudative diathesis in chicks	30
Experimental chick diets, 1A, 2A, 3A						
Denmark	—	0.04	—	1985	Test diets for chicks to determine available Se, prepared using varying amounts of fishmeal 1, 2, or 3 containing 1.25, 2.00, and 2.54 mg Se/kg, varying amounts of single-cell protein yeast and casein and common ingredients: soybean meal (solvent ex-	126
Experimental chick diets, 1B, 2B, 3B						
Denmark	—	0.08	—			
Experimental chick diets, 1C, 2C, 3C						

Table 2 (continued)
CONCENTRATION OF SELENIUM IN ANIMAL FEED INGREDIENTS AND RATIONS[a]

Geographical location[b]	Selenium concentration (mg/kg)[c] Range	Mean	N[d]	Date[e]	Remarks[f]	Ref.
Denmark	—	0.12	—		tracted), soybean oil, finely ground oat hulls, L-arginine, DL-methionine, dicalcium phosphate, calcium carbonate, sodium chloride, vitamin and mineral mixture, and corn starch	
Experimental diet for chicks (low selenium)						
Ithaca, NY, U.S.	0.017—0.023	0.019	4	1971	Low-Se diet containing corn, corn oil, torula yeast, DL-methionine, L-arginine-HCl, mineral mixture, and vitamin mixture	149
Experimental diet for chicks (high selenium)						
Ithaca, NY, U.S.	0.620—0.780	0.730	4	1971	High-Se diet containing corn, corn oil, torula yeast, DL-methionine, L-arginine-HCl, mineral mixture, and vitamin mixture	149
Experimental chick diet						
Ithaca, NY, U.S.	0.0009—0.0013	—	2	1965	Composition: torula yeast, glucose, stripped lard, cellulose, L-arginine-HCl, glycine, DL-methionine, DL-phenylalanine, vitamin mix, mineral mix, and ground alfalfa; this ration resulted in a high incidence of exudative diatheses in chicks	30
Experimental diet containing casein A						
The Netherlands	—	0.106 ± 0.013	—	1976	Calculated Se level; diet also contains maizena, lard, cellulose powder, salt mixture, and vitamin mixture; failed to induce myopathy in vitamin E-deficient ducks	130
Experimental diet containing casein B						
The Netherlands	—	0.054 ± 0.005	6	1976	Measured Se level; diet also contains maizena, lard, cellulose powder, salt mixture, and vitamin mixture; caused myopathy in vitamin E-deficient ducks	130
Experimental diet containing casein C						
The Netherlands	—	0.117 ± 0.014	—	1976	Calculated Se level; diet also contains maizena, lard, cellulose powder, salt mixture, and vitamin mixture, failed to induce myopathy in vitamin E-deficient ducks	130

Experimental diets for pigs						
People's Republic of China	—	0.02	—	1982	Low-Se yeast diet	150
People's Republic of China	—	0.01	—	1982	From staple cereals grown in Keshan disease-affected province	150
People's Republic of China	—	0.15	—	1982	From staple cereals grown in nonaffected area (control diet)	150
Experimental diets for rats						
People's Republic of China	—	0.007	—	1982	From staple cereals grown in Keshan disease-affected province	150
People's Republic of China	—	0.030	—	1982	From staple cereals grown in nonaffected area (control diet)	150
Experimental diets for rats/swine						
South Dakota, U.S.	0.54—0.47	0.51	2	1984	Composition: seleniferous wheat (20 mg Se/kg), wheat, seleniferous oats (7 mg Se/kg), oats, corn soybean meal, dicalcium phosphate, calcium carbonate, trace mineralized salt, and vitamin-antibiotic premix; wheat and oats were replaced with seleniferous wheat and oats to give desired Se levels	151
	2.63—2.58	2.61	2			
	5.69—5.60	5.65	2			
	8.33—8.40	8.37	2			
Experimental duck rations						
Ontario, Canada	0.06—0.08	0.07	3	1974	Composition: corn, soybean, meal, limestone, dicalcium phosphate, vitamin mixture, trace element mixture, and antibiotics	152
Experimental horse feeds						
Finland	0.03—0.14	—	—	1980	Composition: hay, oats, bran, promeca w or w/o trace element mixture	153
Experimental swine basal diet (cull peas)						
Washington, U.S.	—	0.087	—	1974	Composition: peas cull grade (*Pisum sativum*), barley, yellow corn, soybean meal, herring fish meal, partially delactosed dried whey, DL-methionine, dicalcium phosphate, limestone, sodium chloride, trace mineral mix, vitamin mix, and antibiotic premix	154
Experimental swine control diet						
Washington, U.S.	—	0.157	—	1974	Composition of non-pea control diet: barley, yellow corn, soybean meal, herring fish meal, partially delactosed dried whey, DL-methionine, dicalcium phosphate, limestone, sodium chloride, trace mineral mix, vitamin mix, and antibiotic premix	154
Experimental rat diet						
Ithaca, NY, U.S.	0.0004—0.0005	0.0005	—	1965	Composition: torula yeast, sucrose, cellulose, stripped lard, vitamin mixture, mineral mixture, and alfalfa; this diet induced liver necrosis in rats	30
Feeds for laying hens and chickens						
Norway	—	0.36 ± 0.07	20	1985		141

Table 2 (continued)
CONCENTRATION OF SELENIUM IN ANIMAL FEED INGREDIENTS AND RATIONS[a]

Geographical location[b]	Selenium concentration (mg/kg)[c] Range	Mean	N[d]	Date[e]	Remarks[f]	Ref.
Low-selenium practical-type diet						
People's Republic of China	0.007—0.008	—	—	1984	A practical-type diet, as opposed to crystalline amino acid experimental diet, produced using very low-Se ground yellow corn and dehulled soybean meal from one of the most severely Se-deficient regions of the world, Heilongjiang Province, People's Republic of China; also contains calcium phosphate, limestone, salt, DL-methionine L-lysine-HCl, and a vitamin-mineral premix	31
Poultry finisher feed						
Queensland, Australia	0.10—0.34	0.21 ± 0.07	15	1982		118
Poultry grower feed						
Queensland, Australia	0.15—0.31	0.22 ± 0.07	7	1982		118
Poultry growing all-mash feed						
Queensland, Australia	0.14—0.43	0.28	2	1982		118
Poultry laying all-mash feed						
Queensland, Australia	—	0.13	—	1982		118
Poultry starter feed						
Queensland, Australia	0.14—0.32	0.22 ± 0.06	11	1982		118
Swine diets						
Arkansas, U.S.	—	0.152	—	1972	Significant regional differences, Se in pig muscle related to dietary Se levels; highest levels in South Dakota still well below recognized toxic levels of Se for swine	29
Idaho, U.S.	—	0.086	—			
Illinois, U.S.	—	0.036	—			
Indiana, U.S.	—	0.052	—			
Iowa, U.S.	—	0.235	—			
Michigan, U.S.	—	0.040	—			
Nebraska, U.S.	—	0.330	—			
New York, U.S.	—	0.036	—			
North Dakota, U.S.	—	0.412	—			
South Dakota, U.S.	—	0.493	—			
Virginia, U.S.	—	0.027	—			

Wisconsin, U.S.	—	0.178	—			
Wyoming, U.S.	—	0.158	—			
Ohio, U.S.	0.018—0.060	—	—	1973	Actual analyses of several diet samples	128
Turkey grower feed						
Queensland, Australia	—	0.34	—	1982		118
Turkey starter feed						
Queensland, Australia	0.25—0.38	0.32	—	1982		118

[a] Feed ingredients are processed feed and food crops and by-products of food processing, used together with unprocessed feed crops, concentrates, and various premixes to formulate the mixed diet or ration. Nomenclature varies among sources and authors, and generally those names reported in the literature are included.

[b] Commodities are those typically obtained from or imported into geographical locations indicated.

[c] Selenium concentrations refer to total selenium, usually reported on a dry basis by usually acceptable methods of analysis (refer to footnote c, Table 1 for general comment); means are given with standard deviations if reported; medians are indicated by m.

[d] Number of samples as reported in the literature or estimated by the present author from information in the literature; a blank indicates data not reported.

[e] Date of sample collection or publication.

[f] This column contains additional information on the nature or origin of the commodity, composition of diets/rations, and experimental uses of some diets as well as other highlights.

[g] Mean ± standard error.

[h] Mean ± 95% confidence limits.

diet as a factor in the etiology of the disorder. For this reason they prepared a practical-type diet as opposed to a crystalline amino acid experimental diet from very low-selenium ground yellow corn and dehulled soybean meal produced in Heilongjiang Province, People's Republic of China, one of the most severely selenium deficient regions of the world. This diet contained 0.007 to 0.008 mg Se per kilogram.

IV. OTHER PLANTS AND PLANT PRODUCTS

Table 3 lists selenium concentrations in plants and plant products not falling within the categories of feed crops and feed ingredients/rations already treated in the previous sections or selenium accumulator plants to be discussed in the following section. Plants tabulated here are in general nonfood, nonfeed types including wild plants, desert shrubs, lichens, mosses, fungi, trees, and products therefrom such as tobacco, cigarettes, and papers. Again, the odd overlap may occur of an item here with that in the feed crop table.

Several reports have appeared dealing with selenium levels in a variety of tobaccos from various locations around the world. Cigarettes produced from American tobacco contained the lowest selenium levels ranging from 0.010 to 0.075 (mean = 0.038) mg/kg with similar levels of 0.013 to 0.085 (mean = 0.046) mg/kg in American flue-cured and Burley tobaccos.[32] Other published mean or median concentrations for tobaccos from other countries range from 0.11 mg/kg for Rhodesian flue-cured tobacco to 0.82 mg/kg for the air-cured variety from Indonesia.[33] One study[34] of cigarette tobaccos from the U.S., U.K., and Sweden (range of means 0.15 to 0.16 mg/kg) and Mexico and Columbia (range of means 0.48 to 0.49 mg/kg) found a significant difference between the two groups.

Analyses of multiple samples of different paper products obtained in the U.S. indicated a range of individual values of 0.00 to 0.31 mg/kg.[35] The lowest mean value of 0.01 mg/kg was for cigarette paper; the highest mean value of 0.16 mg/kg was observed for corrugated paper.

V. SELENIUM-ACCUMULATING PLANTS

Selenium-accumulating plants and the history of selenium toxicity are intertwined. Reports by Marco Polo in ca. 1295, Stein in 1912 regarding animal disorders in China, and Simon in 1560 relating malformations in chicks and children and loss of hair and nails of people in Columbia, South America, are the earliest reports of diseases later attributed to selenium toxicity.[22] A similar disease afflicting livestock and the populace in the neighborhood of Irapuato, Mexico, was described by Juan Roca[36] to be prevalent since the mid 1700s. In North America the earliest account of this toxicity (later denoted alkali disease when produced by seleniferous grains and grasses, and blind staggers when resulting from the ingestion of seleniferous indicator plants) is in the report of Madison in 1857,[22] describing the symptoms of a fatal disease that afflicted cavalry horses at Fort Randall, South Dakota. In subsequent years numerous incidents, throughout the grazing areas of the western U.S., of acute and chronic poisoning in livestock have been documented and attributed to toxic forage plants. Serious losses to farmers due to livestock deaths and ill thrift led to investigations into the problem by researchers at state and federal agricultural research establishments. Franke and co-workers at the South Dakota Agricultural Experiment Station reported on studies of the selenium toxicity of grains and grasses to livestock.[37] Studies by Beath et al.[38] led to the significant observation that a number of species of *Astragalus, Stanleya, Oonopsis,* and *Xylorhiza* have the ability to absorb vast quantities of selenium from soils or geological deposits and convert it to an available form. The occurrence of these plants is restricted to seleniferous formations and soils. They are known as selenium accumulator, indicator, or converter plants. Trelease and Trelease[39] found that these plants require selenium for growth

Table 3
CONCENTRATION OF SELENIUM IN OTHER PLANTS AND PLANT PRODUCTS[a]

Geographical location[b]	Selenium concentration, (mg/kg)[c] Range	Mean or median	N[d]	Date[e]	Remarks[f]	Ref.
Acacia farnesiana (mimosa)						
NW Queensland, Australia	—	79.4	1	1963	Poison strip area, soil Se 0.15—32.2 (m = 3.3) mg/kg (n = 19)	155
Acacia salicina (doolan)						
NW Queensland, Australia	—	173.7	1	1963	Poison strip area, soil Se 0.15—32.2 (m = 3.3) mg/kg (n = 19)	155
Acer saccharinum (silver maple)						
Montreal, Quebec, Canada	—	141	—	1981	Twigs and leaves, severe injury by Se-polluted air	156
Montreal, Quebec, Canada	—	12	—	1981	Slight injury by Se-polluted air (control)	156
Montreal, Quebec, Canada	—	2.2	—	1981	No injury (control)	156
Aerva persica (kapok bush)						
NW Queensland, Australia	—	115.8	1	1963	Poison strip area, soil Se 0.15—32.2 (m = 3.3) mg/kg (n = 19)	155
Aeschynomene indica (Budda tea)						
Queensland, Australia	—	13.1	1	1963	Landsborough highway traverse, Hughenden, soil Se 0.3—14.5 (m = 3.3) mg/kg (n = 10)	155
Agropyron smithii Rydb. (western wheat grass)						
Colorado, U.S.	4—11	4 m	3	1939	Tops	157
Lemhi, Co., Colorado, U.S.	—	15	1	1939	Tops	157
South Dakota, U.S.	1—79	6 m	32	1942	Soil Se 0.2—384 (m = 4) mg/kg (n = 96)	158
South Dakota, U.S.	3.0—84.0	15.3 m	24	1939	Whole plant	58
South Dakota, U.S.	0.0—84.0	12.1	75	1939	Whole plant	58
South Dakota, U.S.	0.0—60	15.2	20	1939	Whole plant	58
South Dakota, U.S.	1.0—50	8.4	40	1939	Whole plant	58
Lower Brule Indian Reservation, South Dakota, U.S.	0.5—95	2 m	21	1941	Soil Se 0.2—16 (m = 1) mg/kg (n = 80)	134
Alnus incana L.						
Vantaa, Ruskeasanta, Seutula,	—	0.022	—	1974	Needles, soil Se 0.17 mg/kg	159

Table 3 (continued)
CONCENTRATION OF SELENIUM IN OTHER PLANTS AND PLANT PRODUCTS[a]

Geographical location[b]	Selenium concentration, (mg/kg)[c] Range	Mean or median	N[d]	Date[e]	Remarks[f]	Ref.
Finland						
Kaavi, Luikonlahti, Finland	—	0.420	—	1974	Leaves, Soil Se 0.60—5.5 mg/kg	159
Alnus glutinosa L.						
Kaavi, Luikonlahti, Finland	—	0.044	—	1974	Leaves, soil Se 0.60—5.5 mg/kg	159
Amaranthus interruptus						
NW Queensland, Australia	—	65.6	1	1963	Poison strip area, soil Se 0.15—32.2 (m = 3.3) mg/kg (n = 19)	155
Amelanchier alnifolia, (northline serviceberry)						
Alberta, Canada (4)	0.001—0.051	0.009 m	—	1981	Leaves, plants in pots downwind of SO_2 sources	160
Artemisia tridentata Nutt. (big sagebrush)						
Powder River Basin, Wyoming/Montana, U.S. (15)		0.36[g]	15	1977	From same sites as *P. chlorochroa* & *B. gracilis*	161
Washington, U.S.	0.2—1.6	0.11	—	1983	Leaves, at peak litterfall from 4 plants	162
Atriplex spinosa						
Washington, U.S.	—	0.58	—	1983	Leaves, at peak litterfall from 4 plants	162
Betula pubescens Enrh.						
Kaavi, Luikonlahti, Finland	—	0.042	—	1974	Leaves, soil Se 0.60—5.5 mg/kg	159
Bouteloua gracilis (H.B.K.) Lag. (blue grama)						
Powder River Basin, Wyoming/Montana, U.S. (15)		0.19[g]	15	1977	From same sites as *P. chlorochroa* & *A. tridentata*	162
Cigarettes						
Egypt	—	<0.09	—	1985	Filter cigarettes	163
U.S.	0.010—0.075	0.038	13	1984	Different commercial + 1 experimental cigarettes	32
Cigarette ash						
Egypt		0.30 ± 0.005	—	1985		163

Cajanus cajan (Poona pea)						
NW Queensland, Australia	—	30.7	1	1963	Poison strip area, soil Se 0.15—32.2 (m = 3.3) mg/kg (n = 19)	155
Calendula						
Other locations, Israel		0.3	1	1957		65
Calluna vulgaris L.,						
Vantaa, Ruskeasanta, Seutula, Finland	—	<0.010	—	1974	Soil Se 0.17 mg/kg	159
Carex						
Luikonlahti, Patolampi, Finland	—	0.350	—	1974	Soil Se 0.60—5.5 mg/kg	159
Cassia occidentalis (coffee senna)						
Queensland, Australia	—	7.9	1	1963	Landsborough highway traverse, Hughenden, soil Se 0.3—14.5 (m = 3.3) mg/kg (n = 10)	155
NW Queensland, Australia	—	11.1	1	1963	Poison strip area, soil Se 0.15—32.2 (m = 3.3) mg/kg (n = 19)	155
Cenchrus australis (Hillside burr grass)						
NW Queensland, Australia	—	71.1	1	1963	Poison strip area, soil Se 0.15—32.2 (m = 3.3) mg/kg (n = 19)	155
Cenchrus ciliaris (Buffel grass)						
NW Queensland, Australia	6.9—77.6	55 m	4	1963	Poison strip area, soil Se 0.15—32.2 (m = 3.3) mg/kg (n = 19)	155
Chloris barbata (purple-top chloris)						
NW Queensland, Australia	21.2—104.8	63.0	2	1963	Poison strip area, soil Se 0.15—32.2 (m = 3.3) mg/kg (n = 19)	155
Chloris virgata (feathertop Rhodes grass)						
NW Queensland, Australia	—	24.4	1	1963	Poison strip area, soil Se 0.15—32.2 (m = 3.3) mg/kg (n = 19)	155
Crotalaria novaehollandiae (a rattlepod)						
Queensland, Australia	—	18.5	1	1963	Landsborough highway traverse, Hughenden, soil Se 0.3—14.5 (m = 3.3) mg/kg (n = 10)	155
Cynodon dactylon (couch grass)						
NW Queensland, Australia	11.1—263	85 m	6	1963	Poison strip area, soil Se 0.15—32.2 (m = 3.3) mg/kg (n = 19)	155
NW Queensland, Australia	14.9—28.1	21.5	2	1963	Adjacent to poison strip area, soil Se 0.03—385 (m = 2.1) mg/kg (n = 45)	155
Dactyloctenium radulans (button grass)						
NW Queensland, Australia	—	16.2	1	1963	Poison strip area, soil Se 0.15—32.2 (m = 3.3) mg/kg (n = 19)	155

Table 3 (continued)
CONCENTRATION OF SELENIUM IN OTHER PLANTS AND PLANT PRODUCTS[a]

Geographical location[b]	Selenium concentration, (mg/kg)[c] Range	Mean or median	N[d]	Date[e]	Remarks[f]	Ref.
Dichantium-fecundum (gulf blue grass) NW Queensland, Australia	—	5.3	1	1963	Adjacent to poison strip area, soil Se 0.03—385 (m = 2.1) mg/kg (n = 45)	155
Dicranum Vantaa, Ruskeasanta, Seutula, Finland	—	0.150	—	1974	Soil Se 0.17 mg/kg	159
Digitaria ctenantha NW Queensland, Australia	—	<0.5	1	1963	Poison strip area, soil Se 0.15—32.2 (m = 3.3) mg/kg (n = 19)	155
Enneapogon avenaceus (ridge grass) NW Queensland, Australia	—	7.9	1	1963	Poison strip area, soil Se 0.15—32.2 (m = 3.3) mg/kg (n = 19)	155
Enneapogon oblongus (bottle-washer grass) NW Queensland, Australia	—	<0.5	1	1963	Poison strip area, soil Se 0.15—32.2 (m = 3.3) mg/kg (n = 19)	155
Epilobium angustifolium L. Kaavi, Luikonlahti, Finland	—	0.190	—	1974	Soil Se 0.60—5.5 mg/kg	159
Equisetum fluviatile Kaavi, Luikonlahti, Finland	—	0.360	—	1974	Soil Se 0.60—5.5 mg/kg	159
Eragrostis setifolia (neverfail grass) NW Queensland, Australia	<0.5—49.3	25	2	1963	Poison strip area, soil Se 0.15—32.2 (m = 3.3) mg/kg (n = 19)	155
Festuca arundinacea Naot-Mordechai, Israel	—	1.5	1	1957	Soil Se 0.1—6 (m = 0.7) mg/kg (n = 41)	65
Flaveria australasica (speedy weed) NW Queensland, Australia	—	18.4	1	1963	Poison strip area, soil Se 0.15—32.2 (m = 3.3) mg/kg (n = 19)	155

Fungi						
Bela, Yugoslavia; 11 species	0.07—2.0	0.63	11	1976	Whole plants, near sources of contamination	164
Cemsenik, Yugoslavia; 3 species	0.62—1.82	0.62 m	3	1976	Whole plants, near sources of contamination	164
Dvor, Dolenjska, Yugoslavia; 4 species	0.04—2.9	0.60 m	4	1976	Whole plants, noncontaminated rural locations	164
Idrija, Yugoslavia; 3 species	0.18—1.28	0.88	3	1976	Whole plants, near sources of contamination	164
Kurescek, Yugoslavia; 9 species	0.49—4.21	1.76 m	11	1976	Whole plants, noncontaminated rural locations	164
Ljubljana, Yugoslavia; 1 species	—	4.16	1	1976	Whole plants, near source of contamination	164
Smlednik, Yugoslavia; 2 species	0.19—3.24	0.70 m	6	1976	Whole plants, noncontaminated rural locations	164
Glossogyne tenuifolia (native cobbler's peg)						
NW Queensland, Australia	—	31.6	1	1963	Poison strip area, soil Se .5—32.2 (m = 3.3) mg/kg (n = 19)	155
Grass						
Bleiberg-Kreut, Austria	0.03—0.13	0.08 ± 0.03	8	1975	Lead/zinc mining districts	165
Halogeton glomeratus						
Cisco, UT, U.S.	0.5—35.0	5.0 m	6	1962	Leaves, low precipitation, June-Nov. 1959	166
Cisco, UT, U.S.	0.3—6.1	0.9 m	6	1962	Stems, low precipitation, June-Nov. 1959	166
Cisco, UT, U.S.	—	90	—	1962	Leaves, good growth, near normal precipitation, Aug. 1958	166
Crescent Junction, UT, U.S.	0.0—0.3	0.0 m	6	1962	Leaves, low precipitation, June-Nov. 1959	166
Crescent Junction, UT, U.S.	0.0—0.5	0.1 m	6	1962	Stems, low precipitation, June-Nov. 1959	166
Crescent Junction, UT, U.S.	—	20	—	1962	Leaves, good growth, near normal precipitation, Aug. 1958	166
Green River, UT, U.S.	0.5—3.4	0.9 m	6	1962	Leaves, low precipitation, June-Nov. 1959	166
Green River, UT, U.S.	0.0—3.1	0.2 m	6	1962	Stems, low precipitation, June-Nov. 1959	166
Green River, UT, U.S.	—	20	—	1962	Leaves, good growth, near normal precipitation, Aug. 1958	166
Thompson, UT, U.S.	0.5—6.4	2.0 m	6	1962	Leaves, low precipitation, June-Nov. 1959	166
Thompson, UT, U.S.	0.0—1.8	1.1 m	6	1962	Stems, low precipitation, June-Nov. 1959	166
Thompson, UT, U.S.	—	10	—	1962	Leaves, good growth, near normal precipitation, Aug. 1958	166
Yellow Cat, UT, U.S.	—	95	—	1962	Leaves, good growth, near normal precipitation, Aug. 1958	166
Yellow Cat, UT, U.S.	0.0—8.9	2.2 m	6	1962	Leaves, low precipitation, June-Nov. 1959	166
Yellow Cat, UT, U.S.	0.3—9.0	2.0 m	6	1962	Stems, low precipitation, June-Nov. 1959	166
Helminthia echioides						
La Roche-Posay, France	2.05—7.90	4.70 ± 1.48	30	1985	Clay loam soils	167
Hylocomium splendens (forest moss)						
Norway (12)	0.1—1.1	0.3 m	12	1984	Estimates from figures in reference	168
Inula viscosa						
Other locations, Israel		0.2	1	1957		65

Table 3 (continued)
CONCENTRATION OF SELENIUM IN OTHER PLANTS AND PLANT PRODUCTS[a]

Geographical location[b]	Selenium concentration, (mg/kg)[c]		N[d]	Date[e]	Remarks[f]	Ref.
	Range	Mean or median				
Iseilema spp.						
NW Queensland, Australia	—	4.6	1	1963	Adjacent to poison strip area, soil Se 0.03—385 (m = 2.1) mg/kg (n = 45)	155
Jack pine						
Sudbury, Ontario, Canada	0.10—0.63		—	1970—76	Leaves	156
Ledum palustre						
Kaavi, Luikonlahti, Finland	—	0.420	—	1974	Soil Se 0.60—5.5 mg/kg	159
Lichen						
Finland and U.S.S.R.	0.09—0.1	0.113 ± 0.017	—	1980	Whole plant	169
Finland and U.S.S.R.	—	0.088 ± 0.022	—	1980	Plant top	169
Lycoperdon perlatum						
Bela, Yugoslavia	—	7.02	1	1976	Spores	164
Cemsenik, Yugoslavia	—	3.09	1	1976	Spores	164
Malvastrum spicatum (malvastrum)						
NW Queensland, Australia	—	6.2	1	1963	Poison strip area, soil Se 0.15—32.2 (m = 3.3) mg/kg (n = 19)	155
Medicago sativa (alfalfa, lucerne)						
NW Queensland, Australia	—	2.3	1	1963	Adjacent to poison strip area, soil Se 0.03—385 (m = 2.1) mg/kg (n = 45)	155
Neptunia gracilis						
NW Queensland, Australia	<0.5—25.4	8.5 m	2	1963	Adjacent to poison strip area, soil Se 0.03—385 (m = 2.1) mg/kg (n = 45)	155
Queensland, Australia	<0.5—<0.5	<0.5	3	1963	Landsborough highway traverse, Hughenden, soil Se 0.3—14.5 (m = 3.3) mg/kg (n = 10)	155
Nerium oleander (oleander)						
NW Queensland, Australia	—	9.9	1	1963	Poison strip area, soil Se 0.15—32.2 (m = 3.3) mg/kg (n = 19)	155
Paper, book paper (coated)						
U.S.	0.03—0.08	0.04	5	1974	From different sources or brands	35

Paper, cardboard, corrugated						
U.S.	0.03—0.31	0.16	5	1974	From different sources or brands	35
Paper, cigarette						
U.S.	0.00—0.02	0.01	4	1974	From different sources or brands	35
Egypt	—	<0.4	—	1985		163
Paper, filter paper						
U.S.	0.01—0.05	0.02	5	1974	From different sources or brands	35
Paper, newsprint						
U.S.	0.03—0.05	0.04	6	1974	From different sources or brands	35
Paper, office stationery						
U.S.	0.00—0.12	0.03	9	1974	From different sources or brands	35
Paper, pulps						
U.S.	0.005—0.073	0.015 m	17	1974	17 different samples, interlaboratory analyses	170
Paper, tissue						
U.S.	0.01—0.05	0.03	4	1974	From different sources or brands	35
Paper, towels & napkins						
U.S.	0.01—0.04	0.03	5	1974	From different sources or brands	35
Parmelia chlorochroa Tuck						
Powder River Basin Wyoming/Montana, U.S. (15)	—	0.32[g]	15	1977	From same sites as *A. tridentata* & *B. gracilis*	161
Picea excelsa Link.						
Kuhmoinen, Patavesi, Finland	—	0.120	—	1974	Needles, soil Se as high as 9.1 mg/kg	159
Vantaa, Ruskeasanta, Seutula, Finland	—	<0.010	—	1974	Needles, soil Se 0.17 mg/kg	159
Pinus silvestris L.						
Kuhmoinen, Patavesi, Finland	—	0.040	—	1974	Needles, soil Se as high as 9.1 mg/kg	159
Vantaa, Ruskeasanta, Seutula, Finland	—	0.120	—	1974	Soil Se 0.17 mg/kg	159
Kaavi, Luikonlahti, Finland	—	0.140	—	1974	Needles, soil Se 0.60—5.5 mg/kg	159
Polonisia viscosa (tick weed)						
NW Queensland, Australia	—	19.7	1	1963	Poison strip area, soil Se 0.15—32.2 (m = 3.3) mg/kg (n = 19)	155
Polystichum spinulosumm Mull.						
Vantaa, Ruskeasanta, Seutula, Finland	—	0.015	—	1974	Soil Se 0.17 mg/kg	159
Populus balsamifera (cottonwood)						
Montreal, Quebec, Canada	—	550	—	1981	Twigs and leaves, severe injury by Se-polluted air	156

Table 3 (continued)
CONCENTRATION OF SELENIUM IN OTHER PLANTS AND PLANT PRODUCTS[a]

Geographical location[b]	Selenium concentration, (mg/kg)[c] Range	Mean or median	N[d]	Date[e]	Remarks[f]	Ref.
Populus tremula L.						
Kaavi, Luikonlahti, Finland	—	0.340	—	1974	Leaves, soil Se 0.60—5.5 mg/kg	159
Prosopis farcata						
Naot-Mordechai, Israel	136.2—311.5	215.1	6	1957	Soil Se 0.1—6 (m = 0.7) mg/kg (n = 41)	65
Other locations, Israel	1.2—61.1	3.4 m	6	1957		65
Prunas virginiana (choke cherry)						
Montreal, Quebec, Canada	—	54	—	1981	Twigs and leaves, severe injury by Se polluted air	156
Purshia tridentata						
Washington, U.S.	—	0.10	—	1983	Leaves, at peak litterfall from 4 plants	162
Rhynchosia minima						
NW Queensland, Australia	—	34.4	1	1963	Poison strip area, soil Se 0.15—32.2 (m = 3.3) mg/kg (n = 19)	155
Sacrobatus vermiculatus (greasewood)						
Washington, U.S.	—	0.55	—	1983	Leaves, at peak litterfall from 4 plants	162
Salix alba vitellina, (golden willow)						
Alberta, Canada (4)	0.009—0.029	0.022 m	—	1981	Leaves, plants in pots downwind of SO_2 sources	160
Salix bebbiana, (beaked willow)						
Alberta, Canada (1)	—	0.065	—	1981	Leaves, plants in pots downwind of SO_2 sources	160
Scirpus lacustrae L.						
Kaavi, Luikonlahti, Palolampi, Finland	—	0.120	—	1974	Soil Se 0.60—5.5 mg/kg	159
Sesbania aculeata (Sesbania pea)						
NW Queensland, Australia	22.5—55.4	39.0	2	1963	Poison strip area, soil Se 0.15—32.2 (m = 3.3) mg/kg (n = 19)	155
NW Queensland, Australia	—	29.1	1	1963	Adjacent to poison strip area, soil Se 0.03—385 (m = 2.1) mg/kg (n = 45)	155
Setaria oplismenoides (native pigeon grass)						
NW Queensland, Australia	—	23.0	1	1963	Poison strip area, soil Se 0.15—32.2 (m = 3.3) mg/kg	155

Solanum esuriale (quena)						
NW Queensland, Australia	—	<0.5	1	1963	Poison strip area, soil Se 0.15—32.2 (m = 3.3) mg/kg (n = 19)	155
Sorbus aucuparia L.						
Kaavi, Luikonlahti, Finland	—	0.250	—	1974	Leaves, soil Se 0.60—5.5 mg/kg	159
Sorghum almun						
NW Queensland, Australia	5.3—21.0	14.6 m	5	1963	Poison strip area, soil Se 0.15—32.2 (m = 3.3) mg/kg (n = 19)	155
Sorghum halepense						
Other locations, Israel	0.7—3.0	1.3 m	5	1957		65
Sphagnum mosses						
Vantaa, Ruskeasanta, Seutula, Finland	—	<0.010	—	1974	Soil Se 0.17 mg/kg	159
Adirondacks, New York, U.S. (11)	0.3—0.6	0.4 m	11	1979	Probably due to atmospheric transport/deposition	171
Syringia vulgaris (lilac)						
Montreal, Quebec, Canada	—	52	—	1981	Twigs and leaves, severe injury by Se-polluted air	156
Montreal, Quebec, Canada	—	36	—	1981	Intermediate injury by Se-polluted air (control)	156
Montreal, Quebec, Canada	—	1.1	—	1981	No injury (control)	156
Trifolium hybridum L.						
Petajavesi, Puttola, Finland	—	<0.010	—	1974	Soil Se 0.14 mg/kg	159
Tilia americanum (basswood)						
Montreal, Quebec, Canada	—	100	—	1981	Twigs and leaves, severe injury by Se-polluted air	156
Tobacco, air cured						
Belgium	—	0.65	—	1979	Raw tobacco imported into Belgium/Luxembourg	33
Indonesia	—	0.82	—	1979	Raw tobacco imported into Belgium/Luxembourg	33
Tobacco, air cured, fire cured						
Malawi	—	0.18	—	1979	Raw tobacco imported into Belgium/Luxembourg	33
Tobacco, Burley, fire, air, or flue cured						
U.S.	0.08—0.19	0.17 m	—	1979	Raw tobacco imported into Belgium/Luxembourg	33
Tobacco, chewing						
U.S.	0.11—0.14	0.13	4	1974	From different sources or brands	35
Tobacco, cigarette						
Columbia	0.15—0.71	0.48 ± 0.25	5	1981	5 different brands	34
Egypt	—	0.087 ± 0.005	—	1985		163
Mexico	0.30—0.84	0.49 ± 0.22	8	1981	8 different brands	34

Table 3 (continued)
CONCENTRATION OF SELENIUM IN OTHER PLANTS AND PLANT PRODUCTS[a]

Geographical location[b]	Selenium concentration, (mg/kg)[c] Range	Mean or median	N[d]	Date[e]	Remarks[f]	Ref.
Mexico & Columbia	0.15—0.84	0.49 ± 0.22	13	1981	13 different brands	34
Sweden	0.12—0.20	0.15 ± 0.04	4	1981	4 different brands	34
U.K.	0.15—0.22	0.16 ± 0.04	5	1981	5 different brands	34
U.S.	—	0.87 ± 0.22	—	1970	Filter cigarette brand	172
U.S.	—	1.31 ± 0.85	—	1970	Nonfilter cigarette brand	172
U.S.	0.08—0.31	0.16 ± 0.06	17	1981	17 different brands	34
U.S., U.K., & Sweden	0.08—0.31	0.16 ± 0.05	26	1981	26 different brands	34
Tobacco, flue cured						
India	—	0.34	—	1979	Raw tobacco imported into Belgium/Luxembourg	33
Rhodesia	—	0.11	—	1979	Raw tobacco imported into Belgium/Luxembourg	33
Tobacco, flue cured and Burley						
U.S.	0.013—0.085	0.046	9	1976—81	5 types of flue-cured and Burley tobaccos	32
Tobacco, pipe						
U.S.	0.04—0.21	0.11	12	1974	From different sources or brands	35
Tobacco, reference IR-I						
Kentucky, U.S.	—	0.089 ± 0.01	6	1978		173
Tobacco, sun cured						
Greece	—	0.26	—	1979	Raw tobacco imported into Belgium/Luxembourg	33
Turkey	—	0.44	—	1979	Raw tobacco imported into Belgium/Luxembourg	33
Tobacco						
Yugoslavia (8)	—	<0.05	—	1978		173
Tobacco, ash						
U.S.	—	1.91 ± 0.58	—	1970		172
Trembling aspen						
Sudbury, Ontario, Canada	0.04—1.14		—	1970—76	Leaves	156
Typha augustatat						
Other locations, Israel	1.0—4.4	2.0 m	4	1957		65
Urochloa panicoides (Urochloa grass)						
NW Queensland, Australia	—	<0.5	1	1963	Poison strip area, soil Se 0.15—32.2 (m = 3.3) mg/kg (n = 19)	155

Vaccinium myrtillus L.						
Vantaa, Ruskeasanta, Seutula, Finland	—	0.035	—	1974	Soil Se 0.17 mg/kg	159
Vegetation						
Iraputo area, Mexico	1—120	6 m	34	1940	Soil Se 0.3—65 (m = 3.5) mg/kg (n = 50)	36
Juarez & Chihuahua area, Mexico	0.5—2	1 m	29	1940	Soil Se 0.1—9.2 (m = 0.25) mg/kg (n = 42) excluding Rio-Tinto Mine	36
Torreon area, Mexico	0.5—8	2 m	21	1940	Soil Se 0.1—2.5 (m = 0.2) mg/kg (n = 44)	36
Four Corners Power Plant, NM, U.S.	—	0.7	1	1961	Shrubs, herbs, grasses 1—20 km from power plant	174
Four Corners Power Plant, NM, U.S.	0.1—5.04	0.8	8	1961	Shrubs, herbs, grasses 21—40 km from power plant	174
Four Corners Power Plant, NM, U.S.	0.05—0.6	0.22	5	1961	Shrubs, herbs, grasses 41—80 km from power plant	174
Four Corners Power Plant, NM, U.S.	0.1—1.5	0.6	6	1971	Shrubs, herbs, grasses 1—20 km from power plant	174
Four Corners Power Plant, NM, U.S.	0.075—5	1.0	9	1971	Shrubs, herbs, grasses 21—40 km from power plant	174
Four Corners Power Plant, NM, U.S.	0.075—0.4	0.15	6	1971	Shrubs, herbs, grasses 41—80 km from power plant	174
Utah, U.S. (18)	1.4—3.3	—	332	1977	Whole plant, 33 different species, ranges of means for each of 18 locations; soil Se 0—2.3 mg/kg from 3 depths 0—15 cm (n = 408)	175
Vicia cracca L.						
Kaavi, Luikonlahti, Finland	—	0.220	—	1974	Soil Se 0.60—5.5 mg/kg	159
White birch						
Sudbury, Ontario, Canada	0.14—1.06		—	1970—76	Leaves	156

[a] Plants and plant products included are those typically not considered food or feed crops or ingredients, listed generally in order of taxonomic name.
[b] General or specific regions from where plants or products were obtained. Numbers in parenthesis indicate number of different sampling sites.
[c] Selenium concentrations refer to total selenium, usually reported on a dry basis by usually acceptable methods of analysis (refer to footnote c, Table 1 for general comment); means are given with standard deviations if reported; medians are indicated by m.
[d] Number of samples as reported in the literature or estimated by the present author from information in the literature; a blank indicates data not reported.
[e] Date of sample collection or, more typically, date of publication of the report.
[f] This column contains additional information on the nature or origin of the plant or product as well as other highlights. Concentration of selenium in soil (and geological material) from the vicinity of the plant is provided as a range and median (m) together with the number (n) of samples analyzed, all data being estimated by the present author from information provided in the literature.
[g] Geometric mean.

and development and suggested that the element may be an essential constituent. These observations firmly established the important role of selenium indicator plants. Numerous reports of extensive, in-depth surveys on the occurrence and distribution of these plants on seleniferous formations in the semiarid western U.S. and the Great Plains of the midwestern U.S. and results of analyses for total selenium content have appeared and are referred to in Table 4.

Selenium concentrations in selenium indicator plants found in the U.S., Canada, and other countries are presented in Table 4. Division of plants has been made into the two categories of primary and secondary indicator plants according to listings by Trelease and Beath,[40,41] Rosenfeld and Beath,[22] and Munz and Keck,[41,42] and with few exceptions species therein reflect these assignments. The primary group includes plants always associated with seleniferous soil and absorbing toxic amounts of selenium. The secondary group includes plants capable of absorbing selenium to toxic levels when growing on seleniferous soil, but found also growing on nonseleniferous soils. In contrast to selenium concentrations in the sub-milligram-per-kilogram region and frequently closer to the vicinity of 0.1 mg/kg in most plants, levels in selenium indicator plants are usually in the hundreds and occasionally in the thousands of milligrams per kilogram. The highest concentration of 14,900 mg/kg on a dry basis was reported by Beath et al.[43] in a composite sample of *Astragalus racemoseus* growing on Pierre loam in Wyoming. Such elevated levels in indicator plants, even while nonaccumulating plants growing adjacent have more or less low normal levels, stresses the capacity of indicator plants for absorbing the element and substantiates the existence of a selenium-based physiology.

VI. CONCLUDING REMARKS

The approach taken in this report has been a tabular presentation of concentration data and descriptive information. An advantage is the summarizing and archival recording of numerical data. Data in map format also has its advantages in presenting to the reader a pictorial overview of the data, and such a treatment should be considered with the existing (and augmented) data base where sufficient. In addition to the excellent detailed cartographic presentation of concentration information for selenium and other elements in soils and surficial materials of the U.S. by workers at the U.S. Geological Survey (see for example Shacklette et al.,[44] and Chapter 9), several other maps showing the distribution of selenium in vegetation have been published. Rosenfeld and Beath[22] have mapped the distribution of seleniferous vegetation in the western U.S. and southwestern Canada. A generalized regional distribution pattern of selenium concentrations in forage, wheat, and feed grain crops in the U.S. has been published by Kubota et al.[45] and Kubota and Allaway,[46] indicating low, variable, and adequate selenium areas. Local areas in the U.S. Midwest containing selenium accumulator plants with >50 mg Se per kilogram are indicated on the latter map. Muth and Allaway[47] mapped for the U.S. the geographical distribution of seleniferous soils with associated seleniferous vegetation and nonseleniferous soils with associated white muscle disease in sheep caused by selenium deficiency in feedstuffs. These maps have been republished in original or adapted forms on several occasions.[48-54] A similar map including eastern and central Canada has appeared,[55] and the selenium contents of corn samples from midwestern U.S. states have been mapped.[56] A generalized pattern of selenium concentrations in crops from some European countries has been published.[57]

A number of factors influencing the availability of selenium to and its absorption by plants must be considered for complete interpretation of data. These factors include (1) nature (type) of plant, (2) stage of growth when sampled and analyzed, (3) the concentration and chemical form or speciation of the selenium in the material (soil or experimental medium) in which the plant grew, (4) plant associations (presence of other plants making soil selenium

Table 4
CONCENTRATION OF SELENIUM IN SELENIUM ACCUMULATOR PLANTS[a]

Geographical location[b]	Selenium concentration (mg/kg)[c] Range	Median	N[d]	Date[e]	Remarks[f]	Ref.
Primary Accumulators						
Astragalus albulus, Woot. and Standl.						
Arizona, U.S.	—	220 av	1	1941	Whole plant	176
New Mexico, U.S.	23—530	83	4	1941	Whole plant	176
Astragalus argillosus Jones						
Arizona, U.S.	—	36	1	1941	Whole plant	176
Arizona, U.S.	—	446	1	1940	Whole plant	177
Utah, U.S.	495—631	517	3	1941	Whole plant	176
Astragalus artemisiarum						
Clark County, NV, U.S.	0.5—970	123	6	1941	Soil Se 0—1 (m = 0.4) mg/kg (n = 25)	178
Clark County, NV, U.S.	35—970	175	6	1940	Soil Se 0—0.04 (m = 0.6) mg/kg (n = 7)	179
Astragalus beathii Porter						
Arizona, U.S.	180—1034	607	2	1941	Whole plant	176
Arizona, U.S.	1550—3135	2343	2	1940	Whole plant	177
Astragalus bisulcatus (Hook.) Gray (two-groove poisonvetch)						
Alberta, Canada	—	3.0	—	1941	Stalks, Belly River formation, Soil Se 0.4—0.5 mg/kg	180
Alberta, Canada	—	5.0	—	1941	Leaves, Belly River formation, Soil Se 0.4—0.5 mg/kg	180
Alberta, Canada	200—465	330	—	1941	Stalks, Bear Paw formation, Soil Se 0.3—0.7 mg/kg	180
Alberta, Canada	—	600	—	1941	Leaves, Bear Paw formation, Soil Se 0.3—0.7 mg/kg	180
Alberta, Canada	12—470	100	8	1941	Soil Se 0.1—2 (m = 0.5) mg/kg (n = 55)	14
Manitoba, Canada	160—890	420	5	1941	Soil Se 0.3—1.5 (m = 0.6) mg/kg (n = 6)	14
Saskatchewan, Canada	15—3640	640	17	1941	Soil Se 0.1—6 (m = 0.7) mg/kg (n = 72)	14
Colorado, U.S.	19—1620	172	14	1939	Tops	157
Colorado, U.S.	—	4	1	1941	Whole plant	176
Colorado, U.S.	8—1140	98	10	1938	Soil Se 0.2—54 (m = 1.5) mg/kg (n = 160)	181
Lemhi County, ID, U.S.	122—2700	410	5	1939	Tops	157
Graham County, KS, U.S.	5—840	45	10	1936	Soil Se 0.4—20 (m = 1.8) mg/kg (n = 14)	117
Gove County, KS, U.S.	2—570	20	11	1936	Soil Se 0.2—18 (m = 2) mg/kg (n = 67)	117
Logan County, KS, U.S.	8—2680	170	37	1936	Soil Se 0.2 —140 (m = 2) mg/kg (n = 258)	117

Table 4 (continued)
CONCENTRATION OF SELENIUM IN SELENIUM ACCUMULATOR PLANTS[a]

Geographical location[b]	Selenium concentration (mg/kg)[c]		N[d]	Date[e]	Remarks[f]	Ref.
	Range	Median				
Rooks County, KS, U.S.	1—200	63	10	1936	Soil Se 0.2—100 (m = 0.7) mg/kg (n = 52)	117
Trego County, KS, U.S.	2—520	90	7	1936	Soil Se 0.3—20 (m = 1) mg/kg (n = 53)	117
Wallace County, KS, U.S.	45—690	190	5	1936	Soil Se 0.3—8 (m = 1) mg/kg (n = 70)	117
Big Horn County, MT, U.S.	20—1180	270	8	1940	Soil Se 0.1—8 (m = 0.7) mg/kg (n = 72)	36
Cascade County, MT, U.S.	3—1650	300	4	1940	Soil Se 0.4—3 (m = 1) mg/kg (n = 9)	36
Chouteau County, MT, U.S.	20—410	95	8	1940	Soil Se 0.2—5 (m = 0.7) mg/kg (n = 82)	36
Fergus County, MT, U.S.	6—1070	33	18	1940	Soil Se 0.2—4 (m = 0.7) mg/kg (n = 34)	36
Pondera County, MT, U.S.	6—1580	130	40	1940	Soil Se 0.2—3 (m = 0.5) mg/kg (n = 85)	36
Teton County, MT, U.S	80—170	125	2	1941	Soil Se 0.1—12 (m = 1) mg/kg (n = 37)	14
Teton County, MT, U.S.	2—1630	75	30	1940	Soil Se 0.1—5 (m = 0.5) mg/kg (n = 151)	36
Yellowstone County, MT, U.S.	3—530	78	4	1940	Soil Se 0.2—3 (m = 0.4) mg/kg (n = 39)	36
Montana, U.S.	2—1550	118	18	1939	Tops, leaves	157
Montana, U.S.	2—1468	76	25	1941	Whole plant	176
Montana, U.S.	—	380	1	1940	Whole plant	177
Montana, U.S.	7—1520	150	25	1936	Soil Se 0.1—2.5 (m = 0.6) mg/kg (n = 68)	117
S & SW of Black Hills, Nebraska, U.S.	—	570	1	1936	Soil Se 0.2—14 (m = 0.7) mg/kg (n = 206)	117
Springer area, New Mexico, U.S.	—	1321	1	1939	Whole plant	157
New Mexico, U.S.	—	274	1	1941	Whole plant	176
New Mexico, U.S.	147—540	344	2	1940	Whole plant	177
New Mexico, U.S.	2—520	80	12	1938	Soil Se 0.2—10 (m = 1) mg/kg (n = 61)	181
Benville County, ND, U.S.	110—1520	420	7	1948	Soil Se 0.2—2 (m = 1) mg/kg (n = 9)	182
Bottineau County, ND, U.S.	320—4400	810	11	1948	Soil Se 0.6—4 (m = 2) mg/kg (n = 15)	182
Divide County, ND, U.S.	—	160	1	1948	Soil Se 0.1—1.2 (m = 0.6) mg/kg (n = 16)	182
McLean County, ND, U.S.	70—2620	580	4	1948	Soil Se 0.4—2 (m = 0.8) mg/kg (n = 15)	182
McHenry County, ND, U.S.	360—2030	1560	4	1948	Soil Se 0.4—1.6 (m = 1.2) mg/kg (n = 6)	182
Mountrail County, ND, U.S.	—	250	1	1948	Soil Se 0.4—8 (m = 1) mg/kg (n = 23)	182
North Dakota, U.S.	—	255	1	1941	Whole plant	176
North Dakota, U.S.	60—470	230	6	1941	Soil Se 0.1—2 (m = 1) mg/kg (n = 14)	14

North Dakota, U.S.	—	10	1	1940	Whole plant	177
Ward County, ND, U.S.	80—1730	300	7	1948	Soil Se 0.2—3 (m = 0.8) mg/kg (n = 21)	182
S & SW of Black Hills, South Dakota, U.S.	2—3110	160	14	1936	Soil Se 0.1—40 (m = 1.5)mg/kg (n = 76)	117
South Dakota, U.S.	—	2.6	1	1938	Whole plant, Niobrara formation	183
S & SW of Black Hills, Wyoming, U.S.	15—450	80	8	1936	Soil Se 0.2—16 (m = 1) mg/kg (n = 33)	117
Wyoming, U.S.	2—4040	273	7	1941	Whole plant	176
Wyoming, U.S.	102—3533	1479	8	1940	Whole plant	177
Wyoming, U.S.	255—8840	2100	8	1937		43
Astragalus confertiflorus Gray						
Arizona, U.S.	—	27	1	1939	Tops	157
Colorado, U.S.	155—888	236	4	1939	Tops, flowers, seeds	157
Colorado, U.S.	—	240	1	1940	Whole plant	177
Colorado, U.S.	351—474	413	2	1941	Whole plant	176
New Mexico, U.S.	—	143	1	1939	Tops	157
New Mexico, U.S.	14—1361	688	2	1941	Whole plant	176
Thompson, UT, U.S.	—	1322	—	1943	Vanadium-uranium zone	184
Utah, U.S.	—	179	1	1939	Tops	157
Utah, U.S.	12—89	28	4	1941	Whole plant	176
Utah, U.S.	5—131	15	3	1940	Whole plant	177
Astragalus eastwoodiae Jones						
Utah, U.S.	—	162	1	1939	Tops	157
Astragalus grayi Parry						
Montana, U.S.	—	1330	1	1940	Whole plant	177
Wyoming, U.S.	—	4420	—	1946		185
Wyoming, U.S.	—	658	1	1941	Whole plant	176
Astragalus haydenianus Gray						
Colorado, U.S.	25—2148	97	8	1939	Tops	157
Colorado, U.S.	25—1916	143	6	1940	Whole plant	177
New Mexico, U.S.	19—2377	1198	2	1941	Whole plant	176
Utah, U.S.	214—237	226	2	1940	Whole plant	177
Astragalus pattersonii var. *praelongus* (Sheld.) Jones						
Arizona, U.S.	970—4835	2903	2	1941	Whole plant	176
New Mexico, U.S.	212—2370	1923	3	1941	Whole plant	176
Utah, U.S.	—	3090	1	1941	Whole plant	176

Table 4 (continued)
CONCENTRATION OF SELENIUM IN SELENIUM ACCUMULATOR PLANTS[a]

Geographical location[b]	Selenium concentration (mg/kg)[c]		N[d]	Date[e]	Remarks[f]	Ref.
	Range	Median				
Astragalus pectinatus Dougl. (narrow-leaf milk vetch)						
Alberta, Canada	—	30	—	1941	Stalks, Bear Paw formation, Soil Se 0.3—0.7 mg/kg	180
Nevada, U.S.	—	631	1	1939	Tops	157
Springer area, New Mexico, U.S.	—	185	1	1939	Whole plant	157
New Mexico, U.S.	13—53	33	2	1939	Tops	157
New Mexico, U.S.	57—965	228	3	1940	Whole plant	177
Utah, U.S.	—	22	1	1939	Tops	157
Utah, U.S.	10—47	39	2	1940	Whole plant	177
Wyoming, U.S.	—	497	1	1940	Whole plant	177
Astragalus limatus Sheld.						
California, U.S.	183—614	399	2	1939	Tops	157
California, U.S.	—	2175	1	1941	Whole plant	176
California, U.S.	42—250	146	2	1940	Whole plant	177
Astragalus osterhoutti Jones						
Colorado, U.S.	—	1356	1	1941	Whole plant	176
Colorado, U.S.	958—2678	1818	2	1940	Whole plant	177
Astragalus pattersonii Gray						
Arizona, U.S.	62—265	164	1	1941	Whole plant	176
Colorado, U.S.	146—5042	288	7	1939	Tops	157
Colorado, U.S.	36—257	197	3	1941	Whole plant	176
Colorado, U.S.	125—1260	614	5	1940	Whole plant	177
New Mexico, U.S.	11—676	15	3	1939	Tops	157
New Mexico, U.S.	21—1008	700	5	1941	Whole plant	176
New Mexico, U.S.	100—1640	592	6	1940	Whole plant	177
Texas, U.S.	1425—2362	1917	3	1940	Whole plant	177
Thompson, UT, U.S.	—	1478	—	1943	Vanadium-uranium zone	184
Thompson, UT, U.S.	—	8512	—	1943	Old Yellow Cat Camp	184
Utah, U.S.	1—296	199	6	1941	Whole plant	176

Utah, U.S.	149—2154	313	3	1939	Tops	157
Utah, U.S.	214—237	226	2	1940	Whole plant	177
Astragalus pattersonii var. *praelongus* (Sheld.) Jones						
Arizona, U.S.	970—4835	2903	2	1941	Whole plant	176
New Mexico, U.S.	212—2370	1923	3	1941	Whole plant	176
Utah, U.S.	—	3090	1	1941	Whole plant	176
Astragalus pectinatus Dougl. (narrow-leaf milk vetch)						
Alberta, Canada	—	30	—	1941	Stalks, Bear Paw formation, Soil Se 0.3—0.7 mg/kg	180
Alberta, Canada	—	120	—	1941	Leaves, Bear Paw formation, Soil Se 0.3—0.7 mg/kg	180
Alberta, Canada	140—3690	670	19	1941	Soil Se 0.1—2 (m = 0.5) mg/kg (n = 55)	14
Saskatchewan, Canada	120—4190	1125	22	1941	Soil Se 0.1—6 (m = 0.7) mg/kg (n = 72)	14
Colorado, U.S.	502—2114	660	5	1939	Tops	157
Colorado, U.S.	35—3890	790	34	1938	Soil Se 0.2—54 (m = 1.5) mg/kg (n = 160)	181
Gove County, KS, U.S.	110—660	385	2	1936	Soil Se 0.2—18 (m = 2) mg/kg (n = 67)	117
Logan County, KS, U.S.	30—2660	530	39	1936	Soil Se 0.2—140 (m = 2) mg/kg (n = 258)	117
Trego County, KS, U.S.	100—140	120	2	1936	Soil Se 0.3—20 (m = 1) mg/kg (n = 53)	117
Wallace County, KS, U.S.	50—840	510	11	1936	Soil Se 0.3—8 (m = 1) mg/kg (n = 70)	117
Cascade County, MT, U.S.	220—550	460	4	1940	Soil Se 0.4—3 (m = 1) mg/kg (n = 9)	36
Chouteau County, MT, U.S.	4—2090	190	61	1940	Soil Se 0.2—5 (m = 0.7) mg/kg (n = 82)	36
Fergus County, MT, U.S.	6—910	250	13	1940	Soil Se 0.2—4 (m = 0.7) mg/kg (n = 34)	36
Astragalus preussii Gray						
Arizona, U.S.	17—154	44	2	1941	Whole plant	176
Arizona, U.S.	—	18	1	1940	Whole plant	177
Nevada, U.S.	—	27	1	1939	Tops	157
Thompson, UT, U.S.	1846—4188	3017	2	1943	Old Yellow Cat Camp	184
Utah, U.S.	40—1483	410	4	1941	Whole plant	176
Astragalus preussii var. *latus* Jones						
Utah, U.S.	—	45	1	1939	Tops	157
Astragalus racemosus Pursh.						
Colorado, U.S.	0.7—1630	130	23	1938	Soil se 0.2—54 (m = 1.5) mg/kg (n = 160)	181
Colorado, U.S.	74—530	302	2	1939	Tops	157
Montana, U.S.	—	17	1	1941	Whole plant	176
S & SW of Black Hills, Nebraska, U.S.	70—5560	570	15	1936	Soil Se 0.2—14 (m = 0.7) mg/kg (n = 206)	117
Nebraska, U.S.	37—1727	528	15	1941	Whole plant	176
Springer area, New Mexico, U.S.	—	113	1	1939	Whole plant	157
New Mexico, U.S.	—	6	1	1941	Whole plant	176
New Mexico, U.S.	22—262	122	3	1940	Whole plant	177

Table 4 (continued)
CONCENTRATION OF SELENIUM IN SELENIUM ACCUMULATOR PLANTS[a]

Geographical location[b]	Selenium concentration (mg/kg)[c]		N[d]	Date[e]	Remarks[f]	Ref.
	Range	Median				
New Mexico, U.S.	180—1690	180	3	1938	Soil Se 0.2—10 (m = 1) mg/kg (n = 61)	181
North Dakota, U.S.	—	328	1	1940	Whole plant	177
S & SW of Black Hills, South Dakota, U.S.	1—2170	260	14	1936	Soil Se 0.1—40 (m = 1.5) mg/kg (n = 76)	117
South Dakota, U.S.	24—4100	550	24	1938	Whole plant, Niobrara formation	183
Pondera County, MT, U.S.	5—1700	270	33	1940	Soil Se 0.2—3 (m = 0.5) mg/kg (n = 85)	36
Teton County, MT, U.S.	10—2140	1330	7	1941	Soil Se 0.1—12 (m = 1) mg/kg (n = 37)	14
Teton County, MT, U.S.	6—5170	320	83	1940	Soil Se 0.1—5 (m = 0.5) mg/kg (n = 151)	36
Montana, U.S.	10—815	357	26	1939	Tops	157
Montana, U.S.	28—2140	350	27	1941	Whole plant	176
Montana, U.S.	110—730	260	3	1936	Soil Se 0.1—2.5 (m = 0.6) mg/kg (n = 68)	117
Benville County, ND, U.S.	940—960	950	2	1948	Soil Se 0.2—2 (m = 1) mg/kg (n = 9)	182
Bottineau County, ND, U.S.	1070—2590	1830	2	1948	Soil Se 0.6—4 (m = 2) mg/kg (n = 15)	182
Burke County, ND, U.S.	280—3860	1855	10	1948	Soil Se 1—1.6 (m = 1.2) mg/kg (n = 10)	182
Divide County, ND, U.S.	310-4740	1880	15	1948	Soil Se 0.1—1.2 (m = 0.6) mg/kg (n = 16)	182
McLean County, ND, U.S.	340—4950	2250	8	1948	Soil Se 0.4—2 (m = 0.8) mg/kg (n = 15)	182
Mountrail County, ND, U.S.	190—3720	2215	22	1948	Soil Se 0.4—8 (m = 1) mg/kg (n = 23)	182
Ward County, ND, U.S.	1040—3190	1870	11	1948	Soil Se 0.2—3 (m = 0.8) mg/kg (n = 21)	182
Williams County, ND, U.S.	520—3990	2310	21	1948	Soil Se 0.4—8 (m = 0.9) mg/kg (n = 24)	182
North Dakota, U.S.	—	924	—	1939	Composite	157
North Dakota, U.S.	280—1660	890	9	1941	Soil Se 0.1—2 (m = 1) mg/kg (n = 14)	14
South Dakota, U.S.	—	3116	1	1939	Tops	157
Wyoming, U.S.	48—3190	1868	10	1941	Whole plant	176
Wyoming, U.S.	—	1160	1	1940	Whole plant	177
Wyoming, U.S.	124—3250	374	7	1937		43
Astragalus praelongus Sheld.						
Arizona, U.S.	3380—4500	3940	2	1939	Tops	157
Arizona, U.S.	754—3856	1517	5	1940	Whole plant	177
New Mexico, U.S.	262—4474	870	4	1939	Tops	157

New Mexico, U.S.	12—1817	464	3	1940	Whole plant	177
Springer area, New Mexico, U.S.	1030—2600	1815	2	1939	Whole plant	157
Utah, U.S.	178—1284	731	2	1939	Tops	157
Astragalus preussii Gray						
Arizona, U.S.	17—154	44	2	1941	Whole plant	176
Arizona, U.S.		18	1	1940	Whole plant	177
Nevada, U.S.	—	27	1	1939	Tops	157
Thompson, UT, U.S.	1846—4188	3017	2	1943	Old Yellow Cat Camp	184
Utah, U.S.	40—1483	410	4	1941	Whole plant	176
Astragalus preussii var. *latus* Jones						
Utah, U.S.	—	45	1	1939	Tops	157
Astragalus racemosus Pursh.						
Colorado, U.S.	0.7—1630	130	23	1938	Soil se 0.2—54 (m = 1.5) mg/kg (n = 160)	181
Colorado, U.S.	74—530	302	2	1939	Tops	157
Montana, U.S.	—	17	1	1941	Whole plant	176
S & SW of Black Hills, Nebraska, U.S.	70—5560	570	15	1936	Soil Se 0.2—14 (m = 0.7) mg/kg (n = 206)	117
Nebraska, U.S.	37—1727	528	15	1941	Whole plant	176
Springer area, New Mexico, U.S.	—	113	1	1939	Whole plant	157
New Mexico, U.S.	—	6	1	1941	Whole plant	176
New Mexico, U.S.	22—262	122	3	1940	Whole plant	177
New Mexico, U.S.	180—1690	180	3	1938	Soil Se 0.2—10 (m = 1) mg/kg (n = 61)	181
North Dakota, U.S.	—	328	1	1940	Whole plant	177
S & SW of Black Hills, South Dakota, U.S.	1—2170	260	14	1936	Soil Se 0.1—40 (m = 1.5) mg/kg (n = 76)	117
South Dakota, U.S.	24—4100	550	24	1938	Whole plant, Niobrara formation	183
South Dakota, U.S.	0—1100	29	40	1938	Whole plant, Pierre formation	183
South Dakota, U.S.	8—3920	86	20	1939	Tops	157
South Dakota, U.S.	5.5—3134	340	31	1946	Soil Se <0.05—8.90 (m = 1.46) mg/kg (n = 220)	186
South Dakota, U.S.	—	2600	1	1941	Whole plant	176
South Dakota, U.S.	0—4800	186	47	1939	Soil Se 0—52 (m = 4.5) mg/kg (n = 253)	58
South Dakota, U.S.	3.5—4100	550	25	1939	Whole plant	58
Utah, U.S.	2—345	12	4	1941	Whole plant	176
S & SW of Black Hills, Wyoming, U.S.	500—680	590	2	1936	Soil Se 0.2—16 (m = 1) mg/kg (n = 33)	117
South Dakota, U.S.	0—1100	29	40	1938	Whole plant, Pierre formation	183
South Dakota, U.S.	8—3920	86	20	1939	Tops	157
South Dakota, U.S.	5.5—3134	340	31	1946	Soil Se <0.05—8.90 (m = 1.46) mg/kg (n = 220)	186

Table 4 (continued)
CONCENTRATION OF SELENIUM IN SELENIUM ACCUMULATOR PLANTS[a]

Geographical location[b]	Selenium concentration (mg/kg)[c]		N[d]	Date[e]	Remarks[f]	Ref.
	Range	Median				
South Dakota, U.S.	—	2600	1	1941	Whole plant	176
South Dakota, U.S.	0—4800	186	47	1939	Soil Se 0—52 (m = 4.5) mg/kg (n = 253)	58
South Dakota, U.S.	3.5—4100	550	25	1939	Whole plant	58
Utah, U.S.	2—345	12	4	1941	Whole plant	176
S & SW of Black Hills, Wyoming, U.S.	500—680	590	2	1936	Soil Se 0.2—16 (m = 1) mg/kg (n = 33)	117
Wyoming, U.S.						
Wyoming, U.S.	142—14920	3140	3	1937		43
Wyoming, U.S.	—	38	1	1941	Whole plant	176
Astragalus sabulosus Jones						
Arizona, U.S.	—	1734	1	1939	Tops	157
Utah, U.S.	2025—2210	2118	3	1941	Whole plant	176
Astragalus sobinatulus Sheld.						
Nevada, U.S.	25—320	173	2	1948	Soil Se 0.1—4 (m = 0.1) mg/kg (n = 17)	182
Wyoming, U.S.	100—1282	691	2	1941	Whole plant	176
Astragalus toanus Jones						
Idaho, U.S.	280—622	451	2	1939	Tops	157
Idaho, U.S.	100—990	620	3	1948	Soil Se 0—6 (m = 0.2) mg/kg (n = 31)	182
Nevada, U.S.	—	265	1	1939	Tops	157
Nevada, U.S.	—	210	1	1948	Soil Se 0.1—4 (m = 0.1) mg/kg (n = 17)	182
Utah, U.S.	—	200	1	1939	Tops	157
Oonopsis condensata						
Wyoming, U.S.	1585—4800	2290	3	1937		43
Oonopsis engelmannii (Gray) Greene						
Colorado, U.S.	70—219	123	3	1939	Tops	157
Oonopsis foliosa (Gray) Greene						
Colorado, U.S.	140—3630	164	3	1939	Tops	157
Oonopsis wardii (Gray) Greene						
Wyoming, U.S.	—	1422	1	1940		177

Stanleya albescens Jones						
Arizona, U.S.	45—356	200	2	1940	Whole plant	177
Arizona, U.S.	—	6.0	1	1941	Whole plant	176
Colorado, U.S.	—	65	1	1941	Whole plant	176
Stanleya bipinnata Greene						
Idaho, U.S.	304—364	352	3	1939	Tops	157
Idaho, U.S.	—	70	1	1948	Soil Se 0—6 (m = 0.2) mg/kg (n = 31)	182
South Dakota, U.S.	2.0—2380	385	17	1938	Whole plant, Niobrara formation	183
South Dakota, U.S.	—	250	1	1938	Whole plant, Pierre formation	183
South Dakota, U.S.	40—2260	380	18	1939	Soil Se 0—52 (m = 4.5) mg/kg (n = 253)	58
South Dakota, U.S.	136—2380	498	8	1939		58
Wyoming, U.S.	—	2490	1	1940	Whole plant	177
Stanleya integrifolia James						
Colorado, U.S.	—	130	1	1939	Tops	157
Colorado, U.S.	—	38	1	1940	Whole plant	177
Texas, U.S.	—	525		1933	Stems, Herbarium specimens collected in 1933	175
Texas, U.S.	—	99	1	1940	Whole plant	177
Utah, U.S.	87—977	502	4	1941	Whole plant	176
Utah, U.S.	—	1	1	1948	Excluding Provo Canyon; Soil Se 0.1—26 (m = 0.2) mg/kg (n = 19)	182
Utah, U.S.	4—1456	62	10	1941	Whole plant	176
Utah, U.S.	12—463	43	8	1940	Whole plant	177
Wyoming, U.S.	—	64	1	1940	Whole plant	177
Yellowstone County, MT, U.S.	190—550	270	4	1940	Soil Se 0.2—3 (m = 0.4) mg/kg (n = 39)	36
Stanleya pinnata and *Stanleya bipinnata* (Stanleya)						
Colorado, U.S.	2—1390	290	28	1938	Soil Se 0.2—54 (m = 1.5) mg/kg (n = 160)	181
Gove County, KS, U.S.	130—600	260	4	1936	Soil Se 0.2—18 (m = 2) mg/kg (n = 67)	117
Logan County, KS, U.S.	10—1070	230	20	1936	Soil Se 0.2—140 (m = 2) mg/kg (n = 258)	117
Rooks County, KS, U.S.	—	230	1	1936	Soil Se 0.2—100 (m = 0.7) mg/kg (n = 52)	117
Trego County, KS, U.S.	—	1160	1	1936	Soil Se 0.3—20 (m = 1) mg/kg (n = 53)	117
S & SW of Black Hills, Nebraska, U.S.		1080	1	1936	Soil Se 0.2—14 (m = 0.7) mg/kg (n = 206)	117
New Mexico, U.S.	10—70	40	2	1938	Soil Se 0.2—10 (m = 1) mg/kg (n = 61)	181
S & SW of Black Hills, Wyoming, U.S.	370—430	400	2	1936	Soil Se 0.2—16 (m = 1) mg/kg (n = 33)	117
Stanleya viridiflora Nutt.						
Lemhi County, ID, U.S.	<1—51	4	3	1939	Tops	157

Table 4 (continued)
CONCENTRATION OF SELENIUM IN SELENIUM ACCUMULATOR PLANTS[a]

Geographical location[b]	Selenium concentration (mg/kg)[c]		N[d]	Date[e]	Remarks[f]	Ref.
	Range	Median				
Stanleya pinnata (Pursh) Britt.						
Arizona, U.S.	41—485	263	2	1940	Whole plant	177
Arizona, U.S.	15—59	33	5	1939	Leaves, stems, flowers	157
Arizona, U.S.	—	32	1	1941	Whole plant	176
California, U.S.	1—46	6	11	1939	Tops, seeds, stems, leaves	157
Colorado, U.S.	15—320	37	3	1939	Tops	157
Colorado, U.S.	70—150	110	2	1940	Whole plant	177
Colorado, U.S.	—	423	1	1941	Whole plant	176
Idaho, U.S.	53—787	419	3	1939	Tops	157
Big Horn County, MT, U.S.	2—360	50	7	1940	Soil Se 0.1—8 (m = 0.7) mg/kg (n = 72)	36
Montana, U.S.	—	112	1	1939	Tops	157
Clark County, NV, U.S.	25—770	420	4	1940	Soil Se 0.04—1 (m = 0.6) mg/kg (n = 7)	179
Clark County, NV, U.S.	0.5—770	30	15	1941	Soil Se 0—1 (m = 0.4) mg/kg (n = 25)	182
Nevada, U.S.	0.2—140	5	5	1948	Soil Se 0.1—4 (m = 0.1) mg/kg (n = 17)	182
Nevada, U.S.	3—75	8	10	1939	Tops, leaves, seeds	157
New Mexico, U.S.	35—1110	573	2	1939	Tops	157
New Mexico, U.S.	2—17	10	2	1941	Whole plant	176
New Mexico, U.S.	208—560	384	2	1940	Whole plant	177
South Dakota, U.S.	—	299	1	1939	Tops	157
Provo Canyon, Utah, U.S.	300—1440	870	2	1948	Soil Se 1.2—54 (m = 12) mg/kg (n = 6)	182
Utah, U.S.	3—110	44	4	1939	Tops	157
Utah, U.S.	—	1	1	1948	Excluding Provo Canyon; Soil Se 0.1—26 (m = 0.2) mg/kg (n = 19)	182
Utah, U.S.	4—1456	62	10	1941	Whole plant	176
Utah, U.S.	12—463	43	8	1940	Whole plant	177
Wyoming, U.S.	—	64	1	1940	Whole plant	177
Yellowstone County, MT, U.S.	190—550	270	4	1940	Soil Se 0.2—3 (m = 0.4) mg/kg (n = 39)	36

Stanleya pinnata and *Stanleya bipinnata* (Stanleya)						
Colorado, U.S.	2—1390	290	28	1938	Soil Se 0.2—54 (m = 1.5) mg/kg (n = 160)	181
Gove County, KS, U.S.	130—600	260	4	1936	Soil Se 0.2—18 (m = 2) mg/kg (n = 67)	117
Logan County, KS, U.S.	10—1070	230	20	1936	Soil Se 0.2—140 (m = 2) mg/kg (n = 258)	117
Rooks County, KS, U.S.	—	230	1	1936	Soil Se 0.2—100 (m = 0.7) mg/kg (n = 52)	117
Trego County, KS, U.S.	—	1160	1	1936	Soil Se 0.3—20 (m = 1) mg/kg (n = 53)	117
S & SW of Black Hills, Nebraska, U.S.		1080	1	1936	Soil Se 0.2—14 (m = 0.7) mg/kg (n = 206)	117
New Mexico, U.S.	10—70	40	2	1938	Soil Se 0.2—10 (m = 1) mg/kg (n = 61)	181
S & SW of Black Hills, Wyoming, U.S.	370—430	400	2	1936	Soil Se 0.2—16 (m = 1) mg/kg (n = 33)	117
Stanleya viridiflora Nutt.						
Lemhi County, ID, U.S.	<1—51	4	3	1939	Tops	157
Idaho, U.S.	2—2	2	3	1939	Tops	157
Nevada, U.S.	1—28	3	4	1939	Tops	157
Wyoming, U.S.	5—15	5	3	1941	Whole plant	176
Xylorrhiza glabriuscula Nutt.						
Colorado, U.S.	—	4	1	1939	Tops	157
Xylorrhiza parryi (Gray) Greene						
Wyoming, U.S.	—	1850	—	1946		185
Wyoming, U.S.	1—607	304	2	1941	Whole plant	176
Wyoming, U.S.	109—317	213	2	1940	Whole plant	177
Wyoming, U.S.	13—2300	1744	3	1937		43
Xyhorrhiza venusta (Jones) Heller						
Colorado, U.S.	—	206	1	1940	Whole plant	184
Thompson, UT, U.S.	—	648	—	1943	Vanadium-uranium zone	177
Utah, U.S.	—	14	1	1939	Tops	157
Utah, U.S.	29—223	54	3	1941	Whole plant	176
Utah, U.S.	14—35	25	2	1940	Whole plant	177

Table 4 (continued)
CONCENTRATION OF SELENIUM IN SELENIUM ACCUMULATOR PLANTS[a]

Geographical location[b]	Selenium concentration (mg/kg)[c]		N[d]	Date[e]	Remarks[f]	Ref.
	Range	Median				
Secondary Accumulators						
Acacia cana (Boree)						
NW Queensland, Australia	902—1121	1012	2	1963	Poison strip area, soil Se 0.15—32.2 (m = 3.3) mg/kg (n = 19)	155
Amanita muscaria						
New Zealand	8.1—17.8	16.8	3	1964	Whole plant	187
Aster adscendens Lindl.						
Lemhi County, ID, U.S.	18—713	366	2	1939	Tops	157
Wyoming, U.S.	3—20	12	3	1941	Whole plant	176
Aster commutatus						
South Dakota, U.S.	282—419	335	3	1939	Tops	157
South Dakota, U.S.	ND—18	9	2	1941	Whole plant	176
Wyoming, U.S.	15—590	253	4	1937		43
Aster ericoides L.						
Colorado, U.S.	6—100	44	4	1938	Soil Se 0.2—54 (m = 1.5) mg/kg (n = 160)	181
Graham County, KS, U.S.	1—330	22	4	1936	Soil Se 0.4—20 (m = 1.8) mg/kg (n = 14)	117
Gove County, KS, U.S.	2—170	8	5	1936	Soil Se 0.2—18 (m = 2) mg/kg (n = 67)	117
Logan County, KS, U.S.	1—230	80	15	1936	Soil Se 0.2—140 (m = 2) mg/kg (n = 258)	117
Rooks County, KS, U.S.	0—160	2	14	1936	Soil Se 0.2—100 (m = 0.7) mg/kg (n = 52)	117
Trego County, KS, U.S.	1—40	7	6	1936	Soil Se 0.3—20 (m = 1) mg/kg (n = 53)	117
Wallace County, KS, U.S.	1—310	29	6	1936	Soil Se 0.3—8 (m = 1) mg/kg (n = 70)	117
Montana, U.S.	1—12	7	2	1936	Soil Se 0.1—2.5 (m = 0.6) mg/kg (n = 68)	117
S & SW of Black Hills, Nebraska, U.S.	7—180	94	2	1936	Soil Se 0.2—14 (m = 0.7) mg/kg (n = 206)	117
North Dakota, U.S.	—	9	1	1940	Whole plant	177
S & SW of Black Hills, South Dakota, U.S.	4—260	10	8	1936	Soil Se 0.1—40 (m = 1.5) mg/kg (n = 76)	117
South Dakota, U.S.	2—440	80	3	1939	Tops	157
South Dakota, U.S.	—	9	1	1941	Whole plant	176

Aster glaucus, T. and G.						
Colorado, U.S.	25	1	—	1939	Tops	157
Aster multiflorus, Ait.						
South Dakota, U.S.	7.2—366	71	8	1938	Whole plant, Niobrara formation	183
South Dakota, U.S.	0—11	4	17	1938	Whole plant, Pierre formation	183
South Dakota, U.S.	0—180	14.5	16	1939	Soil Se 0—52 (m = 4.5) mg/kg (n = 253)	58
South Dakota, U.S.	14—1800	250	14	1939	Whole plant	58
South Dakota, U.S.	0—320	22	12	1939	Whole plant	58
Atriplex canescens (Pursh) Nutt.						
Colorado, U.S.	<1—213	2	8	1939	Tops, leaves, stems	157
New Mexico, U.S.	—	ND	2	1941	Whole plant	176
South Dakota, U.S.	—	Trace	1	1941	Whole plant	176
Wyoming, U.S.	27—450	227	6	1941	Whole plant	176
Atriplex confertifolia (Torr. and Frém.) Wats.						
Thompson, UT, U.S.	—	1031	—	1943	Vanadium-uranium zone	184
Thompson, UT, U.S.	688—1734	1211	2	1943	Old Yellow Cat Camp	184
Wyoming, U.S.	8—84	46	2	1941	Whole plant	176
Atriplex nuttallii Wats.						
Thompson, UT, U.S.	—	611	—	1943	Vanadium-uranium zone	184
Wyoming, U.S.	—	524	—	1946		185
Wyoming, U.S.	3—244	19	12	1941	Whole plant	176
Wyoming, U.S.	30—202	90	6	1940	Whole plant	177
Castilleja chromosa A. Nels.						
Colorado, U.S.	—	77	1	1941	Whole plant	176
New Mexico, U.S.	—	17	1	1941	Whole plant	176
Utah, U.S.	ND—1	0.5	2	1941	Whole plant	176
Wyoming, U.S.	—	3152	—	1946		185
Wyoming, U.S.	Trace—287	143	6	1941	Whole plant	176
Grindelia squarrosa (Pursh) Dunal. (gumweed)						
Colorado, U.S.	2—420	11	6	1938	Soil Se 0.2—54 (m = 1.5) mg/kg (n = 160)	181
Big Horn County, MT, U.S.	<1—40	1	14	1940	Soil Se 0.1—8 (m = 0.7) mg/kg (n = 72)	36
Pondera County, MT, U.S.	3—80	4	5	1940	Soil Se 0.2—3 (m = 0.5) mg/kg (n = 85)	36
Teton County, MT, U.S.	1—7	3	10	1940	Soil Se 0.1—5 (m = 0.5) mg/kg (n = 151)	36
Yellowstone County, MT, U.S.	0—3	1	10	1940	Soil Se 0.2—3 (m = 0.4) mg/kg (n = 39)	36
S & SW of Black Hills, Nebraska, U.S.	1—6	3	5	1936	Soil Se 0.2—14 (m = 0.7) mg/kg (n = 206)	117

Table 4 (continued)
CONCENTRATION OF SELENIUM IN SELENIUM ACCUMULATOR PLANTS[a]

Geographical location[b]	Selenium concentration (mg/kg)[c] Range	Median	N[d]	Date[e]	Remarks[f]	Ref.
New Mexico, U.S.	0—150	1	3	1938	Soil Se 0.2—10 (m = 1) mg/kg (n = 61)	181
Lower Brule Indian Reservation, South Dakota, U.S.	2—930	100	11	1941	Soil Se 0.2—16 (m = 1) mg/kg (n = 80)	178
S & SW of Black Hills, South Dakota, U.S.	5—330	98	4	1936	Soil Se 0.1—40 (m = 1.5) mg/kg (n = 76)	117
South Dakota, U.S.	2.0—260	9.5	4	1938	Whole plant, Niobrara formation	183
South Dakota, U.S.	0—10	8	5	1938	Whole plant, Pierre formation	183
South Dakota, U.S.	0—12	2.7	8	1939	Soil Se 0—52 (m = 4.5) mg/kg (n = 253)	58
South Dakota, U.S.	11—230	61	8	1939	Whole plant	58
South Dakota, U.S.	0—260	7.2	8	1939	Whole plant	58
Gutierrezia sarothrae (Pursh) B. and R. (broom snakeweed, turpentineweed)						
Gove County, KS, U.S.	1—430	15	11	1936	Soil Se 0.2—18 (m = 2) mg/kg (n = 67)	117
Graham County, KS, U.S.	—	4	1	1936	Soil Se 0.4—20 (m = 1.8) mg/kg (n = 14)	117
Logan County, KS, U.S.	1—610	11	20	1936	Soil Se 0.2—140 (m = 2) mg/kg (n = 258)	117
Rooks County, KS, U.S.	0—8	1	8	1936	Soil Se 0.2—100 (m = 0.7) mg/kg (n = 52)	117
Trego County, KS, U.S.	0—15	1	12	1936	Soil Se 0.3—20 (m = 1) mg/kg (n = 53)	117
Wallace County, KS, U.S.	8—20	14	2	1936	Soil Se 0.3—8 (m = 1) mg/kg (n = 70)	117
Montana, U.S.	5—15	8	4	1936	Soil Se 0.1—2.5 (m = 0.6) mg/kg (n = 68)	117
New Mexico, U.S.	1—350	1	17	1938	Soil Se 0.2—10 (m = 1) mg/kg (n = 61)	181
S & SW of Black Hills, South Dakota, U.S.	1—45	1	11	1936	Soil Se 0.1—40 (m = 1.5) mg/kg (n = 76)	117
South Dakota, U.S.	8.0—220	20	8	1938	Whole plant, Niobrara formation	183
South Dakota, U.S.	0—1	0.5	2	1938	Whole plant, Pierre formation	183
South Dakota, U.S.	8—160	20	8	1939	Soil Se 0—52 (m = 4.5) mg/kg (n = 253)	58
South Dakota, U.S.	8—220	22	9	1939	Whole plant	58
Wyoming, U.S.	48—217	133	2	1941	Whole plant	176

Wyoming, U.S.	—	69	1	1940	Whole plant	177
Machaeranthera ramosa A. Nels.						
Utah, U.S.	ND—5	5	3	1941	Whole plant	176
Wyoming, U.S.	13—500	100	11	1941	Whole plant	176
Wyoming, U.S.	—	740	1	1940	Whole plant	177
Mentzelia decapetala Pursh (stickleaf)						
South Dakota, U.S.	1.2—4.0	2.6	4	1938	Whole plant, Niobrara formation	183
South Dakota, U.S.	1—6	1	5	1938	Whole plant, Pierre formation	183
South Dakota, U.S.	0.0—8	1.8	12	1939	Soil Se 0—52 (m = 4.5) mg/kg (n = 253)	58
Neptunia amplexicaulis						
NW Queensland, Australia	25.7—4334	390	43	1963	Poison strip area, soil Se 0.15—32.2 (m = 3.3) mg/kg (n = 19)	155
NW Queensland, Australia	4.8—227	20.6	14	1963	Adjacent to poison strip, soil Se 0.3—385 (m = 2.1) mg/kg (n = 45)	155
NW Queensland, Australia	34.3—105.4	49	6	1963	Landsborough highway traverse, Hughenden; soil Se 0.3—14.5 (m = 3.3) mg/kg (n = 10)	155
Neptunia gracilis						
NW Queensland, Australia	<5—25.4	8.5	4	1963	Adjacent to poison strip, soil Se 0.3—385 (m = 2.1) mg/kg (n = 45)	155
NW Queensland, Australia	—	<0.5	3	1963	Landsborough highway traverse, Hughenden; soil Se 0.3—14.5 (m = 3.3) mg/kg (n = 10)	155

[a] Division of plants is into two categories, primary and secondary accumulator plants, according to listings by Trelease and Beath,[40,41] Rosenfeld and Beath,[21] and Munz and Keck.[41,42] Several species reported on after these publications have been placed in appropriate categories.

[b] Plants are typically from uncultivated areas within the geographical location specified, collected during detailed survey explorations or reconnaissance forays. Data have been arranged by counties and smallest subdivisions in instances when there was an abundance of information or when an actual or perceived difference in data existed among these regions.

[c] Selenium concentrations refer to total selenium usually reported on a dry matter basis (air dried, bone dry, partly dry), determined by procedures based on acid decomposition-distillation-precipitation-titrimetry or nephelometry-colorimetry. Median concentrations were generally calculated by this author; av refers to average concentration, ND indicates selenium not detected by method used.

[d] Number of samples analyzed as reported in the literature or estimated by the present author from information in the literature.

[e] Date of publication of report. Date of sample collection is generally very close to date of publication with the exception of the data (generated in ca. 1938 to 1940) in the 1948 USDA Technical Bulletin No. 950, the publication of which was delayed by wartime restrictions.

Table 4 (continued)
CONCENTRATION OF SELENIUM IN SELENIUM ACCUMULATOR PLANTS[a]

This column contains additional information regarding parts of the plant analyzed as described in the reference, soil selenium content, and more specific details on the location. Whole plant refers to above ground portion; no data for roots are included here. Soil selenium content is reported as a range and median (m) with an indication of the number (n) of samples analyzed. Soil refers to actual soil as well as material described as shale and mineral in the literature, and selenium levels reflect those typical of the geographic region denoted. Refer to the literature cited for soil selenium levels at specific soil and plant collection sites as well as detailed descriptions of locations, soils, plants, and growth stages.

more available to the plant under study), (5) variations in geological formations supporting plant growth, (6) climate and season, (7) plant component or tissue investigated, (8) thrift and age of plant when sampled, and (9) selenium accumulation in subsoils. Reviews of these variables have been reported,[22,41,54,58] with a particularly thorough exposition by Moxon et al.[58] early in the selenium story. Once in the plant, release of selenium from the food or feed crop to the consumer of the commodity (bioavailability) is influenced by other factors of interest to researchers in nutrition and toxicology.

ACKNOWLEDGMENTS

The author acknowledges the assistance of M. Graham in locating and obtaining copies of some of the references used in this work. The bulk of the work was diligently and efficiently set into typescript from handwritten manuscript by Mrs. Karen Burns whose unfailing efforts are deeply appreciated. Discussions with D. Veira regarding animal feed crops and ingredients and A. McElroy on plant taxonomy were helpful. Land Resource Research Centre, Agriculture Canada, Contribution No. 87-90.

REFERENCES

1. **Muth, O. H., Oldfield, J. E., and Weswig, P. H., Eds.,** *Selenium in Biomedicine,* AVI Publishing, Westport, CT, 1967.
2. **Spallholz, J. E., Martin, J. L., and Ganther, H. E., Eds.,** *Selenium in Biology and Medicine,* AVI Publishing, Westport, CT, 1981.
3. **Klayman, D. L. and Gunther, W. H. H., Eds.,** *Organic Selenium Compounds: Their Chemistry and Biology,* Wiley-Interscience, Toronto, 1973.
4. **Shamberger, R. J.,** *Biochemistry of Selenium,* Plenum Press, New York, 1983.
5. **Ganther, H. E.,** Biochemistry of selenium, in *Selenium,* Zingaro, R. A. and Cooper, W. C., Eds., Van Nostrand Reinhold, New York, 1974, 546.
6. **Cooper, W. C., and Glover, J. R.,** The Toxicology of selenium and its compounds, in *Selenium,* Zingaro, R. A. and Cooper, W. C., Eds., Van Nostrand Reinhold, New York, 1974, 654.
7. **Ganther, H. E., Hafeman, D. G., Lawrence, R. A., Serfass, R. E., and Hoekstra, W. G.,** Selenium and glutathione peroxidase in health and disease — a review, in *Trace Elements in Human Health and Disease,* Vol. 2, Prasad, A. S. and Overleas, D., Eds., Academic Press, New York, 1976, 165.
8. **Taboury, M.,** Sur la présence accidentelle du selenium dans certains végétaux (in French), *C. R.,* 195, 171, 1932.
9. **Robinson, W. O.,** Determination of selenium in wheat and soils, *J. Assoc. Off. Anal. Chem.,* 16, 423, 1933.
10. **Byers, H. G.,** Selenium occurrence in certain soils in the United States with a discussion of related topics, *U.S. Dep. Agric. Tech. Bull.,* 482, 1935.
11. **Byers, H. G. and Lakin, H. W.,** Selenium in Canada, *Can. J. Res. Sect. B,* 17, 364, 1939.
12. **Thorvaldson, T. and Johnson, L. R.,** The selenium content of Saskatchewan wheat, *Can. J. Res. Sect. B,* 18, 138, 1940.
13. **Lakin, H. W. and Byers, H. G.,** Selenium in wheat and wheat products, *Cereal Chem.,* 8, 73, 1941.
14. **Williams, K. T., Lakin, H. W., and Byers, H. G.,** Selenium occurrence in certain soils in the United States, with a discussion of related topics: fifth report, *U.S. Dep. Agric. Tech. Bull.,* 758, 1941.
15. **Moxon, A. L.,** Alkali disease or selenium poisoning, *S. D. Agric. Exp. Stn. Bull.,* 311, 1937.
16. *S. D. Agric. Exp. Stn. 45th Annu. Rep.,* 1932, plus subsequent reports and references therein.
17. **Franke, K. W., Rice, T. D., Johnson, A. G., and Schoening, H. W.,** Preliminary field survey of the so-called "alkali disease" of livestock, *U.S. Dep. Agric. Circ.,* 320, 1934.
18. **Anderson, M. S., Lakin, H. W., Beeson, K. C., Smith, F. F., and Thacker, E.,** Selenium in agriculture, *U.S. Dep. Agric. Handbook,* 200, 1961.
19. **Underwood, E. J.,** Trace elements, in *Toxicants Occurring Naturally in Foods,* National Academy of Sciences, Washington, D.C., 1973, 43.

20. **Moxon, A. L.,** Selenium: its occurrence in rocks and soils, absorption by plants, toxic action in animals, and possible essential role in animal nutrition, in *Trace Elements — Proc. Conf. Wooster, Ohio,* Lamb, C. A., Bentley, O. G., and Beattie, J. M., Eds., Academic Press, New York, 1958, 175.
21. **Moxon, A. L.,** Selenium in Agriculture, in *Selenium,* Zingaro, R. A. and Cooper, W. C., Eds., Van Nostrand Reinhold, New York, 1974, 675.
22. **Rosenfeld, I. and Beath, O. A.,** *Selenium, Geobotany, Biochemistry, Toxicity and Nutrition,* Academic Press, New York, 1964.
23. **Nicholas, D. J. D. and Egan, A. R., Eds.,** *Trace Elements in Soil-Plant-Animal Systems,* Academic Press, New York, 1975.
24. **Mills, C. F., Bremner, I., Chesters, J. K., and Quarterman, J., Eds.,** *Trace Element Metabolism in Animals,* Churchill Livingstone, Edinburgh, 1970.
25. **Hoekstra, W. G., Suttie, J. W., Ganther, H. E., and Mertz, W., Eds.,** *Trace Element Metabolism in Animals-2,* University Park Press, Baltimore, 1974.
26. **Kirchgessner, M., Roth-Maier, D. A., Roth, H.-P., Schwarz, F. J., and Weigand, E., Eds.,** *Trace Element Metabolism in Man and Animals-3,* Arbeitskreis fur Tierernahrungsforschung, Weihenstephan, W. Germany, 1978.
27. **Gawthorne, J. M., Howell, J. McC., and White, C. L., Eds.,** *Trace Element Metabolism in Man and Animals,* TEMA 4, Springer-Verlag, Berlin, 1982.
28. **Mills, C. F., Bremner, I., and Chesters, J. K., Eds.,** *Trace Elements in Man and Animals, TEMA-5,* Commonwealth Agricultural Bureaux, Farnham Common, U.K., 1985.
29. **Ku, P. K., Ely, W. T., Groce, A. W., and Ullrey, D. E.,** Natural dietary selenium, α-tocopherol and effect on tissue selenium, *J. Anim. Sci.,* 34, 208, 1972.
30. **Mathias, M. M., Allaway, W. H., Hogue, D. E., Marion, M. V., and Gardner, R. W.,** Value of selenium in alfalfa for the prevention of selenium deficiencies in chicks and rats, *J. Nutr.,* 86, 213, 1965.
31. **Combs, G. F., Jr., Liu, C. H., Lu, Z. H., and Su, Q.,** Uncomplicated selenium deficiency produced in chicks fed a corn-soy-based diet, *J. Nutr.,* 114, 964, 1984.
32. **Chortyk, O. T. and Schlotzhauer, W. S.,** Increasing selenium in cigarettes and smoke: transfer to smoke, *Arch. Environ. Health,* 39, 419, 1984.
33. **Slegers, G. and Claeys, A.,** Instrumental neutron-activation analysis of raw tobacco leaves imported in the Belgian-Luxemburg Economic Union, *Meded. Fac. Landbouwwet., Rijksuniv. Gent,* 44, 1139, 1979.
34. **Bogden, J. D., Kemp, F. W., Buse, M., Thind, I. S., Louria, D. B., Forgacs, J., Lianos, G., and Terrones, I. M.,** Composition of tobaccos from countries with high and low incidences of lung cancer. I. Selenium, polonium-210, *Alternaria,* tar, and nicotine, *J. Natl. Cancer Inst.,* 66, 27, 1981.
35. **Leisure, O. W. and Olson, O. E.,** Use of the AOAC fluorometric method for selenium in plants for the analysis of papers and tobaccos, *J. Assoc. Off. Anal. Chem.,* 57, 658, 1974.
36. **Williams, K. T. and Lakin, H. W.,** Selenium occurrence in certain soils in the United States, with a discussion of related topics; fourth report, *U.S. Dep. Agric. Tech. Bull.,* 702, 1940.
37. **Franke, K. W.,** A new toxicant occurring naturally in certain samples of plant foodstuffs. I. Results obtained in preliminary feeding trials, *J. Nutr.,* 8, 597, 1934.
38. **Beath, O. A., Eppson, H. F., and Gilbert, C. S.,** Selenium and other toxic minerals in soils and vegetation, *Wyo. Agric. Exp. Stn. Bull.,* 206, 1935.
39. **Trelease, S. F. and Trelease, H. M.,** Selenium as a stimulating and possibly essential element for indicator plants, *Science,* 87, 70, 1938.
40. **Trelease, S. F. and Beath, O. A.,** *Selenium,* published by the authors, New York, 1949.
41. **Shrift, A.,** Selenium compounds in nature and medicine. E. Metabolism of selenium by plants and organisms, in *Organic Selenium Compounds: Their Chemistry and Biology,* Klayman, D. L. and Gunther, W. H. H., Eds., Wiley-Interscience, Toronto, 1973, chap. 13E.
42. **Munz, P. A. and Keck, D. D.,** *A California Flora,* University of California Press, Berkeley, 1959.
43. **Beath, O. A., Eppson, H. F., and Gilbert, C. S.,** Selenium distribution in and seasonal variation of type vegetation occurring on seleniferous soils, *J. Am. Pharm. Assoc.,* 26, 394, 1937.
44. **Shacklette, H. T., Boerngen, J. G., and Keith, J. R.,** Selenium, fluorine and arsenic in surficial materials of the conterminous United States, *U.S. Geol. Surv. Circ.,* 692, 1974.
45. **Kubota, J., Allaway, W. H., Carter, D. L., Cary, E. E., and Lazar, V. A.,** Selenium in crops in the United States in relation to selenium-responsive diseases of animals, *J. Agric. Food Chem.,* 15, 448, 1967.
46. **Kubota, J. and Allaway, W. H.,** Geographic distribution of trace element problems, in *Micronutrients in Agriculture,* Soil Science Society of America, Madison, WI, 1972, chap. 21.
47. **Muth, O. H. and Allaway, W. H.,** The relationship of white muscle disease to the distribution of naturally occurring selenium, *J. Am. Vet. Med. Assoc.,* 142, 1379, 1963.
48. **Shrift, A.,** Aspects of selenium metabolism in higher plants, *Annu. Rev. Plant Physiol.,* 20, 475, 1969.
49. **Allaway, W. H.,** Selenium in the food chain, *Cornell Vet.,* 63, 151, 1973.
50. **Johnson, C. M.,** Selenium in soils and plants: contrasts in conditions providing safe but adequate amounts of selenium in the food chain, in *Trace Elements in Soil-Plant-Animal Systems,* Nicholas, D. J. D. and Egan, A. R., Eds., Academic Press, New York, 1975, 165.

51. **Shearer, T. R. and Hadjimarkos, M.,** Geographic distribution of selenium in human milk, *Arch. Environ. Health,* 30, 230, 1975.
52. **Brown, T. A. and Shrift, A.,** Selenium: toxicity and tolerance in higher plants, *Biol. Rev. Cambridge Philos. Soc.,* 57, 59, 1982.
53. **Shamberger, R. J.,** *Biochemistry of Selenium,* Plenum Press, New York, 1983, 167.
54. **Gissel-Nielsen, G., Gupta, U. C., Lamand, M., and Westermarck, T.,** Selenium in soils and plants and its importance in livestock and human nutrition, *Adv. Agron.,* 37, 397, 1984.
55. **Ullrey, D. E.,** Selenium in the soil-plant food chain, in *Selenium in Biology and Medicine,* Spallholz, J. E., Martin, J. L., and Ganther, H. E., Eds., AVI Publishing, Westport, CT, 1981, 176.
56. **Patrias, G. and Olson, O. E.,** Selenium contents of samples of corn from midwestern states, *Feedstuffs,* 41(43), 32, 1969.
57. **Gissel-Nielsen, G.,** Selenium in soils and plants, *Proc. of the Symposium on Selenium-Tellurium in the Environment,* Industrial Health Foundation, Pittsburgh, 1976, 10.
58. **Moxon, A. L., Olson, O. E., and Searight, W. V.,** Selenium in rocks, soils and plants, *S.D. Agric. Exp. Stn. Tech. Bull.,* 2, 1939.
59. **Van Ryswyk, A. L., Broersma, K., and Kalnin, C. M.,** Selenium content of alfalfa grown on orthic gray luvisolic and carbonated orthic gleysolic soils, *Can. J. Plant Sci.,* 56, 753, 1976.
60. **Lévesque, M.,** Some aspects of selenium relationships in eastern Canadian soils and plants, *Can. J. Soil Sci.,* 54, 205, 1974.
61. **Bisbjerg, B.,** *Studies on Selenium in Plants and Soils,* Riso Rep. No. 200, Danish Atomic Energy Commission, Roskilde, Denmark, 1972.
62. **Gissel-Nielsen, G.,** Selenium concentration in Danish forage crops, *Acta Agric. Scand.,* 25, 216, 1975.
63. **Arvy, M. P.,** Teneur en sélénium de la luzerne (*Medicago sativa* L.), *Lett. Bot.,* 5, 419, 1980.
64. **Patel, C. A. and Mehta, B. V.,** Selenium status of soils and common fodders in Gujarat, *Ind. J. Agric. Sci.,* 40, 389, 1970.
65. **Ravikovitch, S. and Margolin, M.,** Selenium in soils and plants, *Isr. J. Agric. Res. (Ktavim [Engl. Ed.]),* 7, 41, 1957.
66. **Grant, A. B. and Sheppard, A. D.,** Selenium in New Zealand pastures, *N.Z. Vet. J.,* 31, 131, 1983.
67. **Lindberg, P. and Lannek, N.,** Amounts of selenium in Swedish forages, soils, and animal tissues, in *Trace Element Metabolism in Animals (TEMA-1969),* Mills, C. F., Bremner, I., Chesters, J. K., and Quarterman, J., Eds., Churchill Livingstone, Edinburgh, 1970, 421.
68. **Carter, D. L., Brown, M. J., and Robbins, C. W.,** Selenium concentrations in alfalfa from several sources applied to a low selenium, alkaline soil, *Proc. Soil Sci. Soc. Am.,* 33, 715, 1969.
69. **Carter, D. L., Robbins, C. W., and Brown, M. J.,** Effect of phosphorus fertilization on the selenium concentration in alfalfa *(Medicago sativa), Proc. Soil Sci. Soc. Am.,* 36, 624, 1972.
70. **Gutierrez, J. L., Smith, G. S., Wallace, J. D., and Nelson, A. B.,** Selenium in plants, water and blood: New Mexico and Chihuahua, *J. Anim. Sci.,* 38, 1330, 1974.
71. **Allaway, W. H., Moore, D. P., Oldfield, J. E., and Muth, O. H.,** Movement of physiological levels of selenium from soils through plants to animals, *J. Nutr.,* 88, 411, 1966.
72. **Bell, M. C. and Bacon, J. A.,** Selenium in forages and the need for selenium for livestock, *Tenn. Farm Home Sci. Prog. Rep.,* 99, 16, 1976.
73. **Cathcart, E. B., Shelford, J. A., and Peterson, R. G.,** Mineral analyses of dairy cattle feed in the upper Fraser Valley of British Columbia, *Can. J. Anim. Sci.,* 60, 177, 1980.
74. **Gissel-Nielsen, G.,** *Control of Selenium in Plants,* Riso Rep. No. 370, Riso National Laboratory, Roskilde, Denmark, 1977.
75. **Redshaw, E. S., Martin, P. J., and Laverty, D. H.,** Iron, manganese, copper, zinc and selenium concentrations in Alberta grains and roughages, *Can. J. Anim. Sci.,* 58, 553, 1978.
76. **Miltimore, J. E., van Ryswyk, A. L., Pringle, W. L., Chapman, F. M., and Kalnin, C. M.,** Selenium concentrations in British Columbia forages, grains, and processed feeds, *Can. J. Anim. Sci.,* 55, 101, 1975.
77. **Arthur, D.,** Selenium content of some feed ingredients available in Canada, *Can. J. Anim. Sci.,* 51, 71, 1971.
78. **Winter, K. A., Gupta, U. C., Nass, H. G., and Kunelius, H. T.,** Selenium content of feedstuffs produced in Prince Edward Island, *Can. J. Anim. Sci.,* 53, 113, 1973.
79. **Gupta, U. C. and Winter, K. A.,** Selenium content of soils and crops and the effects of lime and sulfur on plant selenium, *Can. J. Soil Sci.,* 55, 161, 1975.
80. **Owen, B. D., Sosulski, F., Wu, K. K., and Farmer, M. J.,** Variation in mineral content of Saskatchewan feed grains, *Can. J. Anim. Sci.,* 57, 679, 1977.
81. **Gissel-Nielsen, G.,** The fate of selenium added to agricultural crops, in *Comparative Studies of Food and Environmental Contamination,* IAEA STI/PUB/348, International Atomic Energy Agency, Vienna, 1974, 333.
82. **Oksanen, H. E. and Sandholm, M.,** The selenium content of Finnish forage crops, *Maataloustieteelinen Aikak,* 42, 250, 1970.

83. **Gissel-Nielsen, G.,** Foliar application and pre-sowing treatment of cereals with selenite, *Z. Pflanzenernaehr. Bodenkd.*, 1, 97, 1975.
84. **Moré, E., Coppenet, M. M., and Le Corre, L.,** Teneurs en sélénium des plantes fourragères récoltées dans le Finistère enrichissement du ray-grass et du mais par apport de sélénite au sol, *C. R. Seances Acad. Agric. Fr.*, 61, 870, 1975.
85. **Moré, E., Coppenet, M., and Le Corre, L.,** Teneurs en sélénium des plantes fourragères influence de la fertilisation et des apports de sélénite, *Ann. Agron.*, 31, 297, 1980.
86. **Kappel, L. C., Morgan, E. B., Kilgore, L., Ingraham, R. H., and Babcock, D. K.,** Seasonal changes of mineral content of southern forages, *J. Dairy Sci.*, 68, 1822, 1985.
87. **Dhillon, K. S., Randhawa, N. S., and Sinha, M. K.,** Selenium status of some common fodders and natural grasses of Punjab, *Indian J. Dairy Sci.*, 30, 218, 1977.
88. **Berg, H., Woien, T., and Aasehaug, B.,** Selenium in East African feeding materials, *Nord. Veterinaermed.*, 25, 521, 1973.
89. **Lessard, J. R., Hidiroglou, M., Carson, R. B., and Dermine, P.,** Intra-seasonal variations in the selenium content of various forage crops at Kapuskasing, Ontario, *Can. J. Plant Sci.*, 48, 581, 1968.
90. **Davies, E. B. and Watkinson, J. H.,** Uptake of native and applied selenium by pasture species. I. Uptake of Se by browntop, ryegrass, cocksfoot, and white clover from Atiamuri sand, *N.Z. J. Agric. Res.*, 9, 317, 1966.
91. **Watkinson, J. H. and Davies, E. B.,** Uptake of native and applied selenium by pasture species. IV. Relative uptake through foliage and roots by white clover and browntop. Distribution of selenium in white clover, *N.Z. J. Agric. Res.*, 10, 122, 1967.
92. **Spencer, K.,** Effect of sulfur application on selenium content of subterranean clover plants grown at different levels of selenium supply, *Aust. J. Exp. Agric. Anim. Husb.*, 22, 420, 1982.
93. **Gardiner, M. R. and Nairn, M. E.,** Studies on the effect of cobalt and selenium in clover disease of ewes, *Aust. Vet. J.*, 45, 215, 1969.
94. **Nelson, K. W. and Bundy, S. D.,** Environmental aspects of selenium and tellurium, in *Proc. Int. Symp. Industrial Uses Selenium Tellurium,* 1980, 18.
95. **Williams, C. and Thornton, I.,** The effect of soil additives on the uptake of molybdenum and selenium from soils from different environments, *Plant Soil,* 36, 395, 1972.
96. **Watkinson, J. H. and Davies, E. B.,** Uptake of native and applied selenium by pasture species. III. Uptake of selenium from various carriers, *N.Z. J. Agric. Res.*, 10, 116, 1967.
97. **Davies, E. B. and Watkinson, J. H.,** Uptake of native and applied selenium by pasture species. II. Effects of sulphate and of soil type on uptake by clover, *N.Z. J. Agric. Res.*, 9, 641, 1966.
98. **Millar, K. R.,** The estimation of α-tocopherol in pasture samples and a comparison of α-tocopherol levels in samples from areas differing in selenium status, *N.Z. J. Agric. Res.*, 14, 142, 1971.
99. **Moxon, A. L.,** Selenium deficiency in cattle, *S. Afr. J. Anim. Sci.*, 11, 183, 1981.
100. **Lindberg, P. and Jacobsson, S. O.,** Relationship between selenium content of forage, blood and organs of sheep, and lamb mortality rate, *Acta Vet. Scand.*, 11, 49, 1970.
101. **Walsh, T. and Fleming, G. A.,** Selenium levels in rocks, soils and herbage from a high selenium locality in Ireland, *Trans. Int. Soc. Soil Sci. Comm II and IV*, 2, 1978, 1952 (1953).
102. **Cook, K. A. and Graham, E. R.,** A neutron activation method for determining submicrogram selenium in forage grasses, *J. Soil Sci. Soc. Am.*, 42, 57, 1978.
103. **Bisbjerg, B., Jochumsen, P., and Rasbech, N. O.,** Selenium content in organs, milk and fodder of the cow, *Nord. Veterinaermed.*, 22, 532, 1970.
104. **Ettala, E. and Kossila, V.,** Mineral contents in heavy nitrogen fertilized grass and its silage, *Ann. Agric. Fenn.*, 18, 252, 1979.
105. **Périgand, S.,** Les carences en oligo-éléments chez les ruminants en France. Leur diagnostic. Les problèmes soulevés par l'intensification fourragère, *Ann. Agron.*, 21, 635, 1970.
106. **Gissel-Nielsen, G.,** Improvement of selenium status of pasture crops, *Biol. Trace Elem. Res.*, 6, 281, 1984.
107. **Saunders, W. M. H.,** Mineral composition of soil and pasture from areas of grazed paddocks, affected and unaffected by dung and urine, *N.Z. J. Agric. Res.*, 27, 405, 1984.
108. **Twomey, T., Crinion, R. A. P., and Glazier, D. B.,** Selenium toxicity in cattle in Co. Meath, *Ir. Vet. J.*, 31, 41, 1977.
109. **Andrews, E. D., Hogan, K. G. and Sheppard, A. D.,** Selenium in soils, pastures and animal tissues in relation to the growth of young sheep on a marginally selenium-deficient area, *N.Z. Vet. J.*, 24, 111, 1976.
110. **Long, M. I. E. and Marshall, B.,** The selenium status of pastures in Uganda, *Trop. Agric. (Trinidad),* 50, 121, 1973.
111. **Collier, R. E. and Parker-Sutton, J.,** A measure of the effect of drying temperature on the selenium content of herbage, *J. Sci. Food Agric.*, 27, 743, 1976.
112. **Robberecht, H., Vanden Berghe, D., Deelstra, H., and Van Grieken, R.,** Selenium in the Belgian soils and its uptake by rye-grass, *Sci. Total Environ.*, 25, 61, 1982.

113. **Ylaranta, Y.,** Effect of liming and sulphate on the selenium content of Italian rye grass *(Lolium multiflorum), Ann. Agric. Fenn.,* 22, 152, 1983.
114. **Ylaranta, Y.,** Effect of added selenite and selenate on the selenium content of Italian rye grass *(Lolium multiflorum)* in different soils, *Ann. Agric. Fenn.,* 22, 139, 1983.
115. **Joblin, K. N. and Pritchard, M. W.,** Selenium in a ryegrass pasture, in *Proc. New Zealand Workshop on Trace Elements in New Zealand,* University of Otago, Dunedin, 1981, 93.
116. **Ylaranta, T.,** Effect of selenite and selenate fertilization and foliar spraying on selenium content of timothy grass, *Ann. Agric. Fenn.,* 23, 96, 1984.
117. **Byers, H. G.,** Selenium occurrence in certain soils in the United States with a discussion of related topics, second report, *U.S. Dep. Agric. Tech. Bull.,* 530, 1936.
118. **Noble, R. M. and Barry, G. A.,** Survey of selenium concentrations in wheat, sorghum and soybean grains, prepared poultry feeds and feed ingredients from Queensland, *Queensl. J. Agric. Anim. Sci.,* 39, 1, 1982.
119. **Perry, T. W., Caldwell, D. M., and Peterson, R. C.,** Selenium content of feeds and effect of dietary selenium on hair and blood serum, *J. Dairy Sci.,* 59, 760, 1976.
120. **Cantor, A. H., Scott, M. L., and Noguchi, T.,** Biological availability of selenium in feedstuffs and selenium compounds for prevention of exudative diathesis in chicks, *J. Nutr.,* 105, 96, 1975.
121. **Martello, M. A. and Latshaw, J. D.,** Utilization of dietary selenium as indicated by prevention of selenium deficiency and by retention in eggs, *Nutr. Rep. Int.,* 26, 43, 1982.
122. **Kifer, R. R., Payne, W. L., and Ambrose, M. E.,** Selenium content of fish meals II, *Feedstuffs,* 41(51), 24, 1969.
123. **Lunde, G.,** Trace metal contents of fish meal and of the lipid phase extracted from fish meal, *J. Sci. Food Agric.,* 24, 413, 1973.
124. **Heffron, C. L., Reid, J. T., Haschek, W. M., Furr, A. K., Parkinson, T. F., Bache, C. A., Gutenmann, W. H., St. John, L. E., Jr., and Lisk, D. J.,** Chemical composition and acceptability of aquatic plants in diets of sheep and pregnant goats, *J. Anim. Sci.,* 45, 1166, 1977.
125. **Poley, W. E. and Moxon, A. L.,** Tolerance levels of seleniferous grains in laying rations, *Poult. Sci.,* 17, 72, 1938.
126. **Hugue, Q. M. E. and Jensen, J. F.,** Biological availability of selenium and phosphorus in fish meal as affected by condition of fish and type of meal, *Br. Poult. Sci.,* 26, 289, 1985.
127. **Kelleher, W. J. and Johnson, M. J.,** Determination of traces of selenium in organic matter. Combined spectrophotometric and isotope dilation method, *Anal. Chem.,* 33, 1429, 1961.
128. **Mahan, D. C.,** Selenium — an essential dietary supplement for swine, *Distill. Feed Res. Counc. Conf. Proc.,* 28, 6, 1973.
129. **Gabrielsen, B. O. and Opstvedt, J.,** Availability of selenium in fish meal in comparison with soybean meal, corn gluten meal and selenomethionine relative to selenium in sodium selenite for restoring glutathione peroxidase activity in selenium-depleted chicks, *J. Nutr.,* 110, 1096, 1980.
130. **Hulstaert, C. E., Molenaar, I., de Goeij, J. J. M., Zegers, C., and van Pijpen, P. L.,** Selenium in vitamin-E-deficient diets and the occurrence of myopathy as a symptom of vitamin E deficiency, *Nutr. Metab.,* 20, 91, 1976.
131. **Boyer, C. I., Jr., Andrews, E. J., deLahunta, A., Bache, C. A., Gutenmann, W. H., and Lisk, D. J.,** Accumulation of mercury and selenium in tissues of kittens fed commercial cat food, *Cornell Vet.,* 68, 365, 1978.
132. **Froslie, A., Karlsen, J. T., and Rygge, J.,** Selenium in animal nutrition in Norway, *Acta Agric. Scand.,* 30, 17, 1980.
133. **Ullrey, D. E.,** The selenium-deficiency problem in animal agriculture in *Trace Element Metabolism in Animals-2,* TEMA 2, Hoekstra, W. G., Suttie, J. W., Ganther, H. E., and Mertz, W., Eds., University Park Press, Baltimore, 1974, 275.
134. **Whitacre, M. and Latshaw, J. D.,** Selenium utilization from Menhaden fish meal as affected by processing, *Poult. Sci.,* 61, 2520, 1982.
135. **Soares, J. H., Jr. and Miller, D.,** Selenium content of Atlantic, Gulf menhaden fish solubles, *Feedstuffs,* 42(37), 22, 1970.
136. **Giovannetti, P. and Bell, J. M.,** Research on Rapeseed Meal, 1st Progress Rep. of the Research Committee, Rapeseed Association of Canada, 1971.
137. **Bragg, D. B. and Seier, L.,** Mineral content and biological activity of selenium in rapeseed meal, *Poult. Sci.,* 53, 22, 1974.
138. **Cantor, A. H., Langevin, M. L., Noguchi, T., and Scott, M. L.,** Efficacy of selenium in selenium compounds and feedstuffs for prevention of pancreatic fibrosis in chicks, *J. Nutr.,* 105, 106, 1975.
139. **Alexander, A. R., Whanger, P. D., and Miller, L. T.,** Bioavailability to rats of selenium in various tuna and wheat products, *J. Nutr.,* 113, 196, 1983.
140. **Kaantee, E. and Kurkela, P.,** Comparative effects of barley feed and sodium selenite on selenium levels in hen eggs and tissues, *J. Sci. Agric. Soc. Finl.,* 52, 357, 1980.

141. **Froslie, A., Moksnes, K., and Overnes, G.,** The effect of selenium supplementation of animal feeds in Norway, *Acta Agric. Scand.*, 35, 139, 1985.
142. **Scott, M. L. and Thompson, J. N.,** Selenium content of feedstuffs and effects of dietary selenium levels upon tissue selenium in chicks and poults, *Poult. Sci.*, 50, 1742, 1971.
143. **Maus, R. W., Martz, F. A., Belyea, R. L., and Weiss, M. F.,** Relationship of dietary selenium to selenium in plasma and milk from dairy cows, *J. Dairy Sci.*, 63, 532, 1980.
144. **Groce, A. W., Miller, E. R., Keahey, K. K., Ullrey, D. E., and Ellis, D. J.,** Selenium supplementation of practical diets for growing-finishing swine, *J. Anim. Sci.*, 32, 905, 1971.
145. **Hitchcock, J. P., Miller, E. R., Keahey, K. K., and Ullrey, D. E.,** Effects of arsanilic acid and vitamin E upon utilization of natural or supplemental selenium by swine, *J. Anim. Sci.*, 46, 425, 1978.
146. **Young, L. G., Miller, R. B., Edmeades, D. E., Lun, A., Smith, G. C., and King, G. J.,** Selenium and vitamin E supplementation of high moisture corn diets for swine reproduction, *J. Anim. Sci.*, 45, 1051, 1977.
147. **Ku, P. K., Miller, E. R., Wahlstrom, R. C., Groce, A. W., Hitchcock, J. P., and Ullrey, D. E.,** Selenium supplementation of naturally high selenium diets for swine, *J. Anim. Sci.*, 37, 501, 1973.
148. **Pehrson, B. and Johnsson, S.,** Addition of selenium to beef cattle given a selenium-deficient diet, *Zentralbl. Veterinaermed. Reihe A*, 32, 428, 1985.
149. **Pond, W. G., Allaway, W. H., Walker, E. F., Jr., and Krook, L.,** Effects of corn selenium content and drying temperature and of supplemental vitamin E on growth, liver selenium and blood vitamin E content of chicks, *J. Anim. Sci.*, 33, 996, 1971.
150. **Zhu, L. Z., Lu, Z. H., et al.,** Effects in pigs fed the crops grown in Keshan disease affected province of China in *Trace Element Metabolism in Man and Animals,* TEMA 4, Gawthorne, J. M., Howell, J. M., and White, C. L., Eds., Springer-Verlag, Berlin, 1982, 360.
151. **Goehring, T. B., Palmer, I. S., Olson, O. E., Libal, G. W., and Wahlstrom, R. C.,** Effects of seleniferous grain and inorganic selenium on tissue and blood composition and growth performance of rats and swine, *J. Anim. Sci.*, 59, 725, 1984.
153. **Kaantee, E. and Kurkela, P.,** The effects of trace element supplements on blood levels of horses, *J. Sci. Agric. Soc. Finl.*, 52, 468, 1980.
154. **McDowell, L. R., Froseth, J. A., Kroening, G. H., and Haller, W. A.,** Effects of dietary vitamin E and oxidized cottonseed oil on SGOT, erythrocyte hemolysis, testicular fatty acids and testicular selenium in swine fed peas *(Pisum sativum), Nutr. Rep. Int.*, 9, 359, 1974.
155. **McCray, C. W. R. and Hurwood, I. S.,** Selenosis in North-Western Queensland associated with a marine Crictaceous formation, *Queensl. J. Agr. Sci.*, 20, 475, 1963.
156. **Burton, M. A. S. and Phillips, M. L.,** Vegetation damage during an episode of selenium pollution, *J. Plant Nutr.*, 3, 503, 1981.
157. **Beath, O. A., Gilbert, C. S., and Eppson, H. F.,** The use of indicator plants in locating seleniferous areas in western United States. II. Correlation studies by states, *Am. J. Bot.*, 26, 296, 1939.
158. **Olson, O. E., Whitehead, E. I., and Moxon, A. L.,** Occurrence of soluble selenium in soils and its availability to plants, *Soil Sci.*, 54, 47, 1942.
159. **Koljonen, T.,** Selenium uptake by plants in Finland, *Oikos*, 25, 353, 1974.
160. **Shaw, G. G. and Cocks, L.,** The effect of SO_2 on food quality for wild herbivores, in *Symp./Workshop Proc: Acid Forming Emissions in Alberta and their Ecological Effects,* Alberta Department of the Environment, Edmonton, 1982, 571.
161. **Erdman, J. A. and Gough, L. P.,** Variation in the element content of *Parmelia chlorochroa* from the Powder River Basin of Wyoming and Montana, *Bryologist,* 80, 292, 1977.
162. **Rickard, W. H. and Garland, T. R.,** Trace element content of leaves of desert shrubs in south-central Washington, *Northwest Sci.*, 57, 57, 1983.
163. **Iskander, F. Y.,** Neutron activation analysis of an Egyptian cigarette and its ash, *J. Radioanal. Nucl. Chem.*, 89, 511, 1985.
164. **Byrne, A. R., Ravnik, V., and Kosta, L.,** Trace element concentrations in higher fungi, *Sci. Total Environ.*, 6, 65, 1976.
165. **Beeker, R. R., Veglia, A., and Schmid, E. R.,** Determination of trace elements in feeding stuffs of a mine district by instrumental neutron activation analysis (in German), *Bodenkultur,* 26, 312, 1975.
166. **Williams, M. C., Binns, W., and James, L. F.,** Occurrence and Toxicology of selenium in Halogeton and associated species, *J. Range Manage.*, 15, 17, 1962.
167. **Arvy, M. P.,** Relationship between selenium and sixteen element concentrations in *Helminthia echioides* and *Trifolium repens* growing on clay loam soils, *Plant Soil,* 84, 437, 1985.
168. **Froslie, A., Norheim, G., Rambaek, J. P., and Steinnes, E.,** Levels of trace elements in liver from Norwegian moose, reindeer and red deer in relation to atmospheric deposition, *Acta Vet. Scand.*, 25, 333, 1984.

169. **Westermarck, H. and Kurkela, P.,** Selenium content in lichen in Lapland and South Finland and its effect on the selenium values in reindeer, in *Proc. 2nd Int. Reindeer/Caribou Symp.*, Reimers, E., Gaare, E., and Skjenneberg, S., Eds., Direktoratet for Vilt og Ferskvannsfisk, Trondheim, Norway, 1980, 278.
170. **Anderson, L. W. and Acs, L.,** Selenium in North American paper pulps, *Environ. Sci. Tech.*, 8, 462, 1974.
171. **Furr, A. K., Schofield, C. L., Grandolfo, M. C., Hofstader, R. A., Gutenmann, W. H., St. John, L. E., Jr., and Lisk, D. J.,** Elemental content of mosses as possible indicators of air pollution, *Arch. Environ. Contam. Toxicol.*, 8, 335, 1979.
172. **Nadkarni, R. A., Ehmann, W. D., and Burdick, D.,** Investigations on the relative transference of trace elements from cigarette tobacco into smoke condensate, *Tobacco Sci.*, 14, 37, 1970.
173. **Dermelj, M., Ravnik, V., Byrne, A. R., and Vakselj, A.,** Trace heavy metals in various Yogoslav Tobaccos, *Mikrochim. Acta*, I, 261, 1978.
174. **Cannon, H. L. and Swanson, V. E.,** Contributions of major and minor elements to soils and vegetation by the coal-fired Four Corners power plant, San Juan County, New Mexico, *U.S. Geol. Surv. Prof. Pap.*, 1129-B, 1979.
175. **Sharma, R. P. and Shupe, J. L.,** Trace metals in ecosystems: relationships of the residues of copper, molybdenum, selenium, and zinc in animal tissues to those in vegetation and soil in the surrounding environment, in *Biological Implications of Metals in the Environment*, ERDA Symp. Ser. 42, CONF-750929, 1977, 595.
176. **Beath, O. A., Gilbert, C. S., and Eppson, H. F.,** The use of indicator plants in locating seleniferous areas in western United States. IV. Progress report, *Am. J. Bot.*, 28, 887, 1941.
177. **Beath, O. A., Gilbert, C. S., and Eppson, H. F.,** The use of indicator plants in locating seleniferous areas in western United States. III. Further studies, *Am. J. Bot.*, 27, 564, 1940.
178. **Lakin, H. W. and Byers, H. G.,** Selenium occurrence in certain soils in the United States, with a discussion of related topics: sixth report, *U.S. Dep. Agric. Tech. Bull.*, 783, 1941.
179. **Lakin, H. W. and Hermann, F. J.,** *Astragalus Artemisiarum* Jones as a selenium absorber, *Am. J. Bot.*, 27, 245, 1940.
180. **Walker, O. J., Harris, W. E., and Rossi, M.,** Selenium in soils, grains, and plants in Alberta, *Can. J. Res. Sect. B*, 19, 173, 1941.
181. **Byers, H. B., Miller, J. T., Williams, K. T., and Lakin, H. W.,** Selenium occurrence in certain soils in the United States with a discussion of related topics: third report, *U.S. Dep. Agric. Tech. Bull.*, 601, 1938.
182. **Lakin, H. W. and Byers, H. G.,** Selenium occurrence in certain soils in the United States, with a discussion of related topics: seventh report, *U.S. Dep. Agric. Tech. Bull.*, 950, 1948.
183. **Moxon, A. L., Olson, O. E., Searight, W. V., and Sandals, K. M.,** The stratigraphic distribution of selenium in the Cretaceous formations of South Dakota and the selenium content of some associated vegetation, *Am. J. Bot.*, 25, 794, 1938.
184. **Beath, O. A.,** Toxic vegetation growing on the salt wash sandstone member of the Morrison formation, *Am. J. Bot.*, 30, 698, 1943.
185. **Beath, O. A., Hagner, A. F., and Gilbert, C. S.,** Some rocks and soils of high selenium content, *Wyo. Geol. Surv. Bull.*, 36, 1946.
186. **Searight, W. V., Moxon, A. L., Hilmo, R. J., and Whitehead, E. I.,** Occurrence of selenium in Pleistocene deposits and their derivatives in South Dakota, *Soil Sci.*, 61, 455, 1946.
187. **Watkinson, J. H.,** A selenium-accumulating plant of the humid regions: *Amanita muscaria, Nature (London)*, 202, 1239, 1964.

Chapter 6

FOODS

Wayne R. Wolf and Anita Schubert

TABLE OF CONTENTS

I. INTRODUCTION

The significance of selenium to human and animal health as both a toxic element and essential nutrient has been mentioned in other chapters in this volume and covered in the literature.[1-11] Selenium is of particular concern because of current active interest in its role as a cancer-preventive agent[12-16] and is the subject of many ongoing investigations. The 1980 edition of U.S. Recommended Dietary Allowances provides "estimated safe and adequate daily dietary intakes" for some trace elements not previously included in its recommendations.[17] The recommended dietary intake range of selenium for individuals aged 7 years and older is 50 to 200 μg/d. Although many published reports of composition data on Se in foods based on sound analytical studies are available, no comprehensive table of critically evaluated data on the selenium content of foods consumed in the U.S. has been published prior to this work. Planning for nutrient data studies must include an evaluation of available data as well as an assessment of those foods which are commonly consumed and are significant sources of the nutrient of concern. A systematic evaluation and compilation of published data regarding selenium concentrations in foods consumed in the U.S. has been carried out.[18] This evaluation considered analytical selenium data published between 1960 and 1985 and forms the basis for the presentation of data in this chapter dealing solely with foods consumed in the U.S.

Throughout this chapter, emphasis is on foods available at the retail level, purchased and prepared by the consumer as either fresh or commercially processed products. Such products are processed versions of the raw, natural agricultural plant and animal tissues discussed in Chapters 5 and 7, respectively. Most of the commodities considered here have undergone some sort of processing or treatment (commercial or home) prior to consumption. Although some overlap in material type and selenium content will occur between foodstuffs and the starting raw material, the products treated here comprise a different but important population. Analytical data for selenium contents in U.S. foods were subjected to in-depth, detailed evaluation described below. At the present time, it is not possible to carry out such a detailed evaluation for foods from other countries. The reader is referred to a rather comprehensive review on selenium in foods for further information.[19]

II. SELECTION OF U.S. FOODS

The goal of the food selection process was to include those frequently consumed foods which contribute the bulk of selenium to the diets of the U.S. population. Foods were considered for evaluation according to their frequency of consumption, based on the U.S. Department of Agriculture 1977—1978 Nationwide Food Consumption Survey (NFCS).[20]

The combination of frequency data (frequency per day and average portion size) and selenium concentration yielded a list of foods with the amount of selenium consumed per day by the survey population. These foods were ranked in descending order by selenium contribution, and the top approximately 100 foods were designated the selenium core foods. A table of selenium contents for this core foods list then includes not only the mean value and ranges of published data for the major sources of selenium in the U.S. dietary, but also a confidence code denoting an estimate of the quality of the mean value for each food. This first effort at compiling a table of core foods with evaluated selenium concentration data has been previously published.[18]

Based on this core foods list, additional selected food samples were analyzed for selenium to improve the data table. These additional foods were chosen for several reasons: foods suspected of being important selenium sources based upon frequency of consumption, for which no acceptable data were available, foods which had lower confidence codes, and several foods sampled to validate mean values obtained from the evaluation process. Sampling plans were developed based on the selenium core foods list, and samples were obtained from three U.S. cities (Washington, D.C., Houston, and Los Angeles). Determination of the selenium content of 88 food items from this sampling was carried out at the Beltsville Human Nutrition Research Center.[21] In addition, a number of foods were obtained from the Food and Drug Administration Total Diet Study and analyzed.[21] Analytical data from both of these samplings were added to the data originally evaluated for the core foods list. An updated or second core foods list was generated which contains 130 core foods as eaten plus their raw counterparts.[22] This updated core foods list is presented herein in Table 1. The 22 food items which supply 75% of the 1977—1978 NFCS population's selenium intake are listed in ranked order in Table 2.

Food names listed in these tables represent either a specific food item or an aggregate. Foods were aggregated based on the similarity of items, the availability of selenium data for each food item, and the proximity of their selenium levels. For example, all cuts and methods of cooking beef were aggregated into ''beef, ckd.'' However, when selenium data were available only for one specific food item within an aggregate, that food item is listed although its designation as a core food was based on the frequency data for the aggregate. As an additional example, all fruit yogurts were aggregated for determining frequency of consumption, but selenium data existed for only strawberry yogurt. Table 1, therefore, lists strawberry yogurt as representative of all fruit yogurts.

III. EVALUATION OF ANALYTICAL DATA FOR U.S. FOODS

Procedures for the collection and evaluation of published selenium data are described in detail in a previous publication[55] and summarized in Chapter 4. To evaluate each published report of nutrient data, a set of criteria for five general categories was developed: number of samples, analytical method, sample handling, sampling plan, and analytical quality control. The basis for defining selenium-specific criteria within each of the five general categories was provided by examination of the published reports of determination of selenium in foods together with a knowledge of accepted analytical methodology, sample handling procedures, and quality control measures for this nutrient, as well as knowledge of statistical methods. The data from each study were rated by food item within each category on a scale from zero (unacceptable) to three (most desirable). A confidence code indicating the confidence a user can have in the grand mean selenium value for each food was calculated and assigned to each datum. They are listed in Table 1. Meanings of the codes are a — considerable confidence, b — confidence, although some problems exist with the data, c — less confidence due to limited quantity/quality of data. The grand mean selenium value for each food item, derived from the data evaluation procedure, was used with the consumption frequency data to select the core foods as described above.

IV. SELENIUM CONTENT OF U.S. FOODS

Table 1 presents selenium concentration results for core foods/food aggregates as eaten which provide most of the selenium consumed by the surveyed population. Since more data are available for raw foods than for their cooked counterparts, values for raw or uncooked forms of the foods are also included in the table. This tabulation represents an updated version of the core foods list published previously,[18] and it includes 36 food items which had no prior acceptable data. For the food items from the first core foods list, most of the original mean values changed very little by the addition of the new analytical values. In addition, confidence codes for many foods on the updated list are improved. One quarter of all of the 130 food items in Table 1 have means with an 'a' confidence code; one third have a 'c', and the remainder have a 'b'. The values for the first 22 foods contributing 74% of selenium intake listed in Table 2 have 'a' or 'b' confidence codes.

Table 1
SELENIUM CONTENT OF U.S. CORE FOODS (LIST 2)[a]

Food or aggregate[b]	Selenium concentration[c] (μg/kg) Range	Mean	Confidence code[d]	Ref.[e]
Beef, Lamb, Pork, Veal				
Beef, raw	50—420	220	a	23—33
Beef, ckd.	150—520	260	a	21, 24, 34, 35
Beef liver, raw	180—630	400	a	23—28, 36
Beef liver, ckd.	430—710	560	b	24, 34
Lamb, raw	60—320	210	a	23, 24, 26—28, 37
Lamb, ckd.	—	170	c	34
Meat loaf, beef, ckd.	120—270	170	b	24, 34
Pork/ham, fresh/cured, raw	190—510	330	a	23—26, 28, 38
Pork/ham, fresh/cured, ckd./cnd. (incl. roasted, pan ckd.)	180—920	330	a	21, 24, 34, 39
Veal, raw	200—350	280	c	26
Veal, ckd.	120—123	120	b	34
Processed Meats				
Bacon, raw	250—260	250	b	24, 28
Bacon, ckd.	—	220	c	34
Bologna	100—300	160	a	21, 24, 34, 39
Frankfurter, unckd.	—	230	c	24, 28
Frankfurter, ckd. (beef, beef/pork)	88—120	98	a	21, 34
Luncheon meat (incl. chopped ham, Spam®[f])	210—280	250	b	21, 24, 39
Pork sausage, unckd.	250—360	310	a	24, 28
Pork sausage, ckd.	130—190	160	a	21, 34
Salami	130—330	200	b	24, 34
Fish & Seafood				
Cod, raw	270—860	520	b	23, 24, 25
Cod, ckd.	330—360	350	a	21, 34
Crab, cnd.	220—350	280	b	21, 40
Fish sticks, ckd./unckd.	120—170	140	a	24, 28, 34
Haddock, raw	190—1030	540	b	21, 24, 25, 34
Haddock, ckd.	330—360	340	b	34
Oyster, raw	190—940	570	b	23, 24, 26, 40
Oyster, ckd.	—	790	c	21
Salmon, raw	450—750	600	c	25, 40
Salmon, cnd.	310—1490	640	a	21, 24—26
Shrimp, raw	210—790	530	b	23, 24, 26
Shrimp, ckd./unckd.	210—1610	590	a	21, 25, 26, 34, 40
Tuna, cnd. (incl. light, white, packed in oil or water)	370—1150	730	a	21, 24, 28, 34, 39—42
Poultry, Eggs, & Other High-Protein Foods				
Chicken, raw	10—400	220	a	23—26, 28
Chicken, ckd. (incl. fried, roasted)	168—260	215	a	34, 35
Chicken liver, raw	10—710	360	c	24, 26, 28
Chicken liver, ckd.	—	710	c	21

Table 1 (continued)
SELENIUM CONTENT OF U.S. CORE FOODS (LIST 2)[a]

Food or aggregate[b]	Selenium concentration[c] (μg/kg) Range	Mean	Confidence code[d]	Ref.[e]
Turkey, raw	—	340	c	24
Turkey, ckd.	230—300	250	b	34, 35
Egg white, raw	20—350	140	b	23—26
Egg, whole, raw	390—470	440	b	24, 25, 28
Egg yolk, raw	180—690	410	b	23—26
Eggs, ckd. (incl. scrambled, fried, soft-boiled)	160—380	250	a	21, 24, 34
Beans, pinto, dry	—	190	c	39
Beans, pinto, cnd./dry, ckd.	50—340	200	b	21
Beans, white, dry (incl. navy, great northern)	60—130	110	b	24, 25, 28, 44
Beans, great northern, dry, soaked, ckd.	48—60	54	b	21
Peanut butter	36—120	85	b	24, 28, 39, 44
Brazil nuts	1990—29600	11,430	b	21, 45
Sunflower seeds	610—920	770	b	24, 44
Dairy Products				
American cheese, processed	56—120	92	a	23, 25, 34
Cheddar cheese	80—140	110	b	25, 26, 34
Cheese spread	—	82	c	21
Cottage cheese	52—79	66	b	21, 23, 25
Ice cream	—	9	c	26
Milk, chocolate	—	12	c	26
Milk shake, chocolate, McDonald's®	—	13	c	21
Milk, skim	10—47	27	b	23, 26, 46
Milk, 2% fat	25—26	26	b	21, 46
Milk, whole	11—25	16	b	23, 25, 26, 46
Swiss cheese	62—100	83	c	23, 26
Yogurt, strawberry	17—24	20	b	21
Breads				
Bagels, white	—	320	c	24
Biscuits, refrig. type/from mix, baked	130—180	160	a	21, 34
Bread stuffing	40—510	280	c	24, 25
Cornbread/muffins/toasties	95—150	120	a	21, 34
Cracked wheat bread	250—670	460	b	21, 25
English muffins, incl. toasted	200—270	230	b	21, 24
French bread	—	560	c	25
Italian bread	—	240	c	21
Raisin bread	—	200	c	21
Rolls, white	210—610	340	a	24, 25, 34
Rye bread	240—590	360	a	21, 24—26, 28, 34
White bread	230—540	320	a	23—26, 28, 34
Whole wheat bread	280—670	430	a	21, 23, 34, 39

Table 1 (continued)
SELENIUM CONTENT OF U.S. CORE FOODS (LIST 2)[a]

Food or aggregate[b]	Selenium concentration[c] (μg/kg) Range	Mean	Confidence code[d]	Ref.[e]
Ready-to-Eat Breakfast Cereals				
Alpha Bits®[g]	—	360	c	25
Apple Jacks/Froot Loops®[h]	60—340	180	a	25, 34
Bran flakes cereal	110—300	180	b	24, 25
Cheerios®[i]	260—500	400	a	23,25,26, 29—32,34,47
Corn flakes, plain or sugar coated (incl. Corn Flakes®[h] & Frosted Flakes®[h])	26—120	63	a	23—25, 48
Crisped rice cereal (incl. Rice Krispies®[h])	210—230	220	b	24, 25, 48
Granola	170—220	190	b	34
Life®[j]	—	250	c	21
Raisin bran cereal	—	110	c	24, 25
Special K®[h]	350—940	630	b	24, 25, 27, 44
Sugar Crisp®[g]	510—1040	780	c	24, 25
Wheaties®[i]	40—550	150	a	23—28, 44, 47
Other Grain Products				
Cake, chocolate w/choc. icing, RTE/frzn.	—	32	c	21
Cake, yellow w/white icing, both from mix	—	39	c	21
Coffee cake, RTE/frzn.	150—170	160	b	34
Cookies, chocolate chip	35—49	42	b	21
Cookies, oatmeal	—	90	c	24
Corn chips	70—90	79	b	24, 44
Corn grits, raw	30—300	170	b	48
Corn grits, ckd.	20—95	58	b	21, 39
Doughnuts	73—100	88	b	24, 34
Egg noodles, macaroni, spaghetti, dry	430—1350	660	a	23—26, 28, 44, 47
Egg noodles, macaroni, spaghetti, ckd.	150—200	180	a	11, 34
Farina, cream of wheat, raw	200—240	220	c	24, 48
Farina, cream of wheat, ckd.	100—110	110	b	21
French toast, frzn., toasted	—	170	c	21
Oats, oatmeal, raw	70—680	270	a	23—25, 44, 47, 48
Oatmeal, ckd.	70—100	88	a	21, 34
Pancakes, made from dry mix	80—98	89	b	34
Pancake & waffle mix	90—260	150	b	24, 28, 36
Rice, fried	—	150	c	21
Rice, white, enriched, raw	200—470	290	a	23, 25, 26, 28, 39, 44, 47, 48
Rice, white, enriched, ckd.	78—93	85	b	21, 34
Saltines/soda crackers	61—110	86	c	24, 44
Sweet rolls	—	190	c	34
Tortillas, corn	—	68	c	24
Tortillas, wheat	130—220	180	b	34
Waffles	120—180	150	b	21, 24

Table 1 (continued)
SELENIUM CONTENT OF U.S. CORE FOODS (LIST 2)[a]

Food or aggregate[b]	Selenium concentration[c] (μg/kg) Range	Mean	Confidence code[d]	Ref.[e]
Fruits & Vegetables				
Apples, fresh	3—6	4.3	a	23—26, 44
Bananas, fresh	4—11	8.5	b	23—26
Broccoli, fresh	2—17	9.5	c	24, 25
Broccoli, fresh/frzn., ckd.	15—32	24	b	21
Carrots, fresh	6—29	17	a	23—25, 28, 44
Carrots, cnd.	—	13	c	23
Coleslaw, homemade	—	17	c	21
Corn, cream style, cnd.	—	14	c	21
Corn, raw	4—340	100	a	23—25, 28, 36, 37, 40, 44, 48—52
Corn, fresh/frzn., ckd./cnd., drained	6—16	11	b	21
Cucumber, fresh, with/without peel	6—120	63	c	24, 25
Grapefruit, fresh	9—21	15	c	25, 26
Green beans, fresh	5—20	10	b	23—25, 27, 44
Green beans, fresh/frzn., ckd./cnd., drained	2—5	3	b	21
Green peas, fresh	5—45	20	b	24, 25, 44
Green peas, frzn., boiled/cnd., drained	4—5	5	b	21
Lettuce, fresh	2—48	15	a	23—25, 49, 53
Orange, fresh	13—18	15	b	23, 25, 26
Orange juice, frzn., reconstituted	0.8—1.0	0.9	b	21
Potatoes, french fried	—	42	c	21
Potatoes, white, raw	4—23	13	a	23—25, 31, 44, 49, 53
Potatoes, white, baked/cnd.	9—13	12	a	21, 23
Potatoes, white, mashed from instant mix	13—18	16	b	21
Tomatoes, fresh	1—13	7	b	23—25
Tomato/barbecue sauce, bottled/cnd.	10.3—10.5	10	b	21
Mixed Dishes				
Baked beans in tomato sauce/molasses	18—60	35	b	24, 35
Beef & vegetable stew, cnd.	36—55	45	b	21, 39
Chicken noodle casserole, homemade	110—130	120	b	34
Chicken noodle/rice soup	—	13	c	24
Chicken/turkey pot pie, frzn., heated/unheated	80—160	120	b	21, 54
Chili, beef & beans, cnd.	—	31	c	21
Chili, without beans, cnd.	—	97	c	39
Chow mein, cnd.	61—110	84	b	21
Hamburger, on bun w/condiments	140—150	150	b	34
Lasagna, homemade	—	140	c	34
Macaroni & cheese, ckd., from box mix	130—160	150	a	21, 34
Pizza, comml. frzn., ckd.	150—220	190	a	21, 34
Pork & beans, cnd.	24—40	32	b	21
Ravioli, beef, cnd.	—	92	c	21
Spaghetti with meat sauce, homemade	—	88	b	34
Spaghetti with tomato sauce, cnd.	—	100	c	34

Table 1 (continued)
SELENIUM CONTENT OF U.S. CORE FOODS (LIST 2)[a]

Food or aggregate[b]	Selenium concentration[c] (μg/kg) Range	Mean	Confidence code[d]	Ref.[e]
Spanish rice, cnd./ckd., from box mix	—	77	c	21
Vegetable soup, vegetarian	—	16	c	21
Miscellaneous				
Beef gravy	—	10	c	24
Beer, regular	8—13	10	b	21
Coffee, beverage, decaf./reg., black	0.25—1.1	0.68	a	21
Jams and jellies	—	20	c	24
Mayonnaise	20—600	310[k]	c	21, 39
Mustard, prepared	—	280	c	21

[a] An updated and augmented version of the first study.[18]

[b] Description represents either the specific food item for which acceptable analytical data are available or the aggregate. Designation of a specific food item as a core food, however, may be based on frequency data for the aggregate in which it is included. See text for further explanation. Abbreviations: ckd. — cooked, comml. — commercial, cnd. — canned, decaf. — decaffeinated, frzn. — frozen, incl. — including, refrig. — refrigerated, reg. — regular, RTE — ready-to-eat, unckd. — uncooked.

[c] Values taken from acceptable studies; all data (usually recorded to two significant figures) are on a wet or fresh weight basis of the edible portion; number of digits does not necessarily reflect precision or reliability of data.

[d] a = considerable confidence; b = confidence, although some problems with data; c = less confidence due to limited quantity/quality of data.

[e] References of acceptable studies.

[f] George A. Hormel & Co., Austin, TX.

[g] General Foods Corp., White Plains, NY.

[h] Kellogg Co., Battle Creek, MI.

[i] General Mills, Inc., Minneapolis, MN.

[j] The Quaker Oats Co., Chicago, IL.

[k] Questionable value; see text.

Mean values with a 'c' confidence code which are derived from only one or two studies should be used with caution. The case of mayonnaise provides a good example of a questionable value. The value of 310 μg/kg was derived from one published value of 600 μg/kg[39] obtained from analysis of two samples, and a second value of 19 μg/kg[21] for one sample. Given the selenium content of the individual ingredients, even the resulting mean value of 310 μg/kg appears to be an overly high value. We were not able to obtain information from the authors of the previously published value and thus had no criteria to confirm or disregard the higher values. Obviously more analytical work needs to be carried out to confirm the position on the core foods list for this food.

Table 3 lists non-core, selenium-rich foods for which acceptable data are available. Data on these foods were evaluated as described above. Average portion sizes for food aggregates were determined by averaging the mean portion sizes (as reported in the NFCS) for all of the food codes included in the aggregate. For those raw foods which are not part of the NFCS data base, yield factors from USDA Agriculture Handbook No. 102[58] were used in conjunction with the NFCS cooked food data to calculate the selenium content per average portion size.

Table 2
FIRST 22 SELENIUM CORE FOODS OF CORE FOOD LIST RANKED BY CONTRIBUTION TO DIETS OF THE SURVEY POPULATION

Rank	Food item or aggregate[a]	Selenium concentration[b] (μg/kg)	Confidence code	Selenium intake (μg/person/d)[c]	Cumulative % of daily intake[d]
1	Beef, ckd.	260	a	12.48	16.9
2	White bread	320	a	10.03	30.5
3	Pork, ckd.	330	a	5.47	37.9
4	Chicken ckd.	180	a	3.88	43.2
5	Eggs, ckd.	250	a	3.37	47.7
6	White rolls	340	a	2.82	51.5
7	Whole wheat bread	430	a	2.27	54.6
8	Egg noodles, mac., spag., ckd.	180	a	2.06	57.4
9	Whole milk	16	b	1.99	60.1
10	Tuna fish, cnd.	730	a	1.47	62.1
11	2% fat milk	26	b	1.25	63.8
12	White rice, enriched, ckd.	85	b	1.10	65.3
13	Mac. & cheese, ckd., from box mix	150	a	0.97	66.6
14	Luncheon meat	250	b	0.81	67.7
15	Spag. with meat sauce, hmd.	88	b	0.77	68.7
16	Oatmeal	88	a	0.73	69.7
17	Meat loaf, ckd.	170	b	0.72	70.7
18	Hamburger, on bun with condiments	150	b	0.71	71.7
19	Pinto beans, cnd./dry, ckd.	200	b	0.69	72.6
20	Turkey, ckd.	250	b	0.66	73.5
21	Rye bread	360	a	0.63	74.3
22	Bacon, ckd.	220	c	0.63	75.2

[a] Description represents either the specific food item for which acceptable analytical data are available or the aggregate. Designation of a specific food item as a core food, however, may be based on frequency data for the aggregate in which it is included. See text for further explanation. Abbreviations: ckd. — cooked, cnd. — canned, hmd. — homemade, mac. — macaroni, spag. — spaghetti.

[b] Wet or fresh weight basis of edible portion.

[c] Daily Se intake/person from all core foods = 74 μg.

[d] Cumulative percent of daily Se intake from foods which contribute >0.04 μg Se per person per day.

Foods which are the best sources of selenium have been defined based on content per average portion size equal to or greater than 20 μg of selenium. The first 22 selenium-rich core foods ranked in descending order by their respective per capita contribution to the selenium intake of the U.S. survey population are listed in Table 2. As seen in this table, more than half of the U.S. intake of selenium is provided by only six food items: beef, white bread, pork, chicken, eggs, and white rolls. Food groups which are the best sources of selenium, based on concentration, are beef, lamb, pork, and veal; fish and seafood; breads; ready-to-eat breakfast cereals; poultry, eggs, and other high-protein foods; and processed meats. Foods which qualify as core foods primarily due to their high frequency of consumption rather than due to a high selenium concentration are dairy products, fruits, and vegetables. Some core foods with very high selenium concentrations and low frequencies of consumption are oysters, salmon, Brazil nuts, and sunflower seeds.

It should be noted that intake from these core foods may not accurately estimate total daily Se intake due to (1) lack of selenium data on some food items consumed relatively frequently and/or likely to contain appreciable amounts of selenium, and (2) use of a cutoff

Table 3
SELENIUM CONTENT OF NON-CORE, SELENIUM-RICH FOODS

Food item	Selenium concentration[a] (μg/kg)		Confidence code	Ref.
	Range	Mean		
Swordfish, raw	2540—3440	2840	b	29—31,56
Pork kidney, raw	1900—3220	2290	a	23—26,28
Beef kidney, raw	1450—2320	1700	b	23—25,28
Lamb kidney, raw	930—1430	1280	a	23,24,26, 28,37,57
Lobster, raw/canned	500—1220	790	b	23,25,26
Molasses	20—1280	630	b	24,28,44
Oysters, raw	190—940	570	b	23,24,26,41
Cod, raw/cooked	270—860	450	a	23—25,34
Chicken liver, raw	10—710	360	c	24,26,28
Crab, cooked/canned	—	220	c	40

[a] Values taken from acceptable studies, reported on a wet or fresh weight basis of the edible portion.

point to define core foods which excludes some intake from the sum of minor sources. The per capita intake of selenium contributed by any food or food aggregate is a function of the frequency of consumption and the average portion size as well as the selenium concentration of that food. Cumulative percentages and the number of NFCS food codes included in each food aggregate are also presented in the table.

The total average daily selenium intake per person from the 130 core foods (accounted for in the survey population's diets) was calculated to be 74 μg. This estimate compares favorably with the results of other studies in which daily selenium intake from U.S. diets was determined to be between 49 and 111 μg/1826 kcal[59] (Table 4). In a study in South Dakota, where the selenium content of soil is known to be high, the daily intake of the element was determined to be 146 μg/1826 kcal according to one study.[28]

Many factors influence the level of selenium in foods. These include the selenium content of the soil on which crops are grown and livestock and poultry are raised,[57,67-71] treatment of the soil or animal feed,[19,44,54,72,73] protein content of the food,[24,35,39] degree of processing,[23,48,74-76] and cooking method.[24,47,77,78] These factors have received attention and discussion elsewhere in the literature.

The amount of total selenium in a food does not determine how much of the selenium will actually be utilizable by an individual. The form of the element and its bioavailability as influenced by various other components in a meal ultimately determine the amount of selenium absorbed by the body. Some work in these areas has been done,[40-42,70,73,79-81] but much research is still required.

Table 4
COMPARISON OF DAILY SELENIUM INTAKES IN AMERICAN DIETS

Geographic area	Diet	Selenium intake μg/d[a]	Selenium intake μg/1826 kcal	Method of intake estimation[b]	Ref.
U.S.	This work; daily intake based on Nationwide Food Consumption Survey frequency data and analyses of foods by various investigators	74	74	Calc.	—
U.S.	Food and Drug Administration Total Diet Study; adult diets including beverages based on analysis of samples			Calc.	34
	Female, 25—30, 1576 kcal	60	70		
	Female, 60—65, 1355 kcal	60	81		
	Male, 25—30, 2541 kcal	100	72		
	Male, 60—65, 1944 kcal	80	75		
Vermont	2-d hospital diet, incl. beverages	62.3	—	Anal.	27
3 Canadian cities	4 composite diets, no beverages, 2780 kcal	168.6	111	Anal.	60
South Dakota	7-d menu using analytical data from this study and References 10, 14, and 46; 2700 kcal/d	216.4	146	Calc.	28
San Diego, CA	Free-living subjects, based on analysis of samples			Calc.	61
	30-d diets				
	Male, 2700 kcal/d	107	72		
	Female, 2450 kcal/d	99	74		
	1 week of diets, 8 subjects	139	—		
North Carolina	NC Agricultural and Technical State University, 7-d diets, including beverages			Anal.	62
	Cafeteria prepared, omnivorous	92.7	—		
	Home prepared, lacto-ovo-vegetarian	84.8	—		
Beltsville, MD	6-d diets, 22 free-living subjects, incl. beverages, 2027 kcal/d	81	73	Anal.	63
Washington, D.C.	7-d hospital diets			Anal.	64
	Meat based, 3156 kcal/d	85.5	49		
	Lacto-ovo-vegetarian, 3103 kcal/d	118.1	69		
	Low protein, low calorie, 1506 kcal/d	54.6	66		
30 sites/year in U.S.	Adult male composite diets based on 1965 Household Nationwide Food Consumption Survey, 2850 kcal/d			Anal.	65
	1974, 1975, 50 collections (ADF)[c]	125	80		
	1976—1982, 132 collections (ADHAAS)[c]	102.5	66		
Ohio	Free-living rural and urban subjects, based on analysis of samples and 3 24-h recalls 4—6 months apart			Calc.	66
	222 males, 2813 kcal/d	98.3	64		
	227 females, 1795 kcal/d	67.2	68		

[a] Average energy intake per person per day in 1977—1978 NFCS Rep. No. I-2, p. 154.[59]
[b] Calc. — calculated; Anal. — analyzed.
[c] Methods of analysis, ADF — acid digestion/fluorometry; ADHAAS — acid digestion/hydride generation atomic absorption spectrometry.

REFERENCES

1. **Muth, O. H., Oldfield, J. E., and Weswig, P. H., Eds.,** *Selenium in Biomedicine,* AVI Publishing, Westport, CT, 1967.
2. **Spallholz, J. E., Martin, J. L., and Ganther, H. E., Eds.,** *Selenium in Biology and Medicine,* AVI Publishing, Westport, CT, 1981.
3. **Klayman, D. L. and Gunther, W. H. H., Eds.,** *Organic Selenium Compounds: Their Chemistry and Biology,* Wiley-Interscience, Toronto, 1973.
4. **Shamberger, R. J.,** *Biochemistry of Selenium,* Plenum Press, New York, 1983.
5. **Ganther, H. E., Hafeman, D. G., Lawrence, R. A., Serfass, R. E., and Hoekstra, W. G.,** Selenium and glutathione peroxidase in health and disease — a review, in *Trace Elements in Human Health and Disease,* Vol. 2, Prasad, A. S. and Oberleas, D., Eds., Academic Press, New York, 1976, 165.
6. **Underwood, E. J.,** Trace elements, in *Toxicants Occurring Naturally in Foods,* National Academy of Sciences, Washington, D.C., 1973, 43.
7. **Lo, M.-T. and Sandi, E.,** Selenium: occurrence in foods and its toxicological significance — a review, *J. Environ. Pathol. Toxicol.,* 4, 193, 1980.
8. **Moxon, A. L.,** Selenium: its occurrence in rocks and soils, absorption by plants, toxic action in animals, and possible essential role in animal nutrition, in *Trace Elements — Proc. Conf. Wooster, Ohio,* Lamb, C. A., Bentley, O. G., and Beattie, J. M., Eds., Academic Press, New York, 1958, 175.
9. **Ganther, H. E.,** Biochemistry of selenium, in *Selenium,* Zingaro, R. A. and Cooper, W. C., Eds., Van Nostrand Reinhold, New York, 1974, 546.
10. **Cooper, W. C. and Glover, J. R.,** The toxicology of selenium and its compounds, in *Selenium,* Zingaro, R. A. and Cooper, W. C., Eds., Van Nostrand Reinhold, New York, 1974, 654.
11. **Moxon, A. L.,** Selenium in agriculture, in *Selenium,* Zingaro, R. A. and Cooper, W. C., Eds., Van Nostrand Reinhold, New York, 1974, 675.
12. **Shamberger, R. J., Tytko, S. A., and Willis, C. E.,** Antioxidants and cancer. VI. Selenium and age-adjusted human cancer mortality, *Arch. Environ. Health,* 31, 231, 1976.
13. **Schrauzer, G. N., White, D. A., and Schneider, C. J.,** Cancer mortality correlation studies. III. Statistical associations with dietary selenium intakes, *Bioinorg. Chem.,* 7, 23, 1977.
14. **Schrauzer, G. N., White, D. A., and Schneider, C. J.,** Cancer mortality correlation studies. IV. Associations with dietary intakes and blood levels of certain trace elements, notably Se-antagonists, *Bioinorg. Chem.,* 7, 35, 1977.
15. **Harr, J. R., Exon, J. H., Whanger, P. D., and Weswig, P. H.,** Effect of dietary selenium on *N*-2-fluorenyl-acetamide (FAA)-induced cancer in vitamin E supplemented, selenium depleted rats, *Clin. Toxicol.,* 5, 187, 1972.
16. **Ip, C. and Sinha, D. K.,** Enhancement of mammary tumorigenesis by dietary selenium deficiency in rats with a high polyunsaturated fat intake, *Cancer Res.,* 41, 31, 1981.
17. Food and Nutrition Board, National Research Council, *Recommended Dietary Allowances,* 9th Ed., National Academy of Sciences, Washington, D.C., 1980.
18. **Schubert, A., Holden, J. M., and Wolf, W. R.,** Selenium content of a core group of foods based on a critical evaluation of published analytical data, *J. Am. Diet. Assoc.,* 87, 285, 1987.
19. **Combs, G. F., Jr. and Combs, S. B.,** *The Role of Selenium in Nutrition,* Academic Press, New York, 1986, Chap. 3.
20. Human Nutrition Information Service, Food Intakes: Individuals in 48 States, Year 1977—78. Nationwide Food Consumption Survey 1977—78 Rep. No. I-1, U.S. Department of Agriculture, Washington, D.C., 1983.
21. **Wolf, W. R., Schubert, A. S., Lurie, D. G., Holden, J. M., and Woolson-Doherty, J.,** Selenium content of selected foods important for improved assessment of dietary intake, personal communication.
22. **Lurie, D. G., Schubert, A., Holden, J. M., and Wolf, W. R.,** Assessment and improvement of selenium composition data, Abstr. 19th Annu. Meet. of the Soc. for Nutrition Education, Society for Nutrition Education, Oakland, CA, 1986.
23. **Morris, V. C. and Levander, O. A.,** Selenium content of foods, *J. Nutr.,* 100, 1383, 1970.
24. **Olson, O. E. and Palmer, I. S.,** Selenium in foods purchased or produced in South Dakota, *J. Food Sci.,* 49, 446, 1984.
25. **Arthur, D.,** Selenium content of Canadian foods, *Can. Inst. Food Sci. Technol. J.,* 5, 165, 1972.
26. **Amer, M. A. and Brisson, G. J.,** Selenium in human food stuffs collected at the Ste-Foy (Quebec) food market, *Can. Inst. Food Sci. Technol. J.,* 6, 184, 1973.
27. **Schroeder, H. A., Frost, D. V., and Balassa, J. J.,** Essential trace metals in man: selenium, *J. Chron. Dis.,* 23, 227, 1970.
28. **Olson, O. E., Palmer, I. S., and Howe, M.,** Selenium in foods consumed by South Dakotans, *Proc. S.D. Acad. Sci.,* 57, 113, 1978.

29. **Ihnat, M.,** Atomic absorption spectrophotometric determination of selenium with carbon furnace atomization, *Anal. Chim. Acta,* 82, 293, 1976.
30. **Ihnat, M.,** Selenium in foods: evaluation of atomic absorption spectrophotometric techniques involving hydrogen selenide generation and carbon furnace atomization, *J. Assoc. Off. Anal. Chem.,* 59, 911, 1976.
31. **Ihnat, M.,** Collaborative study of a fluorometric method for determining seleniun in foods, *J. Assoc. Off. Anal. Chem.,* 57, 373, 1974.
32. **Shum, G. T. C., Freeman, H. C., and Uthe, J. F.,** Flameless atomic absorption spectrophotometry of selenium in fish and food products, *J. Assoc. Off. Anal. Chem.,* 60, 1010, 1977.
33. **Ihnat, M.,** Fluorometric determination of selenium in foods, *J. Assoc. Off. Anal. Chem.,* 57, 368, 1974.
34. **Pennington, J. A. T., Young, B. E., Wilson, D. B., Johnson, R. D., and Vanderveen, J. E.,** Mineral content of foods and total diets: the selected minerals in foods survey, 1982—1984, *J. Am. Diet. Assoc.,* 86, 876, 1986.
35. **Ferretti, R. J. and Levander, O. A.,** Selenium content of soybean foods, *J. Agric. Food Chem.,* 24, 54, 1976.
36. **McCarthy, T. P., Brodie, B., Milner, J. A., and Bevill, R. F.,** Improved method for selenium determination in biological samples by gas chromatography, *J. Chromatogr.,* 225, 9, 1981.
37. **Whetter, P. A. and Ullrey, D. E.,** Improved fluorometric method for determining selenium, *J. Assoc. Off. Anal. Chem.,* 61, 927, 1978.
38. **Ku, P. K., Ely, W. T., Groce, A. W., and Ullrey, D. E.,** Natural dietary selenium, alpha-tocopherol and effect on tissue selenium, *J. Anim. Sci.,* 34, 208, 1972.
39. **Lane, H. W., Taylor, B. J., Stool, E., Servance, D., and Warren, D. C.,** Selenium content of selected foods, *J. Am. Diet. Assoc.,* 82, 24, 1983.
40. **Cappon, C. J. and Smith, J. C.,** Chemical form and distribution of mercury and selenium in edible seafood, *J. Anal. Toxicol.,* 6, 10, 1982.
41. **Cappon, C. J. and Smith, J. C.,** Chemical form and distribution of mercury and selenium in canned tuna, *J. Appl. Toxicol.,* 2, 181, 1982.
42. **Cappon, C. J. and Smith, J. C.,** Mercury and selenium content and chemical form in fish muscle, *Arch. Environ. Contam. Toxicol.,* 10, 305, 1981.
43. **Holak, W.,** Analysis of foods for lead, cadmium, copper, zinc, arsenic and selenium using closed system sample digestion: collaborative study, *J. Assoc. Off. Anal. Chem.,* 63, 485, 1980.
44. **Moxon, A. L. and Palmquist, D. L.,** Selenium content of foods grown or sold in Ohio, *Ohio Rep.,* 65(1), 13, 1980.
45. **Palmer, I. S., Herr, A., and Nelson, T.,** Toxicity of selenium in Brazil nuts to rats, *J. Food Sci.,* 47, 1595, 1982.
46. **Bruhn, J. C. and Franke, A. A.,** Trace metal and protein concentrations in California market milks, *J. Food Protect.,* 40, 170, 1977.
47. **Higgs, D. J., Morris, V. C., and Levander, O. A.,** Effect of cooking on selenium content of foods, *J. Agric. Food Chem.,* 20, 678, 1972.
48. **Ferretti, R. J. and Levander, O. A.,** Effect of milling and processing on the selenium content of grains and cereal products, *J. Agric. Food Chem.,* 22, 1049, 1974.
49. **Hahn, M. H., Kuennen, R. W., Caruso, J. A., and Fricke, F. L.,** Determination of trace amounts of selenium in corn, lettuce, potatoes, soybeans, and wheat by hydride generation/condensation and flame atomic absorption spectrometry, *J. Agric. Food Chem.,* 29, 792, 1981.
50. **Patrias, G. and Olson, O. E.,** Selenium contents of samples of corn from midwestern states, *Feedstuffs,* 41(43), 32, 1969.
51. **Olson, O. E.,** Fluorometric analysis of selenium in plants, *J. Assoc. Off. Anal. Chem.,* 52, 627, 1969.
52. **Olson, O. E., Palmer, I. S., and Cary, E. E.,** Modification of the official fluorometric method for selenium in plants, *J. Assoc. Off. Anal. Chem.,* 58, 117, 1975.
53. **Wolnik, K. A., Fricke, F. L., Capar, S. G., Braude, G. L., Meyer, M. W., Satzger, R. D., and Kuennen, R. W.,** Elements in major raw agricultural crops in the United States. II. Other elements in lettuce, peanuts, potatoes, soybeans, sweet corn, and wheat, *J. Agric. Food Chem.,* 31, 1244, 1983.
54. **Levander, O. A.,** A review: selenium and chromium in human nutrition, *J. Am. Diet. Assoc.,* 66, 338, 1975.
55. **Holden, J. M., Schubert, A., Wolf, W. R., and Beecher, G. R.,** A system for evaluating the quality of published nutrient data: selenium, a test case, *Food and Nutrition Bulletin* (Suppl. 12, Food Composition Data: A Users Perspective), Rand, W. T., Windham, C. T., Wyse, B., and Young, V. R., Eds., United Nations University, Tokyo, 1987, 177.
56. **Ihnat, M. and Miller, H. J.,** Acid digestion, hydride evolution atomic absorption spectrophotometric method for determining arsenic and selenium in foods. Collaborative study. I, *J. Assoc. Off. Anal. Chem.,* 60, 1414, 1977.
57. **Paulson, G. D., Broderick, G. A., Baumann, C. A., and Pope, A. L.,** Effect of feeding sheep selenium fortified trace mineralized salt: effect of tocopherol, *J. Anim. Sci.,* 27, 195, 1968.

58. **Matthews, R. H. and Garrison, Y. J.,** Food Yields Summarized by Different Stages of Preparation, Revised USDA Agriculture Handbook No. 102, U.S. Department of Agriculture, Washington, D.C., 1975.
59. Human Nutrition Information Service, Food Intakes: Individuals in 48 States, Year 1977—78, Nationwide Food Consumption Survey 1977—78 Rep. No. I-2, U.S. Department of Agriculture, Washington, D.C., 1983.
60. **Thompson, J. N., Erdody, P., and Smith, D. C.,** Selenium content of food consumed by Canadians, *J. Nutr.,* 105, 274, 1975.
61. **Schrauzer, G. N. and White, D. A.,** Selenium in human nutrition: dietary intakes and effects of supplementation, *Bioinorg. Chem.,* 8, 303, 1978.
62. **Ganapathy, S. N. and Dhanda, R.,** Selenium content of omnivorous and vegetarian diets, *Indian J. Nutr. Diet.,* 17, 53, 1980.
63. **Welsh, S. O., Holden, J. M., Wolf, W. R., and Levander, O. A.,** Selenium in self-selected diets of Maryland residents, *J. Am. Diet. Assoc.,* 79, 227, 1981.
64. **McConnell, K. P., Smith, J. C., Jr., Higgins, P. J., and Blotcky, A. J.,** Selenium content of selected hospital diets, *Nutr. Res.,* 1, 235, 1981.
65. **Pennington, J. A. T., Wilson, D. B., Newell, R. F., Harland, B. F., Johnson, R. D., and Vanderveen, J. E.,** Selected minerals in food surveys, 1974 to 1981/82, *J. Am. Diet. Assoc.,* 84, 771, 1984.
66. **Snook, J. T., Palmquist, D. L., Moxon, A. L., Cantor, A. H., and Vivian, V. M.,** Selenium status of a rural (predominantly Amish) community living in a low-selenium area, *Am. J. Clin. Nutr.,* 38, 620, 1983.
67. **Allaway, W. H.,** The Effect of Soils and Fertilizers on Human and Animal Nutrition, *U.S. Dep. Agric. Inf. Bull.,* 378, 1975.
68. **Ullrey, D. E.,** Selenium in the soil-plant-food chain, in *Selenium in Biology and Medicine,* Spallholz, J. E., Martin, J. L., and Ganther, H. E., Eds., AVI Publishing, Westport, CT, 1981, 176.
69. **Varo, P. and Koivistoinen, P.,** Annual variations in the average selenium intake in Finland: cereal products and milk as sources of selenium in 1979/80, *Int. J. Vitam. Nutr. Res.,* 51, 79, 1981.
70. **Douglass, J. S., Morris, V. C., Soares, J. H., Jr., and Levander, O. A.,** Nutritional availability to rats of selenium in tuna, beef kidney, and wheat, *J. Nutr.,* 111, 2180, 1981.
71. **Allaway, W. H.,** Selenium in the food chain, *Cornell Vet.,* 63, 151, 1973.
72. **Furr, A. K., Stoewsand, G. S., Bache, C. A., and Lisk, D. J.,** Study of guinea pigs fed Swiss chard grown on municipal sludge-amended soil, *Arch. Environ. Health,* 31, 87, 1976.
73. **Cappon, C. J.,** Mercury and selenium content and chemical form in vegetable crops grown on sludge-amended soil, *Arch. Environ. Contam. Toxicol.,* 10, 673, 1981.
74. **Fink, H.,** Selenium content of skim milk powders and their inclination for producing dietetic liver necrosis, Abstr. 5th Int. Congr. on Nutrition, International Congress on Nutrition, Washington, D.C., 1960, 19.
75. **Lorenz, K.,** Selenium in wheats and commercial wheat flours, *Cereal Chem.,* 55, 287, 1978.
76. **Lorenz, K.,** Selenium in U.S. and Canadian wheats and flours, in *Selenium in Biology and Medicine,* Spallholz, J. E., Martin, J. L., and Ganther, H. E., Eds., AVI Publishing, Westport, CT, 1981, 449.
77. **Ganapathy, S. N., Joyner, B. T., Sawyer, D. R., and Hafner, K. M.,** Selenium content of selected foods, in *Trace Element Metabolism in Man and Animals -3,* TEMA-3, Kirchgessner, M., Roth-Maier, D. A., Roth, H.-P., Schwarz, F. J., and Weigand, E., Eds., Aebeitskreis fur Tierernahrungsforschung, Weihenstephan, W. Germany, 1978, 322.
78. **Ganapathy, S. N., Joyner, B. T., and Hafner, K. M.,** Effect of baking, broiling and frying on the selenium content of selected beef, pork, chicken and fish foods, Proc. 10th Int. Congr. on Nutrition, Koishi, H., Chm. Ed. Comm., Victory-sha Press, Kyoto, 1975, 676.
79. **Cappon, C. J. and Smith, J. C.,** Determination of selenium in biological materials by gas chromatography, *J. Anal. Toxicol.,* 2, 114, 1978.
80. **Levander, O. A., Alfthan, G., Arvilommi, H., Gref, C. G., Huttunen, J. K., Kataja, M., Koivistoinen, P., and Pikkarainen, J.,** Bioavailability of selenium to Finnish men as assessed by platelet glutathione peroxidase activity and other blood parameters, *Am. J. Clin. Nutr.,* 37, 887, 1983.
81. **Lorus, J. E., Catshaw, J. D. and Biggert, M.,** Se bioavailability in foods and feeds, *Nutr. Rep. Int.,* 33, 13, 1986.

Chapter 7

ANIMAL TISSUES

Richard C. Ewan

TABLE OF CONTENTS

I. INTRODUCTION

Selenium was identified as a toxic element for animals by the efforts of Franke and others[1-3] at the South Dakota Station during the mid-1930s. The nutritional essentiality of selenium was suggested by Schwarz and Foltz[4] in 1957. Subsequent investigations established that selenium prevented vitamin E deficiency conditions such as exudative diathesis in chicks,[5] white muscle disease in lambs,[6] and hepatosis dietetica in pigs.[7] The essentiality of selenium in the presence of excess vitamin E was demonstrated in chicks by Thompson and Scott[8] and in rats by McCoy and Weswig.[9]

The selenium content of the tissues of animals is dependent on the selenium content of the food consumed and duration of the consumption of the food. The selenium content of the food is dependent on the selenium content of the soil and the type of plant that is grown in the soil. While some plants will concentrate selenium from the soil and accumulate high levels, the common forage crops and grains contain relatively low levels of selenium.[10] Several authors have used the selenium content of plants to establish regions where the soils contain high or low selenium levels.[11] With a change in the selenium content of the diet, time is required to allow establishment of a stable selenium content of animal tissues.

The data summarized in this chapter are restricted primarily to the tissues of domestic animals that had been fed diets containing naturally occurring selenium provided by the feed ingredients. Tissue levels of animals fed diets containing purposely added inorganic selenium (selenite or selenate) are not included. For this reason, much of the data are from experiments in which animals were fed diets low in selenium.

II. SELENIUM METABOLISM BY ANIMALS

Selenium in plant materials is primarily organic, occurring in the compounds selenocystine, selenocysteine, Se-methylselenocysteine, selenohomocystine, selenomethionine, Se-methylselenomethionine, selenomethionine selenoxide, selenocystathionine, and dimethyl diselenide (Shrift,[12] see also Chapter 3). Inorganic forms such as selenite and selenate may be present,[13,14] but selenocystine and selenomethionine appear to be the major forms of selenium in natural animal feeds.[12-14]

Selenium substitutes for sulfur in organic compounds and has some similarities with the chemistry of sulfur. The metabolism of selenium in animal tissues, however, differs from the metabolism of sulfur because sulfur is oxidized to sulfate for excretion, while selenium is reduced to selenide. Thus, selenate is absorbed by the rat by mechanisms similar to those of sulfate[15] and is reduced to selenite. Selenite is absorbed by diffusion[16] and is reduced by glutathione and NADPH to selenide.[17,18] Selenomethionine is absorbed by the same mechanism as methionine, but selenocysteine is absorbed by diffusion.[16]

Absorbed selenite is bound to plasma proteins for transport to tissues.[19] It has been suggested that selenite is taken up by red blood cells, reduced to selenide,[19,20] released into the plasma, and rapidly bound by plasma proteins.[19-22] Small amounts of selenate and selenomethionine are taken up by red blood cells or plasma proteins, and the majority of the selenium from these sources occurs in the protein-free plasma.[19]

The only known function of selenium in animal tissues was discovered by Rotruck et al.[23,24] and is as a component of the enzyme, glutathione peroxidase. Selenium is present in each of the four subunits of the enzyme as selenocysteine.[25] Sunde and Hoekstra[26] suggested that an inorganic form of selenium, possibly selenide, is incorporated in a posttranslational reaction with a serine residue in the protein subunit of glutathione peroxidase. Hawkes and Tappel[27] presented results to suggest the presence of a transfer RNA for selenocysteine and that selenocysteine was incorporated into the protein during synthesis.

Selenomethionine can bind to the methionine transfer RNA[28] and be incorporated into proteins in place of methionine.[29] Substitutions of selenomethionine for methionine in functional proteins do not appear to affect the biological function of the protein.[30]

Selenomethionine can be metabolized by pathways similar to the sulfur containing counterparts with the synthesis of selenocysteine.[31] Selenocysteine is metabolized by the enzyme, selenocysteine lyase, to alanine and hydrogen selenide.[32] Further metabolism of selenide is by methylation utilizing methyl groups from *S*-adenosyl methionine.[18,33] Selenium is excreted in urine as trimethyl selenide.[34,35] Dimethyl selenide can be produced if the level of selenium in the animal approaches a toxic level and this compound is excreted by the lungs in the expired air.[36] Some selenium may be excreted by the liver into bile.

Generally the selenium in glutathione peroxidase does not account for all of the selenium found in tissues. A small selenoprotein has been described in the muscle tissue of selenium-adequate lambs and absent in the muscle of lambs with nutritional muscular dystrophy.[37-39] A selenoprotein has been identified in bovine and rat sperm.[40-43] Tappel et al.[44] reported that in rats receiving selenium from selenite, about 36% of the selenium in tissues was associated with glutathione peroxidase and an estimated 9 to 23 other selenoproteins were identified. Selenocysteine accounted for over 80% of the total selenium in the tissues. Thus, it is probable that additional biologically active, selenium-containing proteins or compounds will be identified and provide a field of fruitful research.

III. SELENIUM CONTENT OF ANIMAL TISSUES

The data summarized in the tables are from animals fed diets containing no added inorganic selenium. Much of the data is, therefore, associated with diets that are low in selenium, and increased dietary selenium content was obtained by inclusion of seleniferous feed ingredients. The predominant analytical method used was fluorometry, with only a few studies utilizing colorimetric, atomic absorption, or neutron activation analysis methods.

While differences in serum selenium concentrations due to genetic selection have been reported in pigs,[45] no specific comparisons have been made between breeds within a species, and, therefore, breed within species was not considered as a factor that would influence tissue selenium levels. Within the species summarized in this chapter, no effect of sex on tissue selenium levels has been noted, so sex was not considered as a factor affecting tissue selenium levels. Body weight or age were also not considered as major factors affecting tissue selenium levels.

The major factors that influence tissue selenium levels are the level of dietary selenium and the length of time that a specific dietary level of selenium is fed. The accumulation or depletion of selenium in tissues after a change in dietary selenium levels is not considered in this chapter. Thus, repeated measures of selenium levels in tissues within a constant dietary selenium level were averaged to obtain a value that represented the selenium concentration associated with a specific dietary level of selenium. Tissue concentrations have been summarized by averaging the reported values within a tissue and fraction, where applicable, and recording the overall mean and the range of individual (averaged) values. The means were not weighted in any way to reflect the number of animals used or the number of observations made in a study. Values have been used on the same basis as reported by the original authors, and generally no attempt was made to convert values reported on a wet basis to a dry matter basis.

A. Swine

The selenium contents of tissues from swine are summarized in Table 1. In all species, blood has frequently been analyzed as an indicator of selenium status. In the study of Groce et al.,[46] analysis of erythrocytes, serum, and whole blood from the same samples indicated that erythrocytes contained more selenium than whole blood and that whole blood contained

Table 1
SELENIUM CONTENT OF SWINE TISSUES[a]

Geographical location	Mean selenium concentration[b] Diet (mg/kg)	Tissue	Weight of animal[c] (kg)	Duration of feeding[c] (d)	Date[d]	Ref.
Blood: Erythrocytes (mg/l)						
Michigan, U.S.	0.041	0.096 W	—[e]	9	1973	46
Blood: Plasma (mg/l)						
Michigan, U.S.	0.06	0.120 W	—	114	1986	48
Tennessee, U.S.	0.080	0.188 W	100	76	1977	49
Mean/median	0.07	0.154 W				
Blood: Serum (mg/l)						
Iowa, U.S.	0.020	0.013 W	9	28	1984	50
Michigan, U.S.	0.047	0.033 W	8	14	1978	51
Michigan, U.S.	0.041	0.051 W	—	9	1973	46
North Carolina, U.S.	0.025	0.030 W	—	100	1981	47
Ohio, U.S.	0.020	0.027 W	32	49	1980	52
Ohio, U.S.	0.02	0.048 W	125	105	1979	53
Ohio, U.S.	0.050	0.025 W	—	24	1975	54
Ohio, U.S.	0.050	0.030 W	—	60	1975	54
Ohio, U.S.	0.070	0.035 W	14	35	1985	55
Ohio, U.S.	0.050	0.028 W	60	147	1977	56
Ohio, U.S.	0.07	0.013 W	18	35	1978	57
Ohio, U.S.	0.17	0.053 W	21	35	1978	57
Ohio, U.S.	0.17	0.084 W	20	35	1978	57
Ohio, U.S.	0.46	0.151 W	9	35	1978	57
Ohio, U.S.	0.48	0.135 W	21	35	1978	57
Ohio, U.S.	0.07	0.035 W	52	—	1978	57
Ohio, U.S.	0.17	0.129 W	52	—	1978	57
Ohio, U.S.	0.041	0.039 W	95	56	1978	58
Ontario, Canada	0.030	0.032 W	—	56	1977	59
Ontario, Canada	0.035	0.032 W	—	365	1978	60
Ontario, Canada	0.030	0.027 W	—	21	1977	61
Minimum	0.020	0.013 W				
Maximum	0.48	0.151 W				
Mean	0.101	0.050 W				
Median	0.050	0.033 W				
Blood: Whole (mg/l)						
Michigan, U.S.	0.045	0.051 W	8	14	1978	51
Michigan, U.S.	0.041	0.068 W	—	9	1973	46
New York, U.S.	0.05	0.064 W	92	118	1979	62
New York, U.S.	0.100	0.093 W	92	135	1979	62
New York, U.S.	0.180	0.119 W	92	126	1979	62
New York, U.S.	0.310	0.168 W	92	145	1979	62
New York, U.S.	0.510	0.193 W	92	127	1979	62
New York, U.S.	3.4	0.96 W	47	61	1979	62
New York, U.S.	6.0	2.11 W	47	61	1979	62
New York, U.S.	0.035	0.076 W	25	56	1980	63
New York, U.S.	0.100	0.112 W	25	56	1980	63
New York, U.S.	0.130	0.114 W	25	56	1980	63

Table 1 (continued)
SELENIUM CONTENT OF SWINE TISSUES[a]

Geographical location	Mean selenium concentration[b] Diet (mg/kg)	Tissue	Weight of animal[c] (kg)	Duration of feeding[c] (d)	Date[d]	Ref.
New York, U.S.	0.210	0.215 W	25	56	1980	63
New York, U.S.	0.640	0.375 W	25	56	1980	63
South Dakota, U.S.	0.47	0.24 W	33	42	1984	64
South Dakota, U.S.	2.58	0.88 W	33	42	1984	64
South Dakota, U.S.	5.60	1.79 W	33	42	1984	64
South Dakota, U.S.	8.40	2.42 W	33	42	1984	64
South Dakota, U.S.	0.400	0.320 W	24	35	1984	65
Tennessee, U.S.	0.080	0.245 W	100	76	1977	49
Ontario, Canada	0.780	0.350 W	91	—	1973	66
Ontario, Canada	0.020	0.047 W	—	140	1979	67
Quebec, Canada	0.020	0.031 W	15	49	1981	68
Sweden	0.009	0.016 W	16	12	1974	69
Sweden	0.008	0.023 W	35	56	1978	70
Sweden	0.008	0.030 W	12	13	1978	71
Sweden	0.008	0.017 W	18	22	1978	72
Minimum	0.008	0.016 W				
Maximum	8.40	2.42 W				
Mean	1.116	0.412 W				
Median	0.13	0.119 W				
		Brain (mg/kg)				
Quebec, Canada	0.020	0.439 D	19	63	1981	68
		Bulbourethra (mg/kg)				
North Carolina, U.S.	0.025	0.04 W	—	100	1981	47
		Carcass (mg/kg)				
Alberta, Canada	0.34	0.141 W	73	—	1975	73
Alberta, Canada	0.69	0.140 W	10	—	1975	73
Denmark	0.11	0.100 W	84	160	1979	74
Minimum	0.11	0.10 W				
Maximum	0.69	0.141 W				
Mean	0.38	0.127 W				
Median	0.34	0.140 W				
		Carcass (mg/kg)				
Denmark	0.21	0.332 D	—	21	1974	75
Denmark	0.08	0.224 D	85	—	1974	75
Mean/median	0.15	0.278 D				
		Empty Body (mg/kg)				
Alberta, Canada	0.34	0.153 W	73	—	1975	73
Alberta, Canada	0.69	0.122 W	10	—	1975	73
Mean/median	0.52	0.138 W				
		Epididymis (mg/kg)				
North Carolina, U.S.	0.025	0.32 W	—	100	1981	47

Table 1 (continued)
SELENIUM CONTENT OF SWINE TISSUES[a]

Geographical location	Mean selenium concentration[b] Diet (mg/kg)	Tissue	Weight of animal[c] (kg)	Duration of feeding[c] (d)	Date[d]	Ref.
		Fat (mg/kg)				
Indiana, U.S.	0.07	0.04 W	9	—	1975	76
		Fetus (mg/kg)				
Ohio, U.S.	0.020	0.041 W	—	55	1979	53
		Hair (mg/kg)				
Ohio, U.S.	0.030	0.200 D	—	—	1977	56
South Dakota, U.S.	0.47	0.80 D	33	42	1984	68
South Dakota, U.S.	2.58	4.07 D	33	42	1984	64
South Dakota, U.S.	5.60	8.53 D	33	42	1984	64
South Dakota, U.S.	8.40	10.5 D	33	42	1984	64
South Dakota, U.S.	0.400	1.90 D	24	35	1984	65
Ontario, Canada	0.030	0.193 D	—	85	1972	77
Minimum	0.030	0.193 D				
Maximum	8.40	10.5 D				
Mean	2.50	3.74 D				
Median	0.47	1.90 D				
		Kidney: Fetal (mg/kg)				
Ohio, U.S.	0.020	0.278 W	—	105	1979	53
		Kidney: Cortex (mg/kg)				
Indiana, U.S.	0.07	0.70 W	9	—	1975	76
Indiana, U.S.	0.04	1.48 W	—	—	1973	78
Ohio, U.S.	0.020	0.542 W	32	49	1980	52
Ohio, U.S.	0.020	1.32 W	125	105	1979	53
Ohio, U.S.	0.07	0.355 W	18	35	1978	57
Ohio, U.S.	0.17	0.829 W	18	35	1978	57
Ohio, U.S.	0.17	0.904 W	21	35	1978	57
Ohio, U.S.	0.17	1.04 W	20	35	1978	57
Ohio, U.S.	0.46	1.46 W	9	35	1978	57
Ohio, U.S.	0.48	1.34 W	21	35	1978	57
Ohio, U.S.	0.07	1.08 W	52	—	1978	57
Ohio, U.S.	0.17	1.78 W	52	—	1978	57
Minimum	0.020	0.355 W				
Maximum	0.48	1.78 W				
Mean	0.16	1.07 W				
Median	0.12	1.06 W				
		Kidney: Cortex (mg/kg)				
Ohio, U.S.	0.055	3.44 D	60	147	1977	56
Ohio, U.S.	0.053	3.02 D	7	28	1978	58
Oregon, U.S.	0.060	2.89 D	—	116	1977	79
Minimum	0.053	2.89 D				
Maximum	0.060	3.44 D				
Mean	0.056	3.12 D				
Median	0.055	3.02 D				

Table 1 (continued)
SELENIUM CONTENT OF SWINE TISSUES[a]

Geographical location	Mean selenium concentration[b] Diet (mg/kg)	Tissue	Weight of animal[c] (kg)	Duration of feeding[c] (d)	Date[d]	Ref.
		Kidney (mg/kg)				
Michigan, U.S.	0.020	0.743 W	40	84	1973	80
North Carolina, U.S.	0.025	0.72 W	—	100	1981	47
South Dakota, U.S.	0.47	1.72 W	33	42	1984	64
South Dakota, U.S.	2.58	3.44 W	33	42	1984	64
South Dakota, U.S.	5.60	5.12 W	33	42	1984	64
South Dakota, U.S.	8.40	6.62 W	33	42	1984	64
Denmark	0.03	1.04 W	86	160	1979	74
Denmark	0.06	1.90 W	86	160	1979	74
Denmark	0.11	1.57 W	84	160	1979	74
Minimum	0.020	0.720 W				
Maximum	8.40	6.62 W				
Mean	1.92	2.54 W				
Median	0.11	0.172 W				
		Kidney (mg/kg)				
Iowa, U.S.	0.020	1.33 D	24	56	1971	81
Michigan, U.S.	0.036	4.69 D	93	—	1978	51
Michigan, U.S.	0.443	11.2 D	100	141	1973	82
New York, U.S.	0.057	5.18 D	92	118	1979	62
New York, U.S.	0.100	8.02 D	92	135	1979	62
New York, U.S.	0.180	8.49 D	92	126	1979	62
New York, U.S.	0.310	8.75 D	92	145	1979	62
New York, U.S.	0.510	9.17 D	92	145	1979	62
New York, U.S.	3.4	12.4 D	47	61	1979	62
New York, U.S.	6.0	17.4 D	47	61	1979	62
New York, U.S.	0.030	3.60 D	25	56	1980	63
New York, U.S.	0.040	4.15 D	25	56	1980	63
New York, U.S.	0.100	5.55 D	25	56	1980	63
New York, U.S.	0.130	5.43 D	25	56	1980	63
New York, U.S.	0.210	6.90 D	25	56	1980	63
New York, U.S.	0.640	8.92 D	25	56	1980	63
Ohio, U.S.	0.050	2.61 D	—	24	1975	54
Ohio, U.S.	0.050	6.28 D	95	56	1978	58
Oregon, U.S.	0.067	4.40 D	44	56	1978	83
South Dakota, U.S.	0.267	11.5 D	105	117	1973	82
Ontario, Canada	0.030	1.87 D	19	73	1972	84
Ontario, Canada	0.780	10.4 D	91	—	1973	66
Ontario, Canada	0.010	2.54 D	26	63	1976	85
Ontario, Canada	0.035	2.88 D	—	34	1978	60
Ontario, Canada	0.030	2.93 D	70	120	1977	61
Ontario, Canada	0.030	1.87 D	8	—	1972	77
Quebec, Canada	0.020	1.70 D	19	63	1981	68
Denmark	0.080	7.96 D	85	—	1974	75
Sweden	—	9.44 D	—	—	1965	86
Sweden	—	2.98 D	—	—	1965	86
Sweden	0.126	11.4 D	70	78	1965	87
Minimum	0.010	1.33 D				
Maximum	6.00	17.4 D				
Mean	0.445	6.51 D				
Median	0.080	5.55 D				

Table 1 (continued)
SELENIUM CONTENT OF SWINE TISSUES[a]

Geographical location	Mean selenium concentration[b] Diet (mg/kg)	Tissue	Weight of animal[c] (kg)	Duration of feeding[c] (d)	Date[d]	Ref.
		Liver: Fetal (mg/kg)				
Ohio, U.S.	0.020	0.196 W	—	105	1979	53
		Liver (mg/kg)				
Indiana, U.S.	0.07	0.19 W	9	—	1975	76
Indiana, U.S.	0.04	0.178 W	—	—	1973	78
Indiana, U.S.	—	0.220 W	—	—	1970	88
Indiana, U.S.	0.055	0.080 W	—	—	1970	88
Indiana, U.S.	0.150	0.241 W	—	—	1970	88
Michigan, U.S.	0.020	0.150 W	7	17	1973	80
North Carolina, U.S.	0.025	0.08 W	—	100	1981	47
Ohio, U.S.	0.020	0.089 W	24	42	1980	52
Ohio, U.S.	0.020	0.102 W	125	105	1979	53
Ohio, U.S.	0.07	0.053 W	18	35	1978	57
Ohio, U.S.	0.17	0.174 W	18	35	1978	57
Ohio, U.S.	0.17	0.167 W	21	35	1978	57
Ohio, U.S.	0.17	0.276 W	20	35	1978	57
Ohio, U.S.	0.46	0.554 W	9	35	1978	57
Ohio, U.S.	0.48	0.436 W	21	35	1978	57
Ohio, U.S.	0.07	0.108 W	52	—	1978	57
Ohio, U.S.	0.17	0.437 W	52	—	1978	57
South Dakota, U.S.	0.47	0.82 W	33	42	1984	64
South Dakota, U.S.	2.58	3.17 W	33	42	1984	64
South Dakota, U.S.	5.60	5.12 W	33	42	1984	64
South Dakota, U.S.	8.40	6.79 W	33	42	1984	64
Denmark	0.03	0.10 W	87	160	1979	74
Denmark	0.06	0.21 W	87	160	1979	74
Denmark	0.11	0.39 W	84	160	1979	74
Denmark	—	0.068 W	—	—	1975	89
Denmark	—	0.300 W	—	—	1975	89
Norway	0.028	0.070 W	45	114	1982	90
Norway	0.045	0.13 W	95	84	1982	91
Norway	0.010	0.370 W	—	—	1980	92
Minimum	0.010	0.068 W				
Maximum	8.40	6.79 W				
Mean	0.750	0.727 W				
Median	0.07	0.19 W				
		Liver (mg/kg)				
Iowa, U.S.	0.020	0.085 D	24	56	1971	81
Michigan, U.S.	0.036	0.468 D	93	—	1978	51
Michigan, U.S.	0.443	2.89 D	100	141	1973	82
New York, U.S.	0.057	0.617 D	92	118	1979	62
New York, U.S.	0.100	0.97 D	92	135	1979	62
New York, U.S.	0.180	1.31 D	92	126	1979	62
New York, U.S.	0.310	1.55 D	92	145	1979	62
New York, U.S.	0.510	1.83 D	92	145	1979	62
New York, U.S.	3.40	9.16 D	47	61	1979	62
New York, U.S.	06.00	16.1 D	47	61	1979	62

Table 1 (continued)
SELENIUM CONTENT OF SWINE TISSUES[a]

Geographical location	Mean selenium concentration[b] Diet (mg/kg)	Tissue	Weight of animal[c] (kg)	Duration of feeding[c] (d)	Date[d]	Ref.
New York, U.S.	0.030	0.360 D	25	56	1980	63
New York, U.S.	0.040	0.460 D	25	56	1980	63
New York, U.S.	0.100	1.15 D	25	56	1980	63
New York, U.S.	0.130	1.21 D	25	56	1980	63
New York, U.S.	0.210	2.52 D	25	56	1980	63
New York, U.S.	0.640	3.59 D	25	56	1980	63
Ohio, U.S.	0.050	0.413 D	—	—	1975	54
Ohio, U.S.	0.055	0.364 D	3	—	1977	56
Ohio, U.S.	0.053	0.502 D	15	28	1978	58
Oregon, U.S.	0.067	2.63 D	44	56	1978	83
Oregon, U.S.	0.067	0.342 D	44	56	1978	83
Oregon, U.S.	0.060	0.180 D	—	160	1977	79
South Dakota, U.S.	0.267	2.62 D	105	117	1973	82
British Columbia, Canada	0.223	2.22 D	89	—	1977	93
Ontario, Canada	0.030	0.247 D	11	18	1972	84
Ontario, Canada	0.049	0.636 D	20	58	1972	94
Ontario, Canada	0.030	0.235 D	—	85	1972	77
Ontario, Canada	0.780	3.20 D	91	—	1973	66
Ontario, Canada	0.010	0.260 D	26	63	1976	85
Ontario, Canada	0.066	1.26 D	90	—	1977	93
Ontario, Canada	0.035	0.225 D	95	—	1978	60
Ontario, Canada	0.030	0.318 D	—	14	1977	61
Quebec, Canada	0.525	2.53 D	89	—	1977	93
Quebec, Canada	0.020	0.122 D	19	63	1981	68
Canada	0.030	0.247 D	8	91	1972	77
Denmark	0.080	1.39 D	85	—	1974	75
Sweden	—	0.197 D	—	—	1965	86
Sweden	—	1.22 D	—	—	1965	86
Sweden	0.126	1.77 D	70	78	1965	87
Minimum	0.010	0.085 D				
Maximum	6.00	16.1 D				
Mean	0.381	1.73 D				
Median	0.067	0.97 D				
Lung (mg/kg)						
Quebec, Canada	0.020	0.219 D	19	63	1981	68
Milk: Colostrum (mg/l)						
Michigan, U.S.	0.05	0.10 W	—	—	1986	48
Ontario, Canada	0.035	0.056 W	—	365	1978	60
Ontario, Canada	0.030	0.050 W	—	365	1977	61
Denmark	0.170	0.168 W	—	114	1974	95
Minimum	0.030	0.050 W				
Maximum	0.170	0.168 W				
Mean	0.071	0.094 W				
Median	0.043	0.078 W				

Table 1 (continued) SELENIUM CONTENT OF SWINE TISSUES[a]

Geographical location	Mean selenium concentration[b] Diet (mg/kg)	Tissue	Weight of animal[c] (kg)	Duration of feeding[c] (d)	Date[d]	Ref.
Milk (mg/l)						
Michigan, U.S.	0.05	0.02 W	—	21	1986	48
Ohio, U.S.	0.050	0.018 W	—	—	1977	56
Ohio, U.S.	0.050	0.013 W	—	28	1978	58
Denmark	0.170	0.039 W	—	124	1974	95
Minimum	0.050	0.013 W				
Maximum	0.170	0.039 W				
Mean	0.080	0.023 W				
Median	0.050	0.019 W				
Muscle: Diaphragm (mg/kg)						
South Dakota, U.S.	0.47	0.32 W	33	42	1984	64
South Dakota, U.S.	2.58	1.63 W	33	42	1984	64
South Dakota, U.S.	5.60	2.78 W	33	42	1984	64
South Dakota, U.S.	8.40	3.72 W	33	42	1984	64
Minimum	0.470	0.320 W				
Maximum	8.40	3.72 W				
Mean	4.26	2.11 W				
Median	4.09	2.21 W				
Muscle: Diaphragm (mg/kg)						
Michigan, U.S.	0.036	0.174 D	93	—	1978	51
Muscle: Heart (mg/kg)						
North Carolina, U.S.	0.025	0.06 W	—	100	1981	47
Ohio, U.S.	0.020	0.066 W	24	42	1980	52
Ohio, U.S.	0.020	0.066 W	125	105	1979	53
Ohio, U.S.	0.17	0.117 W	21	35	1978	57
Ohio, U.S.	0.17	0.144 W	21	35	1978	57
Ohio, U.S.	0.46	0.239 W	9	35	1978	57
Ohio, U.S.	0.48	0.214 W	21	35	1978	57
South Dakota, U.S.	0.47	0.43 W	33	42	1984	64
South Dakota, U.S.	2.58	1.85 W	33	42	1984	64
South Dakota, U.S.	5.60	2.88 W	33	42	1984	64
South Dakota, U.S.	8.40	4.10 W	33	42	1984	64
Denmark	—	0.051 W	—	—	1975	89
Denmark	—	0.164 W	—	—	1975	89
Minimum	0.020	0.051 W				
Maximum	8.40	4.10 W				
Mean	1.42	0.799 W				
Median	0.46	0.164 W				
Muscle: Heart (mg/kg)						
Ontario, Canada	0.030	0.267 D	11	18	1972	84
Ontario, Canada	0.780	2.64 D	91	—	1973	66
Ontario, Canada	0.010	0.290 D	26	63	1976	85
Ontario, Canada	0.035	0.286 D	95	—	1978	60

Table 1 (continued)
SELENIUM CONTENT OF SWINE TISSUES[a]

Geographical location	Mean selenium concentration[b] Diet (mg/kg)	Tissue	Weight of animal[c] (kg)	Duration of feeding[c] (d)	Date[d]	Ref.
Ontario, Canada	0.030	0.410 D	—	120	1977	61
Ontario, Canada	0.030	0.267 D	8	91	1972	77
Quebec, Canada	0.020	0.135 D	19	63	1981	68
Minimum	0.010	0.135 D				
Maximum	0.780	2.64 D				
Mean	0.134	0.614 D				
Median	0.030	0.286 D				
Muscle: Skeletal (mg/kg)						
Indiana, U.S.	0.07	0.06 W	9	—	1975	76
Indiana, U.S.	0.040	0.098 W	—	—	1973	78
North Carolina, U.S.	0.025	0.04 W	—	100	1981	47
Ohio, U.S.	0.020	0.031 W	32	49	1980	52
Ohio, U.S.	0.020	0.030 W	125	105	1979	53
Ohio, U.S.	0.07	0.04 W	18	35	1978	57
Ohio, U.S.	0.17	0.074 W	18	35	1978	57
Ohio, U.S.	0.17	0.058 W	21	35	1978	57
Ohio, U.S.	0.17	0.093 W	20	35	1978	57
Ohio, U.S.	0.46	0.116 W	9	35	1978	57
Ohio, U.S.	0.48	0.122 W	21	35	1978	57
Ohio, U.S.	0.07	0.057 W	52	—	1978	57
Ohio, U.S.	0.17	0.085 W	52	—	1978	57
Tennessee, U.S.	0.080	0.078 W	100	76	1977	49
Denmark	0.045	0.05 W	87	160	1979	74
Norway	0.045	0.05 W	95	84	1982	91
Minimum	0.020	0.030 W				
Maximum	0.48	0.122 W				
Mean	0.132	0.068 W				
Median	0.07	0.059 W				
Muscle: Skeletal (mg/kg)						
Arkansas, U.S.	0.152	0.817 D	100	—	1972	96
Idaho, U.S.	0.086	0.392 D	100	—	1972	96
Illinois, U.S.	0.036	0.223 D	100	—	1972	96
Indiana, U.S.	0.052	0.232 D	100	—	1972	96
Iowa, U.S.	0.235	0.977 D	100	—	1972	96
Iowa, U.S.	0.020	0.075 D	24	56	1971	81
Michigan, U.S.	0.443	1.88 D	100	141	1973	82
Michigan, U.S.	0.040	0.206 D	100	—	1972	96
Michigan, U.S.	0.042	0.182 D	—	98	1971	97
Nebraska, U.S.	0.330	1.18 D	100	—	1972	96
New York, U.S.	0.036	0.163 D	100	—	1972	96
New York, U.S.	0.057	0.302 D	92	118	1979	62
New York, U.S.	0.100	0.300 D	92	135	1979	62
New York, U.S.	0.180	0.400 D	92	126	1979	62
New York, U.S.	0.310	0.480 D	92	145	1979	62
New York, U.S.	0.510	0.540 D	92	145	1979	62
New York, U.S.	3.40	4.58 D	47	61	1979	62
New York, U.S.	6.00	8.57 D	47	61	1979	62
New York, U.S.	0.030	0.210 D	25	56	1980	63

Table 1 (continued)
SELENIUM CONTENT OF SWINE TISSUES[a]

Geographical location	Mean selenium concentration[b] Diet (mg/kg)	Tissue	Weight of animal[c] (kg)	Duration of feeding[c] (d)	Date[d]	Ref.
New York, U.S.	0.040	0.210 D	25	56	1980	63
New York, U.S.	0.100	0.290 D	25	56	1980	63
New York, U.S.	0.130	0.350 D	25	56	1980	63
New York, U.S.	0.210	1.57 D	25	56	1980	63
New York, U.S.	0.640	2.50 D	25	56	1980	63
North Dakota, U.S.	0.412	1.43 D	100	—	1972	96
Ohio, U.S.	0.050	0.193 D	—	60	1975	54
Ohio, U.S.	0.055	0.154 D	3	—	1977	56
Ohio, U.S.	0.055	0.190 D	7	28	1978	58
Oregon, U.S.	0.067	0.172 D	44	56	1978	83
South Dakota, U.S.	0.267	1.17 D	105	117	1973	82
South Dakota, U.S.	0.493	1.89 D	100	—	1972	96
Virginia, U.S.	0.027	0.118 D	100	—	1972	96
Wisconsin, U.S.	0.178	0.501 D	100	—	1972	96
Wyoming, U.S.	0.158	1.10 D	100	—	1972	96
British Columbia, Canada	0.223	1.07 D	89	—	1977	93
Ontario, Canada	0.030	0.157 D	11	18	1972	84
Ontario, Canada	0.780	1.89 D	91	—	1973	66
Ontario, Canada	0.010	0.200 D	26	63	1976	85
Ontario, Canada	0.065	0.483 D	90	—	1977	93
Ontario, Canada	0.035	0.120 D	95	—	1978	60
Ontario, Canada	0.030	0.160 D	70	120	1977	61
Ontario, Canada	0.030	0.157 D	8	91	1972	77
Quebec, Canada	0.410	1.61 D	89	—	1977	93
Quebec, Canada	0.020	0.056 D	19	63	1981	68
Saskatchewan, Canada	0.562	1.62 D	91	—	1977	93
Sweden	0.126	0.520 D	70	78	1965	87
Minimum	0.010	0.056 D				
Maximum	6.00	8.57 D				
Mean	0.375	0.904 D				
Median	0.100	0.371 D				
		Offal (mg/kg)				
Alberta, Canada	0.34	0.199 W	73	—	1975	73
Alberta, Canada	0.69	0.079 W	10	—	1975	73
Mean/Median	0.52	0.140 W				
		Pancreas (mg/kg)				
Oregon, U.S.	0.067	0.88 D	44	56	1978	83
Iowa, U.S.	0.020	0.059 W	9	28	1984	98
		Prostate (mg/kg)				
North Carolina, U.S.	0.025	0.05 W	—	100	1981	47
		Semen: Plasma (mg/l)				
North Carolina, U.S.	0.025	0.007 W	—	120	1981	47

Table 1 (continued)
SELENIUM CONTENT OF SWINE TISSUES[a]

Geographical location	Mean selenium concentration[b] Diet (mg/kg)	Tissue	Weight of animal[c] (kg)	Duration of feeding[c] (d)	Date[d]	Ref.
		Semen: Sperm[f]				
North Carolina, U.S.	0.025	0.199 W	—	120	1981	47
		Semen: Whole (mg/l)[g]				
North Carolina, U.S.	0.025	0.07 W	—	120	1981	47
		Seminal Vesicle (mg/kg)				
North Carolina, U.S.	0.025	0.05 W	—	100	1981	47
		Spleen (mg/kg)				
Quebec, Canada	0.020	0.354 D	19	63	1981	68
		Spleen (mg/kg)				
South Dakota, U.S.	0.47	0.53 W	33	42	1984	64
South Dakota, U.S.	2.58	1.51 W	33	42	1984	64
South Dakota, U.S.	5.60	2.54 W	33	42	1984	64
South Dakota, U.S.	8.40	3.49 W	33	42	1984	64
Minimum	0.47	0.53 W				
Maximum	8.40	3.49 W				
Mean	4.26	2.02 W				
Median	4.10	2.03 W				
		Testis (mg/kg)				
North Carolina, U.S.	0.025	0.26 W	—	100	1981	47
Ohio, U.S.	0.17	0.142 W	21	35	1978	57
Ohio, U.S.	0.17	0.171 W	20	35	1978	57
Ohio, U.S.	0.46	0.231 W	9	35	1978	57
Ohio, U.S.	0.48	0.218 W	21	35	1978	57
Minimum	0.025	0.142 W				
Maximum	0.48	0.26 W				
Mean	0.26	0.20 W				
Median	0.17	0.218 W				
		Testis				
Oregon, U.S.	0.062	0.393 D	31	35	1978	83
Oregon, U.S.	0.087	1.35 D	80	84	1974	99
Oregon, U.S.	0.157	2.15 D	80	84	1974	99
Oregon, U.S.	0.060	0.61 D	—	109	1977	79
Minimum	0.060	0.393 D				
Maximum	0.157	2.15 D				
Mean	0.092	1.12 D				
Median	0.075	0.78 D				

Table 1 (continued)
SELENIUM CONTENT OF SWINE TISSUES[a]

[a] Total selenium concentrations.

[b] Concentrations are presented as means in diet fed and tissues sampled as reported in the literature or calculated by the author of this chapter. Basis of tissue concentration data are as reported in the literature or as interpreted by this author from information provided. D — dry weight, W — wet weight; diet concentration data are on a fresh (or fed) or dry basis. Concentration units are as reported in the literature or as interpreted by this author from the information provided, with all units reported as parts per million, denoted mg/kg or mg/l as deemed appropriate; concentration units for diets are mg/kg; concentration units for tissues are mg/kg for solid tissues and mg/l for fluids, as indicated adjacent to the tissue listing; units for erythrocytes are mg/l of wet packed cells. The number of significant figures is as reported, as rounded off by this author or expressed to three decimal places, and is not necessarily related to the precision or accuracy of the data. Both means and medians of individual (mean) values for each tissue are reported; the latter reduces effects of extreme values and presents an indication of the typical concentration.

[c] Animal weights and durations of feeding are as reported in the literature or as estimated by this author from information provided.

[d] Date of report.

[e] Information not reported.

[f] Selenium concentration adjusted and expressed as μg Se/10^9 spermatozoa.

[g] Selenium concentration adjusted to reflect mean concentration of spermatozoa (363×10^6/ml).

more selenium than serum. The other data for whole blood and serum support the observation that whole blood contains more selenium than serum. The data for plasma suggest that plasma contains more selenium than serum, but that observation is not consistent with the data for other species.

Analysis of the empty body or carcasses of pigs indicated a lower selenium level on a wet basis than was present in the diet. Fat contained little selenium, while hair had higher levels of selenium than was present in the diet. The kidney of the pig generally contains higher selenium concentrations than the liver. Values have been reported for the cortex of the kidney because the levels in the cortex have been reported to be higher than the levels in the medulla. No direct comparison has been reported, however, in the pig. The selenium level in sow milk is relatively low, but the levels of selenium in colostrum are greater than the level of selenium in mature milk.

Selenium concentrations in various muscle types (diaphragm, heart, and skeletal) are similar and responsive to the dietary level of selenium. Muscle selenium levels are lower than in liver or kidney. Pancreas and spleen also tend to contain lower selenium concentrations than does liver. Limited observations on the selenium content of male reproductive tissues (testis, seminal vesicle, and epididymis) suggest that these tissues contain higher levels of selenium than other soft tissues. Segerson et al.[47] fractionated semen and observed that the majority of the selenium in semen was associated with the sperm.

B. Poultry

Selenium concentrations in the tissues of poultry are summarized in Table 2. Data are included for chickens, turkeys, and Japanese quail. Whole blood contains higher selenium concentrations than plasma, as observed for swine. Egg white contains more selenium than the yolk and results in the whole egg containing an amount of the element that is intermediate between that in the white and the yolk. Feathers contain higher selenium levels than are found in the diet, but this tissue does not seem to respond as much to dietary selenium variations as other tissues. The level of selenium in the kidney of birds is similar to the concentrations found in liver, but at elevated levels of dietary selenium, liver tends to contain

Table 2
SELENIUM CONTENT OF POULTRY TISSUES[a]

Geographical location	Mean selenium concentration[b] Diet (mg/kg)	Tissue	Duration of feeding[c] (d)	Date[d]	Ref.
		Blood: Plasma (mg/l)			
New York, U.S.	0.012	0.002 W	28	1975	100
Ohio, U.S.	0.10	0.09 W	180	1975	101
Ohio, U.S.	0.42	0.26 W	180	1975	101
Ohio, U.S.	0.03	<0.01 W	—[e]	1974	102
Ohio, U.S.	0.06	0.05 W	—	1974	102
Ohio, U.S.	0.047	0.012 W	28	1982	103
Minimum	0.012	0.002 W			
Maximum	0.42	0.26 W			
Mean	0.11	0.07 W			
Median	0.05	0.03 W			
		Blood: Whole (mg/l)			
New York, U.S.	0.045	0.088 W	28	1975	104
New York, U.S.	0.095	0.080 W	28	1975	104
New York, U.S.	0.215	0.164 W	28	1975	104
New York, U.S.	0.615	0.414 W	28	1975	104
New York, U.S.	0.165	0.190 W	133	1975	105
New York, U.S.	0.07	0.085 W	—	1971	106
New York, U.S.	0.68	0.289 W	—	1971	106
Minimum	0.045	0.080 W			
Maximum	0.68	0.414 W			
Mean	0.27	0.187 W			
Median	0.165	0.164 W			
		Bone (mg/kg)			
South Dakota, U.S.	1.25	0.7 D	—	1938	107
South Dakota, U.S.	2.5	1.8 D	—	1938	107
South Dakota, U.S.	5.0	2.0 D	—	1938	107
Minimum	1.25	0.7 D			
Maximum	5.0	2.0 D			
Mean	2.9	1.5 D			
Median	2.5	1.8 D			
		Brain (mg/kg)			
Indiana, U.S.	0.010	0.11 W	—	1977	108
Maryland, U.S.	0.10	0.23 W	7	1978	109
Mean/median	0.06	0.17 W			
		Brain (mg/kg)			
New York, USA	0.06	0.8 D	—	1978	110
New York, USA	3.7	3.4 D	112	1978	110
Mean/median	1.9	2.1 D			
		Egg: White (mg/kg)			
Ohio, U.S.	0.11	0.043 W	196	1978	111
Norway	0.30	0.126 W	—	1983	112

Table 2 (continued)
SELENIUM CONTENT OF POULTRY TISSUES[a]

Geographical location	Mean selenium concentration[b] Diet (mg/kg)	Tissue	Duration of feeding[c] (d)	Date[d]	Ref.
Norway	0.17	0.11 W	—	1982	113
Minimum	0.11	0.043 W			
Maximum	0.30	0.126 W			
Mean	0.19	0.09 W			
Median	0.17	0.11 W			
		Egg: White (mg/kg)			
New York, U.S.	0.06	0.25 D	112	1978	110
New York, U.S.	3.7	10.0 D	112	1978	110
Ohio, U.S.	0.10	0.40 D	180	1975	101
Ohio, U.S.	0.42	2.47 D	180	1975	101
Ohio, U.S.	0.03	0.07 D	—	1974	102
Ohio, U.S.	0.06	0.21 D	—	1974	102
Ohio, U.S.	0.05	0.14 D	14	1975	114
Ohio, U.S.	0.075	0.23 D	14	1975	114
Ohio, U.S.	0.10	0.18 D	14	1975	114
Ohio, U.S.	0.15	0.34 D	14	1975	114
Ohio, U.S.	0.18	0.66 D	14	1975	114
Ohio, U.S.	0.20	0.79 D	14	1975	114
Ohio, U.S.	0.28	1.16 D	14	1975	114
Ohio, U.S.	0.30	1.53 D	14	1975	114
South Dakota, U.S.	2.5	11.3 D	—	1938	107
South Dakota, U.S.	5.0	19.0 D	—	1938	107
South Dakota, U.S.	10.0	41.3 D	—	1938	107
Minimum	0.03	0.07 D			
Maximum	10.0	41.3 D			
Mean	1.37	5.30 D			
Median	0.18	0.66 D			
		Egg: Whole (mg/kg)			
New York, U.S.	0.446	0.283 W	—	1982	103
New York, U.S.	0.024	0.030 W	—	1974	115
New York, U.S.	0.038	0.119 W	—	1979	116
South Dakota, U.S.	<0.050	0.095 W	224	1974	117
South Dakota, U.S.	0.48	0.46 W	224	1974	117
South Dakota, U.S.	0.020	0.133 W	—	1973	118
South Dakota, U.S.	0.070	0.21 W	—	1973	118
South Dakota, U.S.	0.45	0.49 W	—	1973	118
South Dakota, U.S.	0.080	0.155 W	—	1969	119
South Dakota, U.S.	0.50	0.43 W	532	1969	119
Norway	0.30	0.296 W	—	1983	112
Norway	0.17	0.20 W	—	1982	113
Minimum	0.020	0.030 W			
Maximum	0.50	0.49 W			
Mean	0.22	0.24 W			
Median	0.13	0.20 W			
		Egg: Whole (mg/kg)			
Ohio, U.S.	0.10	0.32 D	180	1975	101
Ohio, U.S.	0.42	1.23 D	180	1975	101

Table 2 (continued)
SELENIUM CONTENT OF POULTRY TISSUES[a]

Geographical location	Mean selenium concentration[b] Diet (mg/kg)	Tissue	Duration of feeding[c] (d)	Date[d]	Ref.
South Dakota, U.S.	<0.050	0.096 D	—	1974	117
South Dakota, U.S.	0.48	0.46 D	—	1974	117
Minimum	0.050	0.096 D			
Maximum	0.48	1.23 D			
Mean	0.26	0.53 D			
Median	0.26	0.39 D			
		Egg: Yolk (mg/kg)			
Ohio, U.S.	0.11	0.185 W	196	1978	111
Norway	0.30	0.556 W	—	1983	112
Norway	0.17	0.36 W	—	1982	113
Minimum	0.11	0.185 W			
Maximum	0.30	0.556 W			
Mean	0.19	0.37 W			
Median	0.17	0.36 W			
		Egg: Yolk (mg/kg)			
New York, U.S.	0.06	0.5 D	112	1978	110
New York, U.S.	3.7	3.5 D	112	1978	110
Ohio, U.S.	0.03	0.065 D	—	1974	102
Ohio, U.S.	0.06	0.22 D	—	1974	102
Ohio, U.S.	0.05	0.14 D	14	1975	114
Ohio, U.S.	0.075	0.24 D	14	1975	114
Ohio, U.S.	0.10	0.19 D	14	1975	114
Ohio, U.S.	0.15	0.34 D	14	1975	114
Ohio, U.S.	0.18	0.50 D	14	1975	114
Ohio, U.S.	0.20	0.51 D	14	1975	114
Ohio, U.S.	0.28	0.80 D	14	1975	114
Ohio, U.S.	0.30	0.63 D	14	1975	114
South Dakota, U.S.	2.5	3.6 D	—	1938	107
South Dakota, U.S.	5.0	5.9 D	—	1938	107
South Dakota, U.S.	10.0	8.4 D	—	1938	107
Minimum	0.30	0.065 D			
Maximum	10.0	8.4 D			
Mean	1.51	1.70 D			
Median	0.18	0.50 D			
		Feathers (mg/kg)			
South Dakota, U.S.	0.020	0.33 D	—	1973	118
South Dakota, U.S.	0.070	0.27 D	—	1973	118
South Dakota, U.S.	0.45	0.90 D	—	1973	118
South Dakota, U.S.	<0.050	0.76 D	—	1974	117
South Dakota, U.S.	0.48	0.89 D	—	1974	117
Minimum	0.020	0.27 D			
Maximum	0.48	0.90 D			
Mean	0.21	0.63 D			
Median	0.070	0.76 D			

Table 2 (continued) SELENIUM CONTENT OF POULTRY TISSUES[a]

Geographical location	Mean selenium concentration[b] Diet (mg/kg)	Tissue	Duration of feeding[c] (d)	Date[d]	Ref.
		Kidney (mg/kg)			
Indiana, U.S.	0.010	0.11 W	—	1977	108
Maryland, U.S.	0.10	0.76 W	7	1978	109
Ohio, U.S.	0.11	0.47 W	196	1978	111
South Dakota, U.S.	0.020	0.40 W	—	1973	118
South Dakota, U.S.	0.070	0.37 W	—	1973	118
South Dakota, U.S.	0.45	0.96 W	—	1973	118
Norway	0.30	0.63 W	—	1983	112
Norway	0.17	0.50 W	—	1982	113
Minimum	0.010	0.11 W			
Maximum	0.45	0.96 W			
Mean	0.15	0.53 W			
Median	0.11	0.49 W			
		Kidney (mg/kg)			
New York, U.S.	0.070	1.02 D	—	1971	106
New York, U.S.	0.68	3.99 D	—	1971	106
New York, U.S.	0.06	2.5 D	112	1978	110
New York, U.S.	3.7	9.45 D	112	1978	110
South Dakota, U.S.	1.25	0.90 D	—	1938	107
South Dakota, U.S.	2.5	6.9 D	—	1938	107
South Dakota, U.S.	5.0	8.5 D	—	1938	107
South Dakota, U.S.	10.0	9.3 D	—	1938	107
South Dakota, U.S.	<0.050	0.353 D	—	1974	117
South Dakota, U.S.	0.48	0.885 D	—	1974	117
Minimum	0.050	0.353 D			
Maximum	10.0	9.45 D			
Mean	2.38	4.38 D			
Median	0.97	3.25 D			
		Liver (mg/kg)			
Indiana, U.S.	0.010	0.090 W	—	1977	108
Indiana, U.S.	0.13	0.28 W	—	1977	108
Maryland, U.S.	0.10	0.51 W	7	1978	109
New York, U.S.	0.165	0.658 W	133	1975	105
New York, U.S.	0.019	0.065 W	22	1971	120
New York, U.S.	0.73	0.89 W	22	1971	120
Ohio, U.S.	0.11	0.30 W	196	1978	111
South Dakota, U.S.	0.020	0.285 W	—	1973	118
South Dakota, U.S.	0.070	0.21 W	—	1973	118
South Dakota, U.S.	0.45	0.775 W	—	1973	118
South Dakota, U.S.	0.080	0.356 W	532	1969	119
South Dakota, U.S.	0.50	0.52 W	532	1969	119
Norway	0.30	0.58 W	—	1983	112
Norway	0.13	0.58 W	—	1982	113
Norway	0.17	0.42 W	—	1982	113
Minimum	0.010	0.065 W			
Maximum	0.73	0.89 W			
Mean	0.20	0.43 W			
Median	0.13	0.42 W			

Table 2 (continued)
SELENIUM CONTENT OF POULTRY TISSUES[a]

Geographical location	Mean selenium concentration[b] Diet (mg/kg)	Tissue	Duration of feeding[c] (d)	Date[d]	Ref.
		Liver (mg/kg)			
New York, U.S.	0.07	0.58 D	—	1971	106
New York, U.S.	0.68	2.78 D	—	1971	106
New York, U.S.	0.06	1.15 D	112	1978	110
New York, U.S.	3.7	11.3 D	112	1978	110
Ohio, U.S.	0.10	0.43 D	180	1975	101
Ohio, U.S.	0.42	1.92 D	180	1975	101
Ohio, U.S.	0.03	0.23 D	—	1974	102
Ohio U.S.	0.06	0.40 D	—	1974	102
Ohio, U.S.	0.047	0.261 D	28	1982	103
South Dakota, U.S.	1.25	3.1 D	—	1938	107
South Dakota, U.S.	2.5	3.8 D	—	1938	107
South Dakota, U.S.	5.0	5.4 D	—	1938	107
South Dakota, U.S.	10.0	17.0 D	—	1938	107
South Dakota, U.S.	<0.050	0.298 D	—	1974	117
South Dakota, U.S.	0.48	0.72 D	—	1974	117
Minimum	0.030	0.23 D			
Maximum	10.0	17.0 D			
Mean	1.63	3.29 D			
Median	0.42	1.15 D			
		Lung (mg/kg)			
South Dakota, U.S.	1.25	1.2 D	—	1938	107
South Dakota, U.S.	2.5	1.0 D	—	1938	107
South Dakota, U.S.	5.0	5.4 D	—	1938	107
Minimum	1.25	1.0 D			
Maximum	5.0	5.4 D			
Mean	2.9	2.5 D			
Median	2.5	1.2 D			
		Muscle: Breast (mg/kg)			
New York, U.S.	0.165	0.199 W	133	1975	105
Ohio, U.S.	0.11	0.075 W	196	1978	111
South Dakota, U.S.	0.020	0.060 W	—	1973	118
South Dakota, U.S.	0.070	0.090 W	—	1973	118
South Dakota, U.S.	0.45	0.40 W	—	1973	118
South Dakota, U.S.	0.50	0.36 W	532	1969	119
Norway	0.30	0.20 W	—	1983	112
Norway	0.13	0.20 W	—	1982	113
Norway	0.17	0.15 W	—	1982	113
Minimum	0.020	0.060 W			
Maximum	0.45	0.40 W			
Mean	0.21	0.19 W			
Median	0.165	0.199 W			
		Muscle: Gizzard (mg/kg)			
Indiana, U.S.	0.010	0.040 W	—	1977	108

Table 2 (continued)
SELENIUM CONTENT OF POULTRY TISSUES[a]

Geographical location	Mean selenium concentration[b] Diet (mg/kg)	Tissue	Duration of feeding[c] (d)	Date[d]	Ref.
Muscle: Heart (mg/kg)					
Indiana, U.S.	0.010	0.070 W	—	1977	108
South Dakota, U.S.	0.020	0.18 W	—	1973	118
South Dakota, U.S.	0.070	0.16 W	—	1973	118
South Dakota, U.S.	0.45	0.54 W	—	1973	118
Norway	0.30	0.36 W	—	1983	112
Minimum	0.010	0.070 W			
Maximum	0.45	0.54 W			
Mean	0.17	0.26 W			
Median	0.070	0.18 W			
Muscle: Skeletal (mg/kg)					
Indiana, U.S.	0.010	0.060 W	—	1977	108
Muscle: Thigh (mg/kg)					
New York, U.S.	0.165	0.185 W	133	1975	105
South Dakota, U.S.	0.020	0.10 W	—	1973	118
South Dakota, U.S.	0.070	0.11 W	—	1973	118
South Dakota, U.S.	0.45	0.44 W	—	1973	118
Minimum	0.020	0.10 W			
Maximum	0.45	0.44 W			
Mean	0.18	0.21 W			
Median	0.12	0.15 W			
Muscle: Breast (mg/kg)					
Ohio, U.S.	0.10	0.33 D	180	1975	101
Ohio, U.S.	0.42	1.18 D	180	1975	101
Ohio, U.S.	0.03	0.09 D	—	1974	102
Ohio, U.S.	0.06	0.240 D	—	1974	102
Ohio, U.S.	0.047	0.164 D	28	1982	103
South Dakota, U.S.	1.25	2.2 D	—	1938	107
South Dakota, U.S.	2.5	2.5 D	—	1938	107
South Dakota, U.S.	5.0	8.0 D	—	1938	107
South Dakota, U.S.	10.0	8.0 D	—	1938	107
South Dakota, U.S.	<0.050	0.065 D	—	1974	117
South Dakota, U.S.	0.48	0.378 D	—	1974	117
Minimum	0.03	0.065 D			
Maximum	10.0	8.0 D			
Mean	1.81	2.10 D			
Median	0.42	0.378 D			
Muscle: Gizzard (mg/kg)					
Ohio, U.S.	0.047	0.201 D	28	1982	103
South Dakota, U.S.	1.25	2.8 D	—	1938	107
South Dakota, U.S.	2.5	3.3 D	—	1938	107
South Dakota, U.S.	5.0	5.2 D	—	1938	107
Minimum	0.047	0.201 D			
Maximum	5.0	5.2 D			
Mean	2.2	2.9 D			
Median	1.9	3.1 D			

Table 2 (continued)
SELENIUM CONTENT OF POULTRY TISSUES[a]

Geographical location	Mean selenium concentration[b] Diet (mg/kg)	Tissue	Duration of feeding[c] (d)	Date[d]	Ref.
		Muscle: Heart (mg/kg)			
New York, U.S.	0.06	0.55 D	—	1978	110
New York, U.S.	3.7	4.4 D	112	1978	110
Ohio, U.S.	0.047	0.169 D	—	1982	103
South Dakota, U.S.	1.25	0.3 D	—	1938	107
South Dakota, U.S.	5.0	7.6 D	—	1938	107
South Dakota, U.S.	<0.050	0.143 D	—	1974	117
South Dakota, U.S.	0.48	0.488 D	—	1974	117
Minimum	0.050	0.143 D			
Maximum	5.0	7.6 D			
Mean	1.51	1.95 D			
Median	0.48	0.488 D			
		Muscle: Skeletal (mg/kg)			
New York, U.S.	0.060	0.25 D	112	1978	110
		Muscle: Thigh (mg/kg)			
South Dakota, U.S.	1.25	2.8 D	—	1938	107
South Dakota, U.S.	2.5	5.8 D	—	1938	107
South Dakota, U.S.	5.0	6.1 D	—	1938	107
South Dakota, U.S.	10.0	12.7 D	—	1938	107
South Dakota, U.S.	<0.050	0.115 D	—	1974	117
South Dakota, U.S.	0.48	0.398 D	—	1974	117
Minimum	0.050	0.115 D			
Maximum	10.0	12.7 D			
Mean	3.2	4.7 D			
Median	1.9	4.3 D			
		Muscle: Not Specified (mg/kg)			
New York, U.S.	0.07	0.22 D	—	1971	106
New York, U.S.	0.68	1.27 D	—	1971	106
New York, U.S.	3.7	3.9 D	112	1978	110
Minimum	0.07	0.22 D			
Maximum	3.7	3.9 D			
Mean	1.5	1.8 D			
Median	0.68	1.27 D			
		Ovary (mg/kg)			
South Dakota, U.S.	2.5	3.9 D	40	1938	107
South Dakota, U.S.	5.0	4.7 D	40	1938	107
South Dakota, U.S.	10.0	14.5 D	40	1938	107
Minimum	2.5	3.9 D			
Maximum	10.0	14.5 D			
Mean	5.8	7.7 D			
Median	5.0	4.7 D			

Table 2 (continued)
SELENIUM CONTENT OF POULTRY TISSUES[a]

Geographical location	Mean selenium concentration[b] Diet (mg/kg)	Tissue	Duration of feeding[c] (d)	Date[d]	Ref.
		Oviduct (mg/kg)			
South Dakota, U.S.	1.25	2.3 D	40	1938	107
South Dakota, U.S.	2.5	10.8 D	40	1938	107
South Dakota, U.S.	5.0	17.8 D	40	1938	107
South Dakota, U.S.	10.0	26.5 D	40	1938	107
Minimum	1.25	2.3 D			
Maximum	10.0	26.5 D			
Mean	4.7	14.4 D			
Median	3.8	14.3 D			
		Pancreas (mg/kg)			
New York, U.S.	0.012	0.050 D	28	1975	100
Ohio, U.S.	0.047	0.335 D	28	1982	103
Mean/Median	0.030	0.193 D			
		Proventriculus (mg/kg)			
South Dakota, U.S.	1.25	1.7 D	40	1938	107
South Dakota, U.S.	2.5	2.9 D	40	1938	107
South Dakota, U.S.	5.0	4.3 D	40	1938	107
South Dakota, U.S.	10.0	11.0 D	40	1938	107
Minimum	1.25	1.7 D			
Maximum	10.0	11.0 D			
Mean	4.7	5.0 D			
Median	3.8	3.6 D			
		Skin (mg/kg)			
New York, U.S.	0.07	0.20 D	—	1971	106
New York, U.S.	0.68	0.71 D	—	1971	106
Mean/Median	0.37	0.45 D			
		Small Intestine (mg/kg)			
South Dakota, U.S.	2.5	0.6 D	42	1938	107
South Dakota, U.S.	5.0	4 D	42	1938	107
South Dakota, U.S.	10.0	3 D	42	1938	107
Minimum	2.5	0.6 D			
Maximum	10.0	4 D			
Mean	5.8	2.5 D			
Median	5.0	3 D			
		Spleen (mg/kg)			
South Dakota, U.S.	2.5	13.6 D	35	1938	107
South Dakota, U.S.	5.0	7.1 D	35	1938	107
Mean/Median	3.75	10.4 D			

Table 2 (continued)
SELENIUM CONTENT OF POULTRY TISSUES[a]

[a] Total selenium concentrations.

[b] Concentrations are presented as means in diet fed and tissues sampled as reported in the literature or calculated by the author of this chapter. Basis of tissue concentration data are as reported in the literature or as interpreted by this author from information provided. D — dry weight, W — wet weight; diet concentration data are on a fresh (or fed) or dry basis. Concentration units are as reported in the literature or as interpreted by this author from the information provided, with all units reported as parts per million, denoted mg/kg or mg/l as deemed appropriate; concentration units for diets are mg/kg; concentration units for tissues are mg/kg for solid tissues and mg/l for fluids, as indicated adjacent to the tissue listing; units for erythrocytes are mg/l of wet packed cells. The number of significant figures is as reported, as rounded off by this author or expressed to three decimal places, and is not necessarily related to the precision or accuracy of the data. Both means and medians of individual (mean) values for each tissue are reported; the latter reduces effects of extreme values and presents an indication of the typical concentration.

[c] Animal weights and durations of feeding are as reported in the literature or as estimated by this author from information provided.

[d] Date of report.

[e] Information not reported.

Various muscle types (breast, gizzard, heart, proventriculus, and thigh) contain similar concentrations of selenium that are lower than those in liver or kidney. The ovary and oviduct contain relatively high levels of selenium compared to other soft tissues. Because the proteins in the egg white are synthesized by the oviduct, the concentration of selenium in the latter tissue may explain the observation that egg white contains high levels of selenium.

C. Sheep

Selenium concentrations in the tissues of sheep are summarized in Table 3. Because ruminants are generally fed diets containing roughage and concentrates, the selenium contents of both components are listed where reported. Roughage includes pasture, dry roughage, or the average if both were fed. In blood, selenium distribution is similar to that in other species, with erythrocytes containing more selenium than whole blood, and plasma containing less than whole blood. Kidney tissue contains more selenium than liver. The cortex of the kidney has more than twice the level of selenium as the medulla. Ewe milk is relatively low in selenium content and similar to the milk of other species. Heart and skeletal muscle are similar in selenium content and lower in selenium content than liver. Other tissues (lung, pancreas, spleen, and brain) have similar selenium contents. Glandular tissues (adrenal and pituitary) contain more selenium than liver, but less than kidney. The selenium content of the uterus is low and similar to the selenium content of muscle, in contrast to the high levels observed in the uterus of poultry. Wool selenium content is variable in relationship to the dietary selenium content, but methods of sampling or methods of sample treatment before analysis are suspected as a contributing cause.

D. Beef Cattle

Selenium concentrations in the tissues of beef cattle are summarized in Table 4. The distribution is similar to other species. Whole blood contains higher levels of the element than plasma or serum. Little selenium is found in fatty tissue. Hair and horn tend to contain more selenium than the diets fed; higher levels occur in kidney than in liver. Milk is low in selenium content, but colostrum contains higher levels than mature milk. Muscular tissue is generally lower in selenium content than other tissues, whereas other soft tissues tend to be similar to liver in selenium content. While the testis is relatively high in selenium, sperm does not contain the majority of the selenium associated with semen.

Table 3
SELENIUM CONTENT OF SHEEP TISSUES[a]

	Mean selenium concentration[b]						
	Diet (mg/kg)						
Geographical location	Rough.[c]	Conc.[c]	Tissue	Weight of animal[d](kg)	Duration of feeding[d] (d)	Date[e]	Ref.
			Abomasum (mg/kg)				
New Zealand	—[f]	—	0.027 W	—	—	1967	121
			Adrenal (mg/kg)				
Wisconsin, U.S.	—	0.01	0.171 W	—	56	1976	122
New Zealand	—	—	0.14 W	—	—	1967	121
Mean/median	—	0.01	0.16 W				
			Adrenal (mg/kg)				
Canada	0.020	—	0.67 D	—	90	1968	123
Canada	0.053	0.40	0.67 D	—	90	1968	123
Mean/median	0.037	0.40	0.67 D				
			Blood: Erythrocytes (mg/l)				
Wisconsin, U.S.	—	0.010	0.206 W	—	56	1976	122
			Blood: Plasma (mg/l)				
Michigan, U.S.	0.026	0.033	0.038 W	—	—	1978	124
Michigan, U.S.	0.063	0.036	0.036 W	48	137	1978	124
North Dakota, U.S.	0.64	0.19	0.123 W	38	137	1978	124
North Dakota, U.S.	0.785	0.19	0.169 W	—	—	1978	124
Wisconsin, U.S.	—	0.010	0.015 W	—	56	1976	122
Australia	—	0.020	0.061 W	37	—	1977	125
Australia	—	0.020	0.080 W	37	—	1980	126
Australia	—	—	0.024 W	25	—	1980	127
Australia	0.060	—	0.010 W	—	—	1983	128
Australia	0.060	—	0.014 W	—	—	1983	128
Minimum	0.026	0.010	0.010 W				
Maximum	0.785	0.19	0.169 W				
Mean	0.272	0.071	0.057 W				
Median	0.062	0.033	0.037 W				
			Blood: Whole (mg/l)				
New Mexico, U.S.	0.030	—	0.102 W	—	—	1983	129
New Mexico, U.S.	—	0.120	0.121 W	30	—	1978	130
Oregon, U.S.	0.018	0.016	0.017 W	—	180	1978	131
Wisconsin, U.S.	—	0.010	0.100 W	9	56	1976	122
Wisconsin, U.S.	0.026	0.014	0.050 W	—	284	1976	132
Wisconsin, U.S.	—	0.020	0.050 W	—	126	1976	132
Wisconsin, U.S.	—	0.050	0.115 W	—	180	1984	133
Wyoming, U.S.	—	—	1.57 W	—	—	1955	134
Alberta, Canada	—	—	0.74 W	61	720	1961	135
Ontario, Canada	0.018	—	0.034 W	—	290	1974	136
Ontario, Canada	0.050	—	0.061 W	—	290	1974	136
Ontario, Canada	0.015	—	0.027 W	—	—	1971	137
Australia	—	0.020	0.272 W	37	—	1980	126
Australia	—	—	0.068 W	25	—	1980	127

Table 3 (continued)
SELENIUM CONTENT OF SHEEP TISSUES[a]

Geographical location	Mean selenium concentration[b] Diet (mg/kg) Rough.[c]	Conc.[c]	Tissue	Weight of animal[d](kg)	Duration of feeding[d] (d)	Date[e]	Ref.
Australia	0.060	—	0.030 W	—	—	1983	128
Australia	—	—	0.267 W	—	—	1985	138
Australia	0.070	—	0.136 W	35	200	1976	139
Australia	0.050	—	0.027 W	—	720	1970	140
Australia	—	—	0.026 W	33	—	1982	141
New Zealand	—	—	0.035 W	—	—	1967	121
New Zealand	0.030	—	0.014 W	—	365	1976	142
Norway	0.050	0.35	0.13 W	—	70	1983	143
Scotland	0.028	0.117	0.054 W	—	—	1983	144
Minimum	0.015	0.010	0.014 W				
Maximum	0.070	0.35	1.57 W				
Mean	0.037	0.072	0.171 W				
Median	0.029	0.020	0.061 W				
Blood: Whole (mg/kg)							
Wyoming, U.S.	—	—	0.43 D	—	—	1955	134
Canada	0.020	—	0.13 D	—	90	1968	123
Canada	0.053	0.40	1.31 D	—	90	1968	123
Minimum	—	—	0.13 D				
Maximum	—	—	1.31 D				
Mean	0.037	0.40	0.62 D				
Median	0.037	0.40	0.43 D				
Bone (mg/kg)							
New Zealand	—	—	0.01 W	—	—	1967	121
Bone: Marrow (mg/kg)							
New Zealand	—	—	0.02 W	—	—	1967	121
Brain: Cortex (mg/kg)							
New Zealand	—	—	0.07 W	—	—	1967	121
Brain (mg/kg)							
Colorado, U.S.	—	—	0.2 W	—	—	1967	145
Fetus (mg/kg)							
Australia	—	—	0.22 D	33	—	1982	141
Hoof (mg/kg)							
Canada	0.020	—	0.15 D	—	90	1968	123
Canada	0.053	0.40	0.38 D	—	90	1968	123
Mean/median	0.037	0.40	0.27 D				

Table 3 (continued)
SELENIUM CONTENT OF SHEEP TISSUES[a]

Geographical location	Mean selenium concentration[b] Diet (mg/kg) Rough.[c]	Conc.[c]	Tissue	Weight of animal[d](kg)	Duration of feeding[d] (d)	Date[e]	Ref.
Intestine: Upper (mg/kg)							
New Zealand	—	—	0.64 D	—	—	1961	146
New Zealand	—	0.11	1.90 D	—	—	1961	146
Mean/median	—	0.11	1.27 D				
Kidney (mg/kg)							
Colorado, U.S.	—	—	1.4 W	—	—	1967	145
Michigan, U.S.	0.063	0.036	0.741 W	48	137	1978	124
New Mexico, U.S.	—	0.12	0.696 W	30	—	1978	130
North Dakota, U.S.	0.64	0.19	1.04 W	38	137	1978	124
Wisconsin, U.S.	—	0.010	0.233 W	9	56	1976	122
Wisconsin, U.S.	—	0.020	0.574 W	—	126	1976	132
Canada	—	0.005	0.442 W	—	240	1971	147
Norway	0.050	0.35	1.40 W	—	70	1983	143
Minimum	0.050	0.005	0.233 W				
Maximum	0.64	0.35	1.40 W				
Mean	0.251	0.104	0.815 W				
Median	0.063	0.036	0.719 W				
Kidney (mg/kg)							
Michigan, U.S.	0.063	0.036	3.87 D	48	137	1978	124
Oregon, U.S.	0.010	0.010	0.61 D	—	42	1968	148
Oregon, U.S.	0.40	0.010	3.29 D	15	42	1968	148
Oregon, U.S.	0.010	0.030	0.52 D	15	—	1966	149
Oregon, U.S.	0.020	0.030	0.83 D	12	—	1966	149
Oregon, U.S.	0.50	0.030	3.28 D	15	—	1966	149
Oregon, U.S.	2.65	0.030	7.70 D	12	—	1966	149
South Dakota, U.S.	0.64	0.19	5.38 D	38	137	1978	124
Ontario, Canada	0.018	—	2.04 D	—	290	1974	136
Ontario, Canada	0.050	—	3.42 D	—	290	1974	136
Australia	0.070	—	5.62 D	35	200	1976	139
Australia	0.015	—	4.10 D	—	—	1969	150
Minimum	0.010	0.010	0.52 D				
Maximum	2.65	0.19	7.70 D				
Mean	0.371	0.046	3.39 D				
Median	0.057	0.030	3.36 D				
Kidney: Cortex (mg/kg)							
New Zealand	—	—	0.5 W	—	—	1967	121
Kidney: Cortex (mg/kg)							
Canada	0.020	—	1.56 D	—	90	1968	123
Canada	0.053	0.40	4.38 D	—	90	1968	123
New Zealand	—	0.11	3.70 D	—	—	1961	146
New Zealand	—	—	1.69 D	—	—	1961	146
Minimum	—	—	1.56 D				
Maximum	—	—	4.38 D				
Mean	0.037	0.26	2.83 D				
Median	0.037	0.26	2.70 D				

Table 3 (continued)
SELENIUM CONTENT OF SHEEP TISSUES[a]

Geographical location	Mean selenium concentration[b]			Weight of animal[d](kg)	Duration of feeding[d] (d)	Date[e]	Ref.
	Diet (mg/kg)						
	Rough.[c]	Conc.[c]	Tissue				
Kidney: Medulla (mg/kg)							
Canada	0.020	—	0.30 D	—	90	1968	123
Canada	0.053	0.40	1.15 D	—	—	1968	123
Mean/median	0.037	0.40	0.73 D				
Liver (mg/kg)							
Colorado, U.S.	—	—	0.80 W	—	—	1967	145
Michigan, U.S.	0.063	0.036	0.132 W	48	137	1978	124
New Mexico, U.S.	—	0.12	0.389 W	30	—	1978	130
North Dakota, U.S.	0.64	0.19	0.617 W	38	137	1978	124
Wisconsin, U.S.	—	0.010	0.049 W	—	56	1976	122
Wisconsin, U.S.	—	0.020	0.049 W	—	126	1976	132
Canada	—	0.005	0.051 W	—	240	1971	147
New Zealand	—	—	0.081 W	—	—	1967	121
New Zealand	0.030	—	0.026 W	—	365	1976	142
Norway	0.050	0.35	0.28 W	—	70	1983	143
Norway	—	—	0.211 W	—	—	1978	151
Minimum	0.030	0.005	0.026 W				
Maximum	0.64	0.35	0.8 W				
Mean	0.20	0.104	0.24 W				
Median	0.057	0.036	0.132 W				
Liver (mg/kg)							
Michigan, U.S.	0.063	0.036	0.484 D	48	137	1978	124
Montana, U.S.	0.088	—	0.31 D	—	—	1962	152
Montana, U.S.	0.201	—	1.00 D	—	—	1962	152
North Dakota, U.S.	0.64	0.190	2.14 D	38	137	1978	124
Oregon, U.S.	0.010	0.010	0.050 D	—	42	1968	148
Oregon, U.S.	0.40	0.010	3.60 D	15	42	1968	148
Oregon, U.S.	0.010	0.030	0.040 D	15	—	1966	149
Oregon, U.S.	0.020	0.030	0.10 D	12	—	1966	149
Oregon, U.S.	0.50	0.030	4.02 D	15	—	1966	149
Oregon, U.S.	2.65	0.030	14.7 D	12	—	1966	149
Wisconsin, U.S.	—	0.020	0.30 D	—	40	1968	153
Ontario, Canada	0.018	—	0.172 D	—	290	1974	136
Ontario, Canada	0.050	—	0.444 D	—	290	1974	136
Canada	0.020	—	0.17 D	—	90	1968	123
Canada	0.053	0.40	1.32 D	—	90	1968	123
New South Wales, Australia	—	—	0.724 D	—	180	1981	154
Australia	0.070	—	0.55 D	35	200	1976	139
Australia	—	—	0.16 D	33	—	1982	141
Australia	—	—	0.604 D	—	—	1979	155
Australia	—	—	0.365 D	—	—	1977	156
Australia	—	—	0.473 D	—	—	1980	157
Australia	0.060	—	0.31 D	—	—	1983	128
New Zealand	—	0.11	0.80 D	—	—	1961	146
New Zealand	—	—	0.268 D	—	—	1961	146
Minimum	0.010	0.010	0.040 D				
Maximum	2.65	0.40	14.7 D				
Mean	0.303	0.075	1.38 D				
Median	0.062	0.030	0.464 D				

Table 3 (continued)
SELENIUM CONTENT OF SHEEP TISSUES[a]

Geographical location	Mean selenium concentration[b] Diet (mg/kg) Rough.[c]	Conc.[c]	Tissue	Weight of animal[d](kg)	Duration of feeding[d] (d)	Date[e]	Ref.
			Lung (mg/kg)				
Colorado, U.S.	—	—	0.4 W	—	—	1967	145
Wisconsin, U.S.	—	0.010	0.045 W	—	56	1976	122
Wisconsin, U.S.	—	0.020	0.058 W	—	126	1976	132
New Zealand	—	—	0.05 W	—	—	1967	121
Minimum	—	—	0.045 W				
Maximum	—	—	0.4 W				
Mean	—	0.015	0.14 W				
Median	—	0.015	0.054 W				
			Lung (mg/kg)				
Canada	0.020	—	0.20 D	—	90	1968	123
Canada	0.053	0.40	0.89 D	—	90	1968	123
New Zealand	—	0.11	1.41 D	—	—	1961	146
New Zealand	—	—	0.27 D	—	—	1961	146
Minimum	—	—	0.20 D				
Maximum	—	—	1.41 D				
Mean	0.037	0.26	0.69 D				
Median	0.037	0.26	0.58 D				
			Milk (mg/l)				
Wisconsin, U.S.	0.026	0.014	0.014 W	—	284	1976	132
Ontario, Canada	0.018	—	0.007 W	—	290	1974	136
Ontario, Canada	0.050	—	0.010 W	—	290	1974	136
Ontario, Canada	0.015	—	0.008 W	62	—	1971	137
Australia	0.060	—	0.006 W	—	—	1983	138
Minimum	0.015	—	0.006 W				
Maximum	0.060	—	0.014 W				
Mean	0.034	0.014	0.009 W				
Median	0.026	0.014	0.008 W				
			Milk (mg/kg)				
New York, U.S.	0.020	0.030	0.070 D	—	72	1967	158
			Muscle: Heart (mg/kg)				
Wisconsin, U.S.	—	0.010	0.040 W	—	56	1976	122
			Muscle: Skeletal (mg/kg)				
Michigan, U.S.	0.063	0.036	0.032 W	48	137	1978	124
North Dakota, U.S.	0.64	0.19	0.245 W	38	137	1978	124
Wisconsin, U.S.	—	0.020	0.030 W	—	126	1976	132
Canada	—	0.005	0.039 W	—	240	1971	147
New Zealand	—	—	0.02 W	—	—	1967	121
Norway	0.050	0.35	0.12 W	—	70	1983	143
Minimum	0.050	0.005	0.02 W				
Maximum	0.64	0.35	0.245 W				
Mean	0.25	0.12	0.081 W				
Median	0.063	0.036	0.036 W				

Table 3 (continued)
SELENIUM CONTENT OF SHEEP TISSUES[a]

	Mean selenium concentration[b]						
	Diet (mg/kg)						
Geographical location	Rough.[c]	Conc.[c]	Tissue	Weight of animal[d](kg)	Duration of feeding[d] (d)	Date[e]	Ref.
Muscle: Heart (mg/kg)							
Oregon, U.S.	0.40	0.010	1.010 D	15	42	1968	148
Oregon, U.S.	0.010	0.030	0.025 D	15	—	1966	149
Canada	0.020	—	0.09 D	—	90	1968	123
Canada	0.053	0.40	0.76 D	—	90	1968	123
New Zealand	—	0.110	0.71 D	—	—	1961	146
Minimum	0.010	0.010	0.025 D				
Maximum	0.40	0.40	1.010 D				
Mean	0.12	0.14	0.52 D				
Median	0.037	0.070	0.71 D				
Muscle: Skeletal (mg/kg)							
Michigan, U.S.	0.063	0.036	0.111 D	48	137	1978	124
Montana, U.S.	0.088	—	0.187 D	—	—	1962	152
Montana, U.S.	0.201	—	0.536 D	—	—	1962	152
New York, U.S.	0.020	0.030	0.040 D	12	—	1966	149
Oregon, U.S.	0.010	0.010	0.033 D	—	42	1968	148
Oregon, U.S.	0.50	0.030	1.19 D	15	—	1966	149
Oregon, U.S.	2.65	0.030	2.43 D	12	—	1966	149
South Dakota, U.S.	0.64	0.19	0.919 D	38	137	1978	124
Wisconsin, U.S.	—	0.020	0.137 D	—	40	1968	153
Ontario, Canada	0.018	—	0.054 D	—	290	1974	136
Ontario, Canada	0.050	—	0.10 D	—	290	1974	136
Canada	0.020	—	0.075 D	—	90	1968	123
Canada	0.053	0.40	0.42 D	—	90	1968	123
Australia	0.070	—	0.14 D	35	200	1976	139
New Zealand	—	—	0.145 D	—	—	1961	146
Minimum	0.010	0.010	0.033 D				
Maximum	2.65	0.40	2.43 D				
Mean	0.337	0.093	0.434 D				
Median	0.063	0.030	0.14 D				
Pancreas (mg/kg)							
Wisconsin, U.S.	—	0.010	0.054 W	—	—	1976	122
Wisconsin, U.S.	—	0.020	0.162 W	—	126	1976	132
New Zealand	—	—	0.06 W	—	—	1967	121
Minimum	—	—	0.054 W				
Maximum	—	—	0.162 W				
Mean	—	0.015	0.09 W				
Median	—	0.015	0.06 W				
Pancreas (mg/kg)							
Canada	0.020	—	0.32 D	—	90	1968	123
Canada	0.053	0.40	1.40 D	—	90	1968	123
Mean/median	0.037	0.40	0.86 D				
Pituitary (mg/kg)							
New Zealand	—	—	0.22 W	—	—	1967	121

Table 3 (continued)
SELENIUM CONTENT OF SHEEP TISSUES[a]

Geographical location	Mean selenium concentration[b] Diet (mg/kg) Rough.[c]	Conc.[c]	Tissue	Weight of animal[d](kg)	Duration of feeding[d] (d)	Date[e]	Ref.
			Spleen (mg/kg)				
Colorado, U.S.	—	—	0.4 W	—	—	1967	145
New Zealand	—	—	0.06 W	—	—	1967	121
Mean/median	—	—	0.23 W				
			Spleen (mg/kg)				
Canada	0.020	—	0.30 D	—	90	1968	123
Canada	0.053	0.40	1.16 D	—	90	1968	123
New Zealand	—	0.110	1.12 D	—	—	1961	146
New Zealand	—	—	0.48 D	—	—	1961	146
Minimum	—	—	0.30 D				
Maximum	—	0.40	1.16 D				
Mean	0.037	0.26	0.77 D				
Median	0.037	0.26	0.80 D				
			Testis (mg/kg)				
Wisconsin, U.S.	—	0.010	0.125 W	—	56	1976	122
			Thyroid (mg/kg)				
New Zealand	—	—	0.04 W	—	—	1967	121
			Uterus (mg/kg)				
Canada	—	0.005	0.050 W	—	240	1971	147
			Uterus (mg/kg)				
Australia	—	—	0.050 D	33	—	1982	141
			Wool (mg/kg)				
Wyoming, U.S.	—	—	1.45 D	—	—	1955	134
Alberta, Canada	—	—	3.63 D	61	720	1961	135
Canada	0.020	—	0.12 D	—	90	1968	123
Canada	0.053	0.40	0.23 D	—	90	1968	123
Australia	—	0.020	0.485 D	37	—	1977	125
Australia	—	0.020	0.48 D	37	—	1980	126
Scotland	0.028	0.117	0.063 D	—	—	1983	144
New Zealand	0.030	—	0.070 D	—	365	1976	142
New Zealand	—	0.11	0.26 D	—	—	1961	146
Minimum	0.020	0.020	0.063 D				
Maximum	0.053	0.40	3.63 D				
Mean	0.033	0.133	0.75 D				
Median	0.029	0.11	0.26 D				

Table 3 (continued)
SELENIUM CONTENT OF SHEEP TISSUES[a]

[a] Total selenium concentrations.

[b] Concentrations are presented as means in diet fed and tissues sampled as reported in the literature or calculated by the author of this chapter. Basis of tissue concentration data are as reported in the literature or as interpreted by this author from information provided. D — dry weight, W — wet weight; diet concentration data are on a fresh (or fed) or dry basis. Concentration units are as reported in the literature or as interpreted by this author from the information provided, with all units reported as parts per million, denoted mg/kg or mg/l as deemed appropriate; concentration units for diets are mg/kg; concentration units for tissues are mg/kg for solid tissues and mg/l for fluids, as indicated adjacent to the tissue listing; units for erythrocytes are mg/l of wet packed cells. the number of significant figures is as reported, as rounded off by this author or expressed to three decimal places, and is not necessarily related to the precision or accuracy of the data. Both means and medians of individual (mean) values for each tissue are reported; the latter reduces effects of extreme values and presents an indication of the typical concentration.

[c] Rough. and conc. refer to roughage and concentrate fractions of the diet, respectively.

[d] Animal weights and durations of feeding are as reported in the literature or as estimated by this author from information provided.

[e] Date of report.

[f] Information not reported.

Table 4
SELENIUM CONTENT OF BEEF CATTLE TISSUES[a]

Geographical location	Mean selenium concentration[b]			Weight of animal[d] (kg)	Duration of feeding[d] (d)	Date[e]	Ref.
	Diet (mg/kg)						
	Rough.[c]	Conc.[c]	Tissue				
			Blood: Plasma (mg/l)				
Florida, U.S.	0.049	0.098	0.019 W	497	300	1980	159
Florida, U.S.	0.049	1.186	0.029 W	497	300	1980	159
Ohio, U.S.	0.048	—[f]	0.026 W	335	98	1980	160
Ohio, U.S.	0.040	—	0.039 W	—	—	1981	161
Ontario, Canada	0.030	—	0.013 W	—	1080	1985	162
New South Wales, Australia	—	—	0.043 W	—	—	1981	154
Australia	—	—	0.017 W	—	448	1984	163
Minimum	0.030	0.098	0.013 W				
Maximum	0.049	1.186	0.043 W				
Mean	0.043	0.642	0.027 W				
Median	0.048	0.642	0.026 W				
			Blood: Serum (mg/l)				
Florida, U.S.	0.04	—	0.006 W	—	—	1983	164
Indiana, U.S.	0.069	0.152	0.027 W	157	195	1978	165
Indiana, U.S.	0.069	0.152	0.033 W	461	300	1978	165
Indiana, U.S.	0.08	—	0.044 W	442	113	1976	166
Indiana, U.S.	0.039	0.090	0.030 W	—	—	1976	167
North Carolina, U.S.	—	—	0.010 W	231	150	1980	168
Virginia, U.S.	0.064	0.072	0.037 W	540	223	1980	159
Virginia, U.S.	0.064	1.082	0.038 W	499	223	1980	159
Minimum	0.039	0.072	0.006 W				
Maximum	0.08	1.082	0.044 W				
Mean	0.061	0.310	0.028 W				
Median	0.064	0.152	0.032 W				

Table 4 (continued)
SELENIUM CONTENT OF BEEF CATTLE TISSUES[a]

Geographical location	Mean selenium concentration[b] Diet (mg/kg) Rough.[c]	Conc.[c]	Tissue	Weight of animal[d] (kg)	Duration of feeding[d] (d)	Date[e]	Ref.
Blood: Whole (mg/l)							
Indiana, U.S.	0.08	—	0.064 W	442	113	1976	166
Ohio, U.S.	0.040	—	0.047 W	—	—	1981	161
South Dakota, U.S.	7.6	—	1.7 W	—	360	1944	169
South Dakota, U.S.	7.6	—	2.1 W	—	720	1944	169
Ontario, Canada	0.014	—	0.017 W	—	240	1974	136
New South Wales, Australia	—	—	0.107 W	—	—	1981	154
Australia	—	—	0.030 W	—	—	1984	163
Scotland	0.050	—	0.029 W	—	—	1979	170
Minimum	0.014	—	0.017 W				
Maximum	7.6	—	2.1 W				
Mean	2.56	—	0.51 W				
Median	0.065	—	0.056 W				
Brain (mg/kg)							
Colorado, U.S.	0.11	—	0.116 W	246	270	1960	171
Colorado, U.S.	—	1.70	0.215 W	—	—	1981	172
Mean/median	0.11	1.70	0.166 W				
Epididymus: Caput, Cauda, Corpus (mg/kg)							
North Carolina, U.S.	—	—	0.39 W	231	150	1980	168
North Carolina, U.S.	—	—	0.71 W	231	150	1980	168
North Carolina, U.S.	—	—	0.31 W	231	150	1980	168
Minimum	—	—	0.31 W				
Maximum	—	—	0.71 W				
Mean	—	—	0.47 W				
Median	—	—	0.39 W				
Fat: Omental (mg/kg)							
Virginia, U.S.	0.064	0.072	0.011 W	139	210	1980	159
Virginia, U.S.	0.064	1.082	0.010 W	169	210	1980	159
Mean/median	0.064	0.577	0.011 W				
Fat: Omental (mg/kg)							
Virginia, U.S.	0.064	0.072	0.012 D	139	210	1980	159
Virginia, U.S.	0.064	1.082	0.011 D	169	210	1980	159
Mean/median	0.064	0.577	0.012 D				
Hair (mg/kg)							
Colorado, U.S.	0.11	—	1.196 D	246	270	1960	171
Indiana, U.S.	0.08	—	0.30 D	442	113	1976	166
Indiana, U.S.	0.039	0.090	0.378 D	—	—	1976	167
Ontario, Canada	—	—	0.18 D	—	—	1965	173
Minimum	0.039	—	0.18 D				
Maximum	0.11	—	1.196 D				
Mean	0.076	0.090	0.51 D				
Median	0.08	0.090	0.34 D				

Table 4 (continued) SELENIUM CONTENT OF BEEF CATTLE TISSUES[a]

Geographical location	Mean selenium concentration[b] Diet (mg/kg) Rough.[c]	Conc.[c]	Tissue	Weight of animal[d] (kg)	Duration of feeding[d] (d)	Date[e]	Ref.
			Horn (mg/kg)				
Colorado, U.S.	0.11	—	0.947 W	246	270	1960	171
			Kidney (mg/kg)				
Colorado, U.S.	0.11	—	1.17 W	246	270	1960	171
Colorado, U.S.	—	1.70	1.45 W	—	—	1981	172
Florida, U.S.	0.049	0.098	1.013 W	211	240	1980	159
Florida, U.S.	0.049	1.186	1.344 W	208	240	1980	159
Indiana, U.S.	—	—	0.849 W	157	195	1978	165
North Carolina, U.S.	—	—	0.84 W	231	150	1980	168
South Dakota, U.S.	7.6	—	2.2 W	—	360	1944	169
South Dakota, U.S.	7.6	—	3.95 W	—	720	1944	169
Virginia, U.S.	0.064	0.072	0.88 W	139	210	1980	159
Virginia, U.S.	0.064	1.082	0.95 W	169	210	1980	159
Ontario, Canada	0.030	—	0.628 W	—	330	1985	162
Minimum	0.030	0.072	0.628 W				
Maximum	7.6	1.70	3.95 W				
Mean	1.95	0.83	1.39 W				
Median	0.064	1.082	1.013 W				
			Kidney (mg/kg)				
Florida, U.S.	0.049	0.098	4.56 D	211	240	1980	159
Florida, U.S.	0.049	1.186	6.00 D	208	240	1980	159
Virginia, U.S.	0.064	0.072	4.46 D	139	210	1980	159
Virginia, U.S.	0.064	1.082	4.67 D	169	210	1980	159
Ontario, Canada	0.014	—	2.16 D	—	120	1974	136
Minimum	0.014	0.072	2.16 D				
Maximum	0.064	1.186	6.00 D				
Mean	0.048	0.610	4.37 D				
Median	0.049	0.590	4.56 D				
			Liver (mg/kg)				
Colorado, U.S.	0.11	—	1.15 W	246	270	1960	171
Colorado, U.S.	—	1.70	0.735 W	—	—	1981	172
Florida, U.S.	0.049	0.098	0.093 W	497	300	1980	159
Florida, U.S.	0.049	1.186	0.141 W	497	300	1980	159
Florida, U.S.	0.04	—	0.160 W	—	—	1983	164
Indiana, U.S.	—	—	0.116 W	157	195	1978	165
North Carolina, U.S.	—	—	0.1 W	231	150	1980	168
Ohio, U.S.	0.043	—	0.64 W	479	112	1980	160
Ohio, U.S.	0.121	—	1.15 W	482	112	1980	160
South Dakota, U.S.	7.6	—	1.2 W	—	360	1944	169
South Dakota, U.S.	7.6	—	5.6 W	—	720	1944	169
Virginia, U.S.	0.064	0.072	0.116 W	139	210	1980	159
Virginia, U.S.	0.064	1.082	0.171 W	169	210	1980	159
Ontario, Canada	0.030	—	0.047 W	—	330	1985	162
Minimum	0.030	0.072	0.047 W				
Maximum	7.6	1.70	5.6 W				
Mean	1.43	0.83	0.82 W				
Median	0.064	1.082	0.17 W				

Table 4 (continued)
SELENIUM CONTENT OF BEEF CATTLE TISSUES[a]

Geographical location	Mean selenium concentration[b] Diet (mg/kg) Rough.[c]	Conc.[c]	Tissue	Weight of animal[d] (kg)	Duration of feeding[d] (d)	Date[e]	Ref.
			Liver (mg/kg)				
Florida, U.S.	0.049	0.098	0.31 D	211	240	1980	159
Florida, U.S.	0.049	1.186	0.483 D	208	240	1980	159
Virginia, U.S.	0.064	0.072	0.423 D	139	210	1980	159
Virginia, U.S.	0.064	1.082	0.609 D	169	210	1980	159
Ontario, Canada	0.014	—	0.123 D	—	120	1974	136
New South Wales, Australia	—	—	0.575 D	—	—	1981	154
Costa Rica	—	—	0.83 D	—	—	1978	174
Minimum	0.014	0.072	0.123 D				
Maximum	0.064	1.186	0.83 D				
Mean	0.048	0.610	0.479 D				
Median	0.049	0.590	0.483 D				
			Lung (mg/kg)				
Ontario, Canada	0.030	—	0.026 W	—	330	1985	162
			Milk (mg/l)				
Florida, U.S.	0.049	0.098	0.008 W	497	300	1980	159
Florida, U.S.	0.049	1.186	0.013 W	497	300	1980	159
Indiana, U.S.	0.069	0.152	0.013 W	—	—	1977	175
Ohio, U.S.	0.040	—	0.005 W	—	—	1981	161
Virginia, U.S.	0.064	0.072	0.011 W	540	223	1980	159
Virginia, U.S.	0.064	1.082	0.013 W	499	223	1980	159
Ontario, Canada	0.014	—	0.004 W	—	240	1974	136
Ontario, Canada	0.030	—	0.007 W	—	330	1985	162
Minimum	0.014	0.072	0.004 W				
Maximum	0.069	1.186	0.023 W				
Mean	0.044	0.518	0.011 W				
Median	0.049	0.152	0.010 W				
			Milk: Colostrum (mg/l)				
Ontario, Canada	0.014	—	0.012 W	—	240	1974	136
			Muscle: Heart (mg/kg)				
Ontario, Canada	0.030	—	0.048 W	—	330	1985	162
			Muscle: Skeletal (mg/kg)				
Colorado, U.S.	0.11	—	0.09 W	246	270	1960	171
Colorado, U.S.	—	1.70	0.265 W	—	—	1981	172
Florida, U.S.	0.049	0.098	0.032 W	211	240	1980	159
Florida, U.S.	0.049	1.186	0.076 W	208	240	1980	159
Indiana, U.S.	—	—	0.031 W	157	195	1978	165
South Dakota, U.S.	7.6	—	0.9 W	—	360	1944	169
South Dakota, U.S.	7.6	—	3.0 W	—	720	1944	169
Virginia, U.S.	0.064	0.072	0.035 W	139	210	1980	159
Virginia, U.S.	0.064	1.082	0.054 W	169	210	1980	159

Table 4 (continued)
SELENIUM CONTENT OF BEEF CATTLE TISSUES[a]

Geographical location	Mean selenium concentration[b] Diet (mg/kg) Rough.[c]	Conc.[c]	Tissue	Weight of animal[d] (kg)	Duration of feeding[d] (d)	Date[e]	Ref.
Ontario, Canada	0.030	—	0.020 W	—	330	1985	162
Minimum	0.030	0.072	0.020 W				
Maximum	7.6	1.70	3.0 W				
Mean	1.95	0.828	0.450 W				
Median	0.064	1.082	0.065 W				
Muscle: Skeletal (mg/kg)							
Florida, U.S.	0.049	0.098	0.129 D	211	240	1980	159
Florida, U.S.	0.049	1.186	0.29 D	208	240	1980	159
Virginia, U.S.	0.064	0.072	0.156 D	139	210	1980	159
Virginia, U.S.	0.064	1.082	0.23 D	169	210	1980	159
Ontario, Canada	0.014	—	0.050 D	—	120	1974	136
Minimum	0.014	0.072	0.050 D				
Maximum	0.064	1.186	0.29 D				
Mean	0.048	0.610	0.171 D				
Median	0.049	0.590	0.156 D				
Ovary (mg/kg)							
Ontario, Canada	0.030	—	0.047 W	—	330	1985	162
Semen: Plasma (mg/l)[g]							
North Carolina, U.S.	—	—	0.04 W	231	150	1980	168
Semen: Sperm (mg/l)[h]							
North Carolina, U.S.	—	—	0.03 W	231	150	1980	168
Semen: Whole (mg/l)[i]							
North Carolina, U.S.	—	—	0.07 W	231	150	1980	168
Seminal Vesicle (mg/kg)							
North Carolina, U.S.	—	—	0.1 W	231	150	1980	168
Spleen (mg/kg)							
Colorado, U.S.	0.11	—	0.39 W	246	270	1960	171
Colorado, U.S.	—	1.70	0.375 W	—	—	1981	172
South Dakota, U.S.	7.6	—	0.6 W	—	360	1944	169
South Dakota, U.S.	7.6	—	2.5 W	—	720	1944	169
Ontario, Canada	0.030	—	0.058 W	—	330	1985	162
Minimum	0.030	—	0.058 W				
Maximum	7.6	—	2.5 W				
Mean	3.84	1.70	0.78 W				
Median	3.86	—	0.39 W				
Teeth (mg/kg)							
Oregon, U.S.	—	—	0.043 D	—	—	1979	176

Table 4 (continued)
SELENIUM CONTENT OF BEEF CATTLE TISSUES[a]

	Mean selenium concentration[b]						
	Diet (mg/kg)			Weight of animal[d]	Duration of		
Geographical location	Rough.[c]	Conc.[c]	Tissue	(kg)	feeding[d] (d)	Date[e]	Ref.
			Testis (mg/kg)				
North Carolina, U.S.	—	—	0.35 W	231	150	1980	168

[a] Total selenium concentrations.

[b] Concentrations are presented as means in diet fed and tissues sampled as reported in the literature or calculated by the author of this chapter. Basis of tissue concentration data are as reported in the literature or as interpreted by this author from information provided. D—dry weight, W—wet weight; diet concentration data are on a fresh (or fed) or dry basis. Concentration units are as reported in the literature or as interpreted by this author from the information provided, with all units reported as parts per million, denoted mg/kg or mg/l as deemed appropriate; concentration units for diets are mg/kg: concentration units for tissues are mg/kg for solid tissues and mg/l for fluids, as indicated adjacent to the tissue listing; units for erythrocytes are mg/l of wet packed cells, the number of significant figures is as reported, as rounded off by this author or expressed to three decimal places, and is not necessarily related to the precision or accuracy of the data. Both means and medians of individual (mean) values for each tissue reported; the latter educes effects of extreme values and presents and indication of the typical concentration.

[c] Rough. and conc. refer to roughage and concentrate fractions of the diet, respectively.

[d] Animal weights and durations of feeding are as reported in the literature or as estimated by this author from information provided.

[e] Date of report.

[f] Information not reported.

[g] Supernatant fraction from suspension of semen sample in extender solution (5% semen/95% extender, v/v); data not adjusted for the addition of extender.

[h] Measurement on sperm pellet; data represent a 1.0-ml dionized water suspension adjusted to a concentration of 40 $\times 10^6$ sperm cells.

[i] Measurement on suspension of semen samples in extender solution (5% semen/95% extender, v/v); data not adjusted for the addition of extender.

E. Dairy Cattle and Horses

The selenium contents of tissues from dairy cattle and horses are summarized in Tables 5 and 6, respectively. Distributions of selenium in tissues in these species are similar to the distributions described for the other species.

IV. CONCLUSIONS

The following generalizations can be made from the distribution of selenium in the tissues of animals. All tissues respond to variation in the dietary intake of selenium and require a period of time to stabilize after dietary intake is changed. Erythrocytes contain higher selenium concentrations than plasma or serum, and whole blood is intermediate in selenium content. Kidney contains higher selenium levels than liver, except for poultry where the two tissues contain similar selenium concentrations. Reproductive tissues and glandular tissues tend to contain levels of selenium that are similar to liver. Other soft tissues generally contain less selenium than liver. Muscle tissues generally contain the lowest levels of the element, and various types of muscle contain similar concentrations of selenium. Milk contains a low level of selenium, and colostrum contains more selenium than mature milk. Hair, feathers, and wool generally contain more selenium than is present in the diet and are more variable in response to changes in dietary selenium intake than other tissues.

Table 5
SELENIUM CONTENT OF DAIRY CATTLE TISSUES[a]

Geographical location	Mean selenium concentration[b]			Duration of feeding[d] (d)	Date[e]	Ref.
	Diet (mg/kg)					
	Rough.[c]	Conc.[c]	Tissue			
			Adrenal (mg/kg)			
Ohio, U.S.	—[f]	—	0.267 W	—	1984	177
			Blood: Plasma (mg/l)			
Ohio, U.S.	—	—	0.059 W	—	1984	177
Ohio, U.S.	—	0.047	0.031 W	—	1979	178
Ohio, U.S.	—	0.204	0.072 W	—	1979	178
British Columbia, Canada	0.047	0.368	0.132 W	—	1980	179
Canada	—	—	0.036 W	—	1982	180
Minimum	—	0.047	0.031 W			
Maximum	—	0.368	0.132 W			
Mean	0.047	0.206	0.066 W			
Median	0.047	0.204	0.059 W			
			Blood: Serum (mg/l)			
Nebraska, U.S.	—	—	0.092 W	—	1980	181
			Blood: Whole (mg/l)			
New York, U.S.	0.020	0.040	0.093 W	135	1980	182
Ohio, U.S.	—	—	0.135 W	—	1984	177
Pennsylvania, U.S.	—	0.057	0.043 W	—	1979	183
Pennsylvania, U.S.	—	0.074	0.089 W	—	1979	183
Pennsylvania, U.S.	—	0.132	0.116 W	—	1979	183
Pennsylvania, U.S.	—	0.333	0.280 W	—	1979	183
England	—	—	0.033 W	—	1978	184
Denmark	—	—	0.08 W	—	1970	185
Minimum	—	0.040	0.033 W			
Maximum	—	0.333	0.280 W			
Mean	0.020	0.127	0.109 W			
Median	0.020	0.074	0.091 W			
			Blood: Whole (mg/kg)			
Denmark	—	—	0.4 D	—	1970	185
			Follicular Fluid (mg/l)			
Ohio, U.S.	—	—	0.052 W	—	1984	177
			Kidney (mg/kg)			
Denmark	—	—	1.08 W	—	1970	185
			Kidney (mg/kg)			
Denmark	—	—	4.9 D	—	1970	185
Norway	—	—	4.65 D	—	1975	186
Mean/median	—	—	4.8 D			

Table 5 (continued) SELENIUM CONTENT OF DAIRY CATTLE TISSUES[a]

Geographical location	Mean selenium concentration[b] — Diet (mg/kg) Rough.[c]	Conc.[c]	Tissue	Duration of feeding[d] (d)	Date[e]	Ref.
Kidney: Cortex (mg/kg)						
Norway	—	—	5.74 D	—	1975	186
Kidney: Medulla (mg/kg)						
Norway	—	—	2.36 D	—	1975	186
Liver (mg/kg)						
Ohio, U.S.	—	—	0.238 W	—	1984	177
Denmark	—	—	0.24 W	—	1970	185
Mean/median	—	—	0.24 W			
Liver (mg/kg)						
Denmark	—	—	0.7 D	—	1970	185
Norway	—	—	0.44 D	—	1975	186
Mean/median	—	—	0.57 D			
Luteal Tissue (mg/kg)						
Ohio, U.S.	—	—	0.269 W	—	1984	177
Milk (mg/l)						
Ohio, U.S.	—	0.047	0.009 W	—	1979	178
Ohio, U.S.	—	0.204	0.023 W	—	1979	178
British Columbia, Canada	0.047	0.368	0.028 W	—	1980	179
Denmark	—	—	0.023 W	—	1970	185
Japan	—	—	0.064 W	—	1981	187
Netherlands	—	—	0.061 W	—	1979	188
Minimum	—	0.047	0.009 W			
Maximum	—	0.368	0.064 W			
Mean	0.047	0.206	0.035 W			
Median	0.047	0.204	0.026 W			
Milk (mg/kg)						
Denmark	0.14	—	0.37 D	—	1970	185
Denmark	—	—	0.2 D	—	1970	185
Mean/median	0.14	—	0.29 D			
Muscle: Heart (mg/kg)						
Denmark	—	—	0.22 W	—	1970	185
Denmark	—	—	1.0 D	—	1970	185
Muscle: Not Specified (mg/kg)						
Norway	—	—	0.23 D	—	1975	186

Table 5 (continued)
SELENIUM CONTENT OF DAIRY CATTLE TISSUES[a]

Geographical location	Mean selenium concentration[b]			Duration of feeding[d] (d)	Date[e]	Ref.
	Diet (mg/kg)					
	Rough.[c]	Conc.[c]	Tissue			
			Ovary (mg/kg)			
Ohio, U.S.	—	—	0.163 W	—	1984	177
			Uterus (mg/kg)			
Ohio, U.S.	—	—	0.12 W	—	1984	177

[a] Total selenium concentrations.

[b] Concentrations are presented as means in diet fed and tissues sampled as reported in the literature or calculated by the author of this chapter. Basis of tissue concentration data are as reported in the literature or as interpreted by this author from information provided. D—dry weight, W—wet weight; diet concentration data are on a fresh (or fed) or dry basis. Concentration units are as reported in the literature or as interpreted by this author from the information provided, with all units reported as parts per million, denoted mg/kg or mg/l as deemed appropriate; concentration units for diets are mg/kg: concentration units for tissues are mg/kg for solid tissues and mg/l for fluids, as indicated adjacent to the tissue listing; units for erythrocytes are mg/l of wet packed cells, the number of significant figures is as reported, as rounded off by this author or expressed to three decimal places, and is not necessarily related to the precision or accuracy of the data. Both means and medians of individual (mean) values for each tissue reported; the latter educes effects of extreme values and presents and indication of the typical concentration.

[c] Rough, and conc. refer to roughage and concentrate fractions of the diet, respectively.

[d] Animal weights and durations of feeding are as reported in the literature or as estimated by this author from information provided.

[e] Date of report.

[f] Information not reported.

Table 6
SELENIUM CONTENT OF HORSE TISSUES[a]

Geographical location	Mean selenium concentration[b]			Weight of animal[d] (kg)	Duration of feeding[d] (d)	Date[e]	Ref.
	Diet (mg/kg)						
	Rough[c]	Conc.[c]	Tissue				
			Blood: Plasma (mg/l)				
Kentucky, U.S.	0.043	0.077	0.065 W	535	112	1985	189
Sweden	—[f]	0.027	0.070 W	—	—	1970	190
Mean/median	0.043	0.052	0.068 W				
			Blood: Erythrocytes (mg/l)				
Sweden	—	0.027	0.077 W	—	—	1970	190
			Blood: Serum (mg/l)				
Kentucky, U.S.	—	—	0.120 W	—	—	1967	191
Ontario, Canada	0.060	0.039	0.058 W	—	—	1976	192
England	—	—	0.114 W	—	—	1979	193
Minimum	—	—	0.058 W				
Maximum	—	—	0.120 W				
Mean	0.060	0.039	0.097 W				
Median	0.060	0.039	0.114 W				

Table 6 (continued)
SELENIUM CONTENT OF HORSE TISSUES[a]

Geographical location	Mean selenium concentration[b] Diet (mg/kg) Rough[c]	Conc.[c]	Tissue	Weight of animal[d] (kg)	Duration of feeding[d] (d)	Date[e]	Ref.
			Blood: Whole (mg/l)				
Kentucky, U.S.	0.043	0.077	0.105 W	535	—	1985	189
Sweden	—	0.027	0.031 W	—	—	1970	190
Mean/median	0.043	0.052	0.068 W				
			Kidney: Cortex (mg/kg)				
Denmark	—	—	1.58 W	—	—	1984	194

[a] Total selenium concentrations.
[b] Concentrations are presented as means in diet fed and tissues sampled as reported in the literature or calculated by the author of this chapter. Basis of tissue concentration data are as reported in the literature or as interpreted by this author from information provided. D—dry weight, W—wet weight; diet concentration data are on a fresh (or fed) or dry basis. Concentration units are as reported in the literature or as interpreted by this author from the information provided, with all units reported as parts per million, denoted mg/kg or mg/l as deemed appropriate; concentration units for diets are mg/kg: concentration units for tissues are mg/kg for solid tissues and mg/l for fluids, as indicated adjacent to the tissue listing; units for erythrocytes are mg/l of wet packed cells, the number of significant figures is as reported, as rounded off by this author or expressed to three decimal places, and is not necessarily related to the precision or accuracy of the data. Both means and medians of individual (mean) values for each tissue reported; the latter educes effects of extreme values and presents and indication of the typical concentration.

REFERENCES

1. **Franke, K. W.,** A new toxicant occurring naturally in certain samples of plant foodstuffs. I. Results obtained in preliminary feeding trials, *J. Nutr.,* 8, 597, 1934.
2. **Franke, K. W. and Moxon, A. L.,** A comparison of the minimum fatal doses of selenium, tellurium, arsenic, and vanadium, *J. Pharmacol. Exp. Ther.,* 58, 454, 1936.
3. **Franke, K. W. and Painter, E. P.,** Selenium in proteins from toxic foodstuffs. I. Remarks on the occurrence and nature of the selenium present in a number of foodstuffs or their derived products, *Cereal Chem.,* 13, 67, 1936.
4. **Schwarz, K. and Foltz, C. M.,** Selenium as an integral part of factor 3 against dietary necrotic liver degeneration, *J. Am. Chem. Soc.,* 79, 3292, 1957.
5. **Patterson, E. L., Milstrey, R., and Stokstad, E. L. R.,** Effect of selenium in preventing exudative diathesis in chicks, *Proc. Soc. Exp. Biol. Med.,* 95, 617, 1957.
6. **Muth, O. H., Olfield, J. E., Remmert, L. F., and Westwig, P. H.,** Effects of selenium and vitamin E on white muscle disease, *Science,* 128, 1090, 1958.
7. **Pellegrini, L.,** A Study of Vitamin E Deficiency in Pigs Fed a Torula Yeast Diet, Ph.D. thesis, University of Minnesota, Minneapolis, 1958.
8. **Thompson, J. N. and Scott, M. L.,** Role of selenium in the nutrition of the chick, *J. Nutr.,* 97, 335, 1969.
9. **McCoy, K. E. M. and Weswig, P. H.,** Selenium responses in the rat not related to vitamin E, *J. Nutr.,* 98, 383, 1969.

10. **Rosenfeld, I. and Beath, O. A.**, *Selenium, Geobotany, Biochemistry, Toxicity and Nutrition*, Academic Press, New York, 1964.
11. National Research Council (U.S.), *Selenium in Nutrition*, 2nd ed., National Academy of Sciences, Washington, D.C., 1983.
12. **Shrift, A.**, Aspects of selenium metabolism in higher plants, *Annu. Rev. Plant Physiol.*, 25, 648, 1969.
13. **Peterson, P. J. and Butler, G. W.**, The uptake and assimilation of selenite by higher plants, *Aust. J. Biol. Sci.*, 15, 126, 1962.
14. **Olson, O. E., Novacek, E. J., Whitehead, E. T., and Palmer, I. S.**, Investigations on selenium in wheat, *Phytochemistry*, 9, 1181, 1970.
15. **Ardueser, F., Siegfried, W., and Scharrer, E.**, Active absorption of selenate by rat ileum, *J. Nutr.*, 115, 1203, 1985.
16. **Wolffram, S., Ardueser, F., and Scharrer, E.**, In vivo intestinal absorption of selenate and selenite by rats, *J. Nutr.*, 115, 454, 1985.
17. **Ganther, H. E.**, Metabolism of hydrogen selenide and methylated selenides, *Adv. Nutr. Res.*, 2, 107, 1979.
18. **Hseih, H. S. and Ganther, H. E.**, Biosynthesis of dimethyl selenide from sodium selenite in rat liver and kidney cell-free systems, *Biochim. Biophys. Acta*, 497, 205, 1977.
19. **Jenkins, K. J. and Hidiroglou, M.**, Comparative metabolism of ^{75}Se-selenite, ^{75}Se-selenate and ^{75}Se-methionine in bovine erythrocytes, *Can. J. Physiol. Pharmacol.*, 50, 927, 1972.
20. **Gasiewicz, T. A. and Smith, J. C.**, The metabolism of selenite by intact rat erythrocytes in vitro, *Chem. Biol. Interact.*, 21, 299, 1978.
21. **Lee, M., Dong, A., and Yano, J.**, Metabolism of ^{75}Se-selenite by human whole blood in vitro, *Can. J. Biochem.*, 47, 791, 1969.
22. **Sandholm, M.**, Function of erythrocytes in attaching selenite-Se on to specific plasma proteins, *Acta Pharmacol. Toxicol.*, 36, 321, 1975.
23. **Rotruck, J. T., Pope, A. L., Ganther, H. E., and Hoekstra, W. G.**, Prevention of oxidative damage to rat erythrocytes by dietary selenium, *J. Nutr.*, 102, 689, 1972.
24. **Rotruck, J. T., Pope, A. L., Ganther, H. E., Swanson, E., Hafeman, A. B., and Hoekstra, W. G.**, Selenium: biochemical component of glutathione peroxidase, *Science*, 179, 588, 1973.
25. **Forstrom, J. W., Zakowski, J. J., and Tappel, A. L.**, Identification of the catalytic site of rat liver glutathione peroxidase as selenocysteine, *Biochemistry*, 17, 2639, 1978.
26. **Sunde, R. A. and Hoekstra, W. G.**, Incorporation of selenium from selenite and selenocysteine into glutathione peroxidase in the isolated perfused rat liver, *Biochem. Biophys. Res. Commun.*, 93, 1181, 1980.
27. **Hawkes, W. C. and Tappel, A. L.**, In vitro synthesis of glutathione peroxidase from selenite. Translational incorporation of selenocysteine, *Biochim. Biophys. Acta*, 739, 225, 1983.
28. **Hoffman, J. L., McConnell, K. P., and Carpenter, D. R.**, Aminoacylation of *Escherichia coli* methionine tRNA by selenomethionine, *Biochim. Biophys. Acta*, 199, 531, 1970.
29. **McConnell, K. P. and Hoffman, J. L.**, Methionine-selenomethionine parallels in rat liver polypeptide chain synthesis, *FEBS Lett.*, 24, 60, 1972.
30. **Huber, R. E. and Criddle, R. S.**, Comparison of the chemical properties of selenocysteine and selenocystine with their sulfur analogs, *Arch. Biochem. Biophys.*, 122, 164, 1967.
31. **Esaki, N., Nakamura, T., Tanaka, H., Suzuki, T., Morino, Y., and Soda, K.**, Enzymatic synthesis of selenocysteine in rat liver, *Biochemistry*, 20, 4492, 1981.
32. **Esaki, N., Nakamura, T., Tanaka, H., and Soda, K.**, Selenocysteine lyase, a novel enzyme that specifically acts on selenocysteine, *J. Biol. Chem.*, 257, 4386, 1982.
33. **Ganther, H. E.**, Enzymatic synthesis of dimethyl selenide from sodium selenite in mouse liver extracts, *Biochemistry*, 5, 1089, 1966.
34. **Byard, J. L.**, Trimethyl selenide. A urinary metabolite of selenite, *Arch. Biochem. Biophys.*, 130, 556, 1969.
35. **Palmer, I. S., Gunsalus, R P., Halverson, A. W., and Olson, O. E.**, Trimethyl selenium ion as a general excretory product from selenium metabolism in the rat, *Biochim. Biophys. Acta*, 208, 260, 1970.
36. **McConnell, K. P. and Portman, O. W.**, Excretion of dimethyl selenide by the rat, *J. Biol. Chem.*, 195, 277, 1952.
37. **Black, R. S., Tripp, M. J., Whanger, P. D., and Weswig, P. H.**, Selenium proteins in ovine tissues. III. Distribution of selenium and glutathione peroxidase in tissue cytosols, *Bioinorg. Chem.*, 8, 161, 1978.
38. **Pederson, N. D., Whanger, P. D., Weswig, P. H., and Muth, O. H.**, Selenium binding proteins in tissues of normal and selenium-responsive myopathic lambs, *Bioinorg. Chem.*, 2, 33, 1972.
39. **Whanger, P. D., Pederson, N. D., and Weswig, P. H.**, Selenium proteins in ovine tissues. II. Spectral properties of a 10,000 molecular weight selenium protein, *Biochem. Biophys. Res. Commun.*, 53, 1031, 1973.
40. **Calvin, A. I.**, Selective incorporation of selenium-75 into a polypeptide of the rat sperm tail, *J. Exp. Zool.*, 204, 445, 1978.

41. **McConnell, K. P., Burton, R. M., Kute, T., and Higgins, P. J.,** Selenoproteins from rat testis cytosol, *Biochem. Biophys. Acta,* 588, 113, 1979.
42. **Pallini, V. and Bacci, E.,** Bull sperm-selenium is bound to a structural protein of mitochondria, *J. Submicrosc. Cytol.,* 11, 165, 1979.
43. **Wu, A. S. H., Oldfield, J. E., Shull, L. R., and Cheeke, P. R.,** Specific effect of selenium deficiency on rat sperm, *Biol. Reprod.,* 20, 793, 1979.
44. **Tappel, A. L., Hawkes, W. C., Wilhelmsen, E. C., and Motsenbocker, M. A.,** Selenocysteine-containing proteins and glutathione peroxidase, *Methods Enzymol.,* 107, 602, 1984.
45. **Stowe, H. D. and Miller, E. R.,** Genetic predisposition of pigs to hypo- and hyperselenemia, *J. Anim. Sci.,* 60, 200, 1985.
46. **Groce, A. W., Miller, E. R., Hitchcock, J. P., Ullrey, D. E., and Magee, W. T.,** Selenium balance in the pig as affected by selenium source and vitamin E, *J. Anim. Sci.,* 37, 942, 1973.
47. **Segerson, E. C., Getz, W. R., and Johnson, B. H.,** Selenium and reproductive function in boars fed a low selenium diet, *J. Anim. Sci.,* 53, 1360, 1981.
48. **Loudenslager, M. J., Ku, P. K., Whetter, P. A., Ullrey, D. E., Whitehair, C. K., Stowe, H. D., and Miller, E. R.,** Importance of diet of dam and colostrum to the biological antioxidant status and parenteral iron tolerance of the pig, *J. Anim. Sci.,* 63, 1905, 1986.
49. **Wilkinson, J. E., Bell, M. C., Bacon, J. A., and Melton, C. C.,** Effects of supplemental selenium on swine. II. Growing-finishing, *J. Anim. Sci.,* 44, 229, 1977.
50. **Adkins, R. S. and Ewan, R. C.,** Effect of selenium on performance, serum selenium concentration and glutathione peroxidase activity in pigs, *J. Anim. Sci.,* 58, 346, 1984.
51. **Hitchcock, J. P., Miller, E. R., Keahey, K. K., and Ullrey, D. E.,** Effects of arsanilic acid and vitamin E upon utilization of natural or supplemental selenium by swine, *J. Anim. Sci.,* 46, 425, 1978.
52. **Peplowski, M. A., Mahan, D. C., Murray, F. A., Moxon, A. L., Cantor, A. H., and Ekstrom, K. E.,** Effect of dietary and injectable vitamin E and selenium in weanling swine antigenically challenged with sheep red blood cells, *J. Anim. Sci.,* 51, 344, 1980.
53. **Piatkowski, T. L., Mahan, D. C., Cantor, A. H., Moxon, A. L., Cline, J. H., and Grifo, A. P., Jr.,** Selenium and vitamin E in semi-purified diets for gravid and nongravid gilts, *J. Anim. Sci.,* 48, 1357, 1979.
54. **Diehl, J. S., Mahan, D. C., and Moxon, A. L.,** Effects of single intramuscular injections of selenium at various levels to young swine, *J. Anim. Sci.,* 40, 844, 1975.
55. **Mahan, D. C.,** Effect of inorganic selenium supplementation on selenium retention in postweaning swine, *J. Anim. Sci.,* 61, 173, 1985.
56. **Mahan, D. C., Moxon, A. L., and Hubbard, M.,** Efficacy of inorganic selenium supplementation to sow diets on resulting carry-over to their progeny, *J. Anim. Sci.,* 45, 738, 1977.
57. **Mahan, D. C. and Moxon, A. L.,** Effects of adding inorganic or organic selenium sources to the diets of young swine, *J. Anim. Sci.,* 47, 456, 1978.
58. **Mahan, D. C. and Moxon, A. L.,** Effect of increasing the level of inorganic selenium supplementation in the post-weaning diets of swine, *J. Anim. Sci.,* 46, 384, 1978.
59. **Fontaine, M., Valli, V. E. O., Young, L. G., and Lumsden, J. H.,** Studies on vitamin E and selenium deficiency in young pigs. I. Hematological and biochemical changes, *Can. J. Comp. Med.,* 41, 41, 1977.
60. **Young, L. G., Miller, R. B., Edmeades, D. M., Lun, A., Smith, G. C., and King, G. J.,** Influence of method of corn storage and vitamin E and selenium supplementation on pig survival and reproduction, *J. Anim. Sci.,* 47, 639, 1978.
61. **Young, L. G., Miller, R. B., Edmeades, D. E., Lun, A., Smith, G. C., and King, G. J.,** Selenium and vitamin E supplementation of high moisture corn diets for swine reproduction, *J. Anim. Sci.,* 45, 1051, 1977.
62. **Mandisodza, K. T., Pond, W. G., Lisk, D. J., Hogue, D. E., Krook, L., Cary, E. E., and Gutenmann, W. H.,** Tissue retention of Se in growing pigs fed fly ash or white sweet clover grown on fly ash, *J. Anim. Sci.,* 49, 535, 1979.
63. **Mandisodza, K. T., Pond, W. G., Lisk, D. J., Gutenmann, W. H., and Hogue, D. E.,** Selenium retention in tissues of swine fed carcasses of pigs grown on diets containing sodium selenite or high selenium white sweet clover grown on fly ash, *Cornell Vet.,* 70, 193, 1980.
64. **Goehring, T. B., Palmer, I. S., Olson, O. E., Libal, G. W., and Wahlstrom, R. C.,** Effects of seleniferous grains and inorganic selenium on tissue and blood composition and growth performance of rats and swine, *J. Anim. Sci.,* 59, 725, 1984.
65. **Goehring, T. B., Palmer, I. S., Olson, O. E., Libal, G. W., and Wahlstrom, R. C.,** Toxic effects of selenium on growing swine fed corn-soybean meal diets, *J. Anim. Sci.,* 59, 733, 1984.
66. **Jenkins, K. J. and Winter, K. A.,** Effects of selenium supplementation of naturally high selenium swine rations on tissue levels of the element, *Can. J. Anim. Sci.,* 53, 561, 1973.
67. **Chavez, E. R.,** Effect of dietary selenium on glutathione peroxidase activity in piglets, *Can. J. Anim. Sci.,* 59, 67, 1979.

68. **Chavez, E. R.,** Dietary selenium and cadmium interrelationships in weanling pigs, *Can. J. Anim. Sci.*, 61, 713, 1981.
69. **Bengtsson, G., Hakkarainen, J., Jonsson, L., Lannek, N., and Lindberg, P.,** A low-selenium pig diet based on casein from selenium-deficient cows, *Acta Vet. Scand.*, 15, 135, 1974.
70. **Bengtsson, G., Hakkarainen, J., Jonsson, L., Lannek, N., and Lindberg, P.,** Requirement for selenium (as selenite) and vitamin E (as alpha-tocopherol) in weaned pigs. I. The effect of varying alpha-tocopherol levels in a selenium deficient diet on the development of the VESD syndrome, *J. Anim. Sci.*, 46, 143, 1978.
71. **Bengtsson, G., Hakkarainen, J., Jonsson, L., Lannek, N., and Lindberg, P.,** Requirement for selenium (as selenite) and vitamin E (as alpha-tocopherol) in weaned pigs. II. The effect of varying selenium levels in a vitamin E deficient diet on the development of the VESD syndrome, *J. Anim. Sci.*, 46, 153, 1978.
72. **Hakkarainen, J., Lindberg, P., Bengtsson, G., Jonsson, L., and Lannek, N.,** Requirement for selenium (as selenite) and vitamin E (as alpha-tocopherol) in weaned pigs. III. The effect on the development of the VESD syndrome of varying selenium levels in a low-tocopherol diet, *J. Anim. Sci.*, 46, 1001, 1978.
73. **Doornenbal, H.,** Tissue selenium content of the growing pig, *Can. J. Anim. Sci.*, 55, 325, 1975.
74. **Nielsen, H. E. and Rasmussen, O. K.,** The influence of selenium on performance, meat production and the quality of some edible tissues in pigs, *Acta Agric. Scand. Suppl.*, 21, 246, 1979.
75. **Rasmussen, O. K.,** Selenium concentration and deposition. Performance, and carcass quality in pigs fed different levels of sodium selenite, *Acta Agric. Scand.*, 24, 115, 1974.
76. **Van Vleet, J. F.,** Retention of selenium in tissues of calves, lambs, and pigs after parenteral injection of a selenium-vitamin E preparation, *Am. J. Vet. Res.*, 36, 1335, 1975.
77. **Sharp, B. A., van Dreumel, A. A., and Young, L. G.,** Vitamin E, selenium, and methionine supplementation of dystrophogenic diets for pigs, *Can. J. Comp. Med.*, 36, 398, 1972.
78. **Van Vleet, J. F., Meyer, K. B., and Olander, H. J.,** Control of Selenium-vitamin E deficiency in growing swine by parenteral administration of selenium-vitamin E preparations to baby pigs or to pregnant sows and their baby pigs, *J. Am. Vet. Med. Assoc.*, 163, 452, 1973.
79. **McDowell, L. R., Froseth, J. A., Piper, R. C., Dyer, I. A., and Kroening, G. H.,** Tissue selenium and serum tocopherol concentrations in selenium-vitamin E deficient pigs fed peas *(Pisum sativum)*, *J. Anim. Sci.*, 45, 1326, 1977.
80. **Herigstad, R. R., Whitehair, C. K., and Olson, O. E.,** Inorganic and organic selenium toxicosis in young swine: comparison of pathologic changes with those in swine with vitamin E-selenium deficiency, *Am. J. Vet. Res.*, 34, 1227, 1973.
81. **Ewan, R. C.,** Effect of vitamin E and selenium on tissue composition of young pigs, *J. Anim. Sci.*, 32, 883, 1971.
82. **Ku, P. K., Miller, E. R., Wahlstrom, R. C., Groce, A. W., Hitchcock, J. P., and Ullrey, D. E.,** Selenium supplementation of naturally high selenium diets for swine, *J. Anim. Sci.*, 37, 501, 1973.
83. **McDowell, L. R., Froseth, J. A., and Piper, R. C.,** Influence of arsenic, sulfur, cadmium, tellurium, silver and selenium on the selenium-vitamin E deficiency in the pig, *Nutr. Rep. Int.*, 17, 19, 1978.
84. **Sharp, B. A., Young, L. G., and van Dreumel, A. A.,** Effect of supplemental vitamin E and selenium in high moisture corn diets on the incidence of mulberry heart disease and hepatosis dietetica in pigs, *Can. J. Comp. Med.*, 36, 393, 1972.
85. **Young, L. G., Lumsden, J. H., Lun, A., Claxton, J., and Edmeades, D. E.,** Influence of dietary levels of vitamin E and selenium on tissue and blood parameters in pigs, *Can. J. Comp. Med.*, 40, 92, 1976.
86. **Lindberg, P. and Siren, M.,** Fluorometric selenium determinations in the liver of normal pigs and in pigs affected with nutritional muscular dystrophy and liver dystrophy, *Acta Vet. Scand.*, 6, 59, 1965.
87. **Lindberg, P. and Lannek, N.,** Retention of selenium in kidneys, liver and striated muscle after prolonged feeding of therapeutic amounts of sodium selenite to pigs, *Acta Vet. Scand.*, 6, 217, 1965.
88. **Van Vleet, J. F., Carlton, W., and Olander, H. J.,** Hepatosis dietetica and mulberry heart disease associated with selenium deficiency in Indiana swine, *J. Am. Vet. Med. Assoc.*, 157, 1208, 1970.
89. **Simesen, M. G. and Pedersen, K. B.,** Selenium determinations in Danish swine affected with hepatosis dietetica, *Acta Vet. Scand.*, 16, 137, 1975.
90. **Teige, J.,** Swine dysentery: the influence of dietary vitamin E and selenium on the clinical and pathological effects of *Treponema hyodysenteriae* infection in pigs, *Res. Vet. Sci.*, 32, 95, 1982.
91. **Moksnes, K., Tollersrud, S., and Larsen, H. J.,** Influence of dietary sodium selenite on tissue selenium levels of growing pigs, *Acta Vet. Scand.*, 23, 361, 1982.
92. **Froeslie, A., Karlsen, J. T., and Rygge, J.,** Selenium in animal nutrition in Norway, *Acta Agric. Scand.*, 30, 17, 1980.
93. **Young, L. G., Castell, A. G., and Edmeades, D. E.,** Influence of dietary levels of selenium on tissue selenium of growing pigs in Canada, *J. Anim. Sci.*, 44, 590, 1977.
94. **Sharp, B. A., Young, L. G., and van Dreumel, A. A.,** Dietary induction of mulberry heart disease and hepatosis dietetica in pigs. I. Nutritional aspects, *Can. J. Comp. Med.*, 36, 371, 1972.

95. **Rasmussen, O. K.,** Selenium concentration in sow colostrum and in sow milk, *Acta Agric. Scand.,* 24, 175, 1974.
96. **Ku, P. K., Ely, W. T., Groce, A. W., and Ullrey, D. E.,** Natural dietary selenium, alpha-tocopherol and effect on tissue selenium, *J. Anim. Sci.,* 34, 208, 1972.
97. **Groce, A. W., Miller, E. R., Keahey, K. K., Ullrey, D. E., and Ellis, D. J.,** Selenium supplementation of practical diets for growing-finishing swine, *J. Anim. Sci.,* 32, 905, 1971.
98. **Adkins, R. S. and Ewan, R. C.,** Effect of supplemental selenium on pancreatic function and nutrient digestibility in the pig, *J. Anim. Sci.,* 58, 351, 1984.
99. **McDowell, L. R., Froseth, J. A., Kroening, G. H., and Haller, W. A.,** Effects of dietary vitamin E and oxidized cottonseed oil on SGOT, erythrocyte hemolysis, testicular fatty acids and testicular selenium in swine fed peas *(Pisum sativum), Nutr. Rep. Int.,* 9, 359, 1974.
100. **Cantor, A. H., Langevin, M. L., Noguchi, T., and Scott, M. L.,** Efficacy of selenium in selenium compounds and feeds for prevention of pancreatic fibrosis in chicks, *J. Nutr.,* 105, 106, 1975.
101. **Latshaw, J. D.,** Natural and selenite selenium in the hen and egg, *J. Nutr.,* 105, 32, 1975.
102. **Latshaw, J. D. and Osman, M.,** Selenium and vitamin E responsive condition in the laying hen, *Poult. Sci.,* 53, 1704, 1974.
103. **Cantor, A. H., Moorhead, P. D., and Musser, M. A.,** Comparative effects of sodium selenite and selenomethionine upon nutritional muscular dystrophy, selenium-dependent glutathione peroxidase, and tissue selenium concentrations of turkey poults, *Poult. Sci.,* 61, 478, 1982.
104. **Cantor, A. H., Scott, M. L., and Noguchi, T.,** Biological availability of selenium in feeds and selenium compounds for prevention of exudative diathesis in chicks, *J. Nutr.,* 105, 96, 1975.
105. **Cantor, A. H. and Scott, M. L.,** Influence of dietary selenium on tissue selenium levels in turkeys, *Poult. Sci.,* 54, 262, 1975.
106. **Scott, M. L. and Thompson, J. N.,** Selenium content of feeds and effects of dietary selenium levels upon tissue selenium in chicks and poults, *Poult. Sci.,* 50, 1742, 1971.
107. **Moxon, A. L. and Poley, W. E.,** The relation of selenium content of grains in the ration to the selenium content of poultry carcass and eggs, *Poult. Sci.,* 17, 77, 1938.
108. **Van Vleet, J. F.,** An evaluation of protection offered by various dietary supplements against experimentally induced selenium-vitamin E deficiency in ducklings, *Am. J. Vet. Res.,* 38, 1231, 1977.
109. **Kling, L. J. and Soares, J. H., Jr.,** Mercury metabolism in Japanese quail. I. The effect of dietary mercury and selenium on their tissue distribution, *Poult. Sci.,* 57, 1279, 1978.
110. **Stoewsand, G. S., Gutenmann, W. H., and Lisk, D. J.,** Wheat grown on fly ash: high selenium uptake and response when fed to Japanese quail, *J. Agric. Food Chem.,* 26, 757, 1978.
111. **Ort, J. F. and Latshaw, J. D.,** The toxic level of sodium selenite in the diet of laying chickens, *J. Nutr.,* 108, 1114, 1978.
112. **Moksnes, K.,** Selenium deposition in tissues and eggs of laying hens given surplus of selenium as selenomethionine, *Acta Vet. Scand.,* 24, 34, 1983.
113. **Moksnes, K. and Norheim, G.,** Selenium concentrations in tissues and eggs of growing and laying chickens fed sodium selenite at different levels, *Acta Vet. Scand.,* 23, 368, 1982.
114. **Latshaw, J. D. and Osman, M.,** Distribution of selenium in egg white and yolk after feeding natural and synthetic selenium compounds, *Poult. Sci.,* 54, 1244, 1975.
115. **Cantor, A. H. and Scott, M. L.,** Effect of selenium in the hen's diet on egg production, hatchability, performance of progeny and selenium concentration in eggs, *Poult. Sci.,* 53, 1870, 1974.
116. **Combs, G. F., Jr. and Scott, M. L.,** The selenium needs of laying and breeding hens, *Poult. Sci.,* 58, 871, 1979.
117. **Arnold, R. L., Olson, O. E., and Carlson, C. W.,** Tissue selenium content and serum tocopherols as influenced by dietary type, selenium and vitamin E, *Poult. Sci.,* 53, 2185, 1974.
118. **Arnold, R. L., Olson, O. E., and Carlson, C. W.,** Dietary selenium and arsenic additions and their effects on tissue and egg selenium, *Poult. Sci.,* 52, 847, 1973.
119. **Thapar, N. T., Guenthner, E., Carlson, C. W., and Olson, O. E.,** Dietary selenium and arsenic additions to diets for chickens over a life cycle, *Poult. Sci.,* 48, 1988, 1969.
120. **Pond, W. G., Allaway, W. H., Walker, E. F., Jr., and Krook, L.,** Effects of corn selenium content and drying temperature and of supplemental vitamin E on growth, liver selenium, and blood vitamin E content of chicks, *J. Anim. Sci.,* 33, 996, 1971.
121. **Hartley, W. J.,** Levels of selenium in animal tissues and methods of selenium administration, in *Selenium in Biomedicine,* Muth, O. H., Oldfield, J. E., and Weswig, P. H., Eds., AVI Publishing, Westport, CT, 1967, 79.
122. **Oh, S.-H., Sunde, R. A., Pope, A. L., and Hoekstra, W. G.,** Glutathione peroxidase response to selenium intake in lambs fed a torula yeast-based, artificial milk, *J. Anim. Sci.,* 42, 977, 1976.
123. **Hidiroglou, M., Jenkins, K. J., Carson, R. B., and MacKay, R. R.,** Some aspects of selenium metabolism in normal and dystrophic sheep, *Can. J. Anim. Sci.,* 48, 335, 1968.

124. **Ullrey, D. E., Light, M. R., Brady, P. S., Whetter, P. A., Tilton, J. E., Henneman, H. A., and Magee, W. T.,** Selenium supplements in salt for sheep, *J. Anim. Sci.,* 46, 1515, 1978.
125. **White, C. L. and Somers, M.,** Sulphur-selenium studies in sheep. I. The effects of varying dietary sulfate and selenomethionine on sulfur, nitrogen and selenium metabolism in sheep, *Aust. J. Biol. Sci.,* 30, 47, 1977.
126. **White, C. L.,** Sulfur-selenium studies in sheep. II. Effect of a dietary sulfur deficiency on selenium and sulfur metabolism in sheep fed varying levels of selenomethionine, *Aust. J. Biol. Sci.,* 33, 699, 1980.
127. **Peter, D. W., Board, P. G., and Palmer, M. J.,** Selenium supplementation of grazing sheep. I. Effects of selenium drenching and other factors on plasma and erythrocyte glutathione peroxidase activities and blood selenium concentrations of lambs and ewes, *Aust. J. Agric. Res.,* 31, 981, 1980.
128. **Hunter, R. A., Peter, D. W., Quin, M. P., and Seibert, B. D.,** Intake of selenium and other nutrients in relation to selenium status and productivity of grazing sheep, *Aust. J. Agric. Res.,* 33, 637, 1983.
129. **Kott, R. W., Ruttle, J. L., and Southward, G. M.,** Effects of vitamin E and selenium injections on reproduction and preweaning lamb survival in ewes consuming diets marginally deficient in selenium, *J. Anim. Sci.,* 57, 553, 1983.
130. **Kott, R. W., Ruttle, J. L., Stiffler, D. M., and Smith, G. S.,** Effects of vitamin E plus selenium supplementation on live and carcass characteristics of medium wool lambs, *Proc. Annu. Meet. Am. Soc. Anim. Sci. West. Sect.,* 29, 418, 1978.
131. **Whanger, P. D., Weswig, P. H., Schmitz, J. A., and Oldfield, J. E.,** Effects of various methods of selenium administration on white muscle disease, glutathione peroxidase and plasma enzyme activities in sheep, *J. Anim. Sci.,* 47, 1156, 1978.
132. **Oh, S.-H., Pope, A. L., and Hoekstra, W. G.,** Dietary selenium requirement of sheep fed a practical-type diet as assessed by tissue glutathione peroxidase and other criteria, *J. Anim. Sci.,* 42, 984, 1976.
133. **Sword, J. T., Jr., Ataja, A. M., Pope, A. L., and Hoekstra, W. G.,** Effect of calcium phosphates and zinc in salt-mineral mixtures on ad libitum salt-mix intake and on zinc and selenium status of sheep, *J. Anim. Sci.,* 59, 1594, 1984.
134. **Leonard, R. O. and Burns, R. H.,** A preliminary study of selenized wool, *J. Anim. Sci.,* 14, 446, 1955.
135. **Slen, S. B., Deviraren, A. S., and Smith, A. D.,** Note on the effects of selenium on wool growth and body gains in sheep, *Can. J. Anim. Sci.,* 41, 263, 1961.
136. **Jenkins, K. J., Hidiroglou, M., Wauthy, J. M., and Proulx, J. E.,** Prevention of nutritional muscular dystrophy in calves and lambs by selenium and vitamin E additions to the maternal mineral supplement, *Can. J. Anim. Sci.,* 54, 49, 1974.
137. **Hidiroglou, M., Hoffman, I., Jenkins, K. J., and MacKay, R. R.,** Control of nutritional muscular dystrophy in lambs by selenium implantation, *Anim. Prod.,* 13, 315, 1971.
138. **Trengove, C. L. and Judson, G. J.,** Trace element supplementation of sheep: evaluation of various copper supplements and a soluble glass bullet containing copper, cobalt and selenium, *Aust. Vet. J.,* 62, 321, 1985.
139. **Lee, H. J. and Jones, G. B.,** Interactions of selenium, cadmium and copper in sheep, *Aust. J. Agric. Res.,* 27, 447, 1976.
140. **Godwin, K. O., Kuchel, R. E., and Buckley, R. A.,** Effect of selenium on infertility in ewes grazing improved pastures, *Aust. J. Exp. Agric. Anim. Husb.,* 10, 672, 1970.
141. **Langlands, J. P., Bowles, J. E., Donald, G. E., Smith, A. J., Paull, D. R., and Davies, H. I.,** Deposition of copper, manganese, selenium and zinc in the ovine fetus and associated tissues, *Aust. J. Agric. Res.,* 33, 591, 1982.
142. **Andrews, E. D., Hogan, K. G., and Sheppard, A. D.,** Selenium in soils, pastures and animal tissues in relation to the growth of young sheep on marginally selenium-deficient area, *N.Z. Vet. J.,* 24, 111, 1976.
143. **Moksnes, K. and Norheim, G.,** Selenium and glutathione peroxidase levels in lambs receiving feed supplemented with sodium selenite or selenomethionine, *Acta Vet. Scand.,* 24, 45, 1983.
144. **Wiener, G., Woolliams, J. A., and Vagg, M. J.,** Selenium concentration in the blood and wool and glutathione peroxidase activity in the blood of three breeds of sheep, *Res. Vet. Sci.,* 34, 365, 1983.
145. **Maag, D. D. and Glenn, M. W.,** Toxicity of selenium: farm animals, in *Selenium in Biomedicine,* Muth, O. H., Oldfield, J. E., and Weswig, P. H., Eds., AVI Publishing, Westport, CT, 1967, 127.
146. **Cousins, F. B. and Cairney, I. M.,** Some aspects of selenium metabolism in sheep, *Aust. J. Agric. Res.,* 12, 927, 1961.
147. **Buchanan-Smith, J. G., Sharp, B. A., and Tillman, A. D.,** Tissue selenium concentrations in sheep fed a purified diet, *Can. J. Physiol. Pharmacol.,* 49, 619, 1971.
148. **Wise, W. R., Weswig, P. H., Muth, O. H., and Oldfield, J. E.,** Dietary interrelationship of cobalt and selenium in lambs, *J. Anim. Sci.,* 27, 1462, 1968.
149. **Allaway, W. H., Moore, D. P., Oldfield, J. E., and Muth, O. H.,** Movement of physiological levels of selenium from soils through plants to animals, *J. Nutr.,* 88, 411, 1966.

150. **Gardiner, M. R. and Nairn, M. E.,** Studies on the effect of cobalt and selenium in clover disease of ewes, *Aust. Vet. J.*, 45, 215, 1969.
151. **Sivertsen, T., Karlsen, J. T., Norheim, G., and Froeslie, A.,** Concentration of selenium in liver in relation to copper level in normal and copper-poisoned sheep, *Acta Vet. Scand.*, 19, 472, 1978.
152. **Burton, V., Keeler, R. F., Swingle, K. F., and Young, S.,** Nutritional muscular dystrophy in lambs — selenium analysis of maternal, fetal, and juvenile tissues, *Am. J. Vet. Res.*, 23, 962, 1962.
153. **Ewan, R. C., Baumann, C. A., and Pope, A. L.,** Retention of selenium by growing lambs, *Agric. Food Chem.*, 16, 216, 1968.
154. **Langlands, J. P., Wilkins, J. F., Bowles, J. E., Smith, A. J., and Webb, R. F.,** Selenium concentration in the blood of ruminants grazing in northern New South Wales. I. Analysis of samples collected in the National Brucellosis Eradication Scheme, *Aust. J. Agric. Res.*, 32, 511, 1981.
155. **Allen, J. G., Masters, H. G., and Wallace, S. R.,** The effect of lupinosis on liver copper, selenium and zinc concentrations in merino sheep, *Vet. Rec.*, 105, 434, 1979.
156. **Gabbedy, B. J., Masters, H., and Boddington, E. B.,** White muscle disease of sheep and associated tissue selenium levels in western Australia, *Aust. Vet. J.*, 53, 482, 1977.
157. **Steele, P., Peet, R. L., Skirrow, S., Hopkinson, W., and Masters, H. G.,** Low alpha-tocopherol levels in livers of weaner sheep with nutritional myopathy, *Aust. Vet. J.*, 56, 529, 1980.
158. **Gardner, R. W. and Hogue, D. E.,** Milk levels of selenium and vitamin E related to nutritional muscular dystrophy in the suckling lamb, *J. Nutr.*, 93, 418, 1967.
159. **Ammerman, C. B., Chapman, H. L., Bouwman, G. W., Fontenot, J. P., Bagley, C. P., and Moxon, A. L.,** Effect of supplemental selenium for beef cows on the performance and tissue selenium concentrations of cows and suckling calves, *J. Anim. Sci.*, 51, 1381, 1980.
160. **Byers, F. M. and Moxon, A. L.,** Protein and selenium levels for growing and finishing beef cattle, *J. Anim. Sci.*, 50, 1136, 1980.
161. **Moxon, A. L.,** Selenium deficiency in cattle, *S. Afr. J. Anim. Sci.*, 11, 183, 1981.
162. **Hidiroglou, M., Proulx, J., and Jolette, J.,** Intraruminal selenium pellet for control of nutritional muscular dystrophy in cattle, *J. Dairy Sci.*, 68, 57, 1985.
163. **Judson, G. J. and McFarlane, J. D.,** Selenium status of cattle given ruminal selenium bullets, *Aust. Vet. J.*, 61, 333, 1984.
164. **Salih, Y. M., McDowell, L. R., Hentges, J. F., Mason, R. M., Jr., and Conrad, J. H.,** Mineral status of grazing beef cattle in the warm climate region of Florida, *Trop. Anim. Health Prod.*, 15, 245, 1983.
165. **Perry, T. W., Peterson, R. C., Griffin, D. D., and Beeson, W. M.,** Relationship of blood serum selenium levels of pregnant cows to low dietary intake, and effect on tissue selenium levels of their calves, *J. Anim. Sci.*, 16, 562, 1978.
166. **Perry, T. W., Beeson, W. M., Smith, W. H., and Mohler, M. T.,** Effect of supplemental selenium on performance and deposit of selenium in blood and hair of finishing beef cattle, *J. Anim. Sci.*, 42, 192, 1976.
167. **Perry, T. W., Caldwell, D. M., and Peterson, R. C.,** Selenium content of feeds and effect of dietary selenium on hair and blood serum, *J. Dairy Sci.*, 59, 760, 1976.
168. **Segerson, E. C. and Johnson, B. H.,** Selenium and reproductive function in yearling Angus bulls, *J. Anim. Sci.*, 51, 395, 1980.
169. **Moxon, A. L., Rhian, M. A., Anderson, H. D., and Olson, O. E.,** Growth of steers on seleniferous range, *J. Anim. Sci.*, 3, 299, 1944.
170. **Arthur, J. R., Price, J., and Mills, C. F.,** Observations on the selenium status of cattle in the north-east of Scotland, *Vet. Rec.*, 104, 340, 1979.
171. **Maag, D. D., Orsborn, J. S., and Clopton, J. R.,** The effect of sodium selenite on cattle, *Am. J. Vet. Res.*, 21, 1049, 1960.
172. **Boyer, K. W., Jones, J. W., Linscott, D., Wright, S. K., Stroube, W., and Cunningham, W.,** Trace element levels in tissues from cattle fed a sewage sludge-amended diet, *J. Toxicol. Environ. Health*, 8, 281, 1981.
173. **Hidiroglou, M., Carson, R. B. and Brossard, G. A.,** Influence of Se on the Se contents of hair and on the incidence of nutritional muscular disease in beef cattle, *Can. J. Anim. Sci.*, 45, 197, 1965.
174. **McDowell, L. R., Lang, C. E., Conrad, J. H., Martin, F. G., and Fonseca, H.,** Mineral status of beef cattle in Guanacaste, Costa Rica, *Trop. Agric. (Trinidad)*, 55, 343, 1978.
175. **Perry, T. W., Peterson, R. C., and Beeson, W. M.,** Selenium in milk from feeding small supplements, *J. Dairy Sci.*, 60, 1698, 1977.
176. **Johnson, J. R. and Shearer, T. R.,** Selenium uptake into teeth determined by fluorimetry, *J. Dent. Res.*, 58, 1836, 1979.
177. **Harrison, J. H. and Conrad, H. R.,** Selenium content and glutathione peroxidase activity in tissues of the dairy cow after short-term feeding, *J. Dairy Sci.*, 67, 2464, 1984.
178. **Conrad, H. R. and Moxon, A. L.,** Transfer of dietary selenium to milk, *J. Dairy Sci.*, 62, 404, 1979.

179. **Fisher, L. J., Hoogendoorn, C., and Montemurro, J.,** The effect of added dietary selenium on the selenium content of milk, urine and feces, *Can. J. Anim. Sci.*, 60, 79, 1980.
180. **Hidiroglou, M. and Hartin, K. E.,** Vitamins A, E and selenium blood levels in the fat cow syndrome, *Can. Vet. J.*, 23, 255, 1982.
181. **Larson, L. L., Mabruck, H. S., and Lowry, S. R.,** Relationship between early postpartum blood composition and reproductive performance in dairy cattle, *J. Dairy Sci.*, 63, 283, 1980.
182. **Lein, D. H., Maylin, G. A., Braund, D. G., Gutenmann, W. H., Chase, L. E., and Lisk, D. J.,** Increasing selenium in bovine blood by feed supplements or selenium injections, *Cornell Vet.*, 70, 113, 1980.
183. **Backall, K. A. and Scholz, R. W.,** Reference values for a field test to estimate inadequate glutathione peroxidase activity and selenium status in the blood of cattle, *Am. J. Vet. Res.*, 40, 733, 1979.
184. **Cawley, G. D. and Bradley, R.,** Sudden death in calves associated with acute myocardial degeneration and selenium deficiency, *Vet. Rec.*, 103, 239, 1978.
185. **Bisbjerg, B., Jochumsen, P., and Rasbech, N. O.,** Selenium content in organs, milk and fodder of the cow, *Nord. Veterinaermed.*, 22, 532, 1970.
186. **Hellesnes, I., Underdal, B., Lunde, G., and Havre, G. N.,** Selenium and zinc concentrations in kidney, liver and muscle of cattle from different parts of Norway, *Acta Vet. Scand.*, 16, 481, 1975.
187. **Munehiro, Y., Yasumoto, K., Iwami, K., and Tashiro, H.,** Distribution of selenium in bovine milk and selenium deficiency in rats fed casein-based diets, monitored by lipid peroxide level and glutathione peroxidase activity, *Agric. Biol. Chem.*, 45, 1681, 1981.
188. **Binnerts, W. T.,** The selenium status of dairy cows in the Netherlands derived from milk and blood analysis, *Neth. Milk Dairy J.*, 33, 24, 1979.
189. **Shellow, J. S., Jackson, S. G., Baker, J. P., and Cantor, A. H.,** The influence of dietary selenium levels on blood levels of selenium and glutathione peroxidase activity in the horse, *J. Anim. Sci.*, 61, 590, 1985.
190. **Bergsten, G., Holmback, R., and Lindberg, P.,** Blood selenium in naturally fed horses and the effect of selenium administration, *Acta Vet. Scand.*, 11, 571, 1970.
191. **Stowe, H. D.,** Serum selenium and related parameters of naturally and experimentally fed horses, *J. Nutr.*, 93, 60, 1967.
192. **Wilson, T. M., Morrison, H. A., Palmer, N. C., Finley, G. G., and van Dreumel, A. A.,** Myodegeneration and suspected selenium/vitamin E deficiency in horses, *J. Am. Vet. Med. Assoc.*, 169, 213, 1976.
193. **Blackmore, D. J., Willett, K., and Agness, D.,** Selenium and gamma-glutamyl transferase activity in the serum of thoroughbreds, *Res. Vet. Sci.*, 26, 76, 1979.
194. **Teilmann, A. M. and Hansen, J. C.,** Cadmium and selenium levels in kidneys from Danish horses, *Nord. Veterinaermed.*, 36, 49, 1984.

Chapter 8

HUMAN TISSUES

Yngvar Thomassen and Jan Aaseth

TABLE OF CONTENTS

I. INTRODUCTION

The initial interest in selenium arose because of its potential toxicity,[1] but more recently there is increasing interest in its essential functions in the body. A matter of some debate has been the optimal daily intake.

In human tissues, selenium is found to play a role as an essential component of the enzyme glutathione peroxidase.[2] In experimental animals, selenium acts as an anticancer agent in chemically induced tumors, and it has been claimed that variations in dietary selenium intake may explain differences in the incidence of cancers in various human populations.[3] Furthermore, selenium seems to prevent some cardiovascular diseases and myocardial infarction. Inverse epidemiological relationships have been observed between the incidence of heart disease and environmental selenium,[3] and in selenium-poor areas of the People's Republic of China an endemic cardiomyopathy, Keshan disease, affects children.[4]

In 1980, the U.S. National Research Council suggested that a selenium intake of 50 to 200 μg/d is safe and adequate for adult humans.[5] It is clear, however, that due to variations in the amount of selenium in the soil of different geographical areas, the daily intake by a local population may be above or below the recommended limits, and this will give rise to large differences in the concentration found in tissues.

This chapter presents a selective survey of the distribution of selenium in man. Because of the vast range of selenium concentrations tabulated from the literature, we have been obliged to question the reliability of the analytical data and to evaluate it with respect to biological variations, geographic area, and variations arising from the measurement techniques. An attempt has been made to quote all relevant papers, but as the number of references is large and some of the published studies are incomplete, it has been found necessary to select data. This may seen unjust in some cases to those whose data have not been listed.

It will be evident from the data reported here that concentrations of selenium in tissues are influenced by several factors of both biological and environmental origin. Such variables include age, health, and dietary habits. In general, low concentrations are observed in young children, and lower concentrations in elderly than in middle-aged persons. Furthermore, several diseases may affect the amount of selenium in tissues in a complicated way, probably due to changes in metabolism. It should be pointed out that knowledge of this subject with respect to diseased persons is fragmentary.

Accurate analytical data are essential for any retrospective or prospective studies relating selenium status to health and disease, for establishing appropriate selenium intake and/or supplementation guidelines, and for the monitoring of environmental and occupational exposure. Several possible sources of error must be taken into account when considering the reliability of published analytical data.

II. ANALYTICAL CONSIDERATIONS

A. Sample Contamination

In sharp contrast to most elements, the sampling, storage, and preparation steps in the determination of selenium in human samples are essentially free of contamination problems. Although most of the earlier published data for trace elements in general are strongly influenced by contamination,[6,7] there is no evidence of a similar problem for selenium.

B. Sample Preparation

The biotransformation of selenium in man, which is characterized by a stepwise biochemical reduction, leading to the binding to or direct incorporation of the element into proteins, apparently involves the formation of intermediate volatile species. Dimethylselenide as well as many other organic forms of selenium and its halides are relatively volatile, which explains the large losses resulting from oven drying procedures at temperatures over 120°C. However, oven drying at temperatures below 100°C, lyophilization, and oxygen plasma ashing are reported to cause minimal or no losses of selenium.

As a rule, decomposition methods are required in the determination of selenium in human tissues, and most of those reported involve wet ashing procedures. Dry ashing is not favored because complete losses of selenium may result.[8,9] Extreme care and strictly controlled procedures are required to prevent losses of selenium.[10,11] Preferred digestion mixtures involve combination of nitric and perchloric[12,13] or nitric, sulfuric, and perchloric[14-16] acids. The essential common factor for all methods is that reducing conditions are prevented and oxidizing conditions are maintained throughout the decomposition stage.

C. Sample Homogeneity and Storage

Analytical results are significantly affected by the homogeneity of the samples and by storage procedures. In general, small aliquots of a sample taken from the bulk material are used, often after long-term storage. These can be representative only if sampling and storage procedures are systematic. This is discussed extensively in the literature, and for further information on the subject the reader is referred to the volumes by Heydorn.[17]

D. Quality Control and Reference Materials

The continuing interest in the biomedical and environmental effects of selenium with the attendant requirement for improved analytical reliability has increased the need for biological reference materials. Numerous selenium projects were set up by both the analytical and medical communities — unfortunately in numerous instances without elaborate quality control measures. Although a great many investigations have been reported on research into and application of a variety of analytical methods for the estimation of selenium in human tissues and fluids, very few biological and human materials are available with established concentrations of selenium for method verification and data quality assurance. Biological reference materials prepared in the 1960s and 1970s by individual scientists such as Bowen[18] and by organizations such as the National Bureau of Standards and the International Atomic Energy Agency are currently available. Recently, human body fluids have been introduced as certified or proposed reference materials in the determination of selenium. Discussions of the concepts and status of reference materials may be found in recent literature[19] and references therein. Detailed compilations of available reference materials have been made by Muramatsu and Parr[20] and Ihnat.[21] Currently available reference materials for selenium are listed in Table 1.

III. SELENIUM CONCENTRATIONS IN TISSUES AND FLUIDS

Table 2 contains an extensive compilation of selenium concentration information in a variety of human tissues and fluids from healthy and diseased individuals throughout the world. Some of the data given in the table are discussed briefly in the following text, focusing on the most commonly analyzed human tissues: blood components, liver, hair, and urine. Although the table contains expressions for tissue concentrations related to dry weight or wet weight, the concentrations used in the text are based on wet weight, the condition *in situ*, of the tissues subjected to analysis.

A. Blood and Blood Components

Blood consists of various cells, of which the most abundant are erythrocytes ($\sim 5 \times 10^6/\mu l$), platelets ($\sim 3 \times 10^5/\mu l$), granulocytes ($\sim 5 \times 10^3/\mu l$), and lymphocytes ($\sim 2 \times 10^3/\mu l$) suspended in an approximately equal volume of extracellular fluid. Although determinations have been done on both blood plasma and cellular components, selenium has been most frequently measured in whole blood.

Table 1
CURRENTLY AVAILABLE BIOLOGICAL REFERENCE MATERIALS FOR SELENIUM[a]

Code	Material	Source[b]
—	Bowen's kale	Bowen
SRM 1549	Milk powder	NBS
SRM 1566	Oyster tissue	NBS
SRM 1567	Wheat flour	NBS
SRM 1568	Rice flour	NBS
SRM 1577a	Bovine liver	NBS
SRM 2670	Urine (normal and spiked)	NBS
RM 50	Albacore tuna	NBS
RM 8412	Corn (*Zea mays*) stalk	NBS
RM 8413	Corn (*Zea mays*) kernel	NBS
RM 8419	Bovine serum	NBS
RM 8431	Total daily diet (U.S.)	NBS
A-11	Milk powder	IAEA
A-13	Animal blood	IAEA
H-4	Animal muscle	IAEA
H-8	Horse kidney	IAEA
MA-A-1/TM	Copepod	IAEA
MA-A-2/TM	Fish flesh	IAEA
MA-M-2/TM	Mussel tissue	IAEA
TORT-1	Lobster hepatopancreas	NRCC
RS-00111	Shark muscle	IPH
Seronorm 102	Human serum	NycoMed
Seronorm 103	Human serum	NycoMed
Seronorm 105	Human serum	NycoMed
Seronorm 108	Human urine	NycoMed
—	Human serum	Versieck

[a] Materials with certified or recommended concentrations.
[b] Suppliers: Bowen — Dr. H. J. M. Bowen, Department of Chemistry, The University of Reading, Whiteknights, Reading, U.K.; NBS — Office of Standard Reference Materials, Chemistry Building, National Bureau of Standards, Gaithersburg, MD, U.S.; IAEA — International Atomic Energy Agency, Analytical Quality Control Services, Laboratory Seibersdorf, Vienna, Austria; NRCC — National Research Council Canada, Division of Chemistry, Ottawa, Canada; IPH — Institute of Public Health, Department of Radiological Health, Tokyo, Japan; NycoMed — NycoMed AS, Division Scandinavia, Torshov, Oslo, Norway (chemical analysis done under auspices of IUPAC); Versieck — University of Ghent, Ghent, Belgium (information from Dr. J. Versieck, Department of Internal Medicine, Division of Gastroenterology, University Hospital, Ghent, Belgium).

1. Whole Blood

The concentrations of selenium in whole blood are strikingly different in different populations, apparently depending on dietary intake; wide variations appear to be well tolerated by the human body. In reports from China, the upper and lower limits of safety are considered to be 440 μg/l and 27 μg/l, respectively.[22] Based on their studies in Venezuela, Jaffé et al.[23] concluded that selenium excess does not constitute a serious health hazard provided that in the blood of children its concentration does not exceed 813 μg/l. In China, however, endemic intoxication has been reported in an area where blood selenium concentrations were

Table 2
SELENIUM CONCENTRATIONS IN HUMAN TISSUES AND FLUIDS[a]

Geographical location	Mean Se concentration[b]	N[c]	Analytical method, date of sampling or analysis[d]	Health status, description of subjects, and tissues; remarks	Ref.
			Amniotic Fluid (μg/l)		
Europe					
Greece	5.0	25	INAA, 1982	Normal pregnancies	93
Greece	5.8	15	INAA, 1982	Prolonged pregnancies	93
Munich, Fed. Rep. Germany	2.1	15	RNAA, 1980	Normal pregnancies	94
Berlin, Fed. Rep. Germany	2.9 μg/kg[e] W	142	INAA, 1983	Normal pregnancies	95
			Aorta (mg/kg)		
Europe					
Glasgow, U.K.	1.58 D	9	RNAA, 1981	Intima, post mortem	96
Glasgow, U.K.	1.10 D	10	RNAA, 1981	Media, post mortem	96
Glasgow, U.K.	0.96 D	10	RNAA, 1981	Adventitia, post mortem	96
Glasgow, U.K.	1.17 D	3	RNAA, 1981	Plaque, post mortem	96
North America					
U.S.	0.94 D	18	XRF, 1982	Autopsies	97
Asia					
Japan	2.9 D	41	INAA, 1980	Forensic material	51
Oceania					
Australia	0.71 D	16	XRF, 1982	Autopsies	97
Australia	0.62 D	17	XRF, 1982	Uremic patients, nondialyzed	97
Australia	0.59 D	21	XRF, 1982	Uremic patients, dialyzed	97
			Ascitic Fluid (μg/l)		
North America					
Huntington, WV, U.S.	30	13	INAA, 1979	Laennecs cirrhosis	98
			Blood: Erythrocytes (μg/l)		
Europe					
Belgium	496 μg/kg Hb	26	EAAS, 1983	Healthy adults	99
Belgium	404 μg/kg Hb	19	EAAS, 1983	Healthy subjects (5—20 years)	99
Belgium	329 μg/kg Hb	26	EAAS, 1983	Cystic fibrosis (7—19 years)	99
Gent, Belgium	159	108	HAAS, 1983	Healthy males	100
Gent, Belgium	165	52	HAAS, 1983	Healthy females	100
Gent, Belgium	200	31	HAAS, 1983	Healthy elderly people	100
Paris, France	349 μg/kg Hb	32	EAAS, 1984	Healthy subjects	101
Paris, France	355 μg/kg Hb	23	EAAS, 1984	Down's syndrome	101
Pavia, Italy	38.8	16	AAS, 1983	Healthy adults	102
Pavia, Italy	120	20	AAS, 1983	Multiple sclerosis	102
Tuscany, Italy	105	12	INAA, 1976	Healthy subjects	63
Tuscany, Italy	148	20	INAA, 1976	Exposed workers (Hg)	63
Amsterdam, Netherlands	145	7	INAA, 1983	Healthy females	103
Amsterdam, Netherlands	162	7	INAA, 1983	Female breast cancer	103
Oslo, Norway	149	15	INAA, 1978	Healthy females	104
Oslo, Norway	149	31	INAA, 1978	Full term infants, cord blood	104

Table 2 (continued)
SELENIUM CONCENTRATIONS IN HUMAN TISSUES AND FLUIDS[a]

Geographical location	Mean Se concentration[b]	N[c]	Analytical method, date of sampling or analysis[d]	Health status, description of subjects, and tissues; remarks	Ref.
Sweden	55.7 μg/kg W	20	INAA, 1980	Healthy newborns, cord blood	105
Southhampton, U.K.	489 μg/kg Hb	391	HAAS, 1983	Healthy adults	106
Berlin, Fed. Rep. Germany	140 μg/kg W	17	INAA, 1979	Healthy females	107
Tübingen, Fed. Rep. Germany	350 μg/kg Hb	10	F, 1976	Healthy subjects	108
Ulm, Fed. Rep. Germany	401 μg/kg	6	INAA, 1981	Healthy men	109
North America					
Ontario, Canada	236	51	RNAA, 1967	Healthy adults	25
Augusta, GA, U.S.	158	206	F, 1984	Healthy subjects	110
Evans, GA, U.S.	470 μg/kg D	10	NAA, 1980	Healthy males	111
Evans, GA, U.S.	410 μg/kg D	10	NAA, 1980	Cardiovascular disease, males	111
Chatham, GA, U.S.	469 μg/kg D	20	NAA, 1980	Healthy males	111
Chatham, GA, U.S.	440 μg/kg D	20	NAA, 1980	Cardiovascular disease, males	111
Louisville, KY, U.S.	159.2	84	RNAA, 1983	Healthy men	112
Louisville, KY, U.S.	139.8	38	RNAA, 1983	Stable chronic disease	112
Louisville, KY, U.S.	88.8	12	RNAA, 1983	Alcoholic cirrhosis w biopsy	112
Louisville, KY, U.S.	135.8	14	RNAA, 1983	Alcoholic cirrhosis w/o biopsy	112
Louisville, KY, U.S.	110.1	9	RNAA, 1983	Malignancy w hepatic metastases	112
Louisville, KY, U.S.	139.0	11	RNAA, 1983	Malignancy w/o hepatic metastases	112
Louisville, KY, U.S.	149.5	9	RNAA, 1983	Chronic cholecystitis	112
Louisville, KY, U.S.	142.1	4	RNAA, 1983	Chronic active hepatitis	112
Louisville, KY, U.S.	129.8	18	RNAA, 1983	Chronic renal failure	112
Brooklyn, NY, U.S.	520	16	F, 1978	Healthy females	113
Brooklyn, NY, U.S.	520	34	F, 1978	Mothers at delivery	113
Brooklyn, NY, U.S.	390	34	F, 1978	Cord blood	113
Valhalla, NY, U.S.	126	27	F, 1984	Healthy adults	114
Valhalla, NY, U.S.	92	30	F, 1984	Detoxified alcoholics	114
Valhalla, NY, U.S.	66	16	F, 1984	Alcoholic liver disease	114
Columbus, OH, U.S.	218	423	F, 1983	Rural Amish population	115
Houston, TX, U.S.	730 μg/kg Hb	100	F, 1981	Healthy adults	116
Houston, TX, U.S.	200	36	F, 1983	Healthy elderly	117
Houston, TX, U.S.	330 μg/kg Hb	9	F, 1981	Chronic intravenous hyperalimentation	116
Virginia Tidewater, VA, U.S.	141.4	38	EAAS, 1982	Healthy women	118
Africa					
Egypt	49.3	3	INAA, 1972	Healthy adults	119
Egypt	71.4	34	INAA, 1972	Healthy children	119
Egypt	70.4	20	INAA, 1972	Kwashiorkor children, admission	119
Egypt	65.8	15	INAA, 1972	Kwashiorkor children, recovery	119

Table 2 (continued)
SELENIUM CONCENTRATIONS IN HUMAN TISSUES AND FLUIDS[a]

Geographical location	Mean Se concentration[b]	N[c]	Analytical method, date of sampling or analysis[d]	Health status, description of subjects, and tissues; remarks	Ref.
Asia					
Okayama-shi, Japan	139 (74[f])	13	GC, 1980	Healthy adults	120
Okayama-shi, Japan	96 (45[f])	3	GC, 1980	Liver cirrhosis	120
Okayama-shi, Japan	100 (55[f])	6	GC, 1980	Lung cancer	120
Okayama-shi, Japan	93 (43[f])	4	GC, 1980	Rheumatoid arthritis	120
Oceania					
New Zealand	63	12	F, 1976	Intravenous alimentation	121
Auckland, New Zealand	98	131	F, 1979	Healthy adults	122
Dunedin, New Zealand	58	48	F, 1977	Elderly people	26
Milton, New Zealand	71	118	F, 1983	Healthy males	123
Milton, New Zealand	75	112	F, 1983	Healthy females	123
Otago, New Zealand	66	113	F, 1979	Healthy adults	124
Otago, New Zealand	67	80	F, 1979	Cancer patients	124
Blood: Platelets (mg/kg)					
Europe					
Fed. Rep. Germany	0.782 W	7	INAA, 1979	Healthy adults (pure platelets)	30
Fed. Rep. Germany	0.679 W	7	INAA, 1979	Healthy adults (impure platelets)	30
Julich, Fed. Rep. Germany	0.747 W	8	INAA, 1979	Healthy subjects (pure platelets)	125
Blood: Plasma (μg/l)					
Europe					
Gent, Belgium	97	110	HAAS, 1983	Healthy males	100
Gent, Belgium	97	53	HAAS, 1983	Healthy females	100
Gent, Belgium	73	31	HAAS, 1983	Healthy elderly people	100
Gent, Belgium	66	9	HAAS, 1983	Cord blood	34
Gent, Belgium	88	9	HAAS, 1983	Healthy mothers, maternal plasma	34
Paris, France	93	33	EAAS, 1984	Healthy subjects	101
Paris, France	70	23	EAAS, 1984	Down's syndrome	101
Pavia, Italy	60.6	16	AAS, 1983	Healthy adults	102
Pavia, Italy	86.4	20	AAS, 1983	Patients with multiple sclerosis	102
Verona, Italy	64	7	PIXE, 1977	Patients with raised serum iron	126
Verona, Italy	47	7	PIXE, 1977	Patients with iron deficiency	126
Amsterdam, Netherlands	110	7	INAA, 1983	Healthy females	103
Amsterdam, Netherlands	110	7	INAA, 1983	Female breast cancer	103
Rotterdam, Netherlands	79	12	F, 1982	Healthy children	127
Rotterdam, Netherlands	66	13	F, 1982	Children, cystic fibrosis	127
Hampshire, U.K.	115	300	HAAS, 1982	Healthy subjects	128
Ulm, Fed. Rep. Germany	103	6	INAA, 1981	Healthy men	109
North America					
Ontario, Canada	144	253	INAA, 1967	Healthy adults	25
San Diego, CA, U.S.	102	7	F, 1972	Healthy adults	129

Table 2 (continued) SELENIUM CONCENTRATIONS IN HUMAN TISSUES AND FLUIDS[a]

Geographical location	Mean Se concentration[b]	N[c]	Analytical method, date of sampling or analysis[d]	Health status, description of subjects, and tissues; remarks	Ref.
San Diego, CA, U.S.	94	26	F, 1972	Full term mothers	129
San Diego, CA, U.S.	73	17	F, 1972	Cord blood	129
San Diego, CA, U.S.	69	18	F, 1972	Sudden infant death syndrome	129
Augusta, GA, U.S.	104	206	F, 1984	Healthy subjects	110
Kentucky, U.S.	260	20	GC, 1973	Healthy subjects	130
Brooklyn, NY, U.S.	210	16	F, 1978	Healthy females	113
Brooklyn, NY, U.S.	190	25	F, 1978	Mother at delivery	113
Brooklyn, NY, U.S.	140	25	F, 1978	Cord blood	113
Valhalla, NY, U.S.	109	27	F, 1984	Healthy adults	114
Valhalla, NY, U.S.	65	30	F, 1984	Detoxified alcoholics	114
Valhalla, NY, U.S.	38	16	F, 1984	Alcoholic liver disease	114
Columbus, OH, U.S.	119	423	F, 1983	Rural Amish population	115
Houston, TX, U.S.	96	100	F, 1981	Healthy adults	116
Houston, TX, U.S.	100	36	F, 1983	Healthy elderly	117
Houston, TX, U.S.	43	9	F, 1981	Chronic intravenous hyperalimentation	116
Virginia Tidewater, VA, U.S.	96.3	38	EAAS, 1982	Healthy nonpregnant women	118
Africa					
Egypt	52	3	INAA, 1972	Healthy adults	119
Egypt	59.4	33	INAA, 1972	Healthy children	119
Egypt	55.6	20	INAA, 1972	Kwashiorkor children, admission	119
Egypt	70.2	13	INAA, 1972	Kwashiorkor children, recovery	119
Asia					
Hiroshima, Japan	51.0	22	EAAS, 1984	Healthy children (1—6 months)	131
Hiroshima, Japan	63.0	11	EAAS, 1984	Healthy children (7—12 months)	131
Hiroshima, Japan	73.2	17	EAAS, 1984	Healthy children (1 year)	131
Hiroshima, Japan	82.2	29	EAAS, 1984	Healthy children (2—5 years)	131
Hiroshima, Japan	85.6	28	EAAS, 1984	Healthy children (6—10 years)	131
Hiroshima, Japan	93.5	25	EAAS, 1984	Healthy children (11—15 years)	131
Hiroshima, Japan	99.4	65	EAAS, 1984	Healthy adults	131
Okayama-shi, Japan	119 (34[f])	13	GC, 1980	Healthy adults	120
Okayama-shi, Japan	106 (26[f])	3	GC, 1980	Liver cirrhosis	120
Okayama-shi, Japan	126 (38[f])	6	GC, 1980	Lung cancer	120
Okayama-shi, Japan	105 (31[f])	4	GC, 1980	Rheumatoid arthritis	120
Chiang Mai, Thailand	83	9	F, 1970	Well-nourished infants (1—4 years)	132
Chiang Mai, Thailand	39	8	F, 1970	Malnourished infants (1—4 years)	132
Oceania					
Auckland, New Zealand	64	131	F, 1979	Healthy adults	122
Otago, New Zealand	43	113	F, 1979	Healthy adults	124
Otago, New Zealand	36	80	F, 1979	Cancer patients	124

Table 2 (continued)
SELENIUM CONCENTRATIONS IN HUMAN TISSUES AND FLUIDS[a]

Geographical location	Mean Se concentration[b]	N[c]	Analytical method, date of sampling or analysis[d]	Health status, description of subjects, and tissues; remarks	Ref.
Blood: Proteins (μg/l)					
North America					
Berkeley, CA, U.S.	0.5	11	INAA, 1982	Healthy adults, very low-density lipoprotein (VLDL)	133
Berkeley, CA, U.S.	3	11	INAA, 1982	Healthy adults, low-density lipoprotein (LDL)	133
Berkely, CA, U.S.	3	11	INAA, 1982	Healthy adults, high-density lipoprotein (HDL)	133
Blood: Serum (μg/l)					
Europe					
Belgium	99	83	EAAS, 1983	Healthy adults	99
Belgium	82	19	EAAS, 1983	Healthy subjects (5—20 years)	99
Belgium	63	24	EAAS, 1983	Cystic fibrosis (7—19 years)	99
Gent, Belgium	130	36	INAA, 1977	Healthy adults	134
Copenhagen, Denmark	98	11	NAA, 1974	Healthy subjects	46
Zealand, Denmark	97	12	F, 1980	Healthy subjects	135
Zealand, Denmark	85	14	F, 1980	Multiple sclerosis	135
Finland	72.9	64	EAAS, 1983	Healthy adults	136
Finland	54.3	128	EAAS, 1972	Healthy matched adults	137
Finland	50.5	128	EAAS, 1972	Cancer patients	137
Finland	74	40	HAAS, 1984	Ovarian cancer	138
Finland	71.6	33	EAAS, 1983	Coronary heart disease	136
S and SE Finland	43.6	18	F, 1976	Healthy subjects	139
S and SE Finland	46.4	27	F, 1976	Multiple sclerosis	139
Helsinki, Finland	66	21	F, 1977	Healthy adults	47
Helsinki, Finland	66	8	F, 1977	Healthy males	141
Helsinki, Finland	66	13	F, 1977	Healthy females	141
Helsinki, Finland	50	9	F, 1977	Healthy children (1—6 years)	141
Helsinki, Finland	51	14	F, 1977	Healthy children (7—12 years)	141
Helsinki, Finland	37	4	F, 1977	Acrodermatitis enteropathica	47
Helsinki, Finland	48	13	F, 1977	Myocardial infarction	47
Helsinki, Finland	37	10	F, 1977	Neuronal ceroid lipofuscinosis	47
Helsinki, Finland	83.1	45	EAAS, 1983	Healthy adults	140
Helsinki, Finland	68.7	34	EAAS, 1983	Alcoholics w/o liver cirrhosis	140
Helsinki, Finland	43.3	13	EAAS, 1983	Alcoholics with liver cirrhosis	140
Helsinki, Finland	47.3	17	EAAS, 1983	Primary biliary cirrhosis	140
Helsinki, Finland	74.6	9	EAAS, 1983	Hypoalbuminaemia-renal origin	140
Kuusamo, Finland	75	20	F, 1977	Healthy males	141

Table 2 (continued)
SELENIUM CONCENTRATIONS IN HUMAN TISSUES AND FLUIDS[a]

Geographical location	Mean Se concentration[b]	N[c]	Analytical method, date of sampling or analysis[d]	Health status, description of subjects, and tissues; remarks	Ref.
Lappeenranta, Finland	42	10	F, 1977	Healthy adults	141
Lieksa, Finland	60	8	F, 1977	Healthy males	141
Lieksa, Finland	54	10	F, 1977	Healthy females	141
Mariehavn, Finland	99	11	F, 1977	Healthy adults	141
Mariehavn, Finland	73	9	F, 1977	Healthy children (4—15 years)	141
Oulu, Finland	98.7	56	HAAS, 1984	Healthy adults	142
Oulu, Finland	90.8	44	HAAS, 1984	Gynecological cancer	142
Oulu, Finland	65	37	HAAS, 1984	Cervical cancer	37
Oulu, Finland	80	64	HAAS, 1984	Endometrial cancer	37
Oulu, Finland	109	137	HAAS, 1984	Middle aged-old women	37
Pihtipudas, Finland	50	10	F, 1977	Healthy adults	141
Savukoski, Finland	94	10	F, 1977	Healthy males	141
Seinajoki, Finland	46	8	F, 1977	Healthy adults	141
Utsjoki, Finland	109	12	F, 1977	Healthy men	141
Utsjoki, Finland	106	11	F, 1977	Healthy females	141
Paris, France	118	32	EAAS, 1983	Healthy subjects	143
Paris, France	96	32	EAAS, 1983	Healthy subjects	144
Paris, France	83	17	EAAS, 1983	Nonobstructive cardiomy-opathies	143
Paris, France	71	28	EAAS, 1983	Down's syndrome	144
Greece	42	25	INAA, 1982	Normal pregnancies	93
Greece	37	25	INAA, 1982	Normal pregnancies, umbil-ical cord serum	93
Greece	36	15	INAA, 1982	Prolonged pregnancies	93
Greece	29	15	INAA, 1982	Prolonged pregnancies, um-bilical cord serum	93
Padova, Italy	79	34	PIXE, 1980	Healthy subjects	145
Padova, Italy	52	38	PIXE, 1980	Malignant lymphoprolifera-tive diseases	145
Tuscany, Italy	46	12	INAA, 1976	Healthy subjects	63
Tuscany, Italy	59	20	INAA, 1976	Exposed workers (Hg)	63
Petten, Netherlands	100 μg/kg	39	INAA, 1984	Healthy adults	146
Petten, Netherlands	89 μg/kg	34	INAA, 1984	Cancer patients	146
Etnedal, Norway	106	34	INAA, 1984	Healthy subjects	147
Gamvik, Norway	114	45	INAA, 1984	Healthy subjects	147
Kautokeino, Norway	114	30	INAA, 1984	Healthy subjects	147
Lom, Norway	106	10	INAA, 1984	Healthy subjects	147
Oslo, Norway	112	16	EAAS, 1982	Healthy adults	148
Oslo, Norway	114	12	EAAS, 1979	Healthy adults	149
Oslo, Norway	130	10	RNAA, 1978	Healthy adults	39
Oslo, Norway	120	40	EAAS, 1982	Healthy adults	38
Oslo, Norway	110	15	INAA, 1978	Healthy females	104
Oslo, Norway	94.0	23	RNAA, 1978	Rheumatoid arthritis	39
Oslo, Norway	96	11	EAAS, 1982	Alcoholic cirrhosis	38
Oslo, Norway	51	11	EAAS, 1980	Alcoholic cirrhosis	74
Oslo, Norway	89	15	EAAS, 1982	Chronic active hepatitis	38
Oslo, Norway	95	4	EAAS, 1982	Chronic persistent hepatitis	38
Oslo, Norway	96	12	EAAS, 1982	Cryptogenic cirrhosis	38
Oslo, Norway	108	12	EAAS, 1982	Primary biliary cirrhosis	38
Oslo, Norway	52	31	INAA, 1978	Full term infants	104

Table 2 (continued)
SELENIUM CONCENTRATIONS IN HUMAN TISSUES AND FLUIDS[a]

Geographical location	Mean Se concentration[b]	N[c]	Analytical method, date of sampling or analysis[d]	Health status, description of subjects, and tissues; remarks	Ref.
Porsanger, Norway	113	38	INAA, 1984	Healthy subjects	147
Vadso, Norway	114	50	INAA, 1984	Healthy subjects	147
Vang, Norway	106	22	INAA, 1984	Healthy subjects	147
Vestre Slidre, Norway	104	30	INAA, 1984	Healthy subjects	147
Ostre Toten, Norway	104	30	INAA, 1984	Healthy subjects	147
Wroclaw, Poland	78	95	INAA, 1982	Healthy adults	150
Galicia, Spain	55	55	F, 1983	Healthy mothers	151
Galicia, Spain	51	73	F, 1983	Healthy school children	151
Galicia, Spain	47	8	F, 1983	Children with phenylketonuria	151
Sweden	20 μg/kg W	21	INAA, 1980	Healthy newborns	105
Stockholm, Sweden	190	8	RNAA, 1973	Healthy subjects	152
Stockholm, Sweden	170	11	RNAA, 1973	Arterial hypertension	152
Uppsala, Sweden	65	13	NAA, 1984	Healthy children	153
Uppsala, Sweden	74	27	NAA, 1984	Diabetic children	153
Southampton, U.K.	115.8	391	HAAS, 1983	Healthy adults	106
Southampton, U.K.	134.3	66	HAAS, 1984	Healthy men, nonsmokers	154
Southampton, U.K.	98.4	41	HAAS, 1984	Healthy men, cigarette smokers	154
Southampton, U.K.	105.9	9	HAAS, 1984	Healthy men, pipe/cigar smokers	154
Fed. Rep. Germany	102	19	INAA, 1977	Healthy adults (20—40 years)	35
Fed. Rep. Germany	34	12	INAA, 1977	Healthy children (1—4 months)	35
Fed. Rep. Germany	58	13	INAA, 1977	Healthy children, later infancy	35
Fed. Rep. Germany	82	30	INAA, 1977	Healthy children, toddlers	35
Fed. Rep. Germany	92	28	INAA, 1977	Healthy school children	35
Berlin, Fed. Rep. Germany	88 μg/kg W	17	INAA, 1979	Healthy females	107
Berlin, Fed. Rep. Germany	78 μg/kg W	37	INAA, 1979	Pregnant women	107
Julich, Fed. Rep. Germany	98	184	INAA, 1972	Healthy subjects	155
Mainz, Fed. Rep. Germany	80.6	99	EAAS, 1982	Healthy adults	156
Mainz, Fed. Rep. Germany	81.1	92	EAAS, 1983	Healthy adults	157
Mainz, Fed. Rep. Germany	50.5	20	EAAS, 1983	Congestive cardiomyopathy	157
North America					
Montreal, Quebec, Canada	142.9	15	F, 1982	Healthy men	158
U.S.	136	210	INAA, 1983	Healthy adults	159
U.S.	129	111	INAA, 1983	Cancer patients	159
U.S.	1.48 mg/kg D	18	INAA, 1976	Healthy subjects	160
U.S.	1.27 mg/kg D	110	INAA, 1976	Cancer patients	160
Davis, CA, U.S.	129	11	INAA, 1972	Healthy adults	133
Palo Alto, CA, U.S.	140	21	F, 1981	Healthy children	161
Palo Alto, CA, U.S.	116	32	F, 1981	Children, cystic fibrosis	161

Table 2 (continued)
SELENIUM CONCENTRATIONS IN HUMAN TISSUES AND FLUIDS[a]

Geographical location	Mean Se concentration[b]	N[c]	Analytical method, date of sampling or analysis[d]	Health status, description of subjects, and tissues; remarks	Ref.
Chicago, IL, U.S.	131	212	XRF, 1981	Healthy adults	162
Kentucky, U.S.	430	20	GC, 1973	Healthy subjects	130
Louisville, KY, U.S.	94.3	92	RNAA, 1983	Healthy men	112
Louisville, KY, U.S.	1.57 mg/kg D	27	INAA, 1980	Healthy females	163
Louisville, KY, U.S.	1.25 mg/kg D	35	INAA, 1980	Female breast cancer	163
Louisville, KY, U.S.	1.61 mg/kg D	—	INAA, 1978	Malignant reticuloendothelial tumors	164
Louisville, KY, U.S.	89.2	52	RNAA, 1983	Chronic stable disease	112
Louisville, KY, U.S.	62.2	18	RNAA, 1983	Alcoholic cirrhosis w biopsy	112
Louisville, KY, U.S.	74.3	18	RNAA, 1983	Alcoholic cirrhosis w/o biopsy	112
Louisville, KY, U.S.	66.6	12	RNAA, 1983	Malignancy with hepatic metastases	112
Louisville, KY, U.S.	76.7	14	RNAA, 1983	Malignancy w/o hepatic metastases	112
Louisville, KY, U.S.	88.9	9	RNAA, 1983	Chronic cholecystitis	112
Louisville, KY, U.S.	98.0	4	RNAA, 1983	Chronic active hepatitis	112
Louisville, KY, U.S.	77.8	49	RNAA, 1983	Chronic renal failure	112
Omaha, NE, U.S.	120	37	RNAA, 1979	Healthy adults	165
Omaha, NE, U.S.	130	22	RNAA, 1979	Chronic obstructive pulmonary disease	165
Omaha, NE, U.S.	90	14	RNAA, 1979	Cirrhosis	165
Omaha, NE, U.S.	110	13	RNAA, 1979	Alcohol/pancreatitis	165
Omaha, NE, U.S.	120	42	RNAA, 1979	Arteriosclerotic heart disease	165
Omaha, NE, U.S.	110	20	RNAA, 1979	Malignancy	165
Omaha, NE, U.S.	130	9	RNAA, 1979	Hypertension	165
Omaha, NE, U.S.	130	19	RNAA, 1979	Diabetes	165
Omaha, NE, U.S.	120	14	RNAA, 1979	Infections	165
Omaha, NE, U.S.	120	10	RNAA, 1979	Arthritis	165
Omaha, NE, U.S.	120	11	RNAA, 1979	Ulcer	165
Omaha, NE, U.S.	120	11	RNAA, 1979	Psychoses	165
Long Island, NY, U.S.	95.4	44	F, 1982	Healthy children (5—18 years)	166
Norfolk, VA, U.S.	96.3	38	EAAS, 1983	Healthy females	112
South America					
Lima, Peru	117	20	PIXE, 1983	Healthy adults	167
Rio Mantaro Valley, Peru	59	18	PIXE, 1983	Healthy adults, high altitude	167
Asia					
Kumamoto, Japan	148	16	F, 1983	Healthy lactating mothers	168
Tokyo, Japan	340	20	INAA, 1979	Healthy men	169
Tokyo, Japan	320	25	INAA, 1979	Down's syndrome	169
Oceania					
Australia	77 μg/kg	40	INAA, 1983	Healthy mothers	170
New Zealand	25	12	F, 1976	Intravenous alimentation	121
Dunedin, New Zealand	37	48	F, 1977	Elderly people	26
Milton, New Zealand	48	118	F, 1983	Healthy men	123
Milton, New Zealand	49	112	F, 1983	Healthy females	123

Table 2 (continued)
SELENIUM CONCENTRATIONS IN HUMAN TISSUES AND FLUIDS[a]

Geographical location	Mean Se concentration[b]	N[c]	Analytical method, date of sampling or analysis[d]	Health status, description of subjects, and tissues; remarks	Ref.
			Blood: Whole (μg/l)		
Europe					
Gent, Belgium	122	109	HAAS, 1983	Healthy males	100
Gent, Belgium	125	53	HAAS, 1983	Healthy females	100
Gent, Belgium	130	31	HAAS, 1983	Healthy elderly people	100
Zealand, Denmark	88.5	20	F, 1980	Healthy subjects	135
Zealand, Denmark	92.8	17	F, 1980	Multiple sclerosis	135
Angmagssalik, Greenland	179	62	HAAS, 1984	Healthy subjects, >6 meals of seals/week	171
Angmagssalik, Greenland	135	20	HAAS, 1984	Healthy subjects, 1—6 meals of seals/week	171
Angmagssalik, Greenland	107	36	HAAS, 1984	Healthy subjects, 0—1 meals of seals/week	171
Angmagssalik, Greenland	70	18	HAAS, 1984	Healthy subjects, Danes	171
Bergholz-Rehbrucke, German Dem. Rep.	220	36	F, 1980	Healthy females	172
Bergholz-Rehbrucke, German Dem. Rep.	180	38	F, 1980	Healthy males	172
Helsinki, Finland	81	8	F, 1977	Healthy males	141
Helsinki, Finland	90	17	F, 1977	Healthy females	141
Helsinki, Finland	40	13	F, 1977	Healthy children (1—7 d)	141
Helsinki, Finland	59	14	F, 1977	Healthy children (1—6 years)	141
Helsinki, Finland	60	13	F, 1977	Healthy children (7—12 years)	141
Helsinki, Finland	59	4	F, 1977	Acrodermatitis enterpathica	47
Helsinki, Finland	45	6	F, 1977	Dystrophia musculorium progressiva	47
Helsinki, Finland	64	13	F, 1977	Myocardial infarction	47
Helsinki, Finland	46	17	F, 1977	Neuronal ceroid lipofuscinosis	47
Lappeenranta, Finland	56	17	F, 1977	Healthy adults	141
Kuopia, Finland	56	80	F, 1977	Healthy adults	141
Seinajoki, Finland	69	8	F, 1977	Healthy adults	141
S and SE Finland	61.8	18	F, 1976	Healthy subjects	139
S and SE Finland	56.0	15	F, 1976	Multiple sclerosis	139
Italy	93	40	INAA, 1977	Healthy adults	67
Tuscany, Italy	77	12	INAA, 1976	Healthy subjects	63
Tuscany, Italy	120	20	INAA, 1976	Exposed workers (Hg)	63
Amsterdam, Netherlands	133	7	INAA, 1981	Healthy females	173
Amsterdam, Netherlands	137	7	INAA, 1981	Female breast cancer	173
Sweden	120 μg/kg	8	RNAA, 1966	Healthy adults	174
Sweden	26 μg/kg W	7	INAA, 1980	Healthy mothers, maternal blood	105
Studsvik, Sweden	120	6	RNAA, 1966	Healthy adults	175
Studsvik, Sweden	110	6	RNAA, 1966	Uremic patients	175
U.K.	80 μg/kg W	2500	SSMS, 1972/73	Healthy subjects, pooled sample	56

Table 2 (continued)
SELENIUM CONCENTRATIONS IN HUMAN TISSUES AND FLUIDS[a]

Geographical location	Mean Se concentration[b]	N[c]	Analytical method, date of sampling or analysis[d]	Health status, description of subjects, and tissues; remarks	Ref.
Aberdeen, U.K.	146	58	F, 1983	healthy subjects	176
Aberdeen, U.K.	140	50	F, 1983	Ulcerative colitis	176
Aberdeen, U.K.	137	70	F, 1983	Crohns disease	176
Aberdeen, U.K.	120	68	F, 1984	Healthy children	177
Aberdeen, U.K.	120	17	F, 1984	Asthmatic children	177
Aberdeen, U.K.	120	23	F, 1984	Epileptic children	177
Aberdeen, U.K.	79	15	F, 1984	Cystic fibrosis, children	177
Aberdeen, U.K.	87	18	F, 1984	Coeliac disease, children	177
Birmingham, U.K.	267	20	GC, 1977	Healthy adults	68
Glasgow, U.K.	770 μg/kg D	25	RNAA, 1977	Healthy subjects	96
Hampshire, U.K.	136	300	HAAS, 1982	Healthy subjects	128
Oxford, U.K.	97	10	INAA, 1982	Healthy adults	62
Oxford, U.K.	84	10	INAA, 1982	Multiple sclerosis	62
Southampton, U.K.	137.2	391	HAAS, 1983	Healthy adults	106
Southampton, U.K.	134.3	66	HAAS, 1984	Healthy men, nonsmokers	154
Southampton, U.K.	115.1	41	HAAS, 1984	Healthy men, cigarette smokers	154
Southampton, U.K.	121.9	9	HAAS, 1984	Healthy men, pipe/cigar smokers	154
North America					
Ontario, Canada	182	253	RNAA, 1967	Healthy adults	25
Phoenix, AZ, U.S.	197	10	F, 1968	Healthy males	24
Little Rock, AR, U.S.	201	10	F, 1968	Healthy males	24
Red Bluff, CA, U.S.	182	10	F, 1968	Healthy males	24
San Diego, CA, U.S.	130	6	F, 1972	Healthy adults	129
San Diego, CA, U.S.	98	7	F, 1972	Cord blood	129
San Diego, CA, U.S.	100	12	F, 1972	Sudden infant death syndrome	129
Jacksonville, FL, U.S.	188	10	F, 1968	Healthy males	24
Muncie, IN, U.S.	158	10	F, 1968	Healthy males	24
Chicago, IL, U.S.	223 μg/kg	19	F, 1980	Healthy children	178
Chicago, IL, U.S.	132 μg/kg	14	F, 1980	Phenylketonuria, children	178
Chicago, IL, U.S.	122 μg/kg	20	F, 1980	Cystic fibrosis, children	178
Lafayette, LA, U.S.	176	10	F, 1968	Healthy males	24
Meridian, MI, U.S.	195	10	F, 1968	Healthy males	24
Billings, MT, U.S.	180	10	F, 1968	Healthy males	24
Missoula, MT, U.S.	194	20	F, 1968	Healthy males	24
Fargo, ND, U.S.	217	10	F, 1968	Healthy males	24
Milan, NM, U.S.	171	43	HAAS, 1978	Healthy adults	179
Canandaigua, NY, U.S.	176	10	F, 1968	Healthy males	24
Geneva, NY, U.S.	182	10	F, 1968	Healthy males	24
Valhalla, NY, U.S.	109	27	F, 1984	Healthy adults	114
Valhalla, NY, U.S.	76	30	F, 1984	Detoxified alcoholics	114
Valhalla, NY, U.S.	47	16	F, 1984	Alcoholic liver disease	114
New Orleans, LA, U.S.	119	72	EAAS, 1982	Healthy adults	180
Columbus, OH, U.S.	161	423	F, 1983	Rural Amish Population	115
Lima, OH, U.S.	157	10	F, 1968	Healthy males	24
Oregon, U.S.	100	52	F, 1983	Healthy subjects	181
Oregon, U.S.	101	48	F, 1983	Seventh-Day Adventists, nonvegetarians	181
Oregon, U.S.	76	16	F, 1983	Seventh-Day Adventists, vegetarians	181

Table 2 (continued)
SELENIUM CONCENTRATIONS IN HUMAN TISSUES AND FLUIDS[a]

Geographical location	Mean Se concentration[b]	N[c]	Analytical method, date of sampling or analysis[d]	Health status, description of subjects, and tissues; remarks	Ref.
Oregon, U.S.	110	16	F, 1983	Hormone dependent cancer, women	181
South Dakota, U.S.	265	626	F, 1979	Healthy subjects	182
Rapid City, SD, U.S.	256	20	F, 1968	Healthy males	24
Nashville, TN, U.S.	1100 μg/kg D	717	INAA, 1974	Healthy mothers, maternal blood	183
Nashville, TN, U.S.	1040 μg/kg D	681	INAA, 1974	Fetal blood	183
El Paso, TX, U.S.	192	10	F, 1968	Healthy males	24
Lubbock, TX, U.S.	178	10	F, 1968	Healthy males	24
Montpelier, VT, U.S.	180	10	F, 1968	Healthy males	24
Seattle, WA, U.S.	170	8	INAA, 1976	Long-term total parenteral nutrition	184
Spokane, WA, U.S.	230	10	F, 1968	Healthy males	24
Cheyenne, WY, U.S.	234	10	F, 1968	Healthy males	24
South America					
Caracas, Venezuela	355	50	F, 1972	Healthy school children	23
Villa Bruzual, Venezuela	813	111	F, 1972	Healthy school children	23
Africa					
Egypt	68	8	INAA, 1972	Healthy adults	119
Egypt	44	3	INAA, 1972	Healthy children	119
Egypt	37.9	32	INAA, 1972	Kwashiorkor children, admission	119
Egypt	42.4	32	INAA, 1972	Kwashiorkor children, recovery	119
Asia					
People's Rep. China	3200	72	F, 1983	High-Se area with chronic selenosis	22
People's Rep. China	440	14	F, 1983	High-Se area w/o selenosis	22
People's Rep. China	95	111	F, 1983	Selenium-adequate area	22
People's Rep. China	27	40	F, 1983	Low-Se area	22
People's Rep. China	21	1478	F, 1983	Low-Se area with Keshan disease	22
Chandigarh, India	410 μg/kg D	9	INAA, 1981	Healthy subjects	173
Chandigarh, India	480 μg/kg D	17	INAA, 1981	Leukemia	173
Okayama-shi, Japan	205 (94[f])	13	GC, 1980	Healthy adults	120
Okayama-shi, Japan	158 (60[f])	3	GC, 1980	Liver cirrhosis	120
Okayama-shi, Japan	182 (80[f])	6	GC, 1980	Lung cancer	120
Okayama-shi, Japan	161 (63[f])	4	GC, 1980	Rheumatoid arthritis	120
Kaohsiung, Taiwan	95	80	INAA, 1983	Healthy adults	185
Thailand	187 μg/kg	20	NAA, 1976	Healthy subjects	186
Chiang Mai, Thailand	88	8	F, 1970	Malnourished infants (1—4 years)	132
Chiang Mai, Thailand	120	9	F, 1970	Well-nourished infants (1—4 years)	132
Oceania					
Adelaide, Australia	150	116	F, 1978	Healthy adults	187
New Zealand	68	215	F, 1974	Healthy adults	188
New Zealand	45	12	F, 1976	Intravenous alimentation	121
Auckland, New Zealand	79	131	F, 1979	Healthy adults	122
Auckland, New Zealand	83	122	F, 1978	Healthy adults	189

Table 2 (continued)
SELENIUM CONCENTRATIONS IN HUMAN TISSUES AND FLUIDS[a]

Geographical location	Mean Se concentration[b]	N[c]	Analytical method, date of sampling or analysis[d]	Health status, description of subjects, and tissues; remarks	Ref.
Auckland, New Zealand	62	28	F, 1978	Healthy children (av. 7 years)	189
Christchurch, New Zealand	43	36	F, 1978	Hospitalized children (av. 6 years)	189
Dunedin, New Zealand	62	59	F, 1978	Healthy adults	189
Dunedin, New Zealand	48	48	F, 1977	Elderly people	26
Dunedin, New Zealand	59	18	F, 1978	Healthy children (av. 7 years)	189
Dunedin, New Zealand	48	44	F, 1978	Hospitalized children (av. 5 years)	189
Hamilton, New Zealand	69	24	F, 1974	Healthy adults	190
Milton, New Zealand	60	118	F, 1983	Healthy men	123
Milton, New Zealand	62	112	F, 1983	Healthy females	123
Otago, New Zealand	54	113	F, 1979	Healthy adults	124
Otago, New Zealand	50	80	F, 1979	Cancer patients	124
Tapanui, New Zealand	60	49	F, 1978	Healthy adults	189
Tapanui, New Zealand	48	50	F, 1978	Healthy children (av. 11 years)	189
Wellington, New Zealand	62	20	F, 1978	Hospitalized children (av. 7 years)	189
Bone Tissue (mg/kg)					
Different continents	0.47 D	27	INAA, 1977	Iliac crest, ancient	191
Europe					
Umea, Sweden	0.057 W	5	INAA, 1980	Healthy individuals	192
Umea, Sweden	0.040 W	7	INAA, 1980	Exposed workers	192
Oceania					
Australia	0.56 D	2	XRF, 1982	Autopsies	97
Australia	0.48 D	3	XRF, 1982	Uremic patients, dialyzed	97
Brain (mg/kg)					
Europe					
Helsinki, Finland	0.46 D	3	ADF, 1977	Children (1—6 years)	47
Helsinki, Finland	0.54 D	7	ADF, 1977	Children (1—9 months)	47
Helsinki, Finland	0.64 D	10	ADF, 1977	Cortex, stillborns	47
Helsinki, Finland	0.59 D	11	ADF, 1977	Cortex, full term (1—7 d)	47
Helsinki, Finland	0.63 D	12	ADF, 1977	Cortex, premature (1—7 d)	47
U.K.	0.09 W	10	SSMS, 1972/73	Healthy subjects	56
U.K.	0.05 W	2	SSMS, 1972/73	Healthy subjects, frontal lobe	56
U.K.	0.05 W	2	SSMS, 1972/73	Healthy subjects, basal ganglia	56
Glasgow, U.K.	1.59 D	13	RNAA, 1981	Healthy subjects	96
Cologne, Fed. Rep. Germany	0.57 D	9	INAA, 1974	Tumor free dura	57
Cologne, Fed. Rep. Germany	0.55 D	9	INAA, 1974	Tumor free cerebral cortex	57
Cologne, Fed. Rep. Germany	0.85 D	8	INAA, 1974	Tumor free medulla	57

Table 2 (continued)
SELENIUM CONCENTRATIONS IN HUMAN TISSUES AND FLUIDS[a]

Geographical location	Mean Se concentration[b]	N[c]	Analytical method, date of sampling or analysis[d]	Health status, description of subjects, and tissues; remarks	Ref.
Cologne, Fed. Rep. Germany	1.19 D	4	INAA, 1974	Spongioblastomas	57
Cologne, Fed. Rep. Germany	0.79 D	3	INAA, 1974	Astrocytomas, grade 1	57
Cologne, Fed. Rep. Germany	1.18 D	3	INAA, 1974	Astrocytomas, grade 2	57
Cologne, Fed. Rep. Germany	0.77 D	5	INAA, 1974	Oligodendrogliomas, grade 1	57
Cologne, Fed. Rep. Germany	1.06 D	7	INAA, 1974	Oligodendrogliomas, grade 2	57
Cologne, Fed. Rep. Germany	1.29 D	11	INAA, 1974	Glioblastomas	57
Cologne, Fed. Rep. Germany	1.41 D	7	INAA, 1974	Medulloblastomas	57
Cologne, Fed. Rep. Germany	1.45 D	14	INAA, 1974	Neurinomas	57
Cologne, Fed. Rep. Germany	0.91 D	3	INAA, 1974	Ependymomas	57
Cologne, Fed. Rep. Germany	1.26 D	25	INAA, 1974	Meningiomas	57
Cologne, Fed. Rep. Germany	1.34 D	9	INAA, 1974	Pituitary adenomas	57
North America					
Ontario, Canada	0.423 W	9	GC, 1981	Healthy adults, Se(VI): 24%/total	193
U.S.	1.13 D	21	XRF, 1982	Autopsy material	97
U.S.	1.00 D	96	XRF, 1982	Uremic patients, dialyzed	97
Lexington, KY, U.S.	0.187 W	28	INAA, 1984	Healthy adults	58
Lexington, KY, U.S.	0.107 W	7	INAA, 1984	Infants	58
Asia					
Japan	1.7 D	63	INAA, 1980	Cerebrum, forsenic material	51
Japan	2.1 D	63	INAA, 1980	Cerebellum, forensic material	51
Oceania					
Australia	0.95 D	21	XRF, 1982	Autopsy material	97
Australia	2.54 D	20	XRF, 1982	Uremic patients, nondialyzed	97
Australia	1.78 D	12	XRF, 1982	Uremic patients, dialyzed	97
Auckland & Dunedin, New Zealand	0.66 D	8	F, 1982	Fetuses	50
Breast Tissue (mg/kg)					
North America					
Chicago, IL, U.S.	0.70 D	25	XRF, 1984	Histologically normal	194
Chicago, IL, U.S.	1.02 D	25	XRF, 1984	Neoplastic tissue	194
Asia					
Chandigarh, India	0.059 D	17	INAA, 1984	Healthy individuals	195
Chandigarh, India	0.26 D	17	INAA, 1984	Breast cancer patients	195

Table 2 (continued)
SELENIUM CONCENTRATIONS IN HUMAN TISSUES AND FLUIDS[a]

Geographical location	Mean Se concentration[b]	N[c]	Analytical method, date of sampling or analysis[d]	Health status, description of subjects, and tissues; remarks	Ref.
Cerebrospinal Fluid (μg/l)					
Riyadh, Saudi-Arabia	19.1	10	EAAS, 1984	Various diseases	196
Endometrium (mg/kg)					
Stockholm, Sweden	0.87 D	31	RNAA, 1977	Elective hysterectomy	197
Eye (mg/kg)					
Europe					
Helsinki, Finland	0.45 D	3	INAA, 1978	Clear lenses	198
Helsinki, Finland	0.60 D	77	INAA, 1978	Cataractous lenses w/o pseudoexfoliation	198
Helsinki, Finland	0.56 D	11	INAA, 1978	Cataractous lenses with pseudoexfoliation	198
North America					
U.S.	0.15 W	—	NAA, 1982	Normal lenses	199
U.S.	0.42 W	—	NAA, 1982	Cataractous lenses	199
U.S.	0.38 W	—	NAA, 1982	Early cataractous lenses	199
U.S.	0.30 W	—	NAA, 1982	Nuclear cataract	199
U.S.	0.51 W	—	NAA, 1982	Cortical cataract	199
U.S.	0.34 W	—	NAA, 1982	Mature cataract	199
U.S.	0.52 W	—	NAA, 1982	Hypermature cataract	199
Baltimore, MD, U.S.	0.26 W	6	ADF, 1966	Sclera	200
Baltimore, MD, U.S.	0.12 W	5	ADF, 1966	Retina	200
Baltimore, MD, U.S.	1.4 W	4	ADF, 1966	Uvea	200
Hair (mg/kg)					
Europe					
Angmagssalik, Greenland	0.86	12	HAAS, 1984	Healthy subjects	171
Tuscany, Italy	0.33	12	INAA, 1976	Healthy subjects	63
Tuscany, Italy	0.44	20	INAA, 1976	Occupational exposure to Hg	63
Birmingham, U.K.	1.47	12	GC, 1977	Healthy subjects	68
Glasgow, U.K.	2.24	46	RNAA, 1981	Healthy subjects	96
Oxford, U.K.	0.51 D	10	INAA, 1982	Healthy adults	62
Oxford, U.K.	0.64 D	10	INAA, 1982	Multiple sclerosis	62
Sussex, Essex, U.K.	0.6 D	44	EAAS, 1981	Healthy children	201
Sussex, Essex, U.K.	1.1 D	73	EAAS, 1981	Dyslexic subjects	201
West Sussex, U.K.	2.8 D	156	ICPAES, 1982	Patients, private practice	202
North America					
Montreal, Quebec, Canada	0.36	12	AAS, 1984	Friedreich's disease	203
Montreal, Quebec, Canada	0.41	12	AAS, 1984	Other ataxias	203
Montreal, Quebec, Canada	0.36	15	AAS, 1984	Neurological disease	203

Table 2 (continued)
SELENIUM CONCENTRATIONS IN HUMAN TISSUES AND FLUIDS[a]

Geographical location	Mean Se concentration[b]	N[c]	Analytical method, date of sampling or analysis[d]	Health status, description of subjects, and tissues; remarks	Ref.
Chicago, IL, U.S.	0.72 D	155	XRF, 1983	Healthy males	60
Chicago, IL, U.S.	0.58 D	215	XRF, 1983	Healthy females	60
Greensboro, NC, U.S.	0.64 D	115	ADF, 1982	Healthy subjects	204
Long Island, NY, U.S.	0.765	52	F, 1982	Healthy children (5—18 years)	166
Milan, NM, U.S.	0.46	39	HAAS, 1978	Healthy adults	179
Rapid City, SD, U.S.	0.77	6	ADF, 1979	Healthy subjects	182
Asia					
People's Rep. China	32.2	65	F, 1983	High-Se area with chronic selenosis	22
People's Rep. China	3.7	14	F, 1983	High-Se area w/o selenosis	22
People's Rep. China	0.36	1745	F, 1983	Selenium-adequate area	22
People's Rep. China	0.16	40	F, 1983	Low-Se area	22
People's Rep. China	0.074	1478	F, 1983	Low-Se area with Keshan disease	22
Mao Wu Su Desert Belt, People's Rep. China	0.079 D	2	F, 1982	Healthy boys (5—13 years)	205
Shaanxi Province, People's Rep. China	0.084 D	12	F, 1982	Healthy boys (5—13 years)	205
Qi-Zeo, Shaanxi Province, People's Rep. China	0.212 D	16	F, 1982	Healthy boys (5—13 years)	205
An-Con Basin, People's Rep. China	2.97 D	8	F, 1982	Healthy boys (5—13 years)	205
Heilongjiang, People's Rep. China	0.390	20	F, 1982	City children	4
Heilongjiang, People's Rep. China	0.151	20	F, 1982	Commune children	4
Sichuan, People's Rep. China	0.131	16	F, 1982	City children	4
Sichuan, People's Rep. China	0.069	10	F, 1982	Commune children	4
Bombay, India	1.94	8	INAA, 1979	Healthy students	206
Calcutta, India	7.4	5	INAA, 1979	Healthy subjects	206
Dehli, India	1.4	11	INAA, 1979	Healthy students	206
Hyderabad, India	1.6	8	INAA, 1979	Healthy students	206
Madras, India	1.1	12	INAA, 1979	Healthy students	206
Maharashtra & Gujarat, India	1.2	23	INAA, 1979	Healthy students	206
Eastern India	1.6	26	INAA, 1979	Healthy students	206
Northern India	1.3	66	INAA, 1979	Healthy students	206
Southern India	1.3	49	INAA, 1979	Healthy students	206
Western India	2.0	15	INAA, 1979	Healthy students	206
Chiba, Japan	8.4	5	INAA, 1984	Healthy children (2—7 years)	207
Morioka City, Japan	2.9	37	INAA, 1979	Healthy males	208
Morioka City, Japan	3.0	67	INAA, 1979	Healthy females	208
Tokyo, Japan	3.74	100	INAA, 1979	Healthy males	208
Tokyo, Japan	3.44	102	INAA, 1979	Healthy females	208
Islamabad, Pakistan	1.05	60	INAA, 1982	Healthy males	209
Islamabad, Pakistan	0.98	45	INAA, 1982	Healthy females	209

Table 2 (continued)
SELENIUM CONCENTRATIONS IN HUMAN TISSUES AND FLUIDS[a]

Geographical location	Mean Se concentration[b]	N[c]	Analytical method, date of sampling or analysis[d]	Health status, description of subjects, and tissues; remarks	Ref.
			Heart (mg/kg)		
Europe					
Glasgow, U.K.	1.38 D	9	RNAA, 1981	Healthy subjects	96
Glasgow, U.K.	1.14 D	10	NAA, 1968	Healthy adults	210
Helsinki, Finland	0.57 D	9	F, 1977	Stillborns	47
Helsinki, Finland	0.63 D	8	F, 1977	Premature (1—7 d)	47
Helsinki, Finland	0.67 D	9	F, 1977	Full term (1—7 d)	47
Helsinki, Finland	0.46 D	6	F, 1977	Children (1—9 months)	47
Helsinki, Finland	0.29 D	2	F, 1977	Children (1—6 years)	47
Helsinki, Finland	0.58 D	11	F, 1977	Carcinoma, intestinal	47
Helsinki, Finland	0.65 D	14	F, 1977	Extraintestinal	47
Helsinki, Finland	0.66 D	30	F, 1977	Myocardial infarction	47
Helsinki, Finland	0.71 D	15	F, 1977	Diseased subjects	47
North America					
U.S.	1.60 D	19	XRF, 1982	Autopsy material	97
U.S.	1.54 D	52	XRF, 1982	Uremic patients, dialyzed	97
Asia					
Japan	1.9 D	63	INAA, 1980	Forensic material	51
Oceania					
Australia	1.17 D	19	XRF, 1982	Autopsy material	97
Australia	1.14 D	20	XRF, 1982	Uremic patients, nondialyzed	97
Australia	1.11 D	24	XRF, 1982	Uremic patients, dialyzed	97
Auckland & Dunedin, New Zealand	0.68 D	5	F, 1982	Healthy adults	50
			Kidney (mg/kg)		
Europe					
Belgium	0.78 W	3	INAA, 1979	Nonuremic donors	211
Copenhagen, Denmark	0.58 W	7	NAA, 1974	Healthy subjects	46
Copenhagen, Denmark	0.21 W	6	NAA, 1974	Uremic patients	46
Helsinki, Finland	1.24 D	6	F, 1977	Kidney cortex, stillborns	46
Helsinki, Finland	1.18 D	11	F, 1977	Kidney cortex, premature (1—7 d)	47
Helsinki, Finland	1.38 D	11	F, 1977	Kidney cortex, full-term (1—7 d)	47
Helsinki, Finland	1.32 D	9	F, 1977	Kidney cortex, children (1—9 months)	47
Helsinki, Finland	1.48 D	2	F, 1977	Kidney cortex, children (1—6 years)	47
Helsinki, Finland	1.95 D	12	F, 1977	Diseased subjects	47
U.K.	0.1 W	8	SSMS, 1972/73	Healthy subjects	56
U.K.	0.2 W	8	SSMS, 1972/73	Kidney cortex, healthy subjects	56
Glasgow, U.K.	4.18 D	8	RNAA, 1981	Healthy subjects	96
Burtrask & Jorn, Sweden	0.70 W	8	RNAA, 1980	Healthy subjects	212
Ronnskarsverken, Sweden	0.43 W	21	RNAA, 1980	Exposed workers	212

Table 2 (continued)
SELENIUM CONCENTRATIONS IN HUMAN TISSUES AND FLUIDS[a]

Geographical location	Mean Se concentration[b]	N[c]	Analytical method, date of sampling or analysis[d]	Health status, description of subjects, and tissues; remarks	Ref.
North America					
Kingston, Ontario, Canada	0.31[g] W	42	HAAS, 1982	Healthy subjects, medulla	213
Kingston, Ontario, Canada	0.84[g] W	39	HAAS, 1982	Healthy subjects, kidney cortex	213
U.S.	1.1 W	123	F, 1980	Different diseases	214
U.S.	4.89 D	22	XRF, 1982	Autopsy material	97
U.S.	3.06	32	XRF, 1982	Uremic patients, dialyzed	97
La Jolla, CA, U.S.	0.46 W	4	F, 1976	Healthy subjects	215
La Jolla, CA, U.S.	2.03 W	3	F, 1976	Cystinotic children	215
Cleveland, PA, U.S.	8.0 D	35	F, 1981	Healthy adults, kidney cortex	216
Morgantown, WV, U.S.	4.9 D	32	F, 1984	Caucasian subjects, kidney cortex	217
Asia					
Japan	1.5 D	63	INAA, 1980	Forsenic material	51
Oceania					
Australia	3.51 D	20	XRF, 1982	Autopsy material	97
Australia	1.83 D	16	XRF, 1982	Uremic patients, dialyzed	97
Australia	1.52 D	16	XRF, 1982	Uremic patients, nondialyzed	97
Auckland & Dunedin, New Zealand	3.14 D	33	F, 1982	Healthy adults, kidney cortex	50
Auckland & Dunedin, New Zealand	2.07 D	2	F, 1982	Infants, kidney cortex	50
Liver (mg/kg)					
Europe					
Gent, Belgium	0.26 W	5	RNAA, 1977	Healthy subjects	218
Copenhagen, Denmark	0.39 W	7	NAA, 1974	Healthy subjects	46
Copenhagen, Denmark	0.31 W	7	NAA, 1974	Uremic patients	46
Helsinki, Finland	1.11 D	8	F, 1977	Stillborns	47
Helsinki, Finland	1.21 D	12	F, 1977	Premature (1—7 d)	47
Helsinki, Finland	0.93 D	12	F, 1977	Full term (1—7 d)	47
Helsinki, Finland	0.58 D	8	F, 1977	Children (1—9 months)	47
Helsinki, Finland	0.59 D	3	F, 1977	Children (1—6 years)	47
Helsinki, Finland	0.62 D	8	F, 1977	Carcinoma intestinal	47
Helsinki, Finland	0.69 D	14	F, 1977	Carcinoma extraintestinal	47
Helsinki, Finland	0.70 D	24	F, 1977	Myocardial infarction	47
Helsinki, Finland	0.67 D	8	F, 1977	Diseased subjects	47
Athens, Greece	4.52 D	6	INAA, 1977	Newborns, hypoxia, intracranial hemorrhage	219
Southeastern Norway	0.39 W	31	F, 1980	Healthy adults	45
Oslo, Norway	0.25 W	5	F, 1982	Alcoholic cirrhosis	45
Burtrask & Jorn, Sweden	0.19 W	8	RNAA, 1980	Healthy subjects	212
Ronnskarsverken, Sweden	0.18 W	20	RNAA, 1980	Exposed workers	212
U.K.	0.3 W	11	SSMS, 1972/73	Healthy subjects	56
Glasgow, U.K.	2.47 D	14	RNAA, 1981	Healthy subjects	96

Table 2 (continued)
SELENIUM CONCENTRATIONS IN HUMAN TISSUES AND FLUIDS[a]

Geographical location	Mean Se concentration[b]		N[c]	Analytical method, date of sampling or analysis[d]	Health status, description of subjects, and tissues; remarks	Ref.
Glasgow, U.K.	2.40	D	12	NAA, 1968	Healthy adults	210
Marburg & Lahn, Fed. Rep. Germany	0.33	W	37	PIXE, 1979	Healthy adults	220
Marburg & Lahn, Fed. Rep. Germany	0.41	W	9	PIXE, 1979	Fetal livers, newborns	220
North America						
Ontario, Canada	0.43	W	10	RNAA, 1967	Healthy adults	25
Kingston, Ontario, Canada	0.39[g]	W	44	HAAS, 1982	Healthy subjects	213
U.S.	0.50	W	36	AAS/NAA, 1983	Healthy adults	44
U.S.	2.25	D	25	XRF, 1982	Autopsy material	97
U.S.	2.33	D	52	XRF, 1982	Uremic patients, dialyzed	97
La Jolla, CA, U.S.	2.67		3	F, 1976	Cystinotic children	215
Louisville, KY, U.S.	1.73	D	7	INAA, 1975	Healthy subjects	221
Louisville, KY, U.S.	1.40	D	4	INAA, 1975	Diseased livers	221
Baltimore, MD, U.S.	3.4	D	13	EAAS, 1976	Sudden infant death syndrome	222
Philadelphia, PA, U.S.	3.2	D	14	EAAS, 1976	Autopsy material, children	222
Seattle, WA, U.S.	2.9	D	11	NAA, 1973	Anencephalic fetuses	223
Seattle, WA, U.S.	2.2	D	9	NAA, 1973	Premature fetuses	223
Seattle, WA, U.S.	2.8	D	12	NAA, 1973	Therapeutic abortions	223
Seattle, WA, U.S.	2.3	D	14	NAA, 1973	Spontaneous abortions	223
Asia						
Japan	2.3	D	63	INAA, 1980	Forensic material	51
Oceania						
Australia	1.66	D	21	XRF, 1982	Autopsy material	97
Australia	1.89	D	15	XRF, 1982	Uremic patients, dialyzed	97
Australia	1.59	D	23	XRF, 1982	Uremic patients, nondialyzed	97
New South Wales, Australia	0.259	W	28	F, 1978	Sudden infant death syndrome	224
New Zealand	0.25	W	23	F, 1983	Healthy females	225
New Zealand	0.33	W	73	F, 1983	Healthy males	225
New Zealand	0.159	W	92	F, 1978	Sudden infant death syndrome	224
New Zealand	0.193	W	17	F, 1978	Children, accidental death	224
Auckland & Dunedin, New Zealand	0.72	D	41	F, 1982	Healthy adults	50
Auckland & Dunedin, New Zealand	1.26	D	12	F, 1982	Fetuses	50
Auckland & Dunedin, New Zealand	0.58	D	4	F, 1982	Infants	50
Lung (mg/kg)						
Europe						
Gent, Belgium	0.177	W	7	INAA, 1983	Healthy subjects	226
Copenhagen, Denmark	0.16	W	7	NAA, 1974	Healthy subjects	46
Copenhagen, Denmark	0.12	W	7	NAA, 1974	Uremic patients	46
Burtrask & Jorn, Sweden	0.11	W	9	RNAA, 1980	Healthy subjects	212

Table 2 (continued)
SELENIUM CONCENTRATIONS IN HUMAN TISSUES AND FLUIDS[a]

Geographical location	Mean Se concentration[b]	N[c]	Analytical method, date of sampling or analysis[d]	Health status, description of subjects, and tissues; remarks	Ref.
Burtrask & Jorn, Sweden	0.13 W	36	RNAA, 1980	Exposed workers	212
U.K.	0.1 W	11	SSMS, 1972/73	Healthy subjects	56
Glasgow, U.K.	1.43 D	8	RNAA, 1981	Healthy subjects	96
Cologne, Fed. Rep. Germany	1.02 D	18	INAA, 1976	Healthy subjects	227
Duisburg, Fed. Rep. Germany	2.16 D	5	INAA, 1976	Healthy adults	227
North America					
U.S.	1.30 D	18	XRF, 1982	Autopsy material	97
U.S.	1.39 D	46	XRF, 1982	Uremic patients, dialyzed	97
Asia					
Japan	7.8 D	22	INAA, 1980	Forensic material, female	51
Oceania					
Australia	0.96 D	21	XRF, 1982	Autopsy material	97
Australia	0.77 D	20	XRF, 1982	Uremic patients, non-dialyzed	97
Australia	0.94 D	21	XRF, 1982	Uremic patients, dialyzed	97
Auckland & Dunedin, New Zealand	0.45 D	6	F, 1982	Healthy adults	50
Lymph Nodes (mg/kg)					
Europe					
U.K.	0.05 W	6	SSMS, 1972/73	Healthy subjects	56
Milk (μg/l)					
Europe					
Helsinki, Finland	10.7	13	EAAS, 1983	Healthy mothers, 1st mo. of lactation	228
Helsinki, Finland	5.8	13	EAAS, 1983	Healthy mothers, 3rd month of lactation	228
Athens, Greece	20	24	F, 1973	Healthy mothers	229
Italy	13 μg/kg W	21	INAA, 1982	Healthy mothers (≥15 d after deliv.)	230
Fed. Rep. Germany	128 μg/kg D	3	INAA, 1977	Healthy mothers	35
Dusseldorf, Fed. Rep. Germany	230 μg/kg D	44	INAA, 1978	Healthy mothers	231
Dusseldorf, Fed, Rep. Germany	245 μg/kg D	25	INAA, 1978	Healthy mothers, transitory milk	231
North America					
Champaign & Urbana, IL, U.S.	16.3	72	GC, 1982	Healthy mothers	232
Iowa City, IA, U.S.	20	15	F, 1973	Healthy mothers	229
Portland, OR, U.S.	21	15	F, 1973	Healthy mothers	229
Asia					
Japan	21	34	F, 1983	Healthy mothers	168
Oceania					
Australia	12 μg/kg	40	INAA, 1983	Healthy mothers	170

Table 2 (continued)
SELENIUM CONCENTRATIONS IN HUMAN TISSUES AND FLUIDS[a]

Geographical location	Mean Se concentration[b]	N[c]	Analytical method, date of sampling or analysis[d]	Health status, description of subjects, and tissues; remarks	Ref.
			Nails (mg/kg)		
North America					
Illinois, U.S.	0.9 D	9	INAA, 1979	Healthy females	233
Illinois, U.S.	0.6 D	11	INAA, 1979	Healthy adults	233
Portland, OR, U.S.	1.14	16	F, 1973	Healthy adults	234
Virginia, U.S.	8.7 D	8	SSMS, 1972	Healthy adults	235
Asia					
Bangladesh	1.28 D	51	PIXE, 1984	Healthy adults	236
			Nerves (mg/kg)		
Europe					
Copenhagen, Denmark	0.085 W	7	NAA, 1974	Healthy subjects, N. ischiadicus	46
Copenhagen, Denmark	0.13 W	6	NAA, 1974	Healthy subjects, N. ulnaris	46
Copenhagen, Denmark	0.10 W	6	NAA, 1974	Healthy subjects, N. tibialis	46
Copenhagen, Denmark	0.069 W	7	NAA, 1974	Uremic patients, N. ischiadicus	46
Copenhagen, Denmark	0.099 W	7	NAA, 1974	Uremic patients, N. ulnaris	46
Copenhagen, Denmark	0.12 W	2	NAA, 1974	Uremic patients, N. tibialis	46
			Ovary (mg/kg)		
Europe					
U.K.	0.09 W	6	SSMS, 1972/73	Healthy subjects	56
			Pancreas (mg/kg)		
Europe					
Copenhagen, Denmark	0.22 W	7	NAA, 1974	Healthy subjects	46
Copenhagen, Denmark	0.19 W	3	NAA, 1974	Uremic patients	46
North America					
Louisville, KY, U.S.	0.63 D	12	INAA, 1975	Healthy adults	221
Louisville, KY, U.S.	0.98 D	11	INAA, 1975	Diseased pancreas	221
Asia					
Japan	1.9 D	63	INAA, 1980	Forensic material	51
			Placenta (mg/kg)		
Europe					
Denmark	0.173 W	19	RNAA, 1983	Risk for Menkes disease	237
Athens, Greece	1.90 D	18	INAA, 1977	Healthy mothers	219
Stockholm, Sweden	1.22 D	14	RNAA, 1977	Pregnant women, decidua	197
Galicia, Spain	0.828 D	57	F, 1983	Healthy mothers	151
Birmingham, U.K.	0.34 W	50	GC, 1977	Healthy mothers	68
North America					
Nashville, TN, U.S.	1.70 D	822	INAA, 1974	Healthy mothers	183
Galveston, TX, U.S.	0.373 W	16	INAA, 1968	Healthy mothers	238

Table 2 (continued)
SELENIUM CONCENTRATIONS IN HUMAN TISSUES AND FLUIDS[a]

Geographical location	Mean Se concentration[b]	N[c]	Analytical method, date of sampling or analysis[d]	Health status, description of subjects, and tissues; remarks	Ref.
Saliva (μg/l)					
North America					
Portland, OR, U.S.	3.1	26	F, 1971	Healthy school children	239
Seminal Plasma (μg/l)					
Europe					
Oslo, Norway	35	15	EAAS, 1982	Infertility patients	148
North America					
Montreal, Quebec, Canada	71.3	120	F, 1984	Infertility patients	240
Norfolk, VA, U.S.	21—190[h]	45	EAAS, 1983	Infertility patients	241
Skeletal Muscle (mg/kg)					
Europe					
Copenhagen, Denmark	0.17 W	6	NAA, 1974	Healthy subjects	46
Copenhagen, Denmark	0.14 W	7	NAA, 1974	Uremic patients	46
U.K.	0.11 W	6	SSMS, 1972/73	Healthy subjects	56
North America					
Ontario, Canada	0.37 W	10	RNAA, 1967	Healthy adults	25
U.S.	1.60 D	34	XRF, 1982	Autopsy material	97
U.S.	1.35 D	64	XRF, 1982	Uremic patients, nondialyzed	97
Asia					
Japan	1.7 D	63	INAA, 1980	Forensic material	51
Oceania					
Australia	0.75 D	21	XRF, 1982	Autopsy material	97
Australia	0.84 D	17	XRF, 1982	Uremic patients, nondialyzed	97
Australia	0.89 D	21	XRF, 1982	Uremic patients, dialyzed	97
Auckland & Dunedin, New Zealand	0.29 D	22	F, 1982	Healthy adults	50
Skin (mg/kg)					
Europe					
Manchester, U.K.	0.52 D	17	INAA, 1979	Abdominal skin, epidermis	59
Manchester, U.K.	0.39 D	36	INAA, 1979	Abdominal skin, dermis	59
Manchester, U.K.	0.39 D	9	INAA, 1979	Plantar skin, epidermis	59
Manchester, U.K.	0.46 D	10	INAA, 1979	Plantar skin, dermis	59
North America					
Ontario, Canada	0.27 W	10	RNAA, 1967	Healthy adults	25
Spermatozoa (mg/kg)					
North America					
Norfolk, VA, U.S.	1.80 W	10	EAAS, 1983	Infertility patients	241

Table 2 (continued)
SELENIUM CONCENTRATIONS IN HUMAN TISSUES AND FLUIDS[a]

Geographical location	Mean Se concentration[b]	N[c]	Analytical method, date of sampling or analysis[d]	Health status, description of subjects, and tissues; remarks	Ref.
			Spleen (mg/kg)		
Europe					
Copenhagen, Denmark	0.24 W	7	NAA, 1974	Healthy subjects	46
Copenhagen, Denmark	0.16 W	3	NAA, 1974	Uremic patients	46
Glasgow, U.K.	1.59 D	9	RNAA, 1981	Healthy subjects	96
North America					
U.S.	1.74 D	23	XRF, 1982	Autopsy material	97
U.S.	1.68 D	47	XRF, 1982	Uremic patients, dialyzed	97
Asia					
Japan	1.7 D	63	INAA, 1980	Forensic material	51
Oceania					
Australia	1.13 D	21	XRF, 1982	Autopsy material	97
Australia	1.09 D	15	XRF, 1982	Uremic patients, nondialyzed	97
Australia	1.16 D	22	XRF, 1982	Uremic patients, dialyzed	97
			Synovial Membrane (mg/kg)		
North America					
Louisville, KY, U.S.	0.63 D	28	INAA, 1975	Healthy subjects	221
Louisville, KY, U.S.	0.88 D	94	INAA, 1975	Diseased subjects, osteoarthritis, trauma, inflammation	221
Louisville, KY, U.S.	1.27 D	6	INAA, 1975	Rheumatoid arthritis	221
			Teeth (mg/kg)		
Europe					
Edinburgh, U.K.	0.83	3	RNAA, 1970	Adult permanent teeth, enamel	242
Manchester, U.K.	0.91	3	RNAA, 1970	Adult permanent teeth, enamel	242
North America					
U.S.	0.43 D	56	SSMS, 1973	Healthy boys (av. 14 years), enamel	243
Dunedin, New Zealand & U.S.	1.47	326	SSMS, 1978	Healthy youth (10—20 years), premolars, enamel	244
Virginia, U.S.	0.48 D	173	AAS, 1974	Permanent, sound teeth, enamel	245
Virginia, U.S.	0.28 D	173	AAS, 1974	Permanent, sound teeth, dentin	245
Africa					
South Africa	0.08	10	RNAA, 1976	Permanent, sound teeth, black residents, enamel	246
South Africa	0.012	10	RNAA, 1976	Permanent, south teeth, white residents, enamel	246
Withwatersrand, South Africa	0.08	10	RNAA, 1974	Canine teeth from Bantu patients, enamel	247

Table 2 (continued)
SELENIUM CONCENTRATIONS IN HUMAN TISSUES AND FLUIDS[a]

Geographical location	Mean Se concentration[b]	N[c]	Analytical method, date of sampling or analysis[d]	Health status, description of subjects, and tissues; remarks	Ref.
Thyroid Gland (mg/kg)					
Europe					
Idrija, Yugoslavia	5.6 W	6	RNAA, 1974	Hg-exposed subjects and controls	248
Testis (mg/kg)					
Europe					
U.K.	0.2 W	5	SSMS, 1972/73	Healthy subjects	56
Urine (μg/l)					
Europe					
Lyon, France	12.3	92	F, 1980	Healthy adults	249
Italy	14 μg/24 h	40	INAA, 1977	Healthy adults	67
Tuscany, Italy	7.4	12	INAA, 1976	Healthy adults	63
Tuscany, Italy	7.4	13	INAA, 1976	Exposed workers (Hg)	63
Oslo, Norway	27.3 μg/g C	21	HAAS, 1983	Healthy male workers	72
Oslo, Norway	44.7 μg/g C	28	HAAS, 1983	Exposed workers (Hg)	72
Birmingham, U.K.	5	5	GC, 1977	Healthy adults	68
Stockholm, Sweden	30 μg/24 h	16	RNAA, 1973	Arterial hypertension	152
North America					
Montreal, Quebec, Canada	125 μg/24 h	10	F, 1982	Healthy men	158
Milan, NM, U.S.	19.2	35	HAAS, 1978	Healthy adults	179
Long Island, NY, U.S.	28.7 μg/g C	44	F, 1982	Healthy children (5—18 years)	166
Portland, OR, U.S.	49	16	F, 1973	Healthy subjects	234
Brookings, SD, U.S.	114 μg/g C	8	F, 1983	High animal protein diet	250
Brookings, SD, U.S.	118 μg/g C	8	F, 1983	Ovo-lacto-vegetarian diet	250
Brookings, SD, U.S.	67 μg/g C	8	F, 1983	Vegetarian diet	250
Brookings, SD, U.S.	75 μg/g C	8	F, 1983	Seafood-lacto-vegetarian diet	250
Rapid City, SD, U.S.	63	6	F, 1979	Healthy subjects	182
South America					
Venezuela	152	1055	F, 1971	Healthy school children	69
Villa Bruzual, Venezuela	636 μg/g C	111	F, 1972	Healthy school children	23
Caracas, Venezuela	224 μg/g C	50	F, 1972	Healthy school children	23
Africa					
Yavne, Israel	25	10	RNAA, 1979	Healthy subjects	251
Asia					
People's Rep. China	7	43	F, 1983	Low-Se area with Keshan disease	22
People's Rep. China	26	19	F, 1983	Selenium-adequate area	22
People's Rep. China	140	14	F, 1983	High-Se area w/o selenosis	22
People's Rep. China	2680	17	F, 1983	High-Se area with chronic selenosis	22
Japan	57.9	21	F, 1981	Healthy adults	71

Table 2 (continued)
SELENIUM CONCENTRATIONS IN HUMAN TISSUES AND FLUIDS[a]

Geographical location	Mean Se concentration[b]	N[c]	Analytical method, date of sampling or analysis[d]	Health status, description of subjects, and tissues; remarks	Ref.
Kyoto, Japan	105 (43.4μg/g C)	20	F, 1982	Healthy males (19—27 years)	252
Japan	68.9	22	F, 1981	Exposed workers (Mn)	71
Japan	106.9	14	F, 1981	Exposed workers (Cr)	71
Japan	166.4	5	F, 1981	Exposed workers (Cd)	71
Japan	288	1	F, 1981	Exposed worker (Hg)	71
Japan	44.5	2	F, 1981	Hypertension patients	71
Japan	26.0	23	F, 1981	Cancer patients	71
Japan	36.0	21	F, 1981	Epileptic patients	71
Oceania					
South Otago, New Zealand	21 (17μg/24 h)	39	F, 1960	Healthy school boys (5—14 years)	253
North Canterbury, New Zealand	30 (24μg/24 h)	31	F, 1960	Healthy school boys (5—14 years)	253

[a] Total selenium unless otherwise stated.
[b] Arithmetic mean unless otherwise indicated; basis of concentration data; D — dry weight; W — wet weight; C — creatinine; Hb — hemoglobin.
[c] Number of subjects.
[d] Analytical techniques used: AAS — atomic absorption spectrometry; ADF — acid decomposition fluorometry; EAAS — electrothermal atomic absorption spectrometry; F — fluorometry; GC — gas chromatography; HAAS — hydride generation atomic absorption spectrometry; INAA — instrumental neutron activation analysis; NAA — neutron activation analysis; PIXE — proton-induced X-ray emission; RNAA — radiochemical neutron activation analysis; SSMS — spark source mass spectrometry; XRF — X-ray fluorescence spectrometry. Sample preparation (decomposition) procedures are not indicated. Date of sampling or analysis usually but not necessarily synonymous with date of publication.
[e] Geometric mean.
[f] Reported as the oxidation state Se(IV).
[g] Median.
[h] Range.

in the range of 1300 to 7500 μg/l, in contrast to the endemic Keshan disease (juvenile cardiomyopathy) found in the low-selenium area where blood selenium concentrations were as low as 21 ± 10 μg/l.

Relatively high amounts within the "safe range" (27 to 440 μg/l) are found in Caracas, 355 μg/l,[23] and in general in both the U.S. and Canada. For example, in Rapid City, SD, values reported are 250 to 300 μg/l, and a mean of 206 μg/l for 210 samples collected from several places in the U.S. was reported by Allaway et al.[24] As in most of continental Europe, lower concentrations are also found in Scandinavia, and particularly in Finland where they are below 70 μg/l in a number of regions. In New Zealand, another low-selenium area, blood concentrations are found to be in the same range as or lower than in Finland. The geographical variations can be summarized as follows:
Caracas > South Dakota > Guatemala > Canada and most parts of the U.S. > most of Europe > Finland > New Zealand and Egypt. In general, blood selenium concentration is lower in children than in adults. Furthermore, it has been reported that it decreases further after 60 years of age.[25,26]

2. Erythrocytes

It has been reported that erythrocyte selenium is correlated with plasma selenium content, the cellular concentration (expressed as μg/l packed cells) being about 1.5 to 2.0 times higher than that in plasma.[27] Dickson and Tomlinson[25] found a factor of 1.6, with an average level in erythrocytes for Canadians of 236 μg/l of cells derived from a study of 251 subjects. Regional differences similar to those for whole blood are reported. Quantitative changes in dietary intake are reflected in red blood cell selenium after a period of approximately 4 weeks, slower than in plasma, since incorporation of ingested selenium into red blood cell proteins occurs only during erythropoiesis.[28] Consequently, erythrocyte selenium is considered to provide a long-term indication of general selenium status.

A considerable fraction of the selenium appears to be incorporated in glutathione peroxidase; over 75% of ovine red cell selenium occurs in this enzyme. In humans, there is a correlation between erythrocyte selenium content and glutathione peroxidase activity, at least as long as the former is below a threshold value of approximately 140 μg/l.[28] If this value is exceeded, enzyme activity seems to reach a plateau and an increasing fraction of red cell selenium then exists in a nonenzyme form, probably indicating that the selenium requirement of the enzyme is met. It follows that in humans the percent of red cell selenium existing in glutathione peroxidase will depend at least in part on the selenium status of the whole body.

Intake of organic selenium in excess, e.g., from yeast or wheat, can given rise to very high cellular concentrations of selenium, existing predominantly in a non-glutathione peroxidase form, whereas supplementation by inorganic compounds of the element does not affect the nonenzymatic (inactive) selenium pool to the same extent.[29] Thus, assay of glutathione peroxidase activity may supplement, but not replace, the use of selenium determination for assessment of selenium nutritional status.

3. Platelets and White Blood Cells

It has been found that glutathione peroxidase as well as the selenium concentrations are higher in platelets than in erythrocytes. In a German population, Kasperek et al.[30] found 782 ± 127 μg of Se per kilogram of platelets (wet weight basis), which is six to seven times the erythrocyte concentration in the same population. Furthermore, selenium supplementation given to healthy individuals, as studied in a Finnish group, leads to substantially increased platelet glutathione peroxidase activity.[29] The presence of this large amount of enzyme in blood platelets is presumed to be related to the rapid turnover of prostaglandin precursors with endoperoxide structure in these cells. In contrast to the high amounts seen in platelets of healthy persons, lower amounts of glutathione peroxidase are found in patients with thrombastenia,[31] but the possibility that selenium deficiency plays a role in this disorder is still under debate.

On the other hand, white blood cells, in general, contain rather low concentrations of selenium.[32] The selenium content of mononuclear cells (range: 2.6 to 8.1 μg/g protein) and of neutrophils (range: 1.8 to 3.0 μg/g protein) has been studied by Pleban and McHugh.[33]

4. Plasma and Serum

According to most reports, blood plasma and serum selenium content is approximately 75% of that found in whole blood. It is affected by recent changes in dietary selenium intake to a larger degree than in red cells and fluctuates somewhat more than cellular concentrations. Plasma as well as whole blood concentrations appear to be age dependent with very low amounts being found in small children. Umbilical cord blood plasma contains about 75% as much selenium as is present in the plasma of the mothers. The lowest reported values of ~20 to 35 μg/l were found in Belgian and German babies from 1 to 6 months of age with a gradual rise occurring in toddlers and youngsters up to the age of maturity.[34,35]

Based on studies in China, the concentration of selenium in plasma can be as low as 20 μg/l without risk of cardiomyopathy.[22] However, in German children with phenylketonuria, Lombeck and co-workers[36] found even lower concentrations, although clinical symptoms of heart disease were entirely absent. It is probable that cellular antioxidants, rather than the intracellular concentration of glutathione peroxidase, are involved in protective mechanisms and, therefore, plasma selenium values alone are insufficient to determine the probability of disease.

Low concentrations of selenium in serum may accompany pathological conditions other than Keshan disease. Markedly less selenium was reported in the plasma, but not the cells, of three patients with extensive burns.[26] In another study, lower amounts of plasma selenium were found in children with protein/calorie malnutrition than in controls.[23] Malignant diseases, liver cirrhosis, and rheumatoid arthritis are also associated with lower than normal selenium concentrations in plasma, probably as a result of metabolic disturbances.[37-39]

Normally, only a very small fraction of plasma selenium occurs as glutathione peroxidase. Gel filtration studies have shown that roughly half of the selenium is recovered in fractions corresponding to a molecular weight between 50,000 and 100,000.[40] A smaller peak of selenium enrichment in the chromatogram corresponded to the region $\geqslant$ 700,000. A selenoprotein in monkey plasma, of molecular weight 80,000, has been described and presumed to be a transport agent.[41]

Selenium components circulating in blood plasma can be regarded as transport forms of the element. It is presumed that selenium deposited in internal organs is sooner or later converted to functionally active forms. Distribution among internal organs has been examined by Schroeder et al.[42] in the U.S., who found relative concentrations in tissues as follows: kidney > liver > spleen > testes > heart muscle > lung > brain. The same order of distribution in human tissues also appears in reports of studies carried out in other parts of the world, although the actual concentrations vary considerably.

B. Liver and Spleen

Amounts of trace elements in the liver are generally considered to be good indicators of body trace element status.[43] Therefore, high concentrations in people from Venezuela, Enshi County-Hubei Province (China), and South Dakota, but low concentrations in those from Scandinavia, particularly Finland, as well as New Zealand are anticipated. Data are, however, relatively sparse and completely lacking for Venezuela and Enshi. In the U.S., the National Bureau of Standards (NBS) reported a mean value in liver of 0.5 mg/kg wet weight,[44] which is in good agreement with the mean value of 0.54 mg/kg reported by Schroeder et al.[42] A somewhat lower value of 0.43 ± 0.12 mg/kg was found for Canada.[25] In both Norway[45] and Denmark[46] a mean of 0.39 mg/kg was found, whereas even lower concentrations, viz., 0.15 to 0.25 mg/kg in the range of those from New Zealand were reported for Finland[47] and Sweden.[48]

Although a correlation among different organs could not be found by Dickson and Tomlinson,[25] on a group basis the ratio of the concentration of selenium in the liver (mg/kg wet basis) to that in blood (mg/l) was approximately 2.5.

In hepatocytes, as well as erythrocytes, it is assumed that a significant fraction of the selenium exists as a component of glutathione peroxidase, but a detailed speciation remains to be done. Glutathione peroxidase apparently operates only intracellularly and is dependent on the concentration of glutathione. It is interesting to note that the concentration of both glutathione and selenium is two to three times higher in liver cells than in erythrocytes,[49] indicating much higher enzymatic activity associated with the liver cells, presumably relating to the systems for detoxification of peroxidative chemicals in the liver.

It appears that the selenium concentration in the spleen is 60 to 70% of that found in the liver.[42]

C. Kidneys

Concentrations of selenium in the kidney are usually two to three times higher than those in the liver, with similar regional variations. Thus, low values of 0.75 ± 0.26 mg/kg in the cortex are reported in New Zealand,[50] and about 0.5 mg/kg in Finland.[47] In Sweden, Wester et al.[48] reported a mean value of approximately 0.7 mg/kg in a group of workers. Higher concentrations were found in the U.S. by, among others, Schroeder et al.[42] who reported a mean of 1.09 mg/kg, and even higher concentrations of 1.50 ± 0.96 mg/kg have been reported in Japan.[51] However, since data for Venezuela and China are still lacking and reports on the subject are few in general, the various factors, of which whole body selenium status is apparently one, affecting the concentration of selenium in the kidney are difficult to determine. The larger amounts of glutathione peroxidase and glutathione are an indication of the capacity for detoxification of peroxides.[49] The non-glutathione peroxidase fraction of renal selenium may also be related to detoxification mechanisms, as heavy metal-selenium complexes are transported to and precipitated in renal tissue upon exposure to metals such as silver, mercury, or cadmium.[52,53]

D. Heart and Skeletal Muscle

The concentration of selenium in heart tissue, according to the few available reports, appears to be somewhat lower than in liver. In New Zealand, Casey et al.[50] found 0.19 ± 0.03 mg/kg. Although such information on patients suffering from Keshan disease in China has not been published, if it is assumed that there is a correspondingly lower amount of selenium in the heart as in blood, concentrations in heart tissue may be as low as 0.06 mg/kg. Fleming et al.[54] have reported a case of fatal cardiomyopathy in a patient on total parenteral nutrition with 0.03 mg Se/kg in the heart, whereas they found a mean of 0.20 mg/kg in two control patients. Concentrations of 0.22, 0.28, and 0.26 mg/kg have been reported in hearts of healthy humans in Canada, the U.S., and the Federal Republic of Germany, respectively,[25,42,55] and somewhat lower values in Finland.[47] Information is unavailable for the high-selenium areas of Venezuela and Enshi, China.

The concentration of selenium in skeletal muscle is lower than in heart muscle, judging from most reports. However, since muscular tissue constitutes a large fraction of body weight, roughly 40% of the total amount of selenium in the body is in muscles. Casey et al.[50] found 0.06 ± 0.02 mg/kg in muscular tissue in New Zealand residents. No reports are available for China. Values reported for Europe are in the range of 0.1 to 0.2 mg/kg,[46,56] while those for North America, 0.3 to 0.4 mg/kg, and for Japan[51] are somewhat higher.

E. Brain

Casey et al.[50] found 0.09 ± 0.02 mg/kg in fetal brain tissue (27 to 42 weeks) in New Zealand. Westermarck[47] has reported 0.08 mg/kg in adult humans from Finland, in the same range as in the U.K.[56] and the Federal Republic of Germany,[57] while a range of 0.15 to 0.20 mg/kg is found in Canada,[25] the U.S.,[58] and Japan.[51] Symptoms of neurological disturbances including hemiolegia accompanied the selenium poisoning in a few cases in Enshi, but the authors do not include values for brain selenium concentrations.[22] It is apparent that not only the dietary intake of selenium, but also other factors may play a role. For example, coexposure of selenium and heavy metals such as mercury and cadmium can induce increased transport of metal complexes to the brain.[53]

F. Lung

Casey et al.[50] observed 0.10 ± 0.03 mg/kg for selenium in lung tissue in New Zealand. Similar amounts are reported for Denmark,[46] Sweden,[48] and the U.K.[56] Reported values from the Federal Republic of Germany,[55] Canada,[25] and the U.S.[42] of 0.16, 0.21, and 0.15 mg/kg, respectively, are higher. Dietary intake of selenium is considered a factor contributing to lung selenium concentrations, but it should be pointed out that the effects of occupational exposure to heavy metals or other toxic compounds are insufficiently explored. The amount of selenium in the lungs of patients with various pulmonary diseases is virtually unknown.

G. Skin and Hair

The possible usefulness of skin as an indicator of selenium status of the body has not been fully evaluated. Dickson and Tomlinson[25] found a mean of 0.27 mg/kg in skin (range: 0.12 to 0.62 mg/kg), whereas values reported for Europe are lower.[59] Hair has been used much more extensively to assess body status, although the selenium compounds which are added to certain shampoos are absorbed into the hair as external contaminants.[60] In the selenium-adequate areas of the People's Republic of China a value of 0.36 ± 0.16 mg/kg was found in hair.[22] Mean values of 3.7 and 0.16 mg/kg in hair were reported for the high- and low-selenium areas, respectively, where no symptoms of selenosis or of deficiency were recorded. However, in the high-selenium area, where chronic selenosis is a problem, 32.2 mg/kg (range: 4.1 to 100 mg/kg) in hair has been reported, and where Keshan disease occurs, 0.074 ± 0.050 mg/kg in hair was found.[22] Valentine et al.[61] have reported a mean of 0.46 mg/kg for New Mexico, comparable to the value found in the selenium-adequate area in China. Most of the concentrations in hair reported for Europe are in the range of 0.3 to 0.6 mg/kg.[62,63]

H. Urine

The amount of selenium excreted in the urine has been widely used to estimate the amount ingested. Based on studies by Thompson[64] of New Zealand women, the average daily intake of selenium may be assessed roughly as twice the 24-h urinary excretion rate, provided that no foods unusually high in selenium have been consumed. While it is generally accepted that the concentration of selenium in urine is a reflection of the daily intake of the element, the above relationship may not be valid for regions in which daily intakes are high or average.

Concentrations of selenium in urine samples (Table 2) are expressed in micrograms per liter since most authors use this unit, even though it is a less reliable basis for comparison than the total amount of selenium excreted in a 24-h collection of urine. it should be mentioned, however, that Hojo[65] was able to use single-void urines as substitutes for 24-h urine samples when expressing the concentration in micrograms of Se per gram of creatinine.

One of the lowest values given in Table 2 is 7 μg/l, found in the low-selenium area of China were Keshan disease is endemic. However, values in the same range, viz., 0.9 to 13.6 μg/l, were found in New Guinea[66] and in Italy.[63,67] Low values in New Zealand and in a small group (n = 5) in Great Britain[68] have also been reported. The highest values of 880 to 6630 μg/l come from the high-selenium area of endemic poisoning in Enshi.[22] A high individual value of 3900 μg/l in Venezuela[69] also indicates selenium intoxication. Other data for Venezuela range from 152 to 636 μg/l. In high-selenium areas in China where selenosis does not occur, values in the range of 40 to 330 μg/l are reported.[22]

The normal selenium concentration in urine is below 100 μg/l, and it has been proposed that this value be used as the maximum allowable concentration.[70] The use of this threshold, however, to evaluate the extent of occupational exposure has been questioned for several reasons. First, exposure to dangerous selenium compounds such as hydrogen selenide can cause severe damage to the lungs before the urinary selenium concentration increases. Furthermore, the fraction of urinary selenium derived from pulmonary absorption may be small compared to that arising from dietary selenium intake. Factors other than dietary intake or occupational exposure can also affect urine selenium concentration. Japanese workers exposed to heavy metals (mercury, cadmium) were found to have higher urinary selenium content than control groups,[71] and similar observations have been made in Norway and Italy.[72,73] It may be supposed that these metals can accelerate the excretion of selenium as the metal selenides.

Lower selenium concentrations in urine have been reported for patients with cancer, coeliac disease, and burns, and for alcoholic subjects; in the case of alcoholics it is unknown whether this is a result of reduced intake or reduced resorption of the element.[74,75] It has been suggested that a marginal selenium deficiency for an extended period may aggravate alcoholic cardiomyopathy.[76]

Little is known of the different selenium species in human urine. Trimethylselenonium was first isolated and identified in urine from rats,[77,78] and it is thought to be a normal detoxification metabolite of selenium. The compound is also found in human urine.[79] Oyamada and Ishizaki[80] found that trimethylselenonium made up 10 to 47% of the total selenium content of urine, but the other metabolites of selenium excreted in human urine have not yet been identified. It is reasonable to suppose that a more detailed knowledge of urinary selenium speciation will improve the monitoring of cases of occupational exposure.

IV. UPTAKE, RETENTION, AND EXCRETION OF SELENIUM

It was mentioned previously that the average daily intake of selenium differs much among geographic regions. In New Zealand it is reported to be 28 μg, and even lower at ca. 13 μg in rural parts of Tuscany, Italy.[28,73] A range of 10 to 35 μg/d is ingested in Finland,[81] and as little as 2 to 20 μg/d is ingested in the area of China where Keshan disease occurs.[4] A somewhat higher daily intake is reported for Sweden[82] and ca. 50 to 60 μg/d for Norway.[83] The intake in Japan is approximately 90 μg/d,[84] while in the U.S.[85] and Canada[86] it is almost 150 μg/d. De Mondragón and Jaffé[69] found the intake in Caracas, Venezuela, to be about 325 μg/d, and in a high-selenium area of China 750 μg/d is ingested without symptoms of selenosis.[22] The highest daily intakes, 3200 to 6690 μg/d, were recorded for Enshi and were associated with signs of endemic poisoning such as loss of hair and nails and neurological disturbances.

Almost all ingested selenium (70 to 100%) is absorbed from the intestine,[87] apparently without homeostatic or physiological control. This implies that the percent absorbed will depend primarily on the bioavailability of the selenium in its various dietary forms. No quantitative data on humans with regard to selenium absorption through the lungs are available, but it should be noted that high urinary selenium concentrations have been reported for workers exposed to the element.

It is assumed that the daily intake of selenium is reflected in the total amount of the element found in the body. Rough estimates obtained by extrapolations based on a constant ratio of the amount of selenium in the whole body to that in the liver[42] resulted in values for total body selenium in the U.S., Norway, and New Zealand of 15 (range: 13.0 to 20.3 mg), 11, and 6 mg, respectively. The latter value is in good agreement with the average of 6.1 mg (range: 4.1 to 10.0 mg) reported by Stewart et al. for New Zealand.[87]

The biological half-life of selenium has been calculated to be 98 ± 16 d, based on measurements (in Finland) of the retention in the whole body of trace amounts of administered [^{75}Se]-sodium selenite,[88] but this parameter may depend on present body status and on the chemical form of selenium ingested. Thus, oral doses of [^{75}Se]-selenomethionine were retained longer than oral doses of [^{75}Se]-selenite.[89] The total selenium retention curves of radioactive selenomethionine could be resolved into three components with biological half-lives of 1.7, 18, and 234 d, respectively. Retention of selenium bound in organic form was also longer than that of inorganically bound selenium, as was determined in a bioavailability trial conducted in Finland by Levander et al.[29] who found that selenium present in selenium-rich yeast or wheat resulted in a higher concentration of selenium in the blood than selenium fed as selenate, although there was no difference in the concentration of glutathione peroxidase ultimately achieved during the supplementation. Once the selenium supplements were discontinued, however, the amount of glutathione peroxidase remained somewhat elevated in the groups receiving wheat or yeast compared to those receiving selenate. This indicates that the accumulated selenomethionine could be utilized for synthesis of the enzyme. It appears that the effect of some diseases, e.g., muscular dystrophia, is to shorten the half-life of selenium.[88] Liver diseases caused by alcoholism may also change the retention of and/or the physiological ability of the body to utilize selenium.[74] Biological interferences in selenium metabolism by a large number of toxic substances including heavy metals and carbon tetrachloride have been reported.

Of all excreted selenium, 50 to 70% is in the urine, the exact fraction being dependent on daily intake, the chemical form of the ingested selenium, and probably many other factors.[90] Stewart et al.[87] found this fraction to be 55%, and in their studies, Levander et al.[91] observed 45 to 63%.

An important metabolic step preceding selenium excretion is the biochemical methylation of the reduced form. Whereas trimethylselenonium apparently accounts for a significant fraction of urinary selenium, it is known that humans exposed to a high concentration of the element develop a garlicky breath odor characteristic of dimethylselenide.[92]

V. SUMMARY

The overview presented here includes tables of concentrations of selenium found in human tissues. Substantial regional differences are observed, due to differences in dietary intake. The concentration of selenium found in blood varies as follows: Enshi, China > Caracas > South Dakota > Guatemala > Canada and most parts of the U.S. > most parts of Europe > Finland > New Zealand and Egypt > Keshan, China.

Consistent with concentrations found in blood, the given values for whole body selenium vary within a rather broad range of of 4 to 21 mg. However, relative selenium concentrations in various human tissues seem to be fairly constant, irrespective of the geographical origin of the analyzed material: kidneys > liver > spleen > testes > heart > lungs > brain > blood serum.

A detailed discussion of the selenium concentrations observed, possible sources of the variations, and physiological implications are presented only for the more frequently analyzed tissues, such as blood components, liver, kidneys, muscle, skin, hair, and urine.

REFERENCES

1. **Wilber, C.,** Toxicology of selenium: a review, *Clin. Toxicol.,* 17, 171, 1980.
2. **Awasthi, Y. C., Beutler, E., and Srivastava, S. K.,** Purification and properties of human erythrocyte glutathione peroxidase, *J. Biol. Chem.,* 250, 5144, 1975.
3. **Schrauzer, G. N., White, D. A., and Schneider, C. J.,** Cancer mortality correlation studies. III. Statistical associations with dietary selenium intakes, *Bioinorg. Chem.,* 7, 23, 1977.
4. **Chen, X., Yang, G., Chen, J., Chen, X., Wen, Z., and Ge, K.,** Studies on the relations of selenium and Keshan disease, *Biol. Trace Elem. Res.,* 2, 91, 1980.
5. National Research Council (U.S.), *Recommended Dietary Allowances,* 9th ed., National Academy of Sciences, Washington, D.C., 1980.
6. **Versieck, J.,** Trace elements in human body fluids and tissues, *CRC Crit. Rev. Clin. Lab. Sci.,* 22, 97, 1985.
7. **Versieck, J. and Cornelis, R.,** Normal levels of trace elements in human blood plasma or serum, *Anal. Chim. Acta,* 116, 217, 1980.
8. **Bock, R.,** *A Handbook of Decomposition Methods in Analytical Chemistry,* John Wiley & Sons, New York, 1979, 149.
9. **Fourie, H. O. and Peisach, M.,** Loss of trace elements during dehydration of marine zoological material, *Analyst,* 102, 193, 1977.
10. **Gorsuch, T. T.,** Radiochemical investigations on the recovery for analysis of trace elements in organic and biological materials, *Analyst,* 84, 135, 1959.
11. **Loyd, B., Holt, P., and Delves, H. T.,** Determination of selenium in biological samples by hydride generation and atomic-absorption spectroscopy, *Analyst,* 107, 927, 1982.

12. **Watkinson, J. H.,** Semi-automated fluorimetric determination of nanogram quantities of selenium in biological material, *Anal. Chim. Acta,* 105, 319, 1979.
13. **Olson, O. E., Palmer, I. S., and Cary, E. E.,** Modification of the official fluorometric method for selenium in plants, *J. Assoc. Off. Anal. Chem.,* 58, 117, 1975.
14. **Nève, J., Hanocq, M., Molle, L., and Lefebvre, G.,** Study of some systematic errors during the determination of the total selenium and some of its ionic species in biological materials, *Analyst,* 107, 934, 1982.
15. **Ihnat, M.,** Fluorometric determination of selenium in foods, *J. Assoc. Off. Anal. Chem.,* 57, 368, 1974.
16. **Ihnat, M. and Miller, H. J.,** Analysis of food for arsenic and selenium by acid digestion, hydride evolution atomic absorption spectrophotometry, *J. Assoc. Off. Anal. Chem.,* 60, 813, 1977.
17. **Heydorn, K.,** *Neutron Activation Analysis for Clinical Trace Element Research,* Vol. 1 and 2, CRC Press, Boca Raton, FL, 1984.
18. **Bowen, H. J. M.,** Comparative elemental analyses of a standard plant material, *Analyst,* 92, 124, 1967.
19. **Ihnat, M.,** Biological reference materials for quality control, in *Quantitative Trace Analysis of Biological Materials,* McKenzie, H. A. and Smythe, L. E., Eds., Elsevier, Amsterdam, 1988, 331.
20. **Muramatsu, Y. and Parr, R. M.,** *Survey of Currently Available Reference Materials for Use in Connection with the Determination of Trace Elements in Biological and Environmental Materials,* IAEA/RL/128, International Atomic Energy Agency, Vienna, 1985.
21. **Ihnat, M.,** Biological and related reference materials for determination of elements, in *Quantitative Trace Analysis of Biological Materials,* McKenzie, H. A. and Smythe, L. E., Eds., Elsevier, Amsterdam, 1988, 739.
22. **Yang, G., Wang, S., Zhou, R., and Sun, S.,** Endemic selenium intoxication of humans in China, *Am. J. Clin. Nutr.,* 37, 872, 1983.
23. **Jaffé, W. G., Ruphael, M. D., de Mondragón, M. C. and Cuevas, M. A.,** Estudio clinico y bioquimico en niños escolares de una zona selenifera (in Spanish), *Arch. Latinoam. Nutr.,* 22, 595, 1972.
24. **Allaway, W. H., Kubota, J., Losee, F., and Roth, M.,** Selenium, molybdenum and vanadium in human blood, *Arch. Environ. Health,* 16, 342, 1968.
25. **Dickson, R. C. and Tomlinson, R. H.,** Selenium in blood and human tissues, *Clin. Chim. Acta,* 16, 311, 1967.
26. **Thomson, C. D., Rea, H. M., Robinson, M. F., and Chapman, O. W.,** Low blood selenium concentrations and glutathione peroxidase activities in elderly people, *Proc. Univ. Otago Med. Sch.,* 55(1), 18, 1977.
27. **Yanghorbani, M., Christensen, M. J., Nahapetian, A., and Young, V. R.,** Selenium metabolism in healthy adults: quantitative aspects using the stable isotope $^{74}SeO_3^{2-}$, *Am. J. Clin. Nutr.,* 35, 647, 1982.
28. **Thomson, C. D. and Robinson, M. F.,** Selenium in human health and disease with emphasis on those aspects peculiar to New Zealand, *Am. J. Clin. Nutr.,* 33, 303, 1980.
29. **Levander, O. A., Alfthan, G., Arvilommi, H., Gref, C. G., Huttunen, J. K., Kataja, M., Koivistoinen, P., and Pikkarainen, J.,** Bioavailability of selenium to Finnish men as assessed by platelet glutathione peroxidase activity and other blood parameters, *Am. J. Clin. Nutr.,* 37, 887, 1983.
30. **Kasperek, K., Iyengar, G. V., Kiem, J., Borberg, H., and Feinendegen, L. E.,** Elemental composition of platelets. III. Determination of Ag, Au, Cd, Co, Cr, Cs, Mo, Rb, Sb, and Se in normal human platelets by neutron activation analysis, *Clin. Chem.,* 25, 711, 1979.
31. **Karpatkin, S. and Weiss, H. J.,** Deficiency of glutathione peroxidase associated with high levels of reduced glutathione in Glanzmann's thrombasthenia, *N. Engl. J. Med.,* 287, 1062, 1972.
32. **Verlinden, M., Cooreman, W., and Deelstra, H.,** Atomic absorption spectrometry as a tool to study the distribution pattern of selenium in human blood, in *Trace Element Analytical Chemistry in Medicine and Biology,* Vol. 1, Brätter, P. and Schramel, P., Eds., Walter de Gruyter, Berlin, 1980, 513.
33. **Pleban, P. A. and McHugh, T.,** Copper, magnesium, selenium and zinc in human mononuclear cells and neutrophils, *Clin. Chem.,* 29, 1286, 1983.
34. **Verlinden, M., van Sprundel, M., Van der Auwera, J. C., and Eylenbosch, W. J.,** The selenium status of Belgian population groups. II. Newborns, children and the aged, *Biol. Trace Elem. Res.,* 5, 103, 1983.
35. **Lombeck, I., Kasperek, K., Harbisch, H. D., Feinendegen, L. E., and Bremer, H. J.,** The selenium state of healthy children. I. Serum selenium concentration at different ages; activity of glutathione peroxidase of erythrocytes at different ages; selenium content of food of infants, *Eur. J. Pediatr.,* 125, 81, 1977.
36. **Lombeck, I., Kasperek, K., Harbisch, H. D., Becker, K., Schumann, E., Schröter, W., Feinendegen, L. E., and Bremer, H. J.,** The selenium state of children. II. Selenium content of serum, whole blood, hair and the activity of erythrocyte glutathione peroxidase in dietetically treated patients with phenylketonuria and maple-syrup-urine disease, *Eur. J. Pediatr.,* 128, 213, 1978.
37. **Sundström, H., Yrjänheikki, E., and Kauppila, A.,** Low serum selenium concentration in patients with cervical or endometrial cancer, *Int. J. Gynaecol. Obstet.,* 22, 35, 1984.

38. **Aaseth, J., Alexander, J., Thomassen, Y., Blomhoff, J. P., and Skrede, S.,** Serum selenium levels in liver diseases, *Clin. Biochem.,* 15, 281, 1982.
39. **Aaseth, J., Munthe, E., Forre, O., and Steinnes, E.,** Trace elements in serum and urine of patients with rheumatoid arthritis, *Scand. J. Rheumatol.,* 7, 237, 1978.
40. **Alexander, J., Kofstad, J., Saeed, K., Thomassen, Y., Ovrebo, S., and Aaseth, J.,** The application of direct electrothermal atomic absorption spectrophotometric determination of selenium in clinical chemistry, in *Trace Element Analytical Chemistry in Medicine and Biology,* Vol. 2, Brätter, P. and Schramel, P., Eds., Walter de Gruyter, Berlin, 1983, 729.
41. **Motsenbocker, M. A. and Tappel, A. L.,** Selenocysteine-containing proteins from rat and monkey plasma, *Biochim. Biophys. Acta,* 704, 253, 1982.
42. **Schroeder, H. A., Frost, D. V., and Balassa, J. J.,** Essential trace metals in man: selenium, *J. Chron. Dis.,* 23, 227, 1970.
43. **Underwood, E. J.,** *Trace Elements in Human and Animal Nutrition,* 4th ed., Academic Press, New York, 1977.
44. **Zeisler, R.,** Results of the inorganic analysis of the first year human liver collection, in *The Pilot National Environmental Specimen Bank — Analysis of Human Liver Specimens,* NBS Special Publ. 656, Zeisler, R., Harrison, S. H., and Wise, S. A., Eds., U.S. National Bureau of Standards, Washington, D.C., 1983, 81.
45. **Norheim, G. and Aaseth, J.,** Essential trace elements in human liver in Norway: copper, zinc, molybdenum and selenium, *J. Oslo City Hosp.,* 30, 105, 1980.
46. **Larsen, N. A., Nielsen, B., Pakkenberg, H., Christoffersen, P., Damsgaard, E., and Heydorn, K.,** Concentrations of arsenic, manganese and selenium in organs from normal and uremic persons determined by activation analyses, (in Danish), *Ugeskr. Laeg.,* 136, 2586, 1974.
47. **Westermarck, T.,** Selenium content of tissues in Finnish infants and adults with various diseases, and studies on the effects of selenium supplementation in neuronal ceroid lipofuscinosis patients, *Acta Pharmacol. Toxicol.,* 41, 121, 1977.
48. **Wester, P. O., Brune, D., and Nordberg, G.,** Arsenic and selenium in lung, liver, and kidney tissue from dead smelter workers, *Br. J. Ind. Med.,* 38, 179, 1981.
49. **Jocelyn, P. C.,** *Biochemistry of the SH Group,* Academic Press, London, 1972, 261.
50. **Casey, C. E., Guthrie, B. E., Gaylene, M. F., and Robinson, M. F.,** Selenium in human tissues from New Zealand, *Arch. Environ. Health,* 37, 133, 1982.
51. **Yukawa, M., Suzuki-Yasumoto, M., Amano, K., and Terai, M.,** Distribution of trace elements in the human body determined by neutron activation analysis, *Arch. Environ. Health,* 35, 36, 1980.
52. **Aaseth, J., Olsen, A., Halse, J., and Hovig, T.,** Argyria — tissue deposits of silver as selenide, *Scand. J. Clin. Lab. Invest.,* 41, 247, 1981.
53. **Alexander, J. and Aaseth, J.,** Biliary excretion of copper and zinc in the rat as influenced by diethylmaleate, selenite and diethyldithiocarbamate, *Biochem. Pharmacol.,* 29, 2129, 1980.
54. **Fleming, C. R., Lie, J. T., McCall, J. T., O'Brien, J. F., Baillie, E. E., and Thistle, J. L.,** Selenium deficiency and fatal cardiomyopathy in a patient on home parenteral nutrition, *Gastroenterology,* 83, 689, 1982.
55. **Schicha, H., Kasperek, K., Feinendegen, L. E., Siller, V., and Klein, H. J.,** Aktivierungsanalytische messungen einer inhomogenen teilweise parallel verlaufenden verteilung von kobolt, eisen, selen, zink und antimon in verschiedenen bezirken von leber, lunge, niere, herz und aorta, *Beitr. Pathol.,* 146, 55, 1972.
56. **Hamilton, E. I., Minski, M. J., and Cleary, J. J.,** The concentration and distribution of some stable elements in healthy human tissues from the United Kingdom. An environmental study, *Sci. Total Environ.,* 1, 341, 1972/1973.
57. **Schicha, H., Müller, W., Kasperek, K., and Schröder, R.,** Neutronenaktivierungsanalytische bestimmung der spurenelemente kobalt, eisen, rubidium, selen, zink, chrom, silber, caesium, antimon und scandium in operativ entnommenen hirntumoren des menschen (1. Mitteilung), *Beitr. Pathol.,* 151, 281, 1974.
58. **Markesbery, W. R., Ehmann, W. D., Alauddin, M., and Hossain, T. I. M.,** Brain trace element concentrations in aging, *Neurobiol. Aging,* 5, 19, 1984.
59. **Molokhia, A., Portnoy, B., and Dyer, A.,** Neutron activation analysis of trace elements in skin. VIII. Selenium in normal skin, *Br. J. Dermatol.,* 101, 567, 1979.
60. **Sky-Peck, H. H. and Joseph, B. J.,** The "use" and "misuse" of human hair in trace metal analysis, in *Chemical Toxicology and Clinical Chemistry of Metals,* Brown, S. S. and Savory, J., Eds., Academic Press, London, 1983, 159.
61. **Valentine, J. L., Kang, H. K., Dang, P.-M., and Schluchter, M.,** Selenium concentrations and glutathione peroxidase activities in a population exposed to selenium via drinking water, *J. Toxicol. Environ. Health,* 6, 731, 1980.
62. **Ward, N. I. and Minski, M.,** Comparison of trace elements in whole blood and scalp hair of multiple sclerosis patients and normal individuals, in *Trace Substances in Environmental Health — 16,* Hemphill, D. D., Ed., University of Missouri, Columbia, 1982, 252.

63. **Rossi, L. C., Clemente, G. F., and Santaroni, G.,** Mercury and selenium distribution in a defined area and its population, *Arch. Environ. Health,* 31, 160, 1976.
64. **Thomson, C. D.,** Urinary excretion of selenium in some New Zealand women, *Proc. Univ. Otago Med. Sch.,* 50, 31, 1972.
65. **Hojo, Y.,** Evaluation of the expression of urinary selenium level as ng Se/mg creatinine and the use of single-void urine as a sample for urinary selenium determination, *Bull. Environ. Contam. Toxicol.,* 27, 213, 1981.
66. **Adkins, B. L., Barmes, D. E., and Schamschula, R. G.,** Etiology of caries in Papua New Guinea. The trace element content of urine samples and its relation to individual dental caries experience, *Bull. WHO,* 50, 495, 1974.
67. **Clemente, G. F., Rossi, L. C., and Santaroni, G. P.,** Trace element intake and excretion in the Italian population, *J. Radioanal. Chem.,* 37, 549, 1977.
68. **Poole, C. F., Evans, N. J., and Wibberly, D. G.,** Determination of selenium in biological samples by gas-liquid chromatography with electron-capture detection, *J. Chromatogr.,* 136, 73, 1977.
69. **de Mondragón, M. C. and Jaffé, W. G.,** Selenio en alimentos y en orina de escolares de diferentes zonas de Venezuela, *Arch. Latinoam. Nutr.,* 21, 185, 1971.
70. **Glover, J. R.,** Selenium in human urine: a tentative maximum allowable concentration for industrial and rural populations, *Ann. Occup. Hyg.,* 10, 3, 1967.
71. **Hojo, Y.,** Subject groups high and low in urinary selenium levels, workers exposed to heavy metals and patients with cancer and epilepsy, *Bull. Environ. Contam. Toxicol.,* 26, 466, 1981.
72. **Alexander, J., Thomassen, Y., and Aaseth, J.,** Increased urinary excretion of selenium among workers exposed to elemental mercury vapor, *J. Appl. Toxicol.,* 3, 143, 1983.
73. **Rossi, L. C., Clemente, G. F., and Santaroni, G.,** Mercury and selenium distribution in a defined area and its population, *Arch. Environ. Health,* 31, 160, 1976.
74. **Aaseth, J., Thomassen, Y., Alexander, J., and Norheim, G.,** Decreased serum selenium in alcoholic cirrhosis, *N. Engl. J. Med.,* 303, 944, 1980.
75. **Dutta, S. K., Miller, P. A., Greenberg, L. B., and Levander, O. A.,** Selenium and acute alcoholism, *Am. J. Clin. Nutr.,* 38, 713, 1983.
76. **Goldman, I. S. and Kantrowitz, N. E.,** Cardiomyopathy associated with selenium deficiency, *N. Engl. J. Med.,* 305, 701, 1981.
77. **Byard, J. L.,** Trimethyl selenide. A urinary metabolite of selenite, *Arch. Biochem. Biophys.,* 130, 556, 1969.
78. **Palmer, I. S., Fischer, D. D., Halverson, A. W., and Olson, O. E.,** Identification of a major selenium excretory product in rat urine, *Biochim. Biophys. Acta,* 177, 336, 1969.
79. **Burk, R. F.,** Selenium in man, in *Trace Elements in Human Health and Disease,* Vol. 2, Prasad, A. S. and Oberleas, D., Eds., Academic Press, New York, 1976, 105.
80. **Oyamada, N. and Ishizaki, M.,** Determination of trimethyl-selenonium ion in human urine by graphite furnace atomic absorption spectrometry, in Abstr. 9th Int. Conf. on Atomic Spectroscopy, Tokyo, 1981, 501.
81. **Varo, D. and Koivistoinen, P.,** Annual variations in the average selenium intake in Finland: cereal products and milk as sources of selenium in 1979/80, *Int. J. Vitam. Nutr. Res.,* 51, 79, 1981.
82. **Abdulla, M., Kolar, K., and Svensson, S.,** Selenium, *Scand. J. Gastroenterol. Suppl.,* 14, 181, 1979.
83. **Bibow, K., Riis, G., and Salbu, B.,** A study of trace elements in Norwegian diets by the duplicate portion technique, *Naeringsforskning,* 28, 84, 1984.
84. **Sakurai, K. and Tsuchiya, K.,** A tentative recommendation for the maximum daily intake of selenium, *Environ. Physiol. Biochem.,* 5, 107, 1975.
85. **Morris, V. C. and Levander, O. A.,** Selenium content of foods, *J. Nutr.,* 100, 1333, 1970.
86. **Thomson, J. N., Erdody, P., and Smith, D. C.,** Selenium content of food consumed by Canadians, *J. Nutr.,* 105, 274, 1975.
87. **Stewart, R. D. H., Griffiths, N. M., Thomson, C. D., and Robinson, M. F.,** Quantitative selenium metabolism in normal New Zealand women, *Br. J. Nutr.,* 40, 45, 1978.
88. **Westermarck, T., Rahola, Y., and Suomela, M.,** Selenium (Se) retention and biological half-life (T 1/2) in patients with Dachenne's Muscular Dystrophy, in *Proc. Nordic Symp. Mineral Elements — 1980,* The Academy of Finland, Helsinki, 1981, 1589.
89. **Griffiths, N. M., Stewart, R. D. H., and Robinson, M. F.,** The metabolism of ^{75}Se-selenomethionine in four women, *Br. J. Nutr.,* 35, 373, 1976.
90. **Robberecht, H. J. and Deelstra, H. A.,** Selenium in human urine: concentration levels and medical implications, *Clin. Chim. Acta,* 136, 107, 1984.
91. **Levander, O. A., Sutherland, B., Morris, V. C., and King, J. C.,** Selenium balance in young men during selenium depletion and repletion, *Am. J. Clin. Nutr.,* 34, 2662, 1981.
92. **Glover, J. R.,** Environmental health aspects of selenium and tellurium, in *Proc. Symp. on Selenium-Tellurium in the Environment,* Industrial Health Foundation, Pittsburgh, 1976, 279.

93. **Antoniou, K., Vassilaki-Grimani, M., Lolis, D., and Grimanis, A. P.,** Concentrations of cobalt, rubidium, selenium and zinc in maternal and cord blood serum and amniotic fluid of women with normal and prolonged pregnancies, *J. Radioanal. Chem.*, 70, 77, 1982.
94. **Thieme, R., Schramel, P., and Mahr, W.,** Spurenelemente in menschlichem fruchtwasser, *Geburtshilfe Frauenheilkd.*, 40, 185, 1980.
95. **Rösick, E., Rösick, U., Brätter, P., and Kynast, G.,** Trace element levels of amniotic fluid at term in normal and high risk pregnancies, in *Trace Element Analytical Chemistry in Medicine and Biology,* Vol. 2, Brätter, P. and Schramel, P., Eds., Walter de Gruyter, Berlin, 1983, 463.
96. **Cross, J. D., Raie, R. M., and Smith, H.,** Bromine and selenium in human aorta, *J. Clin. Pathol.*, 34, 393, 1981.
97. **Smythe, W. R., Alfrey, A. C., Craswell, P. W., Crouch, C. A., Ibels, L. S., Kubo, H., Nunnelley, L. L., and Rudolph, H.,** Trace element abnormalities in chronic uremia, *Ann. Intern. Med.*, 96, 302, 1982.
98. **Burch, R. E., Jetton, M. M., Hahn, H. K. J., and Sullivan, J. F.,** Trace element composition of ascitic fluid, *Arch. Intern. Med.*, 139, 680, 1979.
99. **Nève, J., Van Geffel, R., Hanocq, M., and Molle, L.,** Plasma and erythrocyte zinc, copper and selenium in cystic fibrosis, *Acta Paediatr. Scand.*, 72, 437, 1983.
100. **Verlinden, M., van Sprundel, M., Van der Auwera, J. C., and Eylenbosch, W. J.,** The Selenium Status of Belgian population groups. I. Healthy adults, *Biol. Trace Elem. Res.*, 5, 91, 1983.
101. **Sinet, P. M., Nève, J., Nicole, A., and Molle, L.,** Low plasma selenium in Down's Syndrome (Trisomy 21), *Acta Paediatr. Scand.*, 73, 275, 1984.
102. **Mazzella, G. L., Sinforiani, E., Savoldi, F., Allegrini, M., Lanzola, E., and Scelsi, R.,** Blood cells glutathione peroxidase activity and selenium in multiple sclerosis, *Eur. Neurol.*, 22, 442, 1983.
103. **Vernie, L. N., De Vries, M., Benckhuijsen, C., De Goeij, J. J. M., and Zegers, C.,** Selenium levels in blood and plasma, and glutathione peroxidase activity in blood of breast cancer patients during adjuvant treatment with cyclophosphamide, methotrexate and 5-fluorouracil, *Cancer Lett.*, 18, 283, 1983.
and 5-fluorouracil, *Cancer Lett.*, 18, 283, 1983.
104. **Hågå, P. and Lunde, G.,** Selenium and vitamin E in cord blood from preterm and full term infants, *Acta Paediatr. Scand.*, 67, 735, 1978.
105. **Plantin, L.-O. and Meurling, S.,** The concentration of trace elements in blood from healthy newborns, in *Trace Element Analytical Chemistry in Medicine and Biology,* Vol. 1, Brätter, P. and Schramel, P., Eds., Walter de Gruyter, Berlin, 1980, 243.
106. **Lloyd, B., Lloyd, R. S., and Clayton, B. E.,** Effect of smoking, alcohol, and other factors on the selenium status of a healthy population, *J. Epidemiol. Commun. Health,* 37, 213, 1983.
107. **Behne, D. and Wolters, W.,** Selenium content and glutathione peroxidase activity in the plasma and erythrocytes of non-pregnant and pregnant women, *J. Clin. Chem. Clin. Biochem.*, 17, 133, 1979.
108. **Schmidt, K. and Heller, W.,** Selenkonzentration und aktivitat der glutathionperoxydase im Lysat menschlicher erythrozyten, *Blut,* 33, 247, 1976.
109. **Krivan, V., Geiger, H., and Franz, H. E.,** Bestimmung von Fe, Co, Cu, Zn, Se, Rb und Cs in NBS-ochsenleber, blutplasma und erythrocyten durch INAA und AAS, *Fresenius Z. Anal. Chem.*, 305, 399, 1981.
110. **McAdam, P. A., Smith, D. K., Feldman, E. B., and Hames, C.,** Effect of age, sex and race on selenium status of healthy residents of Augusta, Georgia, *Biol. Trace Elem. Res.*, 6, 3, 1984.
111. **Andrews, J. W., Hames, C. G., Metts, J. C., Jr., and Davis, J. M.,** Selenium, cadmium and glutathione peroxidase in blood of cardiovascular diseased and normal subjects from the cardiovascular belt of Southeastern USA, in *Trace Substances in Environmental Health — 14,* Hemphill, D. D., Ed., University of Missouri, Columbia, 1980, 38.
112. **Miller, L., Mills, B. J., Blotcky, A. J., and Lindeman, R. D.,** Red blood cell and serum selenium concentrations as influenced by age and selected diseases, *J. Am. Coll. Nutr.*, 2, 331, 1983.
113. **Rudolph, N. and Wong, S. L.,** Selenium and glutathione peroxidase activity in maternal and cord plasma and red cells, *Pediatr. Res.*, 12, 789, 1978.
114. **Dworkin, B. M. and Rosenthal, W. S.,** Selenium and the alcoholic, *Lancet,* 1, 1015, 1984.
115. **Snook, J. T., Palmquist, D. L., Moxon, A. L., Cantor, A. H., and Vivian, V. M.,** Selenium status of a rural (predominantly Amish) community living in a low-selenium area, *Am. J. Clin. Nutr.*, 38, 620, 1983.
116. **Lane, H. W., Dudrick, S., and Warren, D. C.,** Blood selenium levels and glutathione-peroxidase activities in university and chronic intravenous hyperalimentation subjects, *Proc. Soc. Exp. Biol. Med.*, 167, 383, 1981.
117. **Lane, H. W., Warren, D. C., Taylor, B. J., and Stool, E.,** Blood selenium and glutathione peroxidase levels and dietary selenium of free-living and institutionalized elderly subjects, *Proc. Soc. Exp. Biol. Med.*, 173, 87, 1983.

118. **Pleban, P. A., Munyani, A., and Beachum, J.,** Determination of selenium concentration and glutathione peroxidase activity in plasma and erythrocytes, *Clin. Chem.*, 28, 311, 1982.
119. **Maxia, V., Meloni, S., Rollier, M. A., Brandone, A., Patwardhan, V. N., Waslien, C. I., and Said El Shami,** Selenium and chromium assay in Egyptian foods and in blood of Egyptian children by activation analysis, in *Nuclear Activation Techniques in the Life Sciences 1972*, International Atomic Energy Agency, Vienna, 1972, 527.
120. **Kurahashi, K., Inoue, S., Yonekura, S., Shimoishi, Y., and Tôei, K.,** Determination of selenium in human blood by gas chromatography with electron-capture detection, *Analyst*, 105, 690, 1980.
121. **McKenzie, J. M., Rea, H., Van Rij, A. M., and Robinson, M. F.,** Low blood selenium concentrations during intravenous alimentation, *Proc. Univ. Otago Med. Sch.*, 54, 25, 1976.
122. **Kay, R. G. and Knight, G. S.,** Blood selenium values in an adult Auckland population group, *N. Z. Med. J.*, 90, 11, 1979.
123. **Robinson, M. F., Campbell, D. R., Sutherland, W. H. F., Herbison, G. P., Paulin, J. M., and Simpson, F. O.,** Selenium and risk factors for cardiovascular disease in New Zealand, *N. Z. Med. J.*, 96, 755, 1983.
124. **Robinson, M. F., Godfrey, P. J., Thomson, C. D., Rea, H. M., and van Rij, M.,** Blood selenium and glutathione peroxidase activity in normal subjects and in surgical patients with and without cancer in New Zealand, *Am. J. Clin. Nutr.*, 32, 1477, 1979.
125. **Kiem, J., Iyengar, G. V., Borberg, H., Kasperek, K., Siegers, M., Feinendegen, L. E., and Gross, R.,** Sampling and sample preparation of platelets for trace element analysis and determination of certain selected bulk and trace elements in normal human platelets by means of neutron activation analysis, in *Nuclear Activation Techniques in the Life Sciences 1978*, International Atomic Energy Agency, Vienna, 1979, 143.
126. **Perona, G., Cellerino, R., Guidi, G. C., Moschini, G., Stievano, B. M., and Tregnaghi, C.,** Erythrocytic glutathione peroxidase: its relationship to plasma selenium in man, *Scand. J. Haematol.*, 19, 116, 1977.
127. **Van Caillie-Bertrand, M., De Biéville, F., Neijens, H., Kerrebijn, K., Fernandes, J., and Degenhart, H.,** Trace metals in cystic fibrosis, *Acta Paediatr. Scand.*, 71, 203, 1982.
128. **Lloyd, B., Holt, P., and Delves, H. T.,** Determination of selenium in biological samples by hydride generation and atomic absorption spectroscopy, *Analyst*, 107, 927, 1982.
129. **Rhead, W. J., Cary, E. E., Allaway, W. H., Saltzstein, S. L., and Schrauzer, G. N.,** The vitamin E and selenium status of infants and the sudden infant death syndrome, *Bioinorg. Chem.*, 1, 289, 1972.
130. **Young, J. W. and Christian, G. D.,** Gas-chromatographic determination of selenium, *Anal. Chim. Acta*, 65, 127, 1973.
131. **Hatano, S., Nishi, Y., and Usui, T.,** Plasma selenium concentration in healthy Japanese children and adults determined by flameless atomic absorption spectrophotometry, *J. Pediatr. Gastroenterol. Nutr.*, 3, 426, 1984.
132. **Levine, R. J. and Olson, R. E.,** Blood selenium in Thai children with protein-calorie malnutrition, *Proc. Soc. Exp. Biol. Med.*, 134, 1030, 1970.
133. **Morss, S. G., Ralston, H. R., and Olcott, H. S.,** Selenium determination in human serum lipoprotein fractions by neutron activation analysis, *Anal. Biochem.*, 49, 598, 1972.
134. **Versieck, J., Hoste, J., Barbier, F., Michels, H., and De Rudder, J.,** Simultaneous determination of iron, zinc, selenium, rubidium and cesium in serum and packed blood cells by neutron activation analysis, *Clin. Chem.*, 23, 1301, 1977.
135. **Jensen, G. E., Gissel-Nielsen, G., and Clausen, J.,** Leucocyte glutathione peroxidase activity and selenium level in multiple sclerosis, *J. Neurol. Sci.*, 48, 61, 1980.
136. **Miettinen, T. A., Alfthan, G., Huttunen, J. K., Pikkarainen, J., Naukkarinen, V., Mattila, S., and Kumlin, T.,** Serum selenium concentration related to myocardial infarction and fatty acid content of serum lipids, *Br. Med. J.*, 287, 517, 1983.
137. **Salonen, J. T., Alfthan, G., Huttenen, J. K., and Puska, P.,** Association between serum selenium and the risk of cancer, *Am. J. Epidemiol.*, 120, 342, 1984.
138. **Sundström, H., Yrjänheikki, E., and Kauppila, A.,** Serum selenium in patients with ovarian cancer during and after therapy, *Carcinogenesis*, 5, 731, 1984.
139. **Wikström, J., Westermarck, T., and Palo, J.,** Selenium, vitamin E and copper in multiple sclerosis, *Acta Neurol. Scand.*, 54, 287, 1976.
140. **Välimäki, M. J., Harju, K. J., and Ylikahri, R. H.,** Decreased serum selenium in alcoholics — a consequence of liver dysfunction, *Clin. Chim. Acta*, 130, 291, 1983.
141. **Westermarck, T., Raunu, P., Kirjarinta, M., and Lappalainen, L.,** Selenium content of whole blood and serum in adults and children of different ages from different parts of Finland, *Acta Pharmacol. Toxicol.*, 40, 465, 1977.

142. **Sundström, H., Korpela, H., Viinikka, L., and Kauppila, A.,** Serum selenium and glutathione peroxidase, and plasma lipid peroxides in uterine, ovarian or vulvar cancer, and their responses to antioxidants in patients with ovarian cancer, *Cancer Lett.*, 24, 1, 1984.
143. **Huguet, C., Chappuis, Ph., Peynet, J., Thuillier, F., Legrand, A., and Rousselet, F.,** Dosage du sélénium sérique par spectrophotométrie d'absorption atomique — intérêt dans le diagnostic biologique des cardiomyopathies non obstructives, *Ann. Biol. Clin.*, 41, 277, 1983.
144. **Nève, J., Sinet, P. M., Molle, L., and Nicole, A.,** Selenium, zinc and copper in Down's syndrome (trisomy 21): blood levels and relations with glutathione peroxidase and superoxide dismutase, *Clin Chim. Acta,* 133, 209, 1983.
145. **Calautti, P., Moschini, G., Stievano, B. M., Tomio, L., Calzavara, F., and Perona, G.,** Serum selenium levels in malignant lymphoproliferative diseases, *Scand. J. Haematol.*, 24, 63, 1980.
146. **Woittiez, J. R. W.,** *Elemental Analysis of Human Serum and Serum Protein Fractions by Thermal Neutron Activation,* Rep. ECN-147, Netherlands Energy Research Foundation, Petten, Netherlands, 1984.
147. **Blekastad, V., Jonsen, J., Steinnes, E., and Helgeland, K.,** Concentrations of trace elements in human blood serum from different places in Norway determined by neutron activation analysis, *Acta Med. Scand.*, 216, 25, 1984.
148. **Saeed, K. and Thomassen, Y.,** Electrothermal atomic absorption spectrometric determination of selenium in blood serum and seminal fluid after protein precipitation with trichloroacetic acid, *Anal. Chim. Acta,* 143, 223, 1982.
149. **Saeed, K., Thomassen, Y., and Langmyhr, F. J.,** Direct electrothermal atomic absorption spectrometric determination of selenium in serum, *Anal. Chim. Acta,* 110, 285, 1979.
150. **Masiak, M., Skowron, S., Maleszewska, H., Koziorowski, L., and Herzyk, D.,** Serum levels of certain trace elements (Fe, Rb, Se, Zn) in healthy humans. III, *Acta Physiol. Pol. (Engl. Transl),* 33, 75, 1982.
151. **Fraga, J. M., Cocho, J. A., Alvela, M., Alonso, J. R., Pena, J., and Tojo, R.,** Selenium state of children. The selenium content of the serum of normal children and children with inborn errors of metabolism, *J. Inherit. Metab. Dis.*, 6, (Suppl. 2), 99, 1983.
152. **Wester, P. O.,** Trace elements in serum and urine from hypertensive patients before and during treatment with chlorthalidone, *Acta Med. Scand.*, 194, 505, 1973.
153. **Gebre-Medhin, M., Ewald, U., Plantin, L.-O., and Tuvemo, T.,** Elevated serum selenium in diabetic children, *Acta Paediatr. Scand.*, 73, 109, 1984.
154. **Ellis, N., Lloyd, B., Lloyd, R. S., and Clayton, B. E.,** Selenium and vitamin E in relation to risk factors for coronary heart disease, *J. Clin. Pathol.*, 37, 200, 1984.
155. **Kasperek, K., Schicha, H., Siller, V., and Feinendegen, L. E.,** Normalwerte von spurenelementen im menschlichen serum und korrelation zum lebensalter und zur serum-eiweiss-konzentration, *Strahlentherapie,* 143, 468, 1972.
156. **Oster, O. and Prellwitz, W.,** A methodological comparison of hydride and carbon furnace atomic absorption spectroscopy for the determination of selenium in serum, *Clin. Chim. Acta,* 124, 277, 1982.
157. **Oster, O., Prellwitz, W., Kasper, W., and Meinertz, T.,** Congestive cardiomyopathy and the selenium content of serum, *Clin. Chim. Acta,* 128, 125, 1983.
158. **Lalonde, L., Jean, Y., Roberts, K. D., Chapdelaine, A., and Bleau, G.,** Fluorometry of selenium in serum or urine, *Clin. Chem.*, 28, 172, 1982.
159. **Willett, W. C., Polk, B. F., Morris, J. S., Stampfer, M. J., Pressel, S., Rosner, B., Taylor, J. O., Schneider, K., and Hames, C. G.,** Prediagnostic serum selenium and risk of cancer, *Lancet,* 2, 130, 1983.
160. **Broghamer, W. L., Jr., McConnell, K. P., and Blotcky, A. L.,** Relationship between serum selenium levels and patients with carcinoma, *Cancer,* 37, 1384, 1976.
161. **Castillo, R., Landon, C., Eckhardt, K., Morris, V., Levander, O., and Lewiston, N.,** Selenium and vitamin E status in cystic fibrosis, *J. Pediatr.*, 99, 583, 1981.
162. **Sky-Peck, H. H. and Joseph, B. J.,** Determination of trace elements in human serum by energy dispersive X-ray fluorescence, *Clin. Biochem.*, 14, 126, 1981.
163. **McConnell, K. P., Jager, R. M., Bland, K. I., and Blotcky, A. J.,** The relationship of dietary selenium and breast cancer, *J. Surg. Oncol.*, 15, 67, 1980.
164. **Broghamer, W. L., Jr., McConnell, K. P., Grimaldi, M., and Blotcky, A. J.,** Serum selenium and reticuloendothelial tumors, *Cancer,* 41, 1462, 1978.
165. **Sullivan, J. F., Blotcky, A. J., Jetton, M. M., Hahn, H. K. J., and Burch, R. E.,** Serum levels of selenium, calcium, copper, magnesium, manganese and zinc in various human diseases, *J. Nutr.*, 109, 1432, 1979.
166. **Chen, S. Y., Collipp, P. J., Boasi, L. H., Isenschmid, D. S., Verolla, R. J., San Roman, G. A., and Yeh, J. K.,** Fluorometry of selenium in human hair, urine and blood. A single tube process for submicrogram determination of selenium, *Ann. Nutr. Metab.*, 26, 186, 1982.

167. **Agostoni, A., Gerli, G. C., Beretta, L., Palazzini, G., Buso, G. P., Xusheng, H., and Moschini, G.,** Erythrocyte antioxidant enzymes and selenium serum levels in an Andean population, *Clin. Chim. Acta,* 133, 153, 1983.
168. **Higashi, A., Tamari, H., Kuroki, Y., and Matsuda, I.,** Longitudinal changes in selenium content of breast milk, *Acta Paediatr. Scand.,* 72, 433, 1983.
169. **Nakahara, H., Nagame, Y., Yoshizawa, Y., Oda, H., Gotoh, S., and Murakami, Y.,** Trace element analysis of human blood serum by neutron activation analysis, *J. Radioanal. Chem.,* 54, 183, 1979.
170. **Cumming, F. J., Fardy, J. J., and Briggs, M. H.,** Trace elements in human milk, *Obstet. Gynecol.,* 62, 506, 1983.
171. **Hansen, J. C., Kromann, N., Wulf, H. C., and Alboge, K.,** Selenium and its interrelation with mercury in whole blood and hair in an East Greenlandic population, *Sci. Total Environ.,* 38, 33, 1984.
172. **Macholz, R. M., Kujawa, M., Raab, M., and Engst, R.,** Selengehalt und aktivität der glutathionperoxydase im humanblut, *Zentralbl. Pharm. Pharmakother. Laboratoriumsdiagn.,* 119, 483, 1980.
173. **Mangal, P. C. and Sharma, P.,** Effect of leukaemia on the concentration of some trace elements in human whole blood, *Indian J. Med. Res.,* 74, 559, 1981.
174. **Brune, D., Samsahl, K., and Wester, P. O.,** A comparison between the amounts of As, Au, Br, Cu, Fe, Mo, Se and Zn in normal and uraemic human whole blood by means of neutron activation analysis, *Clin. Chim. Acta,* 13, 285, 1966.
175. **Brune, D., Samsahl, K., and Wester, P. O.,** A comparison between the amounts of As, Au, Br, Cu, Fe, Mo, Se and Zn in normal and uraemic human whole blood by means of neutron activation analysis, *Clin. Chim. Acta,* 13, 285, 1966.
176. **Penny, W. J., Mayberry, J. F., Aggett, P. J., Gilbert, J. O., Newcombe, R. G., and Rhodes, J.,** Relationship between trace elements, sugar consumption, and taste in Crohn's disease, *Gut,* 24, 288, 1983.
177. **Ward, K. P., Arthur, J. R., Russell, G., and Aggett, P. J.,** Blood selenium content and glutathione peroxidase activity in children with cystic fibrosis, coeliac disease, asthma, and epilepsy, *Eur. J. Pediatr.,* 142, 21, 1984.
178. **Lloyd-Still, J. D. and Ganther, H. E.,** Selenium and glutathione peroxidase levels in cystic fibrosis, *Pediatrics,* 65, 1010, 1980.
179. **Valentine, J. L., Kang, H. K., and Spivey, G. H.,** Selenium levels in human blood, urine, and hair in response to exposure via drinking water, *Environ. Res.,* 17, 347, 1978.
180. **Tulley, R. T. and Lehmann, H. P.,** Flameless atomic absorption spectrophotometry of selenium in whole blood, *Clin. Chem.,* 28, 1448, 1982.
181. **Shultz, T. D. and Leklem, J. E.,** Selenium status of vegetarians, nonvegetarians and hormone-dependent cancer subjects, *Am. J. Clin. Nutr.,* 37, 114, 1983.
182. **Howe, S. M.,** Selenium in the blood of South Dakotans, *Arch. Environ. Health,* 34, 444, 1979.
183. **Baglan, R. J., Brill, A. B., Schulert, A., Wilson, D., Larsen, K., Dyer, N., Mansour, M., Schaffner, W., Hoffman, L., and Davies, J.,** Utility of placental tissue as an indicator of trace element exposure to adult and fetus, *Environ. Res.,* 8, 64, 1974.
184. **Hankins, D. A., Riella, M. C., Scribner, B. H., and Babb, A. L.,** Whole blood trace element concentrations during total parenteral nutrition, *Surgery,* 79, 674, 1976.
185. **Lin, S.-M.,** Determination of trace elements in human whole blood by instrumental neutron activation analysis, *Radioisotopes,* 32, 155, 1983.
186. **Meesuk, P.,** The determination of selenium, copper and arsenic in human blood by neutron activation analysis, *J. Natl. Res. Counc. Thailand,* 8(2), 9, 1976.
187. **Judson, G. J., Mattschoss, K. H., and Thomas, D. W.,** Selenium in whole blood of Adelaide residents, *Proc. Nutr. Soc. Aust.,* 3, 105, 1978.
188. **Griffiths, N. M. and Thomson, C. D.,** Selenium in whole blood of New Zealand residents, *N. Z. Med. J.,* 80, 199, 1974.
189. **McKenzie, R. L., Rea, H. M., Thomson, C. D., and Robinson, M. F.,** Selenium concentration and glutathione peroxidase activity in blood of New Zealand infants and children, *Am. J. Clin. Nutr.,* 31, 1413, 1978.
190. **Watkinson, J. H.,** The selenium status of New Zealanders, *N. Z. Med. J.,* 80, 202, 1974.
191. **Brätter, P., Gawlik, D., Lausch, J., and Rosick, U.,** On the distribution of trace elements in human skeletons, *J. Radioanal. Chem.,* 37, 393, 1977.
192. **Lindh, U., Brune, D., Nordberg, G., and Wester, P.-O.,** Levels of antimony, arsenic, cadmium, copper, lead, mercury, selenium, silver, tin and zinc in bone tissue of industrially exposed workers, *Sci. Total Environ.,* 16, 109, 1980.
193. **Cappon, C. J. and Smith, J. C.,** Mercury and selenium content and chemical form in human and animal tissue, *J. Anal. Toxicol.,* 5, 90, 1981.
194. **Rizk, S. L. and Sky-Peck, H. H.,** Comparison between concentrations of trace elements in normal and neoplastic human breast tissue, *Cancer Res.,* 44, 5390, 1984.

195. **Mangal, P. C., and Kumar, S.,** Neutron activation analysis of trace elements in cancerous human breast tissue, *Indian J. Phys.*, 58A, 355, 1984.
196. **El-Yazigi, A., Al-Saleh, I., and Al-Mefty, O.,** Concentration of Ag, Al, Au, Bi, Cd, Cu, Pb, Sb, and Se in cerebrospinal fluid of patients with cerebral neoplasms, *Clin. Chem.*, 30, 1358, 1984.
197. **Hagenfeldt, K., Landgren, B.-M., Plantin, L.-O., and Diczfalusy, E.,** Trace elements in the human endometrium and decidua. A multielement analysis, *Acta Endocrinol.*, 85, 406, 1977.
198. **Lakomaa, E.-L. and Eklund, P.,** Trace element analysis of human cataractous lenses by neutron activation analysis and atomic absorption spectrometry with special reference to pseudo-exfoliation of the lens capsule, *Ophthalmic Res.*, 10, 302, 1978.
199. **Bhuyan, K. C., Baxter, T., and Morris, J. S.,** Selenium status in the eye: increased level in cataract in the human and its distribution in the eye tissues of animals, *Invest. Ophthalmol. Visual Sci.*, 22 (Suppl. to No. 3), 35, 1982.
200. **Christian, G. D. and Michaelis, M.,** The selenium content of tissues of the human eye, *Invest. Ophthalmol.*, 5, 248, 1966.
201. **Capel, I. D., Pinnock, M. H., Dorrel, H. M., Williams, D. C., and Grant, E. C. G.,** Comparison of concentrations of some trace, bulk, and toxic metals in the hair of normal and dyslexic children, *Clin. Chem.*, 27, 879, 1981.
202. **Davies, T. S.,** Hair analysis and selenium shampoos, *Lancet*, 2, 935, 1982.
203. **Barbeau, A., Roy, M., and Paris, S.,** Hair trace elements in Friedreich's disease, *Can. J. Neurol. Sci.*, 11, 620, 1984.
204. **Thimaya, S. and Ganapathy, S. N.,** Selenium in human hair in relation to age, diet, pathological condition and serum levels, *Sci. Total. Environ.*, 24, 41, 1982.
205. **Li, J., Ren, S., and Chen, D.,** A study of Kashin-Beck disease associated with environmental selenium in Shaanxi area (in Chinese), *Acta Sci. Circumstantiae (Huanjing Kexue Xuebao)*, 2, 91, 1982.
206. **Arunachalam, J., Gangadharan, S., and Yegnasubramanian, S.,** Elemental data on human hair sampled from Indian student population and their interpretation for studies in environmental exposure, in *Nuclear Activation Techniques in the Life Sciences 1978*, International Atomic Energy Agency, Vienna, 1979, 499.
207. **Yukawa, M., Suzuki-Yasumoto, M., and Tanaka, S.,** The variation of trace element concentration in human hair: the trace element profile in human long hair by sectional analysis using neutron activation analysis, *Sci. Total Environ.*, 38, 41, 1984.
208. **Imahori, A., Fukushima, I., Shiobara, S., Yanagida, Y., and Tomura, K.,** Multielement neutron activation analysis of human scalp hair. A local population survey in the Tokyo metropolitan area, *J. Radioanal. Chem.*, 52, 167, 1979.
209. **Qureshi, I. H., Chaudhary, M. S., and Ahmad, S.,** Trace element concentration in head hair of the inhabitants of the Rawalpindi-Islamabad area, *J. Radioanal. Chem.*, 68, 209, 1982.
210. **Liebscher, K. and Smith, H.,** Essential and nonessential trace elements, *Arch. Environ. Health*, 17, 881, 1968.
211. **De Reu, L., Cornelis, R., Hoste, J., and Ringoir, S.,** Instrumental neutron activation analysis of minor and trace elements in human kidneys, *Radiochem. Radioanal. Lett.*, 40, 51, 1979.
212. **Brune, D., Nordberg, G., and Wester, P. O.,** Distribution of 23 elements in the kidney, liver and lungs of workers from a smeltery and refinery in North Sweden exposed to a number of elements and of a control group, *Sci. Total. Environ.*, 16, 13, 1980.
213. **Subramanian, K. S. and Méranger, J. C.,** Rapid hydride evolution-electrothermal atomisation atomic-absorption spectrophotometric method for determining arsenic and selenium in human kidney and liver, *Analyst*, 107, 157, 1982.
214. **Shamberger, R. J. and Bratush, C. M.,** Cadmium, selenium and zinc levels in kidneys, in *Trace Substances in Environmental Health — 14*, Hemphill, D. D., Ed., University of Missouri, Columbia, 1980, 203.
215. **Rhead, W. J. and Schneider, J. A.,** Effect of selenium compounds on selenium content, growth and ^{35}S-cystine metabolism of skin fibroblasts from normal and cystinotic individuals, *Bioinorg. Chem.*, 6, 187, 1976.
216. **Pleban, P. A., Kerkay, J., and Pearson, K. H.,** Cadmium, copper, lead, manganese, and selenium levels, and glutathione peroxidase activity in human kidney cortex, *Anal. Lett.*, 14, 1089, 1981.
217. **Horvath, D. J., Barker, F. W., Thayne, W. V., and Frost, J. L.,** Selenium, cadmium, zinc and copper in human kidney cortices and post mortem indices of hypertension, *Biol. Trace Elem. Res.*, 6, 225, 1984.
218. **Lievens, P., Versieck, J., Cornelis, R., and Hoste, J.,** The distribution of trace elements in normal human liver determined by semi-automated radiochemical neutron activation analysis, *J. Radioanal. Chem.*, 37, 483, 1977.
219. **Alexiou, D., Grimanis, A. P., Grimani, M., Papaevangelou, G., Koumantakis, E., and Papadatos, C.,** Trace elements (zinc, cobalt, selenium, rubidium, bromine, gold) in human placenta and newborn liver at birth, *Pediatr. Res.*, 11, 646, 1977.

220. **Meinel, B., Bode, J. C., Koenig, W., and Richter, F.-W.,** Contents of trace elements in the human liver before birth, *Biol. Neonate,* 36, 225, 1979.
221. **McConnell, K. P., Broghamer, W. L., Jr., Blotcky, A. J., and Hurt, O. J.,** Selenium levels in human blood and tissues in health and in disease, *J. Nutr.,* 105, 1026, 1975.
222. **Lapin, C. A., Morrow, G., III, Chvapil, M., Belke, D. P., and Fisher, R. S.,** Hepatic trace elements in the sudden infant death syndrome, *J. Pediatr.,* 89, 607, 1976.
223. **Robkin, M. A., Swanson, D. R., and Shepard, T. H.,** Trace metal concentrations in human fetal livers, *Trans. Am. Nucl. Soc.,* 17, 97, 1973.
224. **Money, D. F. L.,** Vitamin E, selenium, iron and vitamin A content of livers from sudden infant death syndrome cases and control children: interrelationship and possible significance, *N. Z. J. Sci.,* 21, 41, 1978.
225. **Pickston, L., Lewin, J. F., Drysdale, J. M., Smith, J. M., and Bruce, J.,** Determination of potentially toxic metals in human livers in New Zealand, *J. Anal. Toxicol.,* 7, 2, 1983.
226. **Vanoeteren, C., Cornelis, R., Hoste, J., and Haeghen, L. V.,** The regional distribution of trace elements in human lungs with differentiation of the fraction present in the deposited dust, in *Trace Element Analytical Chemistry in Medicine and Biology,* Vol. 2, Brätter, P. and Schramel, P., Eds., Walter de Gruyter, Berlin, 1983, 315.
227. **Persigehl, M., Kasperek, K., Klein, H. J., and Feinendegen, L. E.,** Einfluss der industrialiserung auf den spurenelementgehalt in menschlichen lungen, *Beitr. Pathol.,* 157, 260, 1976.
228. **Kumpulainen, J., Vuori, E., Kuitunen, P., Mäkinen, S., and Kara, R.,** Longitudinal study on the dietary selenium intake of exclusively breast-fed infants and their mothers in Finland, *Int. J. Vitam. Nutr. Res.,* 53, 420, 1983.
229. **Hadjimarkos, D. M. and Shearer, T. R.,** Selenium in mature human milk, *Am. J. Clin. Nutr.,* 26, 583, 1973.
230. **Clemente, G. F. and Ingrao, G.,** The concentration of some trace elements in human milk from Italy, *Sci. Total. Environ.,* 24, 255, 1982.
231. **Lombeck, I., Kasperek, K., Bonnermann, B., Feinendegen, L. E., and Bremer, H. J.,** Selenium content of human milk, cow's milk and cow's milk infant formulas, *Eur. J. Pediatr.,* 129, 139, 1978.
232. **Smith, A. M., Picciano, M. F., and Milner, J. A.,** Selenium intakes and status of human milk and formula fed infants, *Am. J. Clin. Nutr.,* 35, 521, 1982.
233. **Kanabrocki, E. L., Kanabrocki, J. A., Greco, J., Kaplan, E., Oester, Y. T., Brar, S. S., Gustafson, P. S., Nelson, D. M., and Moore, C. E.,** Instrumental analysis of trace elements in thumbnails of human subjects, *Sci. Total. Environ.,* 13, 131, 1979.
234. **Hadjimarkos, D. M. and Shearer, T. R.,** Selenium content of human nails: a new index for epidemiologic studies of dental caries, *J. Dent. Res.,* 52, 389, 1973.
235. **Harrison, W. W. and Clemena, G. G.,** Survey analysis of trace elements in human fingernails by spark source mass spectrometry, *Clin. Chim. Acta,* 36, 485, 1972.
236. **Biswas, S. K., Abdullah, M., Akhter, S., Tarafdar, S. A., Khaliquzzaman, M., and Khan, A. H.,** Trace elements in human fingernails: measurement by proton-induced x-ray emission, *J. Radioanal. Nucl. Chem.,* 82, 111, 1984.
237. **Damsgaard, E., Heydorn, K., and Horn, N.,** Trace elements in the placenta of normal foetuses and male foetuses with Menkes disease determined by neutron activation analysis, in *Trace Element Analytical Chemistry in Medicine and Biology,* Vol. 2, Brätter, P. and Schramel, P., Eds., Walter de Gruyter, Berlin, 1983, 499.
238. **Dawson, E. B., Menon, M. P., Wainerdi, R. E., and McGanity, W. J.,** Activation analysis of placental trace elements, *J. Nucl. Med.,* 9, 160, 1968.
239. **Hadjimarkos, D. M. and Shearer, T. R.,** Selenium concentration in human saliva, *Am. J. Clin. Nutr.,* 24, 1210, 1971.
240. **Bleau, G., Lemarbre, J., Faucher, G., Roberts, K. D., and Chapdelaine, A.,** Semen selenium and human fertility, *Fertil. Steril.,* 42, 890, 1984.
241. **Pleban, P. A. and Mei, D.-S.,** Trace elements in human seminal plasma and spermatozoa, *Clin. Chim. Acta,* 133, 43, 1983.
242. **Nixon, G. S. and Myers, V. B.,** Estimation of selenium in human dental enamel by activation analysis, *Caries Res.,* 4, 179, 1970.
243. **Losee, F., Cutress, T. W., and Brown, R.,** Trace elements in human dental enamel, in *Trace Substances in Environmental Health — 7,* Hemphill, D. D., Ed., University of Missouri, Columbia, 1973, 19.
244. **Curzon, M. E. J. and Crocker, D. C.,** Relationship of trace elements in human tooth enamel to dental caries, *Arch. oral Biol.,* 23, 647, 1978.
245. **Derise, N. L. and Ritchey, S. J.,** Mineral composition of normal human enamel and dentin and the relation of composition to dental caries. II. Microminerals, *J. Dent. Res.,* 53, 853, 1974.
246. **Retief, D. H., Cleaton-Jones, P. E., Turkstra, J., and Beukes, P. J. L.,** Selenium content of tooth enamel obtained from two South African ethnic groups, *J. Dent. Res.,* 55, 701, 1976.

247. **Retief, D. H., Scanes, S., Cleaton-Jones, P. E., Turkstra, J., and Smith, H. J.,** The quantitative analysis of selenium in sound human enamel by neutron activation analysis, *Arch. Oral Biol.*, 19, 517, 1974.
248. **Kosta, L., Zelenko, V., Ravnik, V., Levstek, M., Dermelj, M., and Byrne, A. R.,** Trace elements in human thyroid, with special reference to the observed accumulation of mercury following long-term exposure, in *Comparative Studies of Food and Environmental Contamination,* International Atomic Energy Agency, Vienna, 1974, 541.
249. **Geahchan, A. and Chambon, P.,** Fluorometry of selenium in urine, *Clin. Chem.*, 26, 1272, 1980.
250. **Palmer, I. S., Olson, O. E., Ketterling, L. M., and Shank, C. E.,** Selenium intake and urinary excretion in persons living near a high selenium area, *J. Am. Diet. Assoc.*, 82, 511, 1983.
251. **Weingarten, R., Shamai, Y., and Schlesinger, T.,** Determination of selenium in urine by neutron activation analysis, *Int. J. Appl. Radiat. Isot.*, 30, 585, 1979.
252. **Hojo, Y.,** Single-void urine selenium level expressed in terms of creatinine content as an effective and convenient indicator of human selenium status, *Bull. Environ. Contam. Toxicol.*, 29, 37, 1982.
253. **Cadell, P. B. and Cousins, F. B.,** Urinary selenium and dental caries, *Nature (London),* 185, 863, 1960.

Chapter 9

GEOLOGICAL MATERIALS AND SOILS

Michael L. Berrow and Allan M. Ure

TABLE OF CONTENTS

I. INTRODUCTION

In the following review the contents of total selenium in geological materials, including meteorites, minerals, rocks and sediments, are tabulated and discussed. The contents, sources, forms and distribution of total selenium in soils are then reviewed covering the literature up to 1983. Factors affecting the solubility and mobility of selenium in soils and, thus, its availability to plants are also briefly discussed.

The geochemistry of selenium has been reviewed by Rosenfeld and Beath[1] and Howard,[2] while selenium in the environment has been discussed by Lakin,[3] Moxon,[4] and Swaine.[5] Concentrations of selenium in soils have been tabulated by Swaine,[6] Aubert and Pinta,[7] and Ure and Berrow.[8] The wider geochemical cycle, of which soils form a part, also involves plants and animals. Recent reviews of selenium in soils and plants have been presented by Johnson,[9] Fleming,[10] Peterson et al.,[11] and Kabata-Pendias and Pendias,[12] while selenium in soil-plant-animal systems has been reviewed by Allaway[13] and Sharma and Singh.[14]

II. SELENIUM IN METEORITES AND LUNAR MATERIALS

The selenium content of stony meteorites (chondrites) can vary widely. Pelly and Lipschutz[15] reported the selenium contents of 91 meteorites to range between 0.002 and 34 (mean 8.64) mg/kg, while the values reported in Table 1 vary from 0.0016 to 50 mg/kg. The data in this table have been arranged in chronological order of reference source.

Chondritic meteorites are classified largely on gross petrographic features and can be divided into five compositional groups and six petrologic types. Carbonaceous or C-chondrites can be subdivided into types C1, C2, C3, etc., the type number increasing with increasing degree of ''metamorphism''.[16] The elemental composition of the C1-type chondrites is thought to approximate the composition of the primordial matter of the solar system.[17] Concentrations of selenium in a number of C1-type chondrites have been reported by Krahenbuhl et al.[18] to range from 19.1 to 21.1 (mean 19.5) mg/kg, slightly higher than those reported by Ebihara et al.,[19] which ranged from 7.4 to 19.2 (mean 15.9) mg/kg. The C2-, C3-, and C4-type chondrites appear to contain decreasing amounts of selenium in an order (Table 1) which agrees with the finding of Greenland[20] that the abundance of selenium decreased regularly through the carbonaceous chondrites and was lowest in ordinary chondrites. Greenland concluded that since selenium is the chalcophile element *par excellence*, it was not surprising that its distribution almost exactly parallelled that of sulfur. Troilite, FeS, a variety of pyrrhotite, appears to be the host mineral for selenium in meteorites.[15]

The composition of chondrites varies according to their mineralogy and petrology. Chondrite composition is also affected by the inclusion of chondrules (small rounded bodies composed chiefly of olivine and enstatite), whereas achondrites contain none of these. The 18 achondrites analyzed by Laul et al.[21] contained small amounts of selenium, mean 0.54 mg/kg, and the 9 enstatite-type achondrites analyzed by Biswas et al.[22] contained a mean of 2.84 mg/kg. Achondrites also generally have low sulfide contents. The selenium concentrations in eucrite-chondrites also appear to be very low, with a mean of 0.19 mg/kg.[23]

Lunar basalts generally contain low amounts of selenium relative to chondritic meteorites. Wolf and Anders[24] have reported that selenium had virtually constant and identical abundances in lunar and terrestrial basalts, probably reflecting saturation with iron in the source regions. The selenium contents of rocks affected by meteorite impacts will be considered in a later section.

III. SELENIUM IN MINERALS

Selenium is widely dispersed and occurs in small amounts in nearly all phases of the crust of the earth, although silicates of selenium are not known. It occurs to some extent as the

Table 1
SELENIUM IN METEORITES AND LUNAR MATERIALS

Material	Sample type	Se concentration (mg/kg) Range	Mean	N[a]	Date[b]	Ref.
Meteorites	Ordinary and enstatite	0.96—15.1	7.69	27	1960	25
Meteorites	Chondrites and achondrites	0.0016—13.3	6.43	7	1960	26
Meteorites	Chondrites	4.8—34	12.6	23	1967	20
Meteorites	—	0.002—34	8.64	91	1971	15
Meteorites	Chondrites	7.7—15.5	11.0	13	1971	27
Meteorites	Achondrites	0.078—2.09	0.543	18	1972	21
Chondrites	C1 type	19.1—21.1	19.5	8	1973	18
Chondrites	C2 type	11.3—12.3	11.8	3	1973	18
Chondrite	Supuhee H6 type	6.58—25	9.9	6	1977	28
Chondrite	Supuhee clast 4J	22	22	1	1977	29
Chondrites	—	0.46—23.66	8.84	9	1977	30
Chondrite	Karoonda C4 type	7.7	7.7	1	1977	31
Chondrites	—	3.80—18.4	8.3	10	1978	32
Chondrites	Ordinary	8.8—10	9.4	3	1978	33
Chondrites	Carbonaceous	14.0	14.0	3	1978	33
Chondrites	Eucrites	0.08—0.50	0.19	4	1978	23
Chondrites	C3 type	7.1—19.2	10.2	5	1978	34
Meteorites	Gas-rich light portions	6.13—9.22	7.92	5	1979	35
Meteorites	Gas-rich dark portions	8.00—8.73	8.39	5	1979	35
Chondrites	L-6 type	10.3—11.7	10.8	3	1979	36
Meteorites	Stony iron, basic composition	1.8—13	5.09	7	1979	37
Achondrites	Enstatite type	0.40—6.95	2.84	9	1980	22
Meteorite	Allende	8.26—11.15	9.07	4	1980	38
Chondrites	Carbonaceous	7.3—21.5	11.6	20	1981	16
Chondrites	Carbonaceous	3.2—11.4	7.4	7	1982	39
Chondrites	C-1 type	7.39—19.2	15.9	8	1982	19
Chondrites	L4-6 type	4.4—9.06	7.56	14	1982	40
Meteorites	Forsterite type	3.94—50.0	13.7	16	1983	41
Chondrites	E or enstatite type	6.36—31.2	19.8	8	1983	42
Lunar olivines	—	—	0.15	21	1972	43
Lunar basalts	—	—	0.12	24	1980	24
Lunar materials	Rocks and soils	0.12—1.6	0.47	17	1972	44

[a] N — number of samples.
[b] Date of report.

free element in association with elemental sulfur; selenium contents from traces to 0.2% in native sulfur occur very frequently and similarly, native selenium often contains sulfur.[44] The selenium contents of volcanic sulfur from various sources have been found to vary greatly from 4 mg/kg to 2.9%.[45] In 16 samples of native sulfur of volcanic origin from the Kuriles, U.S.S.R., the contents ranged from 1 to 1040 mg/kg, whereas sedimentary sulfur contained only 0.1 to 10 mg/kg.[46]

Much of the selenium in the crust of the earth occurs in sulfide minerals, although some is found in selenides of silver, copper, lead, mercury, nickel, and other metals.[47] Most of the selenium used in the U.S.S.R. is extracted from chalcopyritic and pyritic deposits.[48] Over 40 selenium-bearing minerals have been identified, but these are not present in sufficient quantity to become commercially viable sources of the element. Isomorphous series such as that between galena, PbS, and clausthalite, PbSe, have been well studied, and the latter compound is indeed the most abundant selenium mineral.[44] Ranges of selenium contents in 16 different minerals of hypogene origin have been previously tabulated by Sindeeva,[49] and the great variability of the selenium contents of various sulfide minerals is illustrated by the data presented in Table 2. Contents generally vary from a few to several hundred milligrams per kilogram, but concentrations of greater than 1% have been found in minerals such as cinnabar, HgS,[50,51] tetradymite, Bi_2Te_2S,[52] and stibnite, Sb_2S_3.[53] Cinnabar may, however, contain much lower concentrations of selenium, in the range of 6 to 280 mg/kg, as found by Vershkovskaya et al.,[54] and a mean of 4.5 mg/kg, reported by Gorovoi and Vershkovskaya.[55] The lower contents in these cinnabars were attributed to the long distances between the ore deposits and the selenium source material. The latter authors pointed out that synthetic cinnabar could contain up to 7.7% selenium.

Selenium occurs in varying concentrations in the sulfide minerals of ore bodies and commonly occurs in pyrite, FeS, chalcopyrite, $CuFeS_2$, pyrrhotite, $Fe_{(n-1)}S_n$, galena, PbS, sphalerite, ZnS, cinnabar, HgS, stibnite Sb_2S_3, molybdenite, MoS_2, arsenopyrite, FeAsS, and others. It can often be present in greater amounts in chalcopyrite than in pyrite and other associated sulfides.[49,56-62] Other authors have reported higher concentration of selenium in pyrrhotite than in associated pyrite or chalcopyrite.[63,64] The concentrations of selenium in the various sulfide phases obviously depends upon the amounts of each mineral forming in a particular ore body. Pokrovskaya[65] has reported that the PbS-PbSe system could accommodate large amounts of selenium, while copper sulfides could contain up to 20% and pyrite only up to 5%. Enikeeva[66] found that selenium accumulated at the end of the high-temperature stage of ore formation and was concentrated mainly in galena, sphalerite, pyrrhotite, and chalcopyrite.

In a study of the main sulfide minerals of the Rudnyi Altai, it was found that selenium was concentrated primarily in galenas formed later than other sulfides such as pyrite, sphalerite, and chalcopyrite.[67] Experiments on the chemical dissolution of galenas showed that selenium was present as an isomorphous substituent for sulfur. Martirosyan and Babaeva,[68] in a study of deposits of a complex ore formation in the Lesser Caucasus, found that most of the selenium was contained in chalcopyrite ore, but was also concentrated in sphalerite. Maximum concentrations, of the order of 1%, were found in arsenopyrite, cinnabar, luzonite, and enargite, Cu_3AsS_4, in an investigation of the selenium contents of 175 hydrothermal sulfide minerals by Nagy.[59] Chalcopyrite contained up to 3900, pyrite 1250, sphalerite 1150, tennantite, $Cu_3(As,Sb)S_3$, 3100, and galena 1000 mg/kg.

In a study of the geochemistry of selenium in sulfate-sulfide mineral paragenesis it was found that sulfides contained 30 to 220 mg/kg, while anhydrite, $CaSO_4$, contained only 0.8 to 3 mg/kg.[69] The average S/Se ratio in sulfides was 8154, whereas in anhydrite it was 179,000. A distinct separation of selenium occurred, therefore, during simultaneous crystallization of sulfide and sulfate minerals. The mineral forms of selenium have been investigated by Aksenov et al.[70] in a micromineralogical study of pyrite-chalcopyrite ores with elevated selenium, bismuth, and tellurium contents. In thin sections, the selenium minerals were present in the form of rare, microscopically small (5 to 50 μm) inclusions in chalcopyrite. Bismuth selenide, galena, and aikinite, $(Pb,Cu,Bi,)S_3$, were identified which contained 20, 8, and 1% selenium, respectively. The natural selenide tiemannite, HgSe, identified in cinnabar from Tien Shan, U.S.S.R. by Finkel'shtein,[50] contained 18.3% selenium.

In endogenic processes, i.e., those occurring within the earth, selenium is a constant and unfailing accessory of sulfur. In exogenic processes, however, i.e., those occurring at or near the surface of the earth, sulfur and selenium take clearly separate geochemical pathways. This is because sulfur is oxidized relatively easily to soluble sulfates which can readily migrate. Selenium, on the other hand, requires strongly oxidizing conditions to form soluble selenates. Selenates can migrate in soluble form, under favorable conditions, but selenium is more commonly present in soils in reduced forms such as selenite or elemental selenium which are relatively immobile. For this reason sulfates which form under exogenic processes contain little or no selenium.[69] Selenium is thus mobile under oxidizing, alkaline conditions and immobile under reducing and neutral-to-acid conditions.

Table 2
SELENIUM IN MINERALS

Location	Sample type	Se concentration (mg/kg)		N[a]	Date[b]	Ref.
		Range	Mean			
Worldwide	Pyrite	<10—300	—	115	1955	76
Worldwide	Sphalerite	<10—900	—	41	1955	76
Worldwide	Chalcopyrite	<10—2100	—	43	1955	76
Worldwide	Pyrrhotite	<10—63	—	20	1955	76
U.S.S.R., Noril'sk region	Chalcopyrite	75—160	120	11	1960	56
U.S.S.R., Noril'sk region	Pyrrhotite	27—70	51	7	1960	56
U.S.S.R.	Pyrite	<8—175	—	74	1964	49
U.S.S.R.	Chalcopyrite	12—200	—	32	1964	49
U.S.S.R., Eastern Siberia	Molybdenites	13—203	70.8	10	1966	77
U.S.S.R.	Gold deposits	10—40	—	—	1966	73
U.S.S.R., Sedimentary rocks	Goethites	3—92	—	—	1967	78
U.S.S.R., Sadon	Galena	Trace—520	—	—	1968	79
U.S.S.R., Kiev	Pyrrhotite	5—21	—	—	1968	80
U.S.S.R., Altai	Biselenide	—	20%	—	1968	70
U.S.S.R., Altai	Galena	—	8%	—	1968	70
U.S.S.R., Altai	Aikinite	—	1%	—	1968	70
U.S.S.R., Karamazar	Sulphides	30—220	88	—	1969	69
U.S.S.R., Karamazar	Anhydrite	0.8—3.0	1.5	—	1969	69
U.S.S.R., Azerbaidzhan	Galena	0.07—89.0	—	80	1969	81
U.S.S.R., Tadzhikistan	Ore samples	7—205	—	20	1970	63
U.S.S.R., Skarn deposits	Molybdenite	—	270	—	1970	82
U.S.S.R., Skarn deposits	Chalcopyrite	—	60	—	1970	82
U.S.S.R., Tien-Shan	Cinnabar	10—14800	—	—	1971	50
U.S.S.R., Tien-Shan	Tiemannite	—	18.3%	—	1971	50
U.S.S.R., Kola peninsula	Pyrrhotite	7—117	35	—	1974	83
U.S.S.R.	Cinnabar	Up to 94000	—	—	1974	51
U.S.S.R., Karamazar	Tetradymite	—	27000	—	1976	52
U.S.S.R., Karamazar	Molybdenite	—	500	—	1976	52
U.S.S.R., Caucasus	Pyrite	3—290	—	—	1978	84
U.S.S.R., Nikitovka	Cinnabar	—	4.5	—	1978	55
U.S.S.R., Kazakhstan	Sulphide ores	10—390	—	—	1979	85
U.S.S.R., Caucasus	Ore minerals	0.1—280	26.5	6	1979	54
U.S.S.R., Azerbaidzhan	Sulfide ores	0.5—18.4	—	—	1979	86
U.S.S.R., Chukota	Stibnite	3200—31800	—	—	1980	53
U.S., western	Sedimentary pyrite	<3—300	22	47	1957	87
U.S., Colorado	Sphalerite	0—323	92	48	1976	61
U.S., Colorado	Chalcopyrite	8—537	275	89	1976	61
U.S., Colorado	Pyrite	37—198	104	8	1976	61
U.S., Colorado	Galena	11—68	49	7	1976	61
Canada, 29 different deposits	Chalcopyrite	8—537	235	89	1975	60
Canada, 29 different deposits	Sphalerite	0—323	92.6	48	1975	60
Canada, 29 different deposits	Pyrite	37—198	104	8	1975	60
Canada, 29 different deposits	Galena	11—68	51.7	7	1975	60
Australia, Hydrothermal	Pyrite	0—132	41	34	1954	88
Australia, Hydrothermal	Chalcopyrite	10—110	30	19	1954	88
Australia, Sedimentary	Pyrite	1—9	3	18	1954	88
Chile, Chuquicamata	Sulfide ores	3.0—25.9	—	—	1974	58
Brazil, Paragominas	Bauxites	0.01—0.1	0.043	8	1982	89
Japan, Mines	Pyrites	0.2—39	12.6	—	1950	90
Japan, Hitachi mine	Cu-bearing ores	0.5—117	—	46	1968	91
China, Tongren-Wanshan	Cinnabar	200—8700	2520	3	1982	92
New Zealand	Phosphatic materials	0.21—7.14	2.1		1966	93
New Zealand	Carbonate minerals	0.04—1.4	0.41		1966	93
New Zealand	Sulfur-rich minerals	0.01—1950	197.5		1966	93

Table 2 (continued)
SELENIUM IN MINERALS

Location	Sample type	Se concentration (mg/kg) Range	Mean	N[a]	Date[b]	Ref.
Czechoslovakia, Smolnik	Galena	440—1160	—		1966 and 1967	94, 95
Czechoslovakia, Smolnik	Sphalerite	198—940	—		1966 and 1967	94, 95
Czechoslovakia, Smolnik	Chalcopyrite	4—415	50	320 samples	1966 and 1967	94, 95
Czechoslovakia, Smolnik	Pyrite	6—600	41		1966 and 1967	94, 95
Czechoslovakia, Smolnik	Ankerite	6—16	—		1966 and 1967	94, 95
Czechoslovakia, Muran plateau	Pyrites	14—49	—	11	1967	96
Czechoslovakia	Pyrargyrite & Miargyrite	100—200	—	15	1969	97
Czechoslovakia, Moravia & Silesia	Pyrite	<5—290	—	32	1974	98
Czechoslovakia, Moravia & Silesia	Pyrrhotite	9—83	—	10	1974	98
Czechoslovakia, Moravia & Silesia	Sphalerite	<5—110	—	43	1974	98
Czechoslovakia, Moravia & Silesia	Galena	<5—3500	—	51	1974	98
Czechoslovakia, Moravia & Silesia	Arsenopyrite	4—42	—	8	1974	98
Czechoslovakia, Moravia & Silesia	Molybdenite	66—650	—	5	1974	98
Czechoslovakia, Moravia & Silesia	Stibnite	5—180	—	5	1974	98
Czechoslovakia, Bohemia	Molybdenite	66—300	—	—	1970	99
Czechoslovakia, Carpathians	Sulfide ores	<4—180	89.4	—	1975	100

[a] N — number of samples.
[b] Date of report.

The strong adsorption of selenium by secondary iron minerals is well illustrated by the concentration of selenium in gossan — a ferruginous deposit filling the upper parts of mineral veins or forming a superficial cover on masses of pyrite. In the gossan profile capping a zinc-lead lode in Queensland, Australia,[71] it was found that selenium and other elements were accumulated in the gossan. This contained goethite, $FeO(OH)$, and hematite, Fe_2O_3, which concentrated in a zone of solution-deposited secondary minerals, possibly as a result of leaching from the surface gossan. Similarly, selenium was increased in all the rocks around a copper ore body in Kaavi, eastern Finland,[72] where anomalously high contents were found in gossans, sediments, and plants in restricted areas immediately adjacent to the sulfide-rich rocks. These rocks could be located by determining selenium in such colloid-rich sediments as gossans, organic soils, and lake sediments.

Selenium is also typical of the rare elements in hydrothermal gold deposits where it enters isomorphously into the crystal lattice of sulfides accompanying gold mineralization.[73] Secondary clausthalite and other selenium minerals have, however, been found associated with unique hydrothermal gold-selenium deposits in Germany.[44]

A comprehensive study on the geochemistry of selenium in relation to the formation of ferroselite, FeS_2, and of the behavior of selenium in the vicinity of oxidizing sulfide and uranium deposits has been made by Howard.[2] He found that the conclusions of the early research of Goldschmidt and co-workers[74,75] are still largely valid. Their conclusions were essentially that (1) in magmatic activity selenium is incorporated into sulfide minerals or volcanic sulfur; (2) selenium is separated from sulfur during weathering by oxidation to elemental selenium or to the selenite ion, SeO_3^{2-}, while sulfur is oxidized to a soluble, mobile SO_4^{2-} ion; (3) of the sedimentary rocks, shales and particularly iron and manganese ores are enriched in selenium; and (4) the selenate ion, SeO_4^{2-} is coprecipitated or strongly adsorbed by freshly precipitated ferric hydroxide, thus explaining its enrichment in sedimentary iron oxide beds.

IV. SELENIUM IN ROCKS

Selenium, like sulfur, is not an essential constituent of rock-forming silicate minerals. The close similarity of the ionic radius of sulfur (S^{2-}, 1.84 Å) and selenium (Se^{2-}, 1.91 Å) means that selenium can readily enter sulfide minerals by isomorphous substitution. The element thus has very strong chalcophile characteristics.[101] This partnership of selenium with sulfur is broken, however, during the weathering cycle as sulfur is more readily oxidized than selenium. Soluble sulfates can then migrate relatively easily following the weathering of rocks, whereas selenium tends to remain in less mobile forms (see Section III).

Selenium can separate from cooling igneous rocks in association with the sulfides of iron, cobalt, and nickel in the early stages of magma crystallization. It also accumulates in the liquids and gases of late-stage processes, and volatile selenium escapes along with volatile sulfur from cooling rocks and volcanoes. It is well known that H_2Se and SeO_2 appear in volcanic gases and that the amount of selenium in sulfur of volcanic origin may vary from traces up to 5%. Thus, igneous rocks tend to have low contents of both selenium and sulfur.[45]

Goldschmidt[101] calculated the mean content of selenium in the lithosphere to be 0.09 mg/kg using the mean sulfur content of 520 mg/kg and an average S/Se ratio of 6000. This value has since been lowered to 0.05 mg/kg by Turekian and Wedepohl[102] who revised the mean sulfur content of igneous rocks downward to 300 mg/kg. Vinogradov[103] has reported the mean content of selenium in Russian crystalline rocks to be 0.01 mg/kg, while Taylor[104] has also calculated the average content of the crust of the earth to be 0.05 mg/kg. Sindeeva[49] found a higher average of 0.14 mg/kg in 18 samples of various magmatic rocks from the U.S.S.R.

The distribution of selenium in rocks of the U.S. formed from the pre-Cambrian to recent times has been extensively reviewed by Rosenfeld and Beath.[1] There appears to be little evidence of large amounts of selenium in rocks formed prior to the Carboniferous age in the U.S. Rocks containing large amounts of selenium, usually black shales, occur in strata formed throughout the periods from the Carboniferous to Quaternary and have been most extensively studied in those of the Cretaceous period. More is known about the distribution of seleniferous plants on Cretaceous formations in the western interior of the U.S. than on rocks of any other period. For example, Moxon et al.[105] have reported the selenium contents of rocks sampled at various depths in South Dakota. A total of 527 samples were taken mainly from the Niobrara and Pierre Formations of Cretaceous age (range 0.0 to 113 [mean 5.8] mg/kg) which were suspected to have given rise to seleniferous vegetation. Preceding and during Cretaceous time, there was extensive volcanic activity throughout the land mass from which the sediments were being derived. It appears that much of the selenium in sedimentary rocks was derived from volcanic dusts and gases emitted into the atmosphere. More recently the volcanic eruption on the island of Heimaey, Iceland, has been investigated by Mroz and Zoller[106] who found 1 mg/kg in lava ash and 265 mg/kg in fumarole deposits. Atmospheric aerosols from active volcanic regions show very great enrichments in selenium. Vulcanism does not satisfactorily explain the relatively large contents found in all sedimentary rocks, however. Erosion and weathering of igneous and seleniferous sedimentary rocks can release selenium which is adsorbed and concentrated in other sediments, particularly those rich in decaying organic matter. Because of its dispersal by vulcanism and its mobility under oxidizing conditions, the selenium contents of rocks vary in different geological formations, in different beds of the same formation, and even in different parts of the same bed. The contents in carbonaceous rocks, lignite, volcanic materials, clays, phosphates, sandstones, limestones, ores, and concretions have been discussed by Rosenfeld and Beath.[1]

In New Zealand, rocks which contain large amounts of selenium are andesites (particularly ashes), Upper Cretaceous argillites, and basalts.[107] On the other hand, small amounts occur in granites, rhyolites, rhyolitic pumices, limestones, and schists. In 106 samples of various rock types, Wells[107] found 0.09 to 2.64 (mean 0.40) mg/kg. On the basis of his results he concluded that the average content of the soil-forming rocks of New Zealand was 0.42 mg/kg, a value strongly influenced by the large areas covered by sedimentary rocks having a clay component, and higher than the average based on igneous rocks. Koljonen[108,109] has investigated the selenium contents of Finnish igneous and sedimentary rocks. The highest contents of selenium were found in clay sediments and sediments containing organic residues. A tentative calculated average content in sandstone and evaporites was <0.01, in shale 0.50, and in limestone 0.03 mg/kg, respectively. This compared with an average content in igneous rocks of 0.058 mg/kg.[108] In Precambrian metamorphic rocks from Finland, Koljonen[110] found 0.02 to 37 mg/kg. The distribution of selenium in these rocks resembled that in sediments being enriched in melanocratic rocks and rocks containing organic matter before metamorphism. The distribution of selenium in the environment including rocks, has been reviewed by Lakin[3] who reported that the element was concentrated in carbonaceous deposits. Selenium contents in U.S. black shales were found to correlate with the corresponding contents of organic carbon. Average contents ranged from <0.5 mg/kg in 8 Cretaceous noncarbonaceous, nonmarine shales to 70 mg/kg in 43 Permian phosphatic shales.

Related findings have been reported in a study of the paragenesis of uranium and selenium in multicoloured Mesozoic-Cenozoic deposits in the U.S.S.R. by Vorob'ev et al.[111] The degree of rock oxidation did not affect the uranium contents, but the distribution of selenium was controlled by the oxidation-reduction conditions of sedimentation. All reduced rocks, black, gray, green, and bluish gray-green in color, had elevated contents of selenium. Sharply reducing geochemical conditions were favorable to the formation of syngenetic selenium-uranium concentrations of both elements fixed mainly by organic substances in the sediment. Savel'ev[112] has also reported a relation between selenium concentrations and the amounts of organic carbon and sulfur in metamorphic rocks from Uzbekistan.

During the weathering of Precambrian granites, gabbro-diorites, and pyroxenites, Boiko et al.[113] found that 30 to 60% of the original selenium was lost. Most of the mobile selenium was derived from accessory minerals in the alteration of granites and from pyroxenes in the alteration of pyroxenites. In the weathering process selenium was mobilized into a water-soluble form. Moxon et al.,[105] in studies of rocks of the Pierre formation and their derived soils in South Dakota, estimated that 60 to 84% of the selenium from the original bedrock was lost during soil-forming processes and felt that the lower figure was probably much too low.

Selenium contents of U.S. coals are also high relative to magmatic rocks, with contents of 86 samples ranging from 0.46 to 10.65 (mean 3.36) mg/kg.[114] These coals contain 10 to 200 times as much selenium as is found in igneous rocks (0.05 mg/kg). Savel'ev and Timofeev[115] have reported analyses of samples from a Russian Jurassic coal seam where the contents ranged from 0.1 to 11.35%. The selenium was present as native selenium, ferroselite, pyrites, pyrobitumens, and adsorbed selenium, the latter form representing approximately 50% of the total content in the coal. Of U.S. crude oil samples, 42 contained 0.06 to 0.35 (mean 0.17) mg/kg.[114] The burning of both forms of fossil fuels would release considerable amounts of selenium into the atmosphere.

Selenium contents in many meteorite samples are considerably greater than those found in most rocks of the crust of the earth. The rocks around meteorite impact craters have been analyzed for selenium by a number of workers.[116-118] At one site examined by Wolf et al.,[116] uniform enrichments corresponding to 1 to 2% Cl chondritic material was found. In the Brent Crater in Ontario,[117] samples from the basal melt zone at 823 to 857 m depth were enriched in selenium and other elements above the content in the basement rocks, the abundance pattern suggesting a chondritic meteorite impact.

The results presented in Table 3 show that igneous rocks generally contain small amounts of selenium, often less than 0.2 mg/kg, but volcanic ashes and tuffs can contain much greater amounts. Contents in metamorphic rocks vary widely, but are generally relatively high in carbonaceous metamorphic rocks. Selenium levels in shales and other sedimentary materials can also vary widely from 0.1 to 675 mg/kg, depending upon the geological origin of the sediment and its conditions of deposition. Black shales and sediments containing much clay-sized colloidal material are generally rich in selenium. Sands and sandstones contain much smaller amounts of selenium than do shales, clays, and mudstones, but some Jurassic sandstones have been found to contain 15 to 20 mg/kg.[119,120] Calcareous rocks, likewise, contain small amounts, but some formations contain up to 20 mg/kg.[121] Coal and oil samples generally contain much more selenium than igneous rocks, presumably because of their elevated contents of organic matter.

V. SELENIUM IN SEDIMENTS

Selenium contents of recent deposits such as marine and river sediments have been determined by a number of workers.

A. Marine Sediments

In recent deposits in the Black Sea, Sokolova and Pilipchuk[152] found that sandy sediments contained 0.1 mg/kg, while argillaceous material contained 2.4 mg/kg. Selenium contents were greatest in the eastern and western peripheral areas. Koljonen,[153] in a study of Finnish sediments, also found that the selenium content depended upon the amount of colloids and fine clay fractions present. Residual sediments containing mostly feldspar and quartz were devoid of selenium. In a study of the accumulation and dispersal of elements in the Bay of Fundy, Loring[154] also found that total selenium contents were low in sandy sediments and higher in those rich in fine-grained material.

Selenium in 66 marine sediments from the northwestern Pacific Ocean[155] ranged from 0.1 to 1.7 mg/kg with a mean of 0.63 mg/kg. Iron-manganese concretions were found to contain 0.4 to 0.8 mg/kg, and it was suggested that organic matter and iron sulfides played a role in the fixation of selenium during sedimentation and early diagenesis. Volkov et al.[156] have examined the distribution of sulfur and selenium in sediments of a Pacific transoceanic profile from Japan to Hawaii to Mexico. Under reducing conditions selenium was found to be concentrated in the sediments as weakly soluble selenides. Accumulation of both sulfur and selenium was related to redox processes of diagenesis and was entirely determined by the decomposition of organic matter in the buried sediments. Thus, total sulfur and selenium, like organic carbon, are accumulated most strongly in the nearshore and hemipelagic lithofacies zones. Surface layer sediments of the Tyrrhenian sea were found by Pilipchuk and Sokolova[157] to contain between 0.1 and 0.4 mg/kg, with maximum concentrations on the continental slope and minimum concentrations in the central parts of the sea. A correlation was again found between selenium and organic carbon contents.

B. River Sediments

In a geochemical survey of Ireland, Kiely and Fleming[158] investigated the relationship between stream sediment trace element contents and those of their related soils in selected

Table 3
SELENIUM IN ROCKS

Location	Sample type	Se concentration (mg/kg) Range	Mean	N[a]	Date[b]	Ref.
U.S.S.R.	Magmatic rocks, various	0.10—0.21	0.14	18	1964	49
New Zealand	Basaltic ashes	0.17—1.50	0.60	6	1967	107
New Zealand	Igneous rocks, various	0.09—1.08	0.35	11	1967	107
Finland	Igneous rocks, various	—	0.058	—	1973	108
U.S.S.R., Uzbekistan	Paleozoic igneous & sedimentary	0.2—0.8	0.40	6	1971	122
U.S.S.R.	Granites	0.02—0.09	0.04	40	1972	43
U.S.S.R.	Granodiorites	0.05—0.08	0.07	22	1972	43
U.S.S.R.	Basalts	0.09—0.34	0.13	26	1972	43
U.S.S.R.	Olivinite	0.09—0.52	0.15	21	1972	43
U.S.	Magmatic standard rocks	0.004—0.11	0.04	8	1967	123
U.S.S.R., Yakutia	Kimberlites	0.08—0.235	0.15	6	1975	124
Ocean floor	Basalts	0.095—0.179	0.15	4	1972	21
Ocean floor	Basalts	—	0.16	10	1980	24
Ocean floor	Basalts	0.013—0.335	0.205	7	1980	125
U.S., Wyoming	Volcanic tuffs	12.5—187	91.5	5	1964	1
New Zealand	Rhyolitic pumice	0.11—0.24	0.17	11	1967	107
New Zealand	Andesitic ashes	0.47—2.64	1.19	7	1967	107
	Metamorphic Rocks					
New Zealand	Schists & gneisses	0.14—0.40	0.23	7	1967	107
U.S.S.R., Uzbekistan	Carbonaceous schists	30—120	—	—	1970	112
U.S.S.R.	Phyllites	10—70	—	—	1970	112
U.S.S.R.	Quartzites	0.3—12	—	—	1970	112
U.S.S.R.	Amphibolites	2.0—3.3	—	—	1970	112
Finland	Precambrian metamorphic	0.02—37	—	—	1973	110
Canada	Meteoric impact crater rocks	0.0004-0.023	0.0062	17	1980	116
Canada, Ontario	Meteoric impact crater rocks	0.069—0.252	0.050	15	1981	117
Finland, Lappajarvi	Impact Melts	0.2—0.77	0.46	7	1982	118
	Shales and Other Sediments					
U.S., Alabama	—	0.15—0.6	0.25	5	1934	126
Europe	Paleozoic	Composite	1.2	36	1935	127
Japan	Paleozoic	Composite	0.24	14	1935	127
Japan	Mesozoic	Composite	0.38	10	1935	127
U.S., South Dakota	Mainly Cretaceous	0.0—113	5.79	527	1939	105
U.S., Wyoming	Permian	100—675	277	23	1961	128
U.S., Nebraska	Cretaceous	1.5—103	20.9	27	1961	129
U.S., South Dakota	Cretaceous	0.0—5.0	1.3	15	1961	129
U.S., Montana, Wyoming, & South Dakota	Cretaceous	1.0—8	2.0	22	1962	130
U.S.S.R., Estonia	Silurian	0.3—9	3.3	3	1964	49
Australia, Queensland	Calcareous clay shales	0.05—385	28.0	36	1963	131
New Zealand	Greywackes	0.11—0.72	0.33	32	1967	107
Finland	Shales	—	0.50	—	1973	109
Poland, Leszczyn deposits	Upper Permian	—	1.31	—	1966	132
Poland, Grodziec syncline	Upper Permian	—	1.5	—	1967	133
U.S.S.R., central Asia	Shales & mudstones	—	0.6	—	1970	121

Table 3 (continued)
SELENIUM IN ROCKS

Location	Sample type	Se concentration (mg/kg) Range	Mean	N[a]	Date[b]	Ref.
U.S.S.R., central Asia	Paleozoic sediments	0.42—2.8	—	—	1970	134
U.S.S.R., Tuva	Sedimentary rocks	—	2.64	—	1978	135
U.S., Missouri & Arkansas	Red cherty clay	—	0.39[c]	126	1980	136
U.S., Missouri	Red cherty clay	—	0.46[c]	21	1980	136
U.S.S.R., Moldavia	Sedimentary rocks	0.06—5.5	1.05	—	1983	137
U.S., western	Coal measure shales	<0.1—6.71	1.17	75	1975	138
U.S.S.R., Tuva	Siltstones, gravelites, & shales	0.17—0.5	—	—	1967	120
U.S.	Black shales	—	24.1	156	1973	3
U.S.	Mudstones	6.0—24.1	—	5	1981	139
U.S.	Opaline mudstones	3.6—48.6	—	5	1981	139
U.S.	Eipidiatomites	1.6—13.7	—	4	1981	139
Australia, Lake Yindarlgooda	Black shales	<2—4	2.4	28	1981	140
Australia	Disseminated ore zone	2—40	7.1	21	1981	140
Australia	Massive ore zone	2—4	2.0	17	1981	140
Turkey, Koprubasi uranium deposits	Neogene sedimentary rocks	1.0—27	2.58	103	1981	141
U.S., northern Great Plains	Siltstones & shales	<0.1—0.79	0.21[c]	24	1982	142
U.S., northern Great Plains	Dark shale	<0.1—0.99	0.30[c]	23	1982	142
U.S., northern Great Plains	Fine-grained rocks	<0.08—0.62	0.16[c]	50	1982	142
U.S., Colorado Plateau	Triassic sandstones	<0.5—15	<1	97	1962	119
U.S., Colorado Plateau	Jurassic sandstones	<0.5—15	<1	96	1962	119
U.S.S.R., Tuva	Cretaceous carbonaceous sandstones	1.6—3.0	2.3	—	1964	143
U.S.S.R., Tuva	Jurassic sandstones	0.69—19.2	—	—	1967	120
U.S.S.R., Central Asia	Sands & sandstones	—	2.0	—	1970	121
U.S.S.R., Central Asia	Gravels & conglomerates	—	0.4	—	1970	121
Finland	Sandstone and evaporites	—	<0.01	—	1973	109
U.S.S.R., Lesser Caucasus	Cretaceous sandstones	—	0.27	—	1974	144
U.S., Missouri	Loess from bluffs	—	0.21[c]	19	1980	136
U.S., Missouri	Loess — dolomite bearing	—	0.39[c]	18	1980	136
U.S., Missouri	Loess — away from bluffs	—	0.46[c]	33	1980	136
U.S., northern Great Plains	Sandstones	<0.10—0.66	0.19[c]	80	1982	142
U.S., Montana	Sandstones	<0.10—0.29	0.042[c]	24	1982	142
U.S., northern Great Plains	Sandstones	<0.2—1.5	0.31[c]	42	1982	142
Calcareous Rocks						
U.S., South Dakota	Cretaceous	0.0—6.0	1.4	9	1938	145
U.S., Utah	Mississippian	0.0—12.0	2.0	10	1939	146
Mexico	—	0.1—2.5	0.5	10	1940	147
New Zealand	Calcareous rocks & mud	0.14—0.66	0.34	10	1967	107
U.S.S.R., Uzbekistan	Limestones & dolomites	1.3—24	—	—	1970	112
U.S.S.R., central Asia	Dolomitic limestones	—	1.3	—	1970	121
Finland	Limestones	—	0.03	—	1973	109
U.S.S.R., Lesser Caucasus	Cretaceous limestones	—	0.43	—	1974	144

Table 3 (continued)
SELENIUM IN ROCKS

Location	Sample type	Se concentration (mg/kg) Range	Mean	N[a]	Date[b]	Ref.
	Coals and Oil					
U.S., 20 different states	Coals	0.46—10.7	3.36	86	1969	114
U.S., Illinois	Coals	1.2—7.7	2.12	25	1973	148
U.S., Montana	Coals	1.49—1.50	1.5	2	1975	149
U.S.	Oil	0.15—1.4	—	4	1970	150
Libya	Oil	0.22—1.1	—	3	1970	150
U.S., Illinois	Oil	<0.04—0.4 mg/l	—	46	1973	151
U.S.	Crude oil	0.06—0.35	0.17	42	1969	114
Australia, New South Wales & Queensland	Bituminous coals	0.21—2.5	0.79	—	1977	5
U.S., Illinois	Basin coals	0.45—7.7	1.99	—	1977	5

[a] N — number of samples.
[b] Date of report.
[c] Geometric mean.

Table 4
SELENIUM IN MARINE AND STREAM SEDIMENTS

Location	Sample Type	Se concentration (mg/kg) Range	Mean	N[a]	Date[b]	Ref.
NW Pacific Ocean	Marine sediments	0.1—1.7	0.63	66	1973	155
Romania	River sediments	0.1—1.4	0.68	36	1973	250
Ireland	Stream sediments	0.2—43.4	—	—	1969	158
U.S., northern Great Plains	Stream sediments	<0.11—0.46	0.19[c]	60	1982	142
U.S., Powder River Basin	Stream sediments >200 μm	<0.1—0.60	0.12[c]	19	1982	142
U.S., Powder River Basin	Stream sediments 100—200 μm	<0.1—0.48	0.13[c]	24	1982	142
U.S., Powder River Basin	Stream sediments 63—100 μm	<0.1—0.50	0.13[c]	24	1982	142
U.S., Powder River Basin	Stream sediments <63 μm	<0.1—0.54	0.17[c]	24	1982	142

[a] N — number of samples.
[b] Date of report.
[c] Geometric mean.

areas. Good correlations were obtained for some elements, but stream sediment analyses were unsatisfactory for predicting the soil contents of others, notably selenium. Stream sediments from the Powder River Basin in the U.S. have been separated into particle size fractions, and these show a general tendency to increase slightly in selenium content with decreasing particle size.[142] The values are reported in Table 4 which contains other data from marine and stream sediment analyses.

Many thousands of samples have been analyzed for trace elements under the National Resource Evaluation Program in the U.S. as part of a hydrogeochemical and stream sediment reconnaissance. Reports such as that by Maassen[159] are available through the National Technical Information Service (NTIS). The limit of detection of the XRF method of analysis used for selenium was 5 mg/kg, insufficient to quantify the natural levels of the element. Thus, contents of 5789 stream sediments listed in Reports GJBX 208-80, 105-81, 215-81, and 216-81 were all less than 5 mg/kg, except for 34 samples which had total contents between 5 and 19 mg/kg.

VI. SELENIUM IN SOILS

Because of the widespread occurrence of seleniferous soils in the western interior of the U.S., particularly in Wyoming and South Dakota, the soils of these areas have been examined in considerable detail. The maximum quantity of selenium found in several thousand soil samples in the U.S. was less than 100 mg/kg, and the majority of the seleniferous soils analyzed contained on average less than 2 mg/kg. In addition to the studies in Wyoming and South Dakota, the distribution of selenium in surface and subsoil samples has also been studied in other western states including Arizona, New Mexico, Colorado, Nebraska, Kansas, North Dakota, Montana, and Utah. Surface and subsoil samples in these areas contain on average less than 2 and 5 mg of Se per kilogram, respectively. All the surface and subsoil samples analyzed from Oklahoma, Texas, Missouri, Nevada, Idaho, and California had low contents.[1] The total contents of selenium in soils from the U.S. and other countries have been tabulated from the literature up to 1953 by Swaine[6] who concluded that most soils contain 0.1 to 2.0 mg/kg. The total selenium contents of 1623 soils from various countries reported in the more recent literature ranged from 0.03 to 2.0 (mean 0.40) mg/kg.[8] The values for total selenium in soils reported in Table 5 are from the literature up to 1983.

The selenium contents of soils listed in Table 5 range widely from 0.005 mg/kg in a selenium-deficient area in Finland to 8000 mg/kg in soils from a biogeochemical province enriched in selenium in the Tuva area of the U.S.S.R. This latter value appears to be the highest content reported in a soil. Selenium deficiencies in livestock have been recognized on soils in the U.S., Canada, New Zealand, Australia, Scotland, Finland, Sweden, Denmark, France, Germany, Greece, Turkey, the U.S.S.R., and the U.K.[160] Selenium-toxic soils, on the other hand, have been reported in the U.S., Canada, Columbia, Ireland, South Africa, Australia, the U.K., the U.S.S.R., and Mexico.[10] In a study with sheep, Watkinson[161] found that responses to selenium were frequently obtained when the total content in topsoils was less than 0.45 mg/kg. A rating for selenium in New Zealand topsoils has been proposed by Wells[162] as follows: <0.3, very low; 0.3 to 0.5, low; 0.5 to 0.9, average; 0.9 to 1.5, high and >1.5 mg/kg, very high. Soils containing greater than 2 mg/kg are likely to be seleniferous.[163]

A. Sources of Selenium in Soils

The amount of selenium in soils is determined mainly by geochemical processes, although anthropogenic sources can contribute. Selenium in soils can be derived from various sources which include: (1) the parent rocks, (2) formations underlying the soil mantle, (3) percolating ground or surface waters, (4) plant and animal residues, (5) mining and other industrial activities, (6) volcanic exhalations, (7) industrial smokes and fumes, (8) fertilizers, (9) wastes such as sewage sludges and fly ash applied to soils, and (10) marine aerosol. Bisbjerg[163] has concluded that the selenium concentration in soil mostly depends upon the parent rock, the climate, the topography and the age of the soil.

Parent materials which give rise to soils with low selenium contents include highly siliceous rocks such as granites, rhyolites, sandstones, and cherts and also limestones and dolomites.[162,164] Soils derived from argillaceous rocks, andesitic and basaltic ashes, and highly organic soils can contain high amounts of the element.[1] The total selenium in 517 surface soils from England and Wales increased in order of the following parent material groups: chalk, sandstone, limestone, clay, mudstone, shale, mineralized rocks, and peat.[165] Formations underlying the topsoil can also influence soil selenium contents.[166]

Percolating ground or surface waters may also enrich or deplete soil contents. Selenium released by the weathering of pyritiferous shales can be carried in drainage water and subsequently adsorbed by clay, iron compounds, or organic matter, as was the case in Ireland where organic soils were enriched to levels as high as 1200 mg/kg.[167] On the other hand,

Table 5
SELENIUM IN SOILS

Location	Sample type	Se concentration (mg/kg)				
		Range	Mean	N[a]	Date[b]	Ref.
U.S., various states	11 profiles	0.01—2.5	0.37	51	1937	220
U.S., Montana	Surface	0.1—5.0	0.8	448	1940	147
U.S., Montana	Subsoils	0.2—5.0	1.4	14	1940	147
U.S., New Mexico	Various depths	0—20	1.03	107	1940	147
U.S., South Dakota	32 profiles	0.2—38.4	4.95	96	1942	221
U.S., western half	Seleniferous topsoils	1—80	4.5	500	1945	222
U.S., New York & Nevada	2 profiles	<0.01—0.32	0.18	12	1972	223
U.S., Powder River Basin	0—2.5 cm	0.018—0.55	—	48	1975	224
U.S., Powder River Basin	15—20 cm	0.006—0.64	—	48	1975	224
U.S., Missouri	Cultivated 0—15 cm	0.1—1.5	0.28—0.74[c]	115	1975	164
U.S., Missouri	Uncultivated B horizons	<0.1—3.4	0.27—0.73[c]	300	1975	164
U.S., Colorado	Cultivated & uncultivated 0—15 cm	<0.1—1.5	0.23[c]	168	1975	164
U.S., Piceance Creek, Colorado	A horizons or 0—10 cm	<0.11—1.2	0.28[c]	97	1976	225
U.S., New York & South Dakota	Topsoil	0.92—35.8	18.36	2	1977	226
U.S.	28 profiles	0.08—0.35	0.18	—	1978	227
U.S.	Cultivated topsoils	<0.10—1.86	0.15	322	1980	228
U.S., south Texas	A & B horizons	0.01—0.90	0.15	256	1980	229
U.S., south Texas	A & B horizons	0.08—16	1.21	95	1980	229
U.S., Missouri	Plow zone	<0.1—2.7	0.39	1140	1981	230
U.S., Colorado & Utah	Alluvial, 0—40 cm	<0.1—0.57	0.079[c]	30	1982	142
U.S., Wyoming & Montana	A horizons	<0.1—1.1	0.25[c]	64	1982	142
U.S., Wyoming & Montana	B horizons	<0.1—2.2	0.30[c]	64	1982	142
U.S., Wyoming & Montana	C horizons	<0.1—1.6	0.23[c]	64	1982	142
U.S., Montana	A horizons	<0.12—0.37	0.17[c]	16	1982	142
U.S., Montana	C horizons	<0.11—0.45	0.18[c]	16	1982	142
U.S., northern Great Plains	A horizons	<0.10—20	0.45[c]	136	1982	142
U.S., northern Great Plains	C horizons	<0.10—26	0.34[c]	136	1982	142
U.S., Wyoming	0—40 cm	<0.10—1.1	0.11[c]	36	1982	142
U.S., Wyoming	0—40 cm	<0.10—0.38	0.098[c]	36	1982	142
U.S., New Mexico	A horizons	<0.20—0.80	0.14[c]	47	1982	142
U.S., New Mexico	C horizons	<0.20—0.50	0.13[c]	47	1982	142
U.S., New Mexico	A horizons	<0.20—0.50	0.23[c]	30	1982	142
U.S., New Mexico	C horizons	<0.20—0.70	0.20[c]	30	1982	142
U.S., conterminous	At 20 cm depth	0.1—4.3	0.39	1267	1984	231
Canada, eastern	0—15 cm	0.155—0.540	0.23	10	1971	232
Canada, Yukon	C horizons	0.5—15.0	4.7	16	1973	233
Canada, Toronto	1 profile	0.53—1.43	0.96	8	1974	234
Canada	54 profiles	0.073—2.090	0.38	234	1974	193
Canada	Surface	0.197—0.744	0.44	10	1974	208
Canada, Prince Edward Island	Topsoils	0.035—0.330	0.15	35	1975	235
Canada, Saskatchewan	C horizons	0.07—1.92	—	61	1977	236
Canada	81 profiles	0.02—3.7	0.30	188	1979	200
Canada, Ontario	0—15 cm	0.10—1.67	0.37	228	1979	202
Canada, Ontario	0—15 cm sludge treated	0.21—0.59	0.37	30	1979	202
Canada	Reference soils	0.05—0.4	0.21	4	1980	237
Ireland, Limerick County	6 profiles	0.4—850	72.1	22	1957	191
Ireland, Limerick County	6 profiles	0.56—175	24.4	26	1957	191
Ireland	0—15 cm	0.3—360	24.7	18	1962	167
Ireland, Meath County	7 profiles	0.8—1200	78.3	26	1962	167
England & Wales	0—30 cm	0.2—7.0	3.13	16	1966	238

Table 5 (continued)
SELENIUM IN SOILS

Location	Sample type	Se concentration (mg/kg) Range	Mean	N[a]	Date[b]	Ref.
Ireland	0—15 cm	0.6—2.8	1.44	7	1969	158
Britain	0—15 cm	0.9—91.4	29.7	4	1975	190
Scotland	0—25 cm	0.31—0.39	0.35	2	1979	239
Scotland	Topsoils	0.02—0.36	0.18	10	1979	240
England & Wales	0—15 cm	0.2—1.8	0.60[c]	114	1980	241
England & Wales	0—15 cm	<0.01—4.6	0.48	517	1983	165
Denmark	Plough layer	0.16—1.5	0.416	7	1964	242
Denmark	Topsoils	0.11—1.6	0.39	6	1969	243
Sweden	Topsoils	0.165—0.976	0.39	24	1970	214
Denmark	—	0.21—1.44	0.57	11	1972	163
Finland	8 podzol profile horizons	<0.01—1.25	0.191	29	1975	199
Finland	Miscellaneous	<0.01—0.49	0.155	46	1975	199
Finland	Topsoils, mainly peaty	0.043—2.30	0.72	24	1975	199
Finland	0—20 cm clay soils	0.131—0.633	0.290	29	1983	194
Finland	0—20 cm coarse mineral soils	0.050—0.489	0.172	64	1983	194
Finland	0—20 cm organogenic soils	0.212—1.281	0.464	19	1983	194
Finland	10—40 cm clay soils	0.140—0.654	0.274	26	1983	194
Finland	10—40 cm coarse mineral soils	0.040—0.560	0.159	54	1983	194
Finland	10—40 cm organogenic soils	0.329—0.982	0.575	13	1983	194
Denmark	Topsoils	0.137—0.415	0.30	3	1976	195
Denmark	Surface	0.14—0.52	0.34	28	1976	209
Norway	Topsoils	0.16—0.39	0.27	9	1977	244
Norway	Topsoils	0.90—7.35	2.98	49	1977	244
Denmark	Topsoils	0.14—0.58	0.32	6	1978	245
Norway, north	5—10 cm	0.08—1.70	0.63	122	1978	170
Norway, east	5—10 cm	0.07—1.35	0.42	117	1978	170
Finland	Topsoils	0.005—1.24	0.201	250	1979	203
Belgium	0—25 cm	0.04—0.27	0.11	10	1982	215
India, Gujarat	Surface	0.142—0.678	0.375	18	1970	246
India, Kaira	Surface	0.262—0.444	0.329	10	1970	246
India, Kaira	3 profiles	0.025—0.615	0.23	14	1970	246
India, various states	Surface	0.158—0.710	0.41	106	1972	210
India, Haryana	12 profiles	1.0—10.5	4.58	45	1976	166
Japan	Surface	0.4—0.9	0.7	7	1950	90
Japan	Subsoil	0.8—1.2	1.0	3	1950	90
Japan	—	0.129—2.821	—	—	1977	211
China	—	0.12—0.45	—	—	1983	247
Australia	Surface	0.02—0.4	—	—	1958	248
New Zealand	0—8 cm	0.07—4.3	0.72	28	1962	161
Australia, Queensland	Affected area	0.15—32.2	6.27	19	1963	131
Australia, Queensland	Nonaffected area	0.03—6.12	1.59	19	1963	131
New Zealand	64 profiles	0.06—17.2	1.69	320	1967	162
Israel	—	Traces — 6.0	—	—	1957	206
U.S.S.R., Tuva	Soils	12—8000	—	—	1967	120
Bulgaria	14 profiles	0.015—0.470	0.13	55	1973	249
Romania	11 profiles	0.1—1.8	0.64	45	1973	250
U.S.S.R., Azerbaidzhan	Mountain & cinnamonic	≤0.095	—	—	1974	251
U.S.S.R., Azerbaidzhan	Yellow & chestnut	0.096—0.176	—	—	1974	251
U.S.S.R., Azerbaidzhan	Sierozems	0.177—0.340	—	—	1974	251
U.S.S.R., Azerbaidzhan	Solonchak	0.341—0.430	—	—	1974	251
Egypt, Nile delta	0—30 cm	0.18—0.85	0.45	55	1980	252
Egypt, Nile delta	3 profiles	0.16—0.78	0.41	12	1980	252
Worldwide	—	0.06—1.8	0.40	1623	1982	8

[a] N — number of samples.
[b] Date of report.
[c] Geometric mean.

irrigation can remove substantial quantities of selenium from soils as has been revealed by the analysis of drainage waters.[45]

Plant and animal residues generally tend to enrich soils in selenium. The differences in physical and chemical properties of selenium and sulfur tend to lead to separation of these elements during erosion and weathering, but their similar functions in biological reactions tend to unite them in plant and animal residues which accumulate in topsoils. Rocks tend to lose selenium during weathering,[105] but the average content in topsoils is generally greater than that of their parent rocks,[1,162] partly due to the retention of selenium by soil organic matter.

Mining and other industrial activities can affect soil selenium contents. Mine spoil can contain selenium concentrations similar to those in soils, as found in 92 mine spoil samples from the northern Great Plains and New Mexico which ranged from <0.1 to 0.7 mg/kg.[142] Erosion of mine tailings by floodwater has, however, resulted in the production of selenium-toxic soils in Mexico. In this case, selenosis was recorded both in humans and in animals.[147]

Selenium is associated with sulfur in volcanic activity which can increase soil selenium contents. Unlike sulfur, selenium is likely to be present in the atmosphere in a particulate form either as Se^0 or as SeO_2 and is subject to removal by rain relatively close to its point of origin. Thus, soils and sediments near volcanic activity are relatively rich in selenium.[3,168,169] The burning of fossil fuels rich in selenium (Table 3) introduces selenium as a pollutant to the atmosphere and can also increase soil contents. In the U.S. it has been estimated that this process introduces 8 million pounds (3.6×10^6 kg) of selenium annually into the atmosphere.[3]

In addition to industrial smokes and fumes, there are marine and natural terrestrial sources which add to airborne selenium concentrations. Concentrations in Norwegian soils showed a distinct decrease with increasing distance from the sea, a trend not evident for arsenic, suggesting a considerable input of selenium, but not of arsenic from the marine aerosol.[170] The selenium content of the marine aerosol is also enhanced by the enrichment of selenium in ocean surface waters caused by anthropogenic activities[171] and by the mechanics of seaspray formation.[172] Another minor source is the release of volatile selenium compounds even from nonaccumulator plants such as alfalfa.[173] Soil microorganisms, such as molds and bacteria, can absorb selenium from the soil and also release it into the atmosphere as volatile organic compounds.[174] The factors influencing volatilization from soils have been discussed by Zieve and Peterson.[175] Further treatment of atmospheric selenium is covered in Chapter 12 of this volume.

The application of fertilizers is another source of selenium in soils. Swaine[176] has tabulated the selenium concentrations of fertilizers and found that while many limestone and phosphate rocks contain less than 1 mg/kg, other types can have higher contents. Rader and Hill[177] found up to 55 mg/kg in phosphatic rock, but American superphosphates contained between 0.8 and 4 mg/kg. Wells[93] found that agricultural fertilizers used in New Zealand had selenium contents within the range of values associated with soil-forming rocks, while Gissel-Neilsen[178] has tabulated the contents of some common fertilizers used in Denmark. Other soil additives such as sewage sludge or fly ash can also affect soil contents. Sewage sludges can contain up to about 10 mg/kg with median contents of 2 to 3 mg/kg.[179] Fly ashes from coal burning power stations have been applied to soils, and in the U.S. their contents have been reported by Gutenmann et al.[180] to range from 1.2 to 16.5 mg/kg. The effects of fly ash on selenium uptake by wheat and on the growth of Japanese quail fed the high-selenium wheat obtained has been investigated by Stoewsand et al.[181]

B. Forms of Selenium in Soils

As with many other nutrients, the total amount of selenium present in soil is important, but the forms in which it occurs determine whether problems of deficiency or toxicity will arise. For example, in 26 topsoils from Hawaii that contain 1 to 20 (mean 6.7) mg/kg, with subsoils containing up to 26 mg/kg, there was no evidence of selenium toxicity.[182] On the other hand, soils from South Dakota and Kansas which contain less than 1 mg/kg can produce seleniferous vegetation.[183] The high iron contents of the lateritic soils of Hawaii, in which selenium is firmly fixed in an unavailable form, partly explain these differences in plant availability.

The availability of selenium to plants is a function of the soil pH and Eh as well as the total content. In acid soils (pH 4.5 to 6.5) it is usually bound as a basic ferric selenite of extremely low solubility and is unavailable to plants. In alkaline soils (pH 7.5 to 8.5) selenium may be oxidized to more soluble selenate ions which are readily available to plants.[3]

Selenium occurs in soils in a number of different forms including elemental selenium, selenides, selenites, selenates, and organic selenium. The forms of occurrence in soils have been investigated by Olson,[184] Allaway et al.,[185] and Hamdy and Gissel-Nielsen.[186] Elemental selenium is formed by bacteria, fungi, and algae which are capable of reducing selenites and selenates. In water-logged acid soils, elemental selenium and selenides can form from added selenium salts, but the extent to which elemental selenium is a naturally occurring soil constituent is not well known.[187] Elemental selenium has, however, been reported to occur in sandstones near a uranium deposit by Howard[2] who reports several Eh-pH diagrams illustrating the stability relations between selenium and various iron compounds. Elemental selenium is moderately stable in soils[188] and, thus, not readily available to plants. Selenides are also largely insoluble and occur in association with pyrites in soils of arid regions where extensive weathering has not taken place.[189] As many seleniferous soils are immature, much of the selenium probably exists in the same form as that in which it occurs in the unweathered parent material. Although selenides are generally unavailable to plants, they may on weathering release some selenium in soluble form.

Selenites represent the most important source of selenium in many soils and largely control its availability to plants. The element often occurs in soils in association with iron as basic ferric selenite, $Fe_2(OH)_4SeO_3$. Geering et al.[188] found that the selenium concentration in solution in soils is governed primarily by this ferric oxide-selenite complex where the element is in the Se^{4+} form. Under certain conditions, controlled largely by redox potential and soil pH, selenium may also exist in $+6$, 0, and -2 oxidation states. Nye and Peterson[190] have shown that selenite is the predominant form extracted from four seleniferous British soils and established a linear relationship between selenium in plants and both total and free selenite-selenium in the soils.

The formation of selenates is favored by alkaline soil conditions. The seleniferous soils of South Dakota are alkaline in reaction and generally located in low rainfall areas. Under these conditions selenium oxidizes to the soluble selenate form which accumulates in the lower part of the soil profile.[4]

Selenium in organic combination occurs in varying quantities in soils, and this form can be increased by the accumulation of decaying plant residues. Organic selenium is also subject to microbiological breakdown, resulting in the liberation of alkylselenium compounds, mainly dimethylselenide.[11] In humid temperate regions with the relatively greater accumulation of soil organic matter, organic-selenium forms obviously assume greater importance. In the toxic soils in Ireland, selenium has clearly concentrated in organic horizons within the soil profile in a plant-available form.[191] Indications that organic soils retain selenium more strongly than mineral soils have been reported by a number of workers.[186,192-194] Studies by Hamdy and Gissel-Nielsen[195,196] showed that the addition of organic matter greatly diminished the evolution of volatile selenium compounds as well as the movement and leaching of the element through soil columns. According to these observations organic matter apparently concentrates selenium in soils because all the processes leading to selenium losses are impeded, thus explaining the marked association between native total selenium in soils and their organic matter contents. Soil microorganisms also concentrate selenium within their cells as reported by Koval'skii et al.[197] who found concentrations up to 1800 mg/kg in dry cells. Accumulation in wild mushrooms (up to 17.8 mg/kg dry matter) growing on uncontaminated Norwegian soils has also been reported by Allen and Steinnes.[198]

C. Distribution of Selenium in Soil Profiles

Studies of selenium distribution in soil profiles indicate how this element behaves during the processes of soil formation, and this aspect has been investigated by a number of authors.[142,162,166,193,199,200] In New Zealand profiles, Wells[162] found that B2 horizons, with their accumulation of iron and clay-sized colloids, had greatest contents within the profile. An accumulation of selenium in podzolic B horizons and organic surface horizons was found in 54 Canadian profiles by Lévesque.[193] Multiple regression analyses revealed that the predominant factors involved in selenium distribution were the content of the parent materials and the organic carbon content of the upper horizons. The contribution of atmospheric contaminants to soil contents was considered important. A subsequent investigation showed that the association of sulfur with selenium in geological formations held through soil formation processes.[201] In Finnish podzols, Koljonen[199] found that selenium is enriched in the O-A1 horizons rich in organic matter and in the iron-rich B horizons, but depleted in the eluviated A2 horizons. In Indian profiles, however, Singh and Kumar[166] found that selenium moves up the profile in areas of low rainfall and high temperature, while in high-rainfall areas it is leached down the profile. Enrichment in B horizons of profiles has also been reported by Ebens and Shacklette[142] in soils of the Powder River Basin, Wyoming, while McKeague et al.[200] found that selenium was generally in greatest amount in either the A or B horizons of Canadian profiles. The latter authors found that linear regression equations of selenium on organic carbon, iron, and clay accounted for more than 50% of the variability for several groups of samples.

Soil texture is also a factor in selenium distribution. Loamy and clay soils were found to have greater selenium contents than podzols and sandy soils in Ontario,[202] while in Finnish mineral soils the mean total content diminished with increasing particle size.[203] It has also been found that at a given pH, heavier textured soils released less selenium for uptake by ryegrass than lighter textured soils.[204]

D. Solubility of Selenium in Soils

Water-soluble selenium has been used as an index of selenium availability in areas where selenium contents of plants are high. Williams and Thornton[205] compared water, 0.001 *M* NaOH, and 0.05 *M* EDTA as extractants on three peaty seleniferous soils used in pot experiments. All three extractants reflected the uptake of selenium by ryegrass. In Ireland, water-soluble selenium levels approaching 3 mg/kg have been recorded,[10] while in Israel, water-soluble selenium was found to comprise up to 30% of the total in the soil.[206] In 11 Bulgarian top soils, concentrations of water-soluble selenium ranged from 0.03 to 0.67 (mean 0.19) mg/kg.[207] Lévesque[208] reported that water-soluble selenium represented 1.9 to 6.9% (mean 3.8%) of the total in some Canadian topsoils which is similar to the findings of Hamdy and Gissel-Nielsen[209] who found that water-soluble selenium constituted 1.7 to 5.5% (mean 2.4%) of the total in 28 Danish topsoils. Hot water-extractable selenium in soils sampled at 0 to 20 and 20 to 40 cm depth at some 100 different sites in Finland was on average 2.9 to 4.8% of the total contents.[194] In some Indian topsoils, water-soluble contents ranged between 0.019 and 0.066 mg/kg, representing 6.5 to 11.0% of the total selenium,[210] whereas in horizon samples from soil profiles water-soluble selenium ranged from 0.05 to 0.62 mg/kg, representing 3.1 to 36% of the total.[166] A good correlation between water-soluble and total selenium has been reported by Asakawa et al.[211] in some Japanese grassland soils.

Cary et al.[212] found that 0.02 *M* K_2SO_4-extractable selenium in soils was related to the selenium content of lucerne, presumably because K_2SO_4 dissolves or exchanges selenate in soils. Ammonium acetate-DTPA, pH 4.65, was used as an extractant by Sippola[203] who found that this removed 1.3 to 67.1 μg of Se per liter from 250 Finnish soils which represented an average of 5% of the total contents. The use of ammonium bicarbonate-DTPA as an extractant to assess the plant available selenium in soils has recently been proposed.[213]

E. Availability of Selenium to Plants

In the selenium-poor soils of Scandinavia, the total selenium content in plants is not correlated with either the total or water-soluble selenium in the surface soils.[209,214] On the other hand, Sippola[203] found that the contents in timothy grass growing on 250 different field sites throughout Finland showed a better correlation with total then with extractable soil selenium. Robberecht et al.[215] have also found a highly significant correlation between total selenium in Belgian soils and the contents of ryegrass, as have Nye and Peterson[190] between total selenium in some British soils and plant content. The variation in the findings of different workers probably reflects the fact that selenium is present in soils in different chemical forms having widely different rates of solubility.[186,212] The complex chemistry of soil selenium is governed by the physicochemical properties of the soil and by microbiological activity. Furthermore, its availability generally increases with increase in the soil pH.[204,216,217] Effects of such factors as liming, addition of fertilizers, and interaction with other nutrients on the availability of selenium have been recently reviewed.[10,14]

The interrelationships between selenium in rocks, soils, and plants have been usefully summarized by Allaway,[13] and his conclusions remain valid in the light of more recent research:

1. Where rocks with a high selenium content weather to form well-drained soils in dry areas (less than 500 mm annual rainfall), selenides and other insoluble forms will be converted to selenates and organic selenium compounds. The vegetation will be potentially toxic.
2. Where rocks with a high selenium content weather to form soils in humid areas, relatively insoluble complexes of ferric oxide or hydroxide and selenite will be formed. These soils will be slightly to strongly acidic, and the vegetation will probably contain sufficient selenium to protect livestock from selenium deficiency.
3. Where rocks with a high selenium content weather to form poorly drained soils or where such soils are enriched by alluvial action and the conditions are alkaline, toxic vegetation will probably be produced.
4. Where rocks with a low selenium content weather to form soils under either humid or dry conditions, the vegetation is likely to contain insufficient selenium to protect animals from selenium deficiency. The more humid the area and the more acid the soil, the greater the likelihood of extremely low selenium concentrations in the vegetation.

A map showing selenium concentrations in surficial materials of the conterminous U.S. has been published by Shacklette et al.[218] and is reproduced as Figure 1 by kind permission of the authors and with acknowledgement to the U.S. Geological Survey. The relatively high contents in soils of the midwestern states is evident as are also the relatively low contents along the Pacific coast and in Florida. A similar map showing a generalized regional pattern of selenium concentrations in crops in the U.S. has also been produced by Kubota et al.[219] who concluded that in the Pacific Northwest, northeastern U.S., and southeastern seaboard states, there are extensive areas where crops are generally low in selenium, and selenium-responsive diseases in livestock are most likely to occur.

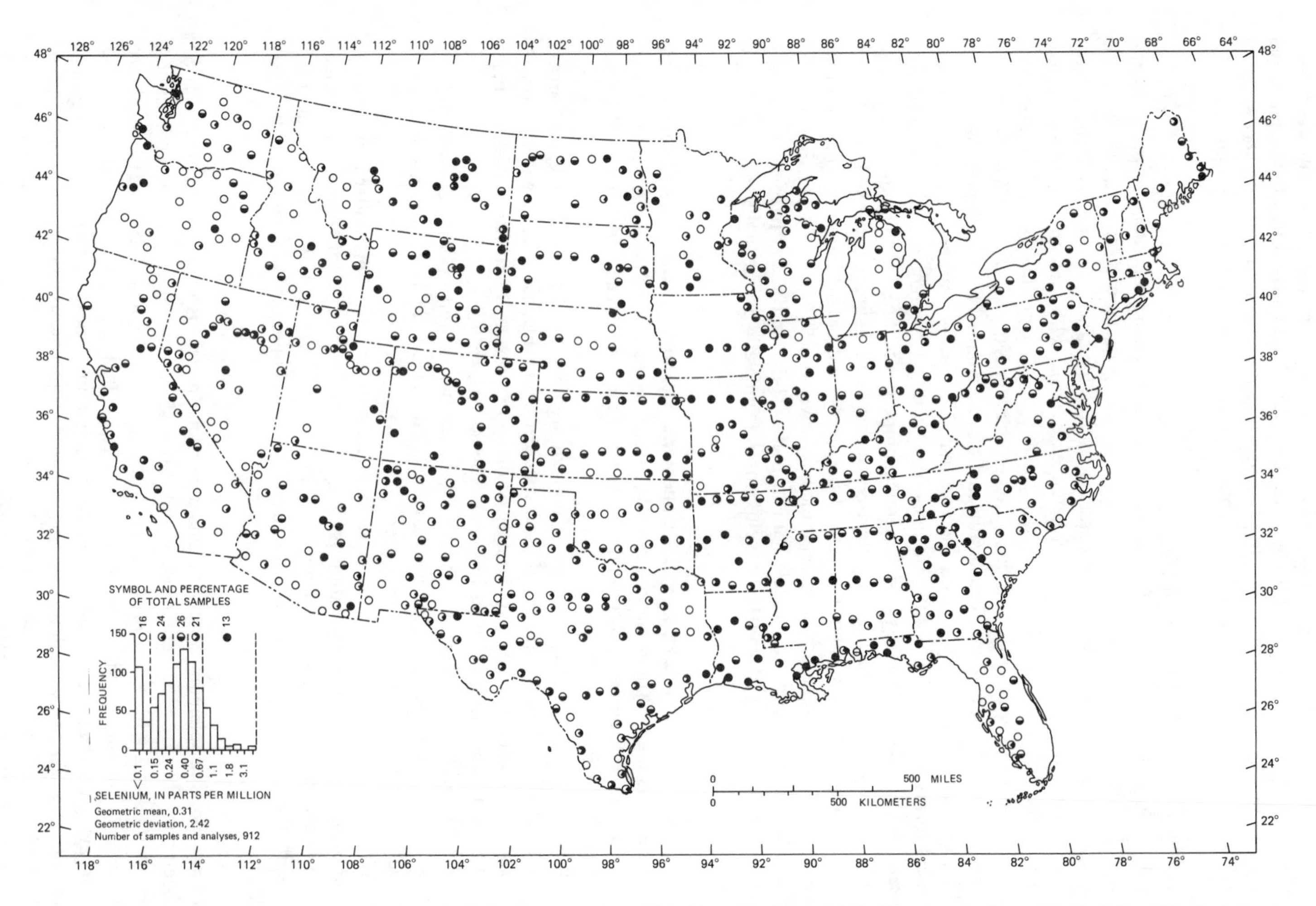

FIGURE 1. Selenium concentrations in surficial materials for the conterminous U.S. (From Shacklette, H. T., Boerngen, J. G., and Keith, J. R., *U.S. Geol. Surv. Circ.*, 692, 1974. With permission.)

Whereas earlier work on selenium in soils was concentrated largely on the problems of selenium toxicity affecting animals, more recent work has concentrated on problems of low selenium in soils. Soils from many parts of the world derived from granites or sandstones can be inherently low in selenium. When these soils become leached and acid, as is the case in temperate latitudes, plant-available selenium is further reduced and problems of selenium deficiency affecting humans and animals can arise as has been found in China, Finland, Scotland, and New Zealand. Attempts to increase the amounts of selenium in crops by the application of selenium-salts to soils present problems because of the dangers of exceeding toxicity thresholds and the fact that the applied selenium appears to be rapidly fixed in many soils. Although much progress has been made in the last 20 years in understanding the complex chemistry of selenium in soils, further study of the interrelated factors controlling the availability of selenium to plants is required. This may be facilitated by recent improvements in the sensitivity of analytical techniques for the accurate determination of the element at the very low concentrations involved in selenium deficiency studies.

REFERENCES

1. **Rosenfeld, I. and Beath, O. A.,** *Selenium, Geobotany, Biochemistry, Toxicity and Nutrition,* Academic Press, New York, 1964, chap. 2.
2. **Howard, J. H., III,** Geochemistry of selenium: formation of ferroselite and selenium behaviour in the vicinity of oxidizing sulfide and uranium deposits, *Geochim. Cosmochim. Acta,* 41, 1665, 1977.
3. **Lakin, H. W.,** Selenium in our environment, in *Trace Elements in the Environment,* Advances in Chemistry Ser., Kothny, E. L., Ed., Vol. 123, American Chemical Society, Washington, D.C., 1973, 96.
4. **Moxon, A. L.,** Natural occurrence of selenium, in *Proc. Symp. Selenium-Tellurium in the Environment,* Industrial Health Foundation, Pittsburgh, 1976, 1.
5. **Swaine, D. J.,** Trace elements in coal, in *Trace Substances in Environmental Health—11,* Hemphill, D. D., Ed., University of Missouri, Columbia, 1977, 107.
6. **Swaine, D. J.,** *The Trace Element Content of Soils,* Tech. Commun. No. 48, Commonwealth Bureau of Soil Science, Commonwealth Agricultural Bureau, Harpenden, England, 1955, 91.
7. **Aubert, H. and Pinta, M.,** *Trace Elements in Soils,* Elsevier, Amsterdam, 1977.
8. **Ure, A. M. and Berrow, M. L.,** The elemental constituents of soils, in *Environmental Chemistry,* Vol. 2, Bowen, H. J. M., Ed., The Royal Society of Chemistry, London, 1982, chap. 3.
9. **Johnson, C. M.,** Selenium in soils and plants: contrasts in conditions providing safe but adequate amounts of selenium in the food chain, in *Trace Elements in Soil-Plant-Animal Systems,* Nicholas, D. J. D. and Egan, A. R., Eds., Academic Press, New York, 1975, 165.
10. **Fleming, G. A.,** Essential micronutrients. II. Iodine and selenium, in *Applied Soil Trace Elements,* Davies, B. E., Ed., John Wiley & Sons, New York, 1980, chap. 6.
11. **Peterson, P. J., Benson, L. M., and Zieve, R.,** Metalloids, in *Effect of Heavy Metal Pollution on Plants,* Vol. 1, Lepp, N. W., Ed., Applied Science, London, 1981, chap. 8.
12. **Kabata-Pendias, A. and Pendias, H.,** *Trace Elements in Soils and Plants,* CRC Press, Boca Raton, FL, 1984, 185.
13. **Allaway, W. H.,** Selenium in Nutrition, *Report Subcommittee on Selenium, Committee on Animal Nutrition,* Agricultural Board, National Research Council, National Academy of Sciences, New York, 1971.
14. **Sharma, S. and Singh, R.,** Selenium in soil, plant and animal systems, *CRC Crit. Rev. Environ. Control,* 13, 23, 1983.
15. **Pelly, I. Z. and Lipschutz, M. E.,** Selenium, in *Handbook of Elemental Abundances in Meteorites,* Mason, B., Ed., Gordon and Breach, New York, 1971, 271.
16. **Kallemeyn, G. W. and Wasson, J. T.,** The compositional classification of chondrites. I. The carbonaceous chondrite groups, *Geochim. Cosmochim. Acta,* 45, 1217, 1981.
17. **Anders, E. and Ebihara, M.,** Solar system abundances of the elements, *Geochim. Cosmochim. Acta,* 46, 2363, 1982.
18. **Krähenbühl, U., Morgan, J. W., Ganapathy, R., and Anders, E.,** Abundance of 17 trace elements in carbonaceous chondrites, *Geochim. Cosmochim. Acta,* 37, 1353, 1973.
19. **Ebihara, M., Wolf, R., and Anders, E.,** Are C1 chondrites chemically fractionated? A trace element study, *Geochim. Cosmochim. Acta,* 46, 1849, 1982.

20. **Greenland, L.,** The abundances of selenium, tellurium, silver, palladium, cadmium, and zinc in chondritic meteorites, *Geochim. Cosmochim. Acta,* 31, 849, 1967.
21. **Laul, J. C., Keays, R. R., Ganapathy, R., Anders, E., and Morgan, J. W.,** Chemical fractionations in meteorites. V. Volatile and siderophile elements in achondrites and ocean ridge basalts, *Geochim. Cosmochim. Acta,* 36, 329, 1972.
22. **Biswas, S., Walsh, T., Bart, G., and Lipschutz, M. E.,** Thermal metamorphism of primitive meteorites. XI. The enstatite meteorites; origin and evolution of a parent body, *Geochim. Cosmochim. Acta,* 44, 2097, 1980.
23. **Morgan, J. W., Higuchi, H., Takahashi, H., and Hertogen, J.,** A ''chondritic'' eucrite parent body: inference from trace elements, *Geochim. Cosmochim. Acta,* 42, 27, 1978.
24. **Wolf, R. and Anders, E.,** Moon and Earth: compositional differences inferred from siderophiles, volatiles, and alkalis in basalts, *Geochim. Cosmochim. Acta,* 44, 2111, 1980.
25. **Dufresne, A.,** Selenium and tellurium in meteorites, *Geochim. Cosmochim. Acta,* 20, 141, 1960.
26. **Schindewolf, U.,** Selenium and tellurium content of stony meteorites by neutron activation, *Geochim. Cosmochim. Acta,* 19, 134, 1960.
27. **Hecht, F. and Kiesl, W.,** Kosmochemische meteoritenuntersuchungen mittels neutronenaktivierungsanalyse, *Chem. Erde,* 30, 145, 1971.
28. **Higuchi, H., Ganapathy, R., Morgan, J. W., and Anders, E.,** ''Mysterite'': a late condensate from the solar nebula, *Geochim. Cosmochim. Acta,* 41, 843, 1977.
29. **Davis, A. M., Grossman, L., and Ganapathy, R.,** Chemical characterization of a ''mysterite''-bearing clast from the Supuhee chondrite, *Geochim. Cosmochim. Acta,* 41, 853, 1977.
30. **Kiesl, W. and Herr, W.,** Die bestimmung der spurenelemente Ru, Pd, Re, Os, Ir, Au, As, Se und Sb in steinmeteoriten, *Chem. Erde,* 36, 324, 1977.
31. **Matza, S. D. and Lipschutz, M. E.,** Volatile/mobile trace elements in Karoonda (C4) chondrite, *Geochim. Cosmochim. Acta,* 41, 1398, 1977.
32. **Takahashi, H., Gros, J., Higuchi, H., Morgan, J. W., and Anders, E.,** Volatile elements in chondrites: metamorphism or nebular fractionation?, *Geochim. Cosmochim. Acta,* 42, 1859, 1978.
33. **Kallemeyn, G. W., Boynton, W. V., Willis, J., and Wasson, J. T.,** Formation of the Bencubbin polymict meteoritic breccia, *Geochim. Cosmochim. Acta,* 42, 507, 1978.
34. **Takahashi, H., Janssens, M.-J., Morgan, J. W., and Anders, E.,** Further studies of trace elements in C3 chondrites, *Geochim. Cosmochim. Acta,* 42, 97, 1978.
35. **Bart, G. and Lipschutz, M. E.,** On volatile element trends in gas-rich meteorites, *Geochim. Cosmochim. Acta,* 43, 1499, 1979.
36. **Curtis, D. B. and Schmitt, R. A.,** The petrogenesis of L-6 chondrites: insights from the chemistry of minerals, *Geochim. Cosmochim. Acta,* 43, 1091, 1979.
37. **Mittlefehldt, D. W.,** Petrographic and chemical characterization of igneous lithic clasts from mesosiderites and howardites and comparison with eucrites and diogenites, *Geochim. Cosmochim. Acta,* 43, 1917, 1979.
38. **Bart, G., Ikramuddin, M., and Lipschutz, M. E.,** Thermal metamorphism of primitive meteorites. IX. On the mechanism of trace element loss from Allende heated up to 1400°C, *Geochim. Cosmochim. Acta,* 44, 719, 1980.
39. **Kallemeyn, G. W. and Wasson, J. T.,** The compositional classification of chondrites. III. Ungrouped carbonaceous chondrites, *Geochim. Cosmochim. Acta,* 46, 2217, 1982.
40. **Walsh, T. M. and Lipschutz, M. E.,** Chemical studies of L chondrites. II. Shock-induced trace element mobilization, *Geochim. Cosmochim. Acta,* 46, 2491, 1982.
41. **Verkouteren, R. M. and Lipschutz, M. E.,** Cumberland Falls chondritic inclusions. II. Trace element contents of forsterite chondrites and meteorites of similar redox state, *Geochim. Cosmochim. Acta,* 47, 1625, 1983.
42. **Hertogen, J., Janssens, M.-J., Takahashi, H., Morgan, J. W., and Anders, E.,** Enstatite chondrites: trace element clues to their origin, *Geochim. Cosmochim. Acta,* 47, 2241, 1983.
43. **Borodin, L. S., Nazarenko, I. I., and Kislova, I. V.,** New data on selenium content in igneous rocks (in Russian), *Dokl. Akad. Nauk S.S.S.R.,* 206, 207, 1972.
44. **Fischer, R. and Leutwein, F.,** Selenium, in *Handbook of Geochemistry,* Vol. 2 (Part 3); Wedepohl, K. H., Ed., Springer-Verlag, Berlin, 1972, chap. 34.
45. **Lakin, H. W. and Davidson, D. F.,** The relation of the geochemistry of selenium to its occurrence in soils, in *Selenium in Biomedicine,* Muth, O. H., Oldfield, J. E., and Weswig, P. H., Eds., AVI Publishing, Westport, CT, 1967, 27.
46. **Srebrodol'skii, B. I. and Sidel'nikova, V. D.,** Selenium in native sulfur (in Russian), *Geokhimiya,* (8), 1013, 1969.
47. **Earley, J. W.,** Description and synthesis of the selenide minerals, *Am. Mineral.,* 35, 337, 1950.
48. **Velikii, A. S., Volgin, V. Yu., and Ivanov, V. S.,** Rare elements in Central Asian antimony-mercury deposits (in Russian), *Formy Nakhozdeniya Osob. Raspredel Redk. Elem. Nekot. Tipakh Gidroterm. Mestorozhd.,* 180, 1967.

49. **Sindeeva, N. D.,** *Mineralogy and Types of Deposits of Selenium and Tellurium* (Engl. Transl.), Ingerson, E., Ed., Interscience Publishers, New York, 1964.
50. **Finkel'shtein, Yu. V.,** Selenium content of the western part of the southern Tien-Shan mercury belt and find of tiemannite in Kyzylkum (in Russian), *Zap. Vses. Mineral. Ova.,* 100, 93, 1971.
51. **Ozerova, N. A., Balitskii, V. S., Komova, V. V., Laputina, I. P., Dobrovol'skaya, M. G., Vorob'ev, Yu. K., Pashkov, Yu, N., Vyal'sov, L. N., Sidel'nikova, V. D. et al.,** Selenium-containing mercury sulfides (according to experimental data and natural observations) (in Russian), in *Problemy Endogennogo Rudoobrazovaniya,* Pavlov, N. V., Ed., Nauka, Moscow, 1974, 150.
52. **Nazirova, R., Mansurov, M., and Stashkov, G. M.,** On the selenious tetradymite from Southern Yangikan (Karamazar) (in Russian), *Uzb. Geol. Zh.,* (2), 70, 1976.
53. **Balitskii, V. S., Ozerova, N. A., Tsepin, A. I., Komova, V. V., Sidel'nikova, V. D., Klientova, G. P., and Vyal'sov, L. N.,** Some features of the inclusion of selenium in antimonite in its recrystallization in selenium-containing hydrothermal solutions (in Russian), in *Novye Dannye Tipomorfizme Minereralov,* Genkin, A. D., Ed., Izdanija Nauka, Moscow, 1980, 89.
54. **Vershkovskaya, O. V., Krapiva, L. Ya., Kislova, I. V., and Shishkunov, M. G.,** Selenium in ore minerals of mercury deposits in the northern Caucasus (in Russian), *Geokhimiya,* (5), 737, 1979.
55. **Gorovoi, A. F. and Vershkovskaya, O. V.,** Selenium content of cinnabar in the Nikitovka ore field (USSR) (in Ukrainian), *Dopov. Akad. Nauk Ukr. RSR, Ser. B,* (9), 774, 1978.
56. **Zainullin, G. G.,** The geochemistry of selenium and tellurium in the copper-nickel ores of the Noril'sk region, *Geokhimiya,* (3), 231, 1960.
57. **Babcan, J., Bluml, A., Kurendova, J., and Tacl, A.,** Geochemistry of selenium, indium and cobalt in the deposit at Tisova u Kraslic, N.W. Bohemia, in *Geochemistry in Czechoslovakia,* Transl. 1st Conf. Geochemistry, Ostrava, September 20 to 24, 1965, Kuehnel, R., Kokta, J., Hak, J., Cadek, J., and Gubac, J., Eds., Vysoka Skola Banska v Ostrave, Ostrava, 1967, 371.
58. **Ambrus, J. W. and Soto, H. P.,** Geology of molybdenum mineralization at Chuquicamata, Chile (in Spanish), *Stud. Geol. Univ. Salamanca,* 8, 45, 1974.
59. **Nagy, B.,** Selenium content of hydrothermal sulfide minerals in Hungary (in Hungarian), *Magy. All. Foldt. Intez. Evi Jel.,* 39, 1974.
60. **Jonasson, I. R. and Sangster, D. F.,** Selenium in Sulfides From Some Canadian Base Metal Deposits, *Geol. Surv. Can. Pap.,* 75-1C, 231, 1975.
61. **Desborough, G. A., Pitman, J. K., and Huffman, C., Jr.,** Concentration and mineralogical residence of elements in rich oil shales of the Green River Formation, Piceance Creek basin, Colorado, and the Uinta Basin, Utah — a preliminary report, *Chem. Geol.,* 17, 13, 1976.
62. **Kashkai, M. A. and Magribi, A. A.,** Geochemistry of chalcopyrite and pyrite ores of the Kashkachai deposit (Dashkesan Region) (in Russian), *Dokl. Akad. Nauk Az. S.S.R.,* 32(5), 48, 1976.
63. **Blokhina, N. A. and Shcheblykina, M. D.,** Selenium and tellurium in sulfide ores of the Maikhura tin-tungsten deposit (central Tadzhikistan) (in Russian), *Dokl. Akad. Nauk Tadzh. S.S.R.,* 13(6), 38, 1970.
64. **Yamamoto, M., Ogushi, N., and Sakai, H.,** Distribution of sulfur isotopes, selenium, and cobalt in the Yanahara ore deposits, Okayama-Ken, Japan, *Geochem. J.,* 2, 137, 1969.
65. **Pokrovskaya, I. V.,** Selenium and tellurium in the Tishinsk and Ridder-Sokol'noe deposits (in Russian), *Izv. Akad. Nauk Kaz. S.S.R. Ser. Geol.,* 26(5), 53, 1970.
66. **Enikeeva, L. N.,** Distribution of rare and trace elements in minerals of lead-zinc deposits of the Kurusai-Turangly ore field (Western Karamazar) (in Russian), *Uzb. Geol. Zh.,* (6), 62, 1981.
67. **Litvinovich, A. N., Gavrilina, K. S., Kalashnikova, G. P., and Lukash, R. I.,** Forms of occurrence of impurities in sulfide minerals (in Russian), *Tr. Inst. Geol. Nauk Akad. Nauk Kaz. S.S.R.,* 33, 122, 1973.
68. **Martirosyan, R. A. and Babaeva, Z. E.,** Selenium and tellurium distribution in deposits of the complex ore formation of the Lesser Caucasus (in Russian), *Geokhimiya,* (3), 418, 1974.
69. **Badalov, S. T., Belopol'skaya, T. L., Prikhid'ko, P. L., and Turesebekov, A.,** Geochemistry of selenium in sulphate-sulfide mineral parageneses (in Russian), *Geokhimiya,* (8), 1007, 1969.
70. **Aksenov, B. S., Inin, V. D., Litvinovich, A. N., Slyusarev, A. P., and Kosyak, E. A.,** Mineral forms of selenium in ores from the Orlovskii pyritic-complex mineral deposit in the Altai (in Russian), *Izv. Akad. Nauk Kaz. S.S.R. Ser. Geol.,* 25, 42, 1968.
71. **Taylor, G. F. and Appleyard, E. C.,** Weathering of the zinc-lead lode, Dugald River, Northwest Queensland. I. The gossan profile, *J. Geochem. Explor.,* 18, 87, 1983.
72. **Koljonen, T.,** Luikonlahti: selenium in the vicinity of copper ore, *J. Geochem. Explor.,* 5(3, Conceptual Models in Exploration Geochemistry: Norden 1975), 263, 415, 1976.
73. **Bezsmertnaya, M. S.,** Rare elements in gold-ore deposits (in Russian), *Geokhim. Mineral. Genet. Tipy Mestorozhd. Redk. Elem.,* 3, 510, 1966.
74. **Goldschmidt, V. M. and Hefter, O.,** Zur Geochemie des Selens. I, *Nachr. Ges. Wiss. Goettingen Math. Physik. Kl. Fachgruppe 1,* 245, 1933.
75. **Goldschmidt, V. M. and Strock, L. W.,** Zur Geochemie des Selens. II, *Nachr. Ges. Wiss. Goettingen, Math. Physik, Kl. Fachgruppe 1,* 123, 1935.

76. **Fleischer, M.,** Minor elements in some sulfide minerals, *Econ. Geol.,* 50th Anniv. vol. (Part II), 970, 1955.
77. **Kazachenko, Yu. A. and Nikonova, L. I.,** Rhenium and selenium in molybdenites of eastern Siberian deposits (in Russian), *Nauchn. Tr. Irkutsk. Gos. Nauchno Issled. Inst. Redk. Tsvetn. Met.,* (14), 79, 1966.
78. **Miralimova, N. M.,** Concentration of thallium, selenium and rare earth elements in the Mesozoic-Cenozoic sedimentary rocks of the Zirabulak-Ziaetdin Mountains (in Russian), *Uzb. Geol. Zh.,* 11, 28, 1967.
79. **Khetagurov, G. V.,** Distribution of selenium and tellurium in complex ores of the Sadon area (in Russian), *Vopr. Soversh. Gorn. Proiz.,* 25, 1968.
80. **Tokarev, V. A. and Yashchenko, A. V.,** Keiv selenium-bearing pyrrhotite (Tyapysh-Manyuk deposit) (in Russian), *Mater. Mineral. Kol'sk. Poluostrova,* 6, 67, 1968.
81. **Babaeva, Z. E. and Efendiev, G. Kh.,** Selenium and tellurium in galenas of polymetallic deposits in the Lesser Caucasus (Azerbaidzhan SSR) (in Russian), in *Mater. Konf. Molodykh Uch. Inst. Neorg. Fiz. Khim. Akad. Nauk Azerb. S.S.R.,* Orudzheva, I. M., Ed., Akademia Nauk Azerbaidzhanskoi S.S.R., Baku, 1969, 241.
82. **Nechelyustov, N. V., Zlenko, B. F., Gubanov, A. M., Kogan, R. I., Razina, I. S., Volkov, B. I., and Pikkat-Ordynskaya, A. P.,** Methods of studying accessory elements (selenium, tellurium, bismuth) in chalcopyrite-molybdenite ores of skarn deposits (in Russian), in *Geokhimiya i Geologiya Nekotorykh Rudnykh Mestorozhdenii,* Velikii, A. S., Ed., Nauka, Moscow, 1970, 154.
83. **Balabonin, N. L. and Astaf'eva, V. V.,** Selenium and tellurium in pyrrhotites from metamorphic rocks of the Allarechensk region (Kola Peninsula) (in Russian), *Vopr. Geol. Metallog. Kol'sk. Poluostrova,* 5, (Part 2), 242, 1974.
84. **Gusev, A. I.,** Selenium and tellurium in pyrites of the copper and lead-zinc deposits of the northern Caucasus (in Russian), *Geokhimiya,* (4), 602, 1978.
85. **Ganzhenko, G. D.,** Selenium and tellurium in the supergene zone of the Kosmurun chalco pyrite deposit in Central Kazakhstan (in Russian), *Geokhimiya,* (8), 1212, 1979.
86. **Babaeva, Z. E.,** Some statistical parameters of selenium and tellurium distribution in complex ores of the Lesser Caucasus (Azerbaidzhan), (in Russian) in *Voprosy Geokhimii Khimi Redkii Elemikh,* Zul'fugarov, Z. G. et al., Eds., Izd. Elm.: Baku, 1979, 118.
87. **Coleman, R. G. and Delevaux, M.,** Occurrence of selenium in sulphides from some sedimentary rocks of the western United States, *Econ. Geol.,* 52, 499, 1957.
88. **Edwards, A. B. and Carlos, G. C.,** The selenium contents of some Australian sulphide deposits, *Aust. Inst. Min. Metal. Proc.,* 172, 31, 1954.
89. **Kronberg, B. I., Fyfe, W. S., McKinnon, B. J., Couston, J. F., Stilianidi-Filho, B., and Nash, R. A.,** Model for bauxite formation: Paragominas (Brazil), *Chem. Geol.,* 35, 311, 1982.
90. **Tsuge, T. and Terada, S.,** The selenium contents of pyrites and soils in Japan (in Japanese/English), *J. Agr. Chem. Soc. Jpn.,* 23, 421, 1950.
91. **Hamada, S., Sato, R., and Shirai, T.,** Distribution of some minor elements in sulfide ores from the Hitachi mine, Ibaraki prefecture, Japan, *Bull. Chem. Soc. Jpn.,* 41, 850, 1968.
92. **Chen, D., Sun, S., and Li, Y.,** Main characteristics of the cinnabar in Tongren-Wanshan region (China) (in Chinese), *Yankuang Ceshi,* 1, 36, 1982.
93. **Wells, N.,** Selenium content of some minerals and fertilisers, *N. Z. J. Sci.,* 9, 409, 1966.
94. **Babcan, J. and Ilavsky, J.,** Geochemistry of selenium from the stratiform pyrite deposit of ferrous sulfide-copper ores in Smolnik (in German), *Sb. Geol. Ved Zapadné Karpaty,* (6), 85, 1966.
95. **Babcan, J. and Ilavsky, J.,** Selenium in the pyrite-chalcopyrite deposit of sedimentary-exhalatative origin near Smolnik, western Carpathians, in *Geochemistry in Czechoslovakia* , Transl. 1st Conf. Geochemistry, Ostrava, September 20 to 24, 1965, Kuehnel, R., Kokta, J., Hak, J., Cadek, J., and Gubac, J., Eds., Vysoka Skola Banska v Ostrave, Ostrava, 1967, 425.
96. **Ivanov, M.,** Metasomatic pyrite and polymetallic mineralization of the Muran Plateau (in Slovak), *Geol. Pr. (Bratislava) Zpr.,* 43, 5, 1967.
97. **Trdlicka, Z.,** Chemistry of pyrargyrite and miargyrite from Czechoslovakia (in German), *Sb. Nar. Muz. Praze Rada B,* 25, 157, 1969.
98. **Kvacek, M.,** Distribution of selenium in sulfides and arsenides from some occurrences in Moravia and Silesia (in Czech), *Cas. Morav. Mus. Vedy Prir.,* 59, 9, 1974.
99. **Kvacek, M. and Trdlicka, Z.,** Die rhenium — und selengehalte in einigen molydäniten der Bohmischen masse (in German), *Acta Univ. Carol. Geol.,* (2), 105, 1970.
100. **Kvacek, M. and Pfeiferova, A.,** Distribution of selenium in sulfides of the Moravian part of the Carpathian system and of the underlying Bohemian massif (in Czech), *Cas. Morav. Mus. Vedy Prir.,* 60, 57, 1975.
101. **Goldschmidt, V. M.,** *Geochemistry,* Muir, A., Ed., Clarendon Press, Oxford, 1954.
102. **Turekian, K. K. and Wedepohl, K. H.,** Distribution of the elements in some major units of the earth's crust, *Geol. Soc. Am. Bull.,* 72, 175, 1961.
103. **Vinogradov, A. P.,** *The Geochemistry of Rare and Dispersed Chemical Elements in Soils,* 2nd ed., (transl. from Russian), Consultants Bureau, New York, 1959.

104. **Taylor, S. R.,** Abundance of chemical elements in the continental crust: a new table, *Geochim. Cosmichim. Acta,* 28, 1273, 1964.
105. **Moxon, A. L., Olson, O. E., and Searight, W. V.,** Selenium in rocks, soils and plants, *S. D. Agric. Exp. Stn. Tech. Bull.,* 2, 1, 1939.
106. **Mroz, E. J. and Zoller, W. H.,** Composition of atmospheric particulate matter from the eruption of Heimaey, Iceland, *Science,* 190, 461, 1975.
107. **Wells, N.,** Selenium content of soil-forming rocks, *N. Z. J. Geol. Geophys.,* 10, 198, 1967.
108. **Koljonen, T.,** Selenium in certain igneous rocks, *Bull. Geol. Soc. Finl.,* 45, (2), 9, 1973.
109. **Koljonen, T.,** Selenium in certain sedimentary rocks, *Bull. Geol. Soc. Finl.,* 45, (2), 119, 1973.
110. **Koljonen, T.,** Selenium in certain metamorphic rocks, *Bull. Geol. Soc. Finl.,* 45, (2), 107, 1973.
111. **Vorob'ev, V. P., Grushevoi, G. V., and Onoshko, I. S.,** Paragenesis of uranium and selenium in multicolored Mesozoic-Cenozoic deposits (in Russian), *Tr. Vses. Nauch Issled. Geol. Inst.,* 142, 114, 1968.
112. **Savel'ev, V. F.,** Selenium in Uzbekistan metamorphic rocks (in Russian), *Nauch. Tr. Tashkent. Gos. Univ.,* (372), 35, 1970.
113. **Boiko, T. F., Nazarenko, I. I., and Kislova, I. V.,** Selenium distribution in weathering profiles of igneous rocks (in Russian), *Geokhimiya,* (3), 399, 1981.
114. **Pillay, K. K. S., Thomas, C. C., Jr., and Kaminski, J. W.,** Neutron activation analysis of the selenium content of fossil fuels, *Nucl. Appl. Technol.,* 7, 478, 1969.
115. **Savel'ev, V. F. and Timofeev, N. I.,** Selenium in coal fields (in Russian), *Sb. Nauch. Tr. Tashk. Un-t,* 530, 101, 1977.
116. **Wolf, R., Woodrow, A. B., and Grieve, R. A. F.,** Meteoric material at four Canadian impact craters, *Geochim. Cosmochim. Acta,* 44, 1015, 1980.
117. **Palme, H., Grieve, R. A. F., and Wolf, R.,** Identification of the projectile at the Brent Crater, Ontario, and further considerations of projectile type at terrestrial craters, *Geochim. Cosmochim. Acta,* 45, 2417, 1981.
118. **Reimold, W. U.,** The Lappäjaervi meteorite crater, Finland: petrography, rubidium-strontium, major and trace element geochemistry of the impact melt and basement rocks, *Geochim. Cosmochim. Acta,* 46, 1203, 1982.
119. **Newman, W. L.,** Distribution of elements in sedimentary rocks of the Colorado Plateau — a preliminary report, *U.S. Geol. Surv. Bull.,* 1107-F, 337, 1962.
120. **Koval'skii, V. V. and Ermakov, V. V.,** Tuva biogeochemical province enriched in selenium, *Geokhimiya,* (1), 86, 1967.
121. **Savel'ev, V. F.,** Selenium content in Upper Cretaceous sedimentary rocks, solonchaks, and lakes in an area of central Asia (in Russian), *Uzb. Geol. Zh.,* 14, 79, 1970.
122. **Bilenskii, M. A.,** Selenium in Paleozoic rocks and waters in southwestern Uzbekistan (in Russian), *Zap. Uzb. Otd. Vses. Mineral. Ova.,* 24, 185, 1971.
123. **Brunfelt, A. O. and Steinnes, E.,** Determination of selenium in standard rocks by neutron activation analysis, *Geochim. Cosmochim. Acta,* 31, 283, 1967.
124. **Sobolev, S. F., Lapin, A. V., Nazarenko, I. I., Kislova, I. V., and Ilupin, I. P.,** Selenium in kimberlites and abyssal peridotite xenoliths of Yakutia (in Russian), *Geokhimiya,* (1), 67, 1975.
125. **Hertogen, J., Janssens, M.-J., and Palme, H.,** Trace elements in ocean ridge basalt glasses: implications for fractionations during mantle evolution and petrogenesis, *Geochim. Cosmochim. Acta,* 44, 2125, 1980.
126. **Williams, K. T. and Byers, H. G.,** Occurrence of selenium in pyrites, *Ind. Eng. Chem.,* 6, 296, 1934.
127. **Minami, E.,** Selenium content of European and Japanese clay shales, *Nachr. Ges. Wiss. Gottingen Math. Physik. Kl. Fachgruppe* 1, 143, 1935.
128. **Davidson, D. F. and Lakin, H. W.,** Metal content of some black shales of the western United States, *U.S. Geol. Surv. Prof. Pap.,* 424C, 329, 1961.
129. **Lakin, H. W.,** Vertical and lateral distribution of selenium in sedimentary rocks of western United States, in Selenium in Agriculture, USDA Agriculture Handbook 200, Anderson, M. S., Lakin, H. W., Beeson, K. C., Smith, F. F., and Thacker, E., Eds., U.S. Department of Agriculture, Washington, D.C., 1961, 12.
130. **Tourtelot, H. A.,** Preliminary investigation of the geologic setting and chemical composition of the Pierre Shale, Great Plains Region, *U.S. Geol. Surv. Prof. Pap.,* 390, 1962.
131. **McCray, C. W. R. and Hurwood, I. S.,** Selenosis in North-western Queensland associated with a marine Cretaceous formation, *Queensl. J. Agric. Sci.,* 20, 475, 1963.
132. **Kanasiewicz, J.,** Geochemical profile of uranium, selenium, and rhenium in the Zechstein of Leszczyna trough (in Polish), *Kwart. Geol.,* 10, 309, 1966.
133. **Kanasiewicz, J.,** Occurrence of rare elements in the copper-bearing marl series of the lower Zechstein in the Grodziec syncline (in Polish), *Kwart. Geol.,* 11, 113, 1967.

134. **Sidel'nikova, V. D. and Shvei, I. V.,** Selenium in Paleozoic formations of Central Asia (in Russian), in *Ocherki po Geologii i Geokhimii Rudnykh Mestorozhdenii,* Vol'fson, F. I., Ed., Nauka, Moscow, 1970, 307.
135. **Ermakov, V. V.,** Subregions and biogeochemical provinces of the USSR with different selenium content (in Russian), *Tr. Biogeokhim. Lab. Akad. Nauk S.S.S.R.,* 15, 54, 1978.
136. **Ebens, R. J. and Connor, J. J.,** Geochemistry of loess and carbonate residuum, *U.S. Geol. Surv. Prof. Pap.,* 954-G, 1980.
137. **Bobrinski, V. M. and Ufnarovskaya, G. P.,** Selenium in sedimentary rocks of central Moldavia, (in Russian), *Izv. Akad. Nauk Mold. S.S.R. Ser. Fiz. Tekh. Mat. Nauk,* (1), 74, 1983.
138. **Mendes, R. V. and Van Trump, G., Jr.,** Miscellaneous geochemical data on late Cretaceous and early Tertiary fine-grained sedimentary rocks from the Western Coal Regions, *U.S. Geol. Surv. Open File Report,* 75-436, 82, 1975.
139. **Bentor, Y. K., Kastner, M., Perlman, I., and Yellin, Y.,** Combustion metamorphism of bituminous sediments and the formation of melts of granitic and sedimentary composition, *Geochim. Cosmochim. Acta,* 45, 2229, 1981.
140. **Christensen, S. M.,** Geochemical parameters of base-metal sulfide-bearing volcano-sedimentary environments, Western Australia, *Chem. Geol.,* 31, 285, 1981.
141. **Yilmaz, H.,** Genesis of uranium deposits in Neogene sedimentary rocks overlying the Menderes metamorphic massif, Turkey, *Chem. Geol.,* 31, 185, 1981.
142. **Ebens, R. J. and Shacklette, H. T.,** Geochemistry of some rocks, mine spoils, stream sediments, soils, plants, and waters in the Western Energy Region of the conterminous United States, *U.S. Geol. Surv. Prof. Pap.,* 1237, 1982.
143. **Savel'ev, V. F.,** Selenium containing coalified plant remnants in the Upper Cretaceous sedimentary rocks of a region of Central Asia (in Russian), *Zap. Uzb. Otd. Vses. Mineral. Ova. Akad. Nauk. Uz. S.S.R.,* 16, 35, 1964.
144. **Zul'fugarly, D. I. and Mekhtieva, R. G.,** Distribution of selenium in sedimentary-volcanic rocks, in *Tezisy Respublikanskoi Nauchnoi Konf. Molodykh Uchenykh-Khimikov Azerbaidzhana, 3.* Akademiya Nauk Azerbaidzhanskoi S.S.R., Instituta Neftekhimii Protsessov, Baku, U.S.S.R., 1974.
145. **Moxon, A. L., Olson, O. E., Searight, W. V., and Sandals, K. M.,** The stratigraphic distribution of selenium in the Cretaceous formations of South Dakota and the selenium content of some associated vegetation, *Am. J. Bot.,* 25, 794, 1938.
146. **Beath, O. A., Gilbert, C. S., and Eppson, H. F.,** The use of indicator plants in locating seleniferous areas in western United States. II. Correlation studies by states, *Am. J. Bot.,* 26, 296, 1939.
147. **Williams, K. T., Lakin, H. W., and Byers, H. G.,** Selenium occurrence in certain soils in the United States with a discussion of related topics: fourth report, *U.S. Dep. Agric. Tech. Bull.,* 702, 1, 1940.
148. **Ruch, R. R., Gluskoter, H. J., and Shrimp, N. F.,** Occurrence and distribution of potentially volatile trace elements in coal: an interim report, *Ill. State Geol. Surv. Environ. Geol. Notes,* 61, 1973.
149. **Chadwick, R. A., Rice, R. C., Bennett, C. M., and Woodriff, R. A.,** Sulfur and trace elements in the Rosebud and McKay coal seams, Colstrip field, Montana, in *Energy Resources of Montana,* Doroshenko, J., Miller, W. R., Thompson, E. E., Jr., and Rawlins, J. H., Eds., Montana Geological Society, Billings, MT, 1975, 167.
150. **Shah, K. R., Filby, R. H., and Haller, W. A.,** Determination of trace elements in petroleum by neutron activation analysis, *J. Radioanal. Chem.,* 6, 413, 1970.
151. **Mast, R. F. and Ruch, R. R.,** Survey of Illinois crude oils for trace concentrations of mercury and selenium, *Ill. State Geol. Surv. Environ. Geol. Notes,* 65, 1973.
152. **Sokolova, E. G. and Pilipchuk, M. F.,** Selenium in recent deposits in the Black Sea (in Russian), *Dokl. Akad. Nauk S.S.S.R.,* 193, 692, 1970.
153. **Koljonen, T.,** Selenium in certain Finnish sediments, *Bull. Geol. Soc. Finl.,* 46, 15, 1974.
154. **Loring, D. H.,** Geochemical factors controlling the accumulation and dispersal of heavy metals in the Bay of Fundy (Canada) sediments, *Can. J. Earth Sci.,* 19, 930, 1982.
155. **Sokolova, E. G. and Pilipchuk, M. F.,** Geochemistry of selenium in deposits in the northwestern part of the Pacific Ocean (in Russian), *Geokhimiya,* (10), 1537, 1973.
156. **Volkov, I. I., Rozanov, A. G., and Zhabina, N. N.,** Distribution of chemical elements in bottom sediments. Elements of the organic material group. Sulfur and selenium (in Russian), *Tr. Geol. Inst. Akad. Nauk S.S.S.R.,* 334, 127 and 231, 1979.
157. **Pilipchuk, M. F. and Sokolova, E. G.,** Selenium in Upper Quaternary formations of the Tyrrhenian Sea (in Russian), *Geokhimiya,* (9), 1374, 1979.
158. **Kiely, P. V. and Fleming, G. A.,** Geochemical survey of Ireland: Meath-Dublin area, *Proc. R. Ir. Acad.,* 68B, 1, 1969.

159. **Maassen, L. W., Sandoval, W. F., Muller, M., Gallimore, D. L., Martell, C. J., Hensley, W. K., and Thomas, G. J.,** Detailed Uranium Hydrogeochemical and Stream Sediment Reconnaissance Data Release for the Eastern Portion of the Montrose NTMS Quadrangle, Colorado, Including Concentrations of Forty-Five Additional Elements, Rep. GJBX-105-81, Los Alamos Scientific Laboratory, Los Alamos, NM, 1981.
160. **Allaway, W. H.,** Control of the environmental levels of selenium, in *Trace Substances in Environmental Health — 2,* Hemphill, D. D., Ed., University of Missouri, Columbia, 1969, 181.
161. **Watkinson, J. H.,** Soil selenium and animal health, in Transactions of Joint Meet. of Commissions IV and V Int. Soc. of Soil Science, Massey University College of Manawatu, Palmerston North, New Zealand, 1962, 149.
162. **Wells, N.,** Selenium in horizons of soil profiles, *N.Z. J. Sci.,* 10, 142, 1967.
163. **Bisbjerg, B.,** *Studies on Selenium in Plants and Soils,* Riso Rep. No. 200, Danish Atomic Energy Commission, Research Establishment Riso, Roskilde, 1972.
164. **Connor, J. J. and Shacklette, H. T.,** Background geochemistry of some rocks, soils, plants, and vegetables in the conterminous United States, *U.S. Geol. Surv. Prof. Pap.,* 574-F, 1975.
165. **Thornton, I., Kinniburgh, D. G., Pullen, G., and Smith, C. A.,** Geochemical aspects of selenium in British soils and implications to animal health, in *Trace Substances in Environmental Health — 17,* Hemphill, D. D., Ed., University of Missouri, Columbia, 1983, 391.
166. **Singh, M. and Kumar, P.,** Selenium distribution in soils of bio-climatic zones of Haryana, *J. Indian Soc. Soil Sci.,* 24, 62, 1976.
167. **Fleming, G. A.,** Selenium in Irish soils and plants, *Soil Sci.,* 94, 28, 1962.
168. **Davidson, D. F. and Powers, H. A.,** Selenium content of some volcanic rocks from the western United States and Hawaiian Islands, *U.S. Geol. Surv. Bull.,* 1084-C, 69, 1959.
169. **Vorob'jev, V. P.,** Two modes of origin of sedimentary selenium concentrations, *Dokl. Akad. Nauk S.S.S.R. Earth Sci. Sect.,* 186, 218, 1969.
170. **Läg, J. and Steinnes, E.,** Regional distribution of selenium and arsenic in humus layers of Norwegian forest soils, *Geoderma,* 20, 3, 1978.
171. **Lantzy, R. J. and Mackenzie, F. T.,** Atmospheric trace metals: global cycles and assessment of man's impact, *Geochim. Cosmochim. Acta,* 43, 511, 1979.
172. **Van Grieken, R. E., Johansson, T. B., and Winchester, J. W.,** Trace metal fractionation effects between sea water and aerosols from bubble bursting, *J. Rech. Atmos.,* 8, 611, 1974.
173. **Lewis, B. G., Johnston, C. M. and Delwiche, C. C.,** Release of volatile selenium compounds by plants. Collection procedures and preliminary observations, *J. Agric. Food Chem.,* 14, 638, 1966.
174. **Ridley, W. P., Dizikes, L. J. and Wood, J. M.,** Biomethylation of toxic elements in the environment, *Science,* 197, 329, 1977.
175. **Zieve, R. and Peterson, P. J.,** Factors influencing the volatilization of selenium from soil, *Sci. Total Environ.,* 19, 277, 1981.
176. **Swaine, D. J.,** *The Trace Element Content of Fertilizers,* Tech. Commun. No. 52, Commonwealth Bureau of Soils, Commonwealth Agricultural Bureaux, Farnham Royal, England, 1962, 215.
177. **Rader, L. F., Jr. and Hill, W. L.,** Occurrence of selenium in natural phosphates, superphosphates and phosphoric acid, *J. Agric. Res.,* 51, 1071, 1935.
178. **Gissel-Nielsen, G.,** Selenium content of some fertilizers and their influence on uptake of selenium in plants, *J. Agric. Food Chem.,* 19, 564, 1971.
179. **Berrow, M. L. and Burridge, J. C.,** Trace elements levels in soils: effects of sewage sludge, in Inorganic Pollution and Agriculture, MAFF Reference Book No. 326, Her Majesty's Stationary Office, London, 1980, 159.
180. **Gutenmann, W. H., Bache, C. A., Youngs, W. D., and Lisk, D. J.,** Selenium in fly ash, *Science,* 191, 966, 1976.
181. **Stoewsand, G. S., Guttenmann, W. H., and Lisk, D. J.,** Wheat grown on fly ash: high selenium uptake and response when fed to Japanese quail, *J. Agric. Food Chem.,* 26, 757, 1978.
182. **Byers, H. G., Miller, J. T., Williams, K. T., and Lakin, H. W.,** Selenium occurrence in certain soils in the United States, with a discussion of related topics. Third report, *U.S. Dep. Agric. Tech. Bull.,* 601, 1938.
183. **Byers, H. G.,** Selenium occurrence in certain soils in the United States, with a discussion of related topics, *U.S. Dep. Agric. Tech. Bull.,* 482, 1935.
184. **Olson, O. E.,** Soil, plant, animal cycling of excessive levels of selenium, in *Selenium in Biomedicine,* Muth, O. H., Oldfield, J. E., and Weswig, P. H., Eds., AVI Publishing, Westport, CT, 1967, chap. 18.
185. **Allaway, W. H., Cary, E. E., and Ehlig, C. F.,** The cycling of low levels of selenium in soils, plants and animals, in *Selenium in Biomedicine,* Muth, O. H., Oldfield, J. E., and Weswig, P. H., Eds., AVI Publishing, Westport, CT, 1967, chap. 17.
186. **Hamdy, A. A. and Gissel-Nielsen, G.,** Fractionation of soil selenium, *Z. Pflanzenernaehr. Bodenkd.,* 139, 697, 1976.

187. **Oldfield, J. E.,** Selenium deficiency in soils and its effect on animal health, in *Geochemical Environment in Relation to Health and Disease,* Cannon, H. L. and Hopps, H. C., Eds., Geological Society of America Spec. Paper 140, Geological Society of America, Boulder, CO, 1972, 57.
188. **Geering, H. R., Cary, E. E., Jones, L. H. P., and Allaway, W. H.,** Solubility and redox criteria for the possible forms of selenium in soils, *Proc. Soil Sci. Soc. Am.,* 32, 35, 1968.
189. **Trelease, S. F. and Beath, O. A.,** *Selenium: its Geological Occurrence and its Biological Effects in Relation to Botany, Chemistry, Agriculture, Nutrition and Medicine,* Trelease, S. F. and Beath, D. A. (Publishers), New York, 1949.
190. **Nye, S. M. and Peterson, P. J.,** The content and distribution of selenium in soils and plants from seleniferous areas in Eire and England, in *Trace Substances in Environmental Health — 9,* Hemphill, D. D., Ed., University of Missouri, Columbia, 1975, 113.
191. **Fleming, G. A. and Walsh, T.,** Selenium occurrence in certain Irish soils and its toxic effects on animals, *Proc. R. Ir. Acad.,* 58B, 151, 1957.
192. **Van Der Elst, F. H. and Tetley, R.,** Selenium studies on peat. I. Selenium uptake of pasture after incorporation of sodium selenite with peat soil, *N. Z. J. Agric. Res.,* 13, 945, 1970.
193. **Lévesque, M.,** Selenium distribution in Canadian soil profiles, *Can. J. Soil Sci.,* 54, 63, 1974.
194. **Yläranta, T.,** Selenium in Finnish agricultural soils, *Ann. Agric. Fenn.,* 22, 122, 1983.
195. **Hamdy, A. A. and Gissel-Nielsen, G.,** Volatilization of selenium from soils, *Z. Pflanzenernaehr. Bodenkd.,* 139, 671, 1976.
196. **Gissel-Nielsen, G. and Hamdy, A. A.,** Leaching of added selenium in soils low in native selenium, *Z. Pflanzenernaehr. Bodenkd.,* 140, 193, 1977.
197. **Koval'skii, V. V., Ermakov, V. V., and Letunova, S. V.,** Geochemical ecology of micro-organisms under conditions of different selenium content in soils, *Mikrobiologiya,* 37, 122, 1968.
198. **Allen, R. O. and Steinnes, E.,** Concentrations of some potentially toxic metals and other trace elements in wild mushrooms from Norway, *Chemosphere,* 7, 371, 1978.
199. **Koljonen, T.,** The behaviour of selenium in Finnish soils, *Ann. Agric. Fenn.,* 14, 240, 1975.
200. **McKeague, J. A., Desjardins, J. G., and Wolynetz, M. S.,** Minor Elements in Canadian Soils, Agriculture Canada, Ottawa, 1979.
201. **Lévesque, M.,** Relationship of sulfur and selenium in some Canadian soil profiles, *Can. J. Soil Sci.,* 54, 333, 1974.
202. **Frank, R., Stonefield, K. I., and Suda, P.,** Metals in agricultural soils of Ontario II, *Can. J. Soil Sci.,* 59, 99, 1979.
203. **Sippola, J.,** Selenium content of soils and Timothy (*Phleum pratense* L.) in Finland, *Ann. Agric. Fenn.,* 18, 182, 1979.
204. **Gissel-Nielsen, G.,** Influence of pH and texture of the soil on plant uptake of added selenium, *J. Agric. Food Chem.,* 19, 1165, 1971.
205. **Williams, C. and Thornton, I.,** The use of soil extractants to estimate plant-available molybdenum and selenium in potentially toxic soils, *Plant Soil,* 39, 149, 1973.
206. **Ravikovitch, S. and Margolin, M.,** Selenium in soils and plants, *Ktavim (Engl. Ed.),* 7, 41, 1957.
207. **Stoyanov, D. V. and Stefanova, V.,** Water-soluble selenium in some Bulgarian soils (in Bulgarian), *Pochvozn. Agrokhim.,* 4(2), 3, 1969.
208. **Lévesque, M.,** Some aspects of selenium relationships in Eastern Canadian soils and plants, *Can. J. Soil Sci.,* 54, 205, 1974.
209. **Hamdy, A. A. and Gissel-Nielsen, G.,** *Relationships Between Soil Factors and Selenium Content of Danish Soils and Plants,* Riso Rep. No. 349, Danish Atomic Energy Commission, Research Establishment Riso, Roskilde, 1976.
210. **Misra, S. G. and Tripathi, N.,** Note on selenium status of surface soils, *Indian J. Agric. Sci.,* 42, 182, 1972.
211. **Asakawa, Y., Kushizaki, M., and Ishizuka, J.,** Soil-plant relation of selenium in grassland. II. Selenium content of soils in Japan (in Japanese), *Nippon Dojo Hiryogaku Zasshi,* 48, 293, 1977.
212. **Cary, E. E., Wieczorek, G. A., and Allaway, W. H.,** Reactions of selenite-selenium added to soils that produce low-selenium forages, *Proc. Soil Sci. Soc. Am.,* 31, 21, 1967.
213. **Soltanpour, P. N. and Workman, S. M.,** Use of NH_4HCO_3-DTPA soil test to assess availability and toxicity of selenium to alfalfa plants, *Commun. Soil Sci. Plant Anal.,* 11, 1147, 1980.
214. **Lindberg, P. and Bingefors, S.,** Selenium levels of forages and soils in different regions of Sweden, *Acta Agric. Scand.,* 20, 133, 1970.
215. **Robberecht, H., Vanden Berghe, D., Deelstra, H., and Van Grieken, R.,** Selenium in Belgian soils and its uptake by ryegrass, *Sci. Total Environ.,* 25, 61, 1982.
216. **Cary, E. E. and Allaway, W. H.,** The stability of different forms of selenium applied to low-selenium soils, *Proc. Soil Sci. Soc. Am.,* 33, 571, 1969.
217. **Van Dorst, S. H. and Peterson, P. J.,** Selenium speciation in the soil solution and its relevance to plant uptake, *J. Sci. Food Agric.,* 35, 601, 1984.

218. **Shacklette, H. T., Boerngen, J. G., and Keith, J. R.,** Selenium, fluorine and arsenic in surficial materials of the conterminous United States, *U.S. Geol. Surv. Circ.*, 692, 1974.
219. **Kubota, J., Allaway, W. H., Carter, D. L., Cary, E. E., and Lazar, V. A.,** Selenium in crops in the United States in relation to selenium responsive diseases of animals, *J. Agric. Food Chem.*, 15, 448, 1967.
220. **Slater, C. S., Holmes, R. S., and Byers, H. G.,** Trace elements in the soils from the erosion Experiment Stations, with supplementary data on other soils, *U.S. Dep. Agric. Tech. Bull.*, 552, 1937.
221. **Olson, O. E., Whitehead, E. I., and Moxon, A. L.,** Occurrence of soluble selenium in soils and its availability to plants, *Soil Sci.*, 54, 47, 1942.
222. **Trelease, S. F.,** Selenium in soils, plants and animals, *Soil Sci.*, 60, 125, 1945.
223. **Kubota, J.,** Sampling of soils for trace element studies, *Ann. N.Y. Acad. Sci.*, 199, 105, 1972.
224. **Anderson, B. M., Keith, J. R., and Connor, J. J.,** Antimony, arsenic, germanium, lithium, mercury, selenium, tin and zinc in soils of the Powder River Basin, *U.S. Geol. Surv. Open File Rep.*, 75-436, 50, 1975.
225. **Ringrose, C. A., Klusman, R. W., and Dean, W. E.,** Geochemical variations in soils in the Piceance Creek Basin, *U.S. Geol. Surv. Open File Rep.*, 76-729, 130, 1976.
226. **Doran, J. W. and Alexander, M.,** Microbial formation of volatile selenium compounds in soil, *J. Soil Sci. Soc. Am.*, 41, 70, 1977.
227. **Heil, R. D. and Mahmoud, K. R.,** Mean concentrations and coefficients of variation of selected trace elements of various soil taxa, in *Forest Soils and Land Use*, Youngberg, C. T., Ed., Colorado State University, Department of Forest and Wood Sciences, Fort Collins, 1979, 198.
228. **Boerngen, J. G. and Shacklette, H. T.,** Chemical analyses of fruits, vegetables, and their associated soils from areas of commercial production in the conterminous United States, *U.S. Geol. Surv. Open File Rep.*, 80-84, 1980.
229. **Henry, C. D. and Kapadia, R. R.,** *Trace Elements in Soils of the South Texas Uranium District: Concentrations, Origin, and Environmental Significance*, Rep. Investigation, University of Texas, Austin, Bur. Economic Geology, No. 101, 1980.
230. **Boerngen, J. G. and Tidball, R. R.,** Chemical analysis of selected agricultural soils of Missouri, *U.S. Geol. Survey, Open-File Rep.*, 81-842, 1981.
231. **Shacklette, H. T. and Boerngen, J. G.,** Element concentrations in soils and other surficial materials of the conterminous United States, *U.S. Geol. Surv. Prof. Pap.*, 1270, 1984.
232. **Lévesque, M. and Vendette, E. D.,** Selenium determination in soil and plant materials, *Can. J. Soil Sci.*, 51, 85, 1971.
233. **Fletcher, K., Doyle, P., and Brink, V. C.,** Seleniferous vegetation and soils in the eastern Yukon, *Can. J. Plant Sci.*, 53, 701, 1973.
234. **Chattopadhyay, A. and Jervis, R. E.,** Multielement determination in market-garden soils by instrumental photon activation analysis, *Anal. Chem.*, 46, 1630, 1974.
235. **Gupta, W. C. and Winter, K. A.,** Selenium content of soils and crops and the effects of lime and sulfur on plant selenium, *Can. J. Soil Sci.*, 55, 161, 1975.
236. **Doyle, P. J. and Fletcher, W. K.,** Influence of soil parent material on the selenium content of wheat from west-central Saskatchewan, *Can. J. Plant Sci.*, 57, 859, 1977.
237. **Agemian, H. and Bedek, E.,** A semi-automated method for the determination of total arsenic and selenium in soils and sediments, *Anal. Chim. Acta*, 119, 323, 1980.
238. **Webb, J. S., Thornton, I., and Fletcher, K.,** Seleniferous soils in parts of England and Wales, *Nature (London)*, 211, 327, 1966.
239. **Forbes, S., Bound, G. P., and West, T. S.,** Determination of selenium in soils and plants by differential pulse cathodic-stripping voltammetry, *Talanta*, 26, 473, 1979.
240. **Ure, A. M., Bacon, J. R., Berrow, M. L., and Watt, J. J.,** The total trace element content of some Scottish soils by spark source mass spectrometry, *Geoderma*, 22, 1, 1979.
241. **Archer, F. C.,** Trace elements in soils in England and Wales, in Inorganic Pollution and Agriculture, MAFF Reference Book No. 326, Her Majesty's Stationary Office, London, 1980, 184.
242. **Gregers-Hansen, B.,** Application of radioactivation analysis for the determination of selenium and cobalt in soils and plants, in *Trans. 8th Int. Congr. Soil Science*, Vol. 3, Publishing House Acad. Bucharest, Romania, 1964, 63.
243. **Bisbjerg, B. and Gissel-Nielsen, G.,** The uptake of applied selenium by agricultural plants. I. The influence of soil type and plant species, *Plant Soil*, 31, 287, 1969.
244. **Elsokkary, I. H. and Oien, A.,** Determination of selenium in soils, *Acta Agric. Scand.*, 27, 285, 1977.
245. **Gissel-Nielsen, G. and Hamdy, A. A.,** Plant uptake of selenium and L_{Se}-values in different soils, *Z. Pflanzenernaehr. Bodenkd.*, 141, 67, 1978.
246. **Patel, C. A. and Mehta, B. V.,** Selenium status of soils and common fodders in Gujarat, *Indian J. Agric. Sci.*, 40, 389, 1970.
247. **Meng, X. and She, Z.,** Background values of certain elements in the soils of the Second Songhua, River basin (in Chinese), *Huanjing Kexue Xuebao*, 3, 25, 1983.

248. **Knott, S. G., McCray, C. W. R., and Hall, W. T. K.,** Selenium poisoning in horses in North Queensland, *Queensl. Dep. Agric. Stock. Div. Anim. Ind. Bull.*, 41, 1958.
249. **Dubikovskii, G. P. and Lebedev, V. N.,** Selenium in soils and meadow plants of the Belorussian SSR (in Russian), *Agrokhimiya,* (5), 113, 1973.
250. **Ababi, V. and Dumitrescu, M.,** Distribution of selenium in the soils and river sediments from the Vatra Moldovitei, Darmanesti, and Lesul Ursului counties (Romania) (in Romanian), *An. Stiint. Univ. "Al. I. Cuza" Iasi, Sect. 1c,* 19, 89, 1973.
251. **Gyul'akhmedov, A. N.,** Content of selenium in zone types of soils of the Azerbaidzhan SSR, in *Selen. Biol. Mater. Nauchn. Konf.*, Gasanov, G. G., Ed., Elm, Baku, U.S.S.R., 1974, 143, 182.
252. **Elsokkary, I. H.,** Selenium distribution, chemical fractionation and adsorption in some Egyptian Alluvial and Lacustrine soils, *Z. Pflanzenernaehr. Bodenkd.*, 143, 74, 1980.

Chapter 10

FRESHWATER SYSTEMS

Gregory A. Cutter

TABLE OF CONTENTS

I. INTRODUCTION

The aquatic chemistry of selenium has become of particular interest in recent years due to the increased mobilization of selenium to the environment via anthropogenic activities, the potential oxidation-reduction behavior of the element, and a realization that selenium can be both an essential and a toxic trace element to biological organisms. With respect to the latter, the different chemical forms, or species, of selenium have differing biological reactivities, as well as chemical and geochemical properties. Therefore, selenium speciation studies are of the utmost importance. It is thus surprising that most of the selenium speciation work has been focused on the marine system. Indeed, in order to fully understand the cycling of an element such as selenium in seawater, the input via fresh waters (e.g., rivers, rain) to the ocean must be evaluated as well. It is the purpose of this chapter to review the current literature on freshwater selenium in order to summarize data presently available. Primary emphasis will be paid to speciation values for the reasons cited above. This is by no means to imply that some current compilations are not useful, the work of Robberecht and Van Grieken[1] being very complete. This chapter will present more recent data and also review available works on selenium in freshwater sediments and biological organisms. However, it will become obvious to the reader that speciation values are sorely lacking.

The data for water will be divided into two general categories, dissolved selenium (in material that passes through filters with 0.45-μm openings) and particulate selenium in material of particle size ≥0.45-μm, typically suspended sediment and other suspended solids. Obviously these are operational definitions, but ones with which aquatic scientists have learned to live. By this definition, selenium in aquatic biota would be deemed "particulate", but in deference to the generally accepted approach, this distinction is reserved in this volume for selenium in particulate matter suspended in water and the atmosphere. The most apparent problem with these definitions occurs with elements in a colloidal phase, not truly dissolved, but passing through the filter. Colloidal selenium in the compilations will be deemed to be in the fraction denoted Se (−II+0). In the dissolved state, selenium may exist in three of its formal oxidation states: Se (−II), Se (IV), and Se (VI); Se (0) would be colloidal. When Se(IV) is present in fresh water, it exists as selenite ($HSeO_3^-$ at pH 6, and 31% $HSeO_3^-$ + 69% SeO_3^{2-} at pH 9).[2] Dissolved Se(VI) is found as selenate, SeO_4^{2-}, in fresh waters from pH 6 to 9.[2] Selenide may be found in the dissolved state as H_2Se + HSe^- (hydrogen selenide) in anoxic waters, or as organic selenides, primarily selenoamino acids bound in soluble peptides,[3] in oxic and anoxic waters. Equilibrium thermodynamic models[2,4] predict that selenate should be the predominant form in oxygenated fresh waters at pH 6 to 9, while hydrogen selenide and Se(0) should be found in anoxic waters. All of the selenium studies in the marine system[5-7] demonstrate that such calculations are questionable since thermodynamically unstable species can be found at significant concentrations due to their biological production and subsequent kinetic stabilization (i.e., slow oxidation rate). Overall, the speciation of dissolved selenium includes different oxidation states and different chemical forms within an oxidation state.

Particulate selenium and the element in biota may exist in the same oxidation states as dissolved selenium and can be found in different phases of the particulate matter (i.e., two levels of speciation: chemical and phase). In sediments, selenium may be within the organic material, iron and manganese oxides, carbonates, or other mineral phases which constitute a sediment particle. The actual chemical forms of selenium may be adsorbed to or coprecipitated with these phases (primarily selenite and selenate). Selenide can be covalently bound in the organic portion of a sediment. In addition, selenium may be found in anoxic sediments as insoluble metal selenide precipitates,[3] as insoluble elemental selenium,[3] or as ferroselite ($FeSe_2$) and selenium-containing pyrite.[8] Within biogenic material (i.e., plankton, fish, etc.), organic selenides are found primarily in the proteins,[9] while intracellular fluids can contain dissolved selenate, selenite, and organic selenides.

II. SAMPLE HANDLING AND STORAGE

Unlike many trace elements which must be sampled with extreme care in order to avoid contamination, selenium does not pose such a problem due to its low abundance in most man-made materials. However, since most samples are analyzed in the laboratory, sample storage procedures must be evaluated in conjunction with the analytical methodology. There have been a number of relevant studies in this respect for dissolved selenium.[10-12] The general consensus is that acidification to ca. pH 2 in large (≥1 l) linear polyethylene or borosilicate glass containers is sufficient to minimize the adsorptive loss of selenium without affecting the speciation of selenite or selenate. The acid used must be nonoxidizing and cannot interfere with the subsequent determination; hydrochloric acid is best suited for this role. Only Robberecht and Van Grieken[11] describe experiments on the storage of elemental selenium; large losses occur. Little can be found in the literature on the storage of organic selenium species (e.g., selenoamino acids, methyl selenides). Cutter[13] reports that dimethyl selenide is completely lost within a day, and therefore must be determined in the field.[3] Recent work in this laboratory[14] has examined the stability of organic selenide species. Reservoir samples, acidified to pH 1.5 with HCl and stored over a period of 7 months lost only 5% total selenium in both borosilicate glass and linear polyethylene bottles. However, during the course of the experiment, fungi grew in the bottles and a 43% increase in the organic selenide fraction was observed. Additional experiments have shown that biologically mediated speciation changes are minimized by refrigerated storage. This problem has not been observed in samples from oligotrophic (low nutrient) waters, but acidified samples should routinely be refrigerated or frozen to prevent biologically induced speciation changes.

A major problem with the storage of particulate and biogenic samples is biodegradation. This process is eliminated if samples are frozen in the field and immediately processed after thawing. The preparation of samples prior to analysis is crucial for an element such as selenium. Samples are usually dried as the first preparative step. However, drying must be at low temperatures (i.e., 40°C) so that losses of selenium through volatilization are eliminated.[15] Studies which did not follow these widely accepted dissolved and particulate storage methods were considered questionable, and data therefrom are not included in this chapter.

III. ANALYTICAL CONSIDERATIONS

The diversity of particulate and dissolved selenium species necessitates ''state of the art'' analytical techniques for examining selenium in any natural water. More specifically, the procedures must be able to accurately and precisely determine the multiple species of selenium, possess low detection limits, and be relatively free from analytical interference (e.g., spectroscopic, effects from other elements). The need for accuracy is obvious, but precision is also crucial since important biological and geochemical processes can result in only minor concentration or speciation changes; these must not be obscured by analytical imprecision. Dissolved selenium concentrations in natural waters are typically in the nanomolar range (parts per trillion). In combination with the fact that multiple species can be present, low detection limits in concert with excellent precision are essential. At the same time, analytical interferences typical with trace element determinations must be minimized. Rather than rigorously describing the possible analytical schemes for determining selenium speciation, a process already covered in Chapter 3, a brief overview of appropriate techniques for freshwater samples should suffice.

Environmental considerations such as low selenium concentrations, multiple species, and potential interferences, combined with analytical constraints (stated above) necessitate the selection of ''acceptable'' procedures prior to the consideration of data which are presented in the following tables and discussion. To begin the selection process, only procedures which have been used for freshwater selenium determinations were considered, and eliminations were made from this group. Techniques capable of presenting only less-than-detection-limit data are of little use, for example. However, techniques which are missing from the list below are not necessarily unacceptable. With the above criteria in mind, the following analytical techniques are considered generally capable of examining freshwater selenium speciation:

1. Reaction of selenite with diaminonaphthalene derivatives and determination with gas chromatography/electron capture detection[10,16] or fluorometry.[17,18]
2. Selective preconcentration/neutron activation analysis.[19]
3. Selective hydride generation/atomic absorption spectrometric detection.[13,20]
4. Electrothermal atomization/atomic absorption spectrometric detection (total selenium only).[21]
5. Selective preconcentration/X-ray emission spectrometry.[11]

Virtually all of the speciation methods determine selenite directly and employ various procedures to selectively convert the other species to selenite. Usually the fractions (species) of selenium measured are total selenium (Se(−II+0+IV+VI)), selenite (SeIV), and selenite + selenate (SeIV+VI). Since selenite is determined directly, selenate is the difference between selenate + selenite and selenite determinations, while Se(−II+0) is the difference between total selenium and selenite + selenate (rigorously defined, Se(0) is not dissolved, but it would be determined in this fraction). Due to such complex schemes, errors in speciation assignments occur. As an example, techniques which reduce selenate to selenite by boiling an HCl acidified (4 to 6 *M*) sample[11,13] determine selenite + selenate, not total dissolved selenium (i.e., which may also include selenide). Indeed, there is strong evidence that organic selenides exist in seawater,[6] and their presence in fresh waters is also shown in the data to be presented below. However, the Se(−II+0) fraction at present can only be operationally defined as organic selenide (similar to Cutter and Bruland[6]).

Most of the acceptable analytical methods are also used for seawater analysis. These techniques have the requisite low detection limits (seawater concentrations of selenium are extremely low), high precision, and the capability of determining multiple species in a complex matrix (e.g., 0.5 *M* NaCl). Neutron activation analysis is usually used only for determination of total selenium in solid materials, as the detection limits are too high for dissolved samples, and the method cannot distinguish between species. However, Massée et al.[19] added species-selective preconcentration steps prior to activation. Thus, the procedure determines total particulate concentration and dissolved selenium speciation if proper care is taken. More specific notes on the analytical techniques used to obtain a data set will be included in the discussions to follow.

The determination of particulate and biogenic selenium involves different methods of sample preparation (solubilization) prior to the use of analytical methods employed for determinations of the dissolved element. No current methods use dry ashing at high temperature to remove organic material due to the volatility of selenium. Digestion procedures (solubilization through oxidative dissolution) typically use combinations of nitric and perchloric acids, or nitric acid and hydrogen peroxide with heated refluxing.[22] However, the loss of selenium through charring must be avoided.[22,23] During an oxidative digest, the original speciation of particulate selenium is lost, and only total selenium values can be reported. Cutter[23] describes a technique which can be used to determine the major speciation of particulate selenium. This procedure uses a sodium hydroxide leach which releases all particulate-bound selenite and selenate without altering speciation. The leach solution is analyzed for selenite and selenate using dissolved sample procedures.[6,13,20] The total selenium concentration of a sample includes selenite and selenate, and the difference between determinations of total selenium and selenite + selenate reveals the concentration of particulate Se(−II+0).

IV. DISSOLVED SELENIUM

A. Rivers, Streams, and Canals

Geochemists studying the cycling of elements in the ocean must consider the input of weathered continental material via rivers and streams. The first thorough examination of river input for selenium was performed by Kharkar et al.[24] in 1968 for seven rivers in the U.S. Neutron activation analysis was employed, and no speciation analyses were attempted. These data as well as those from more recent studies are presented in Table 1. Examining the methods of Kharkar et al.,[24] several observations can be made. Their total selenium values actually represent selenite + selenate since no oxidation step was employed to convert dissolved selenide species to selenite. In addition, the samples were not acidified for storage, and adsorptive losses may have occurred. Nevertheless, these data represent a beginning and are thus presented here.

Speciation data for European rivers have been obtained primarily by Measures and Burton,[25] Burton,[26] and Robberecht and co-workers.[27,28] As the X-ray emission technique of Robberecht and Van Grieken[11] determines selenite and selenite + selenate, their total selenium assignment is therefore misleading. The determination of total selenium by Measures and Burton[10] utilizes UV irradiation which may oxidize any selenide forms present. Although this has not been rigorously evaluated, their "total selenium" fraction is probably identified correctly. However, since the difference between total selenium and selenite determinations may not only represent selenate concentrations, but also any Se(−II+0) present, selenate values are not reported in Table 1.

For the English rivers sampled,[25,26] selenite concentrations are low relative to total selenium. If one assumes organic selenide to be a minor component, then selenate appears to be the dominant species. A similar behavior is observed for the "unpolluted" rivers in Belgium[27,28] (selenate was not listed in the publications, but can be estimated in the ranges provided). The polluted Scheldt River displays elevated dissolved selenium concentrations, with the most notable being an increase in selenite. This can be attributed to the presence of a nearby coal-fired power plant.[29] Data on selenium in Japanese rivers are the most comprehensive available.[30,31] Selenite ranges from approximately 7 to 60% of the total riverine selenium, selenate is 20 to 87% of the total, and organic selenide, isolated via XAD®-2 resin adsorption[31] or bromine oxidation,[30] is less than 10% of the total, with the exception of the Asahi River (70% of total). Overall, large geographic variability in speciation patterns is evident, but selenate concentrations tend to be higher than those of selenite.

Recently, data for several rivers in North America have been obtained (Table 1). In Virginia's James River, selenite and organic selenide (Se(−II+0)) are the dominant species.[32,33] Takayanagi and Wong[33] also determined the amount of colloidal organic and inorganic selenium in the James River. It was found that 77% of the inorganic selenium can be classified as colloidal in the river water, while 70% of the organic selenium is colloidal. Two other rivers in eastern North America display opposite speciation patterns. In the Susquehanna River, which runs through the states of Pennsylvania and Maryland and empties into the Chesapeake Bay, selenate is the predominant form of dissolved selenium (69% of the total).[34] In contrast, samples from the St. Lawrence River in Canada show selenite to be the selenium species of highest concentration (67 to 76% of total).[35]

Cutter[36] has obtained speciation data for the largest river system of California, the San Joaquin and Sacramento Rivers (Table 1). The San Joaquin, which drains the southern portion of the agriculturally important central valley of California, has total selenium concentrations an order of magnitude higher than those of the Sacramento draining the northern

Table 1
DISSOLVED SELENIUM — RIVERS, STREAMS, AND CANALS

Location	Selenium concentration (nmole Se/kg)[a]						
	Total selenium	Selenite	Selenate	Se(−II+0)	Method[b]	Remarks	Ref.
Belgium							
Scheldt River	2.91—22.5	1.65—18.4	—	—	XRF	e	27
Miscellaneous "polluted" rivers	<0.63—7.34	<0.51—2.53	—	—	XRF	e	27
Brazil							
Amazon River	2.66	—	—	—	NAA	c,d,e	24
Canada							
St. Lawrence River	2.00—2.24	1.28—1.70	0.20—0.73	<0.06—0.40	FS		35
France							
Rhone River	1.90	—	—	—	NAA	c,d,e	24
Germany							
Rhine River	2.09	0.82	1.27	—	XRF	e	28
Japan							
Asahi	0.20—0.25	<0.02—0.03	0.04—0.11	0.14	GC		30
Chikugo	1.04	0.28	0.76	<0.06	FS	e	31
Ishikari	0.55	0.25	0.30	<0.06	FS	e	31
Kiso	0.55	0.33	0.22	<0.06	FS	e	31
Kitakami	0.81	0.27	0.54	<0.06	FS	e	31
Mogami	0.39	0.25	0.14	<0.06	FS	e	31
Shinano	0.50	0.25	0.25	<0.06	FS	e	31
Takahashi	0.20	0.05	0.09	0.06	GC		30
Yoshino	1.11	0.32	0.79	<0.06	FS	e	31
Yoshii	2.93	0.20	2.56	0.15	GC		30
Yudo	0.85	0.24	0.61	<0.06	FS	e	31
U.K.							
Hambel	4.94	0.39	—	—	GC		26
Itchen	3.17	0.23	—	—	GC		26
Beaulieu	1.08	<0.03	—	—	GC		26
Meon	3.55	0.25	—	—	GC		26
Test	3.70—4.34	0.20—0.28	—	—	GC		25

U.S.							
Average of 7 rivers	2.65 ± 1.22	—	—	—	NAA	c,d,e	24
California							
Delta-Mendota Canal	2.34—5.08	0.35—1.02	1.42—3.28	0.52—0.72	HAAS		36
Sacramento River	0.65—0.81	<0.02—0.09	0.29—0.35	0.27—0.46	HAAS		36
San Joaquin River	5.61—9.06	0.94—1.22	3.59—8.38	<0.02—0.80	HAAS		36
San Luis Drain Canal	2570—3510	67.6—68.5	2480—2760	21.5—682	HAAS		36
Maryland							
Susquehanna River	2.71	0.79	1.87	0.05	HAAS		34
Virginia							
James River	0.90—1.91	0.54—1.72	0.19—0.36	—	FS	e	32
James River	—	—	—	0.82	FS		33
Overall ranges (excluding San Luis Drain Canal and Scheldt River)	0.20—9.06	<0.02—2.53	<0.02—8.38	<0.02—0.82			

[a] Conversions: nmol/kg × 0.0790 → μg/kg; μmol/kg × 0.0790 → mg/kg.
[b] Analytical techniques: HAAS — atomic absorption spectrometry with hydride generation, EAAS — atomic absorption spectrometry with electrothermal atomization, GC — gas chromatography following derivatization, FS — fluorescence spectroscopy following derivatization, NAA — neutron activation analysis, XRF — X-ray fluorescence spectroscopy.
[c] Possible storage artifacts.
[d] Possible particulate contamination.
[e] Total selenium actually Se (IV + VI).

valley. In the San Joaquin, selenate is the dominant form, while selenate and Se(−II+0) are the primary species in the Sacramento. The Delta-Mendota Canal serves as a source of irrigation water south of the Sacramento/San Joaquin Delta and has a speciation pattern similar to the San Joaquin. The San Luis Drain Canal was constructed to remove subsurface runoff from the agricultural regions of the southern central valley. The San Luis water has amongst the highest dissolved selenium values ever reported (also see Kesterson Reservoir, Table 2). In contrast to other high-selenium waters (e.g., Scheldt River) where selenite is the primary species, selenate constitutes 79 to 97% of the total dissolved selenium in the canal.

The overall riverine speciation data (excluding the San Luis Drain and Scheldt River) show selenate to be the dominant form of selenium, with Se(−II+0) having the lowest concentrations. Nevertheless, the data also clearly indicate that every river is unique with respect to its speciation and concentration of selenium, and broad generalizations are, therefore, inappropriate and misleading. Certainly many more rivers need to be examined, particularly those contributing major inputs of water and suspended material to the sea.

B. Lakes and Reservoirs

The concentration range of total dissolved selenium in lakes varies considerably. Some of this data, however, may be due to analytical uncertainties. Selenium speciation data is presented in Table 2. From the few available results, lakes, like rivers, appear to have variable speciation patterns. Data for two lakes in Belgium[29] indicate that selenite constitutes approximately 65% of the total dissolved selenium. In Lake Constance, Germany, selenate appears to be the predominant species.[37] However, these data can be questioned since no corrections were made[60] for potential interferences due to nitrite.[20] Selenite concentrations, therefore, may be greatly underestimated, and selenate overestimated (refer to analytical discussion above). In Sweden's Lake Erken,[38,39] selenate is clearly the dominant species.

In the U.S., Hyco and Kesterson Reservoirs display the highest concentrations of total dissolved selenium (Table 2). Hyco serves as a cooling lake for a large, coal-fired power plant and receives effluent from coal fly ash leach ponds adjacent to the reservoir. The elevation of selenium is most pronounced in the selenite concentration, with selenite being 68% of the total. The concentrations of Se(−II+0) and selenate are roughly equivalent. Water input from the ash ponds appears to contribute large amounts of selenite.[14] The highest concentration of total dissolved selenium reported for a freshwater system is found at Kesterson Reservoir which receives input from the San Luis Drain (see discussion on rivers and Table 1).[40] In contrast to the Hyco Reservoir, selenate is the major fraction of dissolved selenium in Kesterson. Not only is Kesterson noteworthy for its high selenium concentration, but it is also the only body of freshwater in which methylated selenium species have been documented. Cooke[40] found up to 5.8% of the total selenium as dimethyl selenide and up to 21.7% of the total as an unidentified organic dimethyl selenonium derivative (a cation).

Like Hyco Reservoir, Catfish Lake is associated with a coal-fired power plant, and elevated total selenium concentrations are also displayed.[14] However, in this lake Se(−II+0) is the predominant (66% of total) selenium species. Robinson Impoundment is a blackwater lake with pH values of approximately 4.0. Total selenium is extremely low in this lake even though this reservoir has a coal-fired power plant on its shore.[14] The high tannin and lignin content of this blackwater system may be responsible for the removal of selenium species. Two other lakes, Kentucky[41] and Erie,[42] show total selenium concentrations similar to those in the power plant-associated reservoirs. In contrast, the mountain lake (Arrowhead), has low dissolved selenium species concentrations,[13] perhaps reflecting a difference in selenium input to the lake. The "less than" concentration data for Dufault and Duparquet Lakes[43] are included only to facilitate a comparison with their sediment data in Table 5.

C. Drinking and Ground Waters

The content and speciation of selenium in drinking and ground waters is of the most direct

Table 2
DISSOLVED SELENIUM — LAKES AND RESERVOIRS

Location	Selenium concentration (nmole Se/kg)[a]				Method[b]	Remarks	Ref.
	Total selenium	Selenite	Selenate	Se(−II+0)			
Belgium, Antwerp							
Campus Lake	2.91	1.90	1.01	—	XRF	e	29
Het Broek Pond	1.77	1.14	0.63	—	XRF	e	29
Canada, Quebec							
Lake Dufault	<1.27	—	—	—	HAAS	c	43
Lake Duparquet	<1.27	—	—	—	HAAS	c	43
Germany							
Lake Constance	1.52	0.25	1.27	—	HAAS	e,f	37
Sweden							
Lake Erken	0.82—2.52	0.13—0.51	0.38—2.00	—	FS	e	39
Lake Kinneret	1.29	—	—	—	FS	e	38
U.S.							
California							
Lake Arrowhead	—	0.23	<0.06	—	HAAS		13
Kesterson Reservoir	830—5044	203—456	97.0—4351	50.0—1180	HAAS		40
Kentucky							
Kentucky Lake	15.2—49.4	—	—	—	EAAS	d	41
Michigan							
Lake Erie	12.7—63.3	—	—	—	FS		42
North Carolina							
Catfish Lake	9.12 ± 0.21	2.82 ± 0.10	0.28 ± 0.28	6.02 ± 0.34	HAAS		14
Hyco Reservoir	76.2 ± 2.30	50.5 ± 1.60	14.9 ± 2.80	10.8 ± 3.00	HAAS		14
South Carolina							
Robinson Impoundment	0.68 ± 0.13	0.02 ± 0.005	0.22 ± 0.03	0.44 ± 0.13	HAAS		14

Table 2 (continued)
DISSOLVED SELENIUM — LAKES AND RESERVOIRS

Location	Selenium concentration (nmole Se/kg)[a]				Method[b]	Remarks	Ref.
	Total selenium	Selenite	Selenate	Se(−II+0)			
Overall ranges (excluding Hyco and Kesterson Reservoirs)	0.63—63.3	0.02—2.82	<0.06—1.27	0.44—6.02			

[a] Conversions: nmol/kg × 0.0790 → μg/kg; μmol/kg × 0.0790 → mg/kg
[b] Analytical techniques: HAAS — atomic absorption spectrometry with hydride generation, EAAS — atomic absorption spectrometry with electrothermal atomization, GC — gas chromatography following derivatization, FS — fluorescence spectroscopy following derivatization, NAA — neutron activation analysis, XRF — X-ray fluorescence spectroscopy.
[c] Possible storage artifacts.
[d] Total selenium actually Se IV.
[e] Total selenium actually Se (IV + VI).
[f] Possible sample preparation artifacts

concern to humans (Table 3). In this respect, the high selenium concentrations[44] in the Roche-Posay mineral springs are alarming. Selenate is a major species, but these are old (1963) data, and the analytical methodology can be reasonably questioned. As a comparison, newer data from springs in the U.S.S.R. have total selenium concentrations approximately 30 times lower than those in France.[44]

An important source of drinking water is groundwater, with total selenium concentrations ranging from <0.6 to 342 nmole/kg (Table 3). Although many more data are needed, a worldwide average of 30 nmole/kg can be extrapolated from these results. Like spring waters, all of the available data show selenate to be the major selenium species in ground waters.[27,28,45,46] On an analytical note, Roden and Tallman[46] show that humic substances in groundwater interfere with the determination of groundwater selenium speciation by hydride generation/atomic absorption spectrometry. They remove this interference with XAD®-8 resin. If organically bound selenium (organic selenide?) exists in groundwater, the XAD® treatment precludes its analysis and, therefore, Roden and Tallman's total selenium results are actually selenite + selenate. Drinking water, which is usually obtained from lakes and wells, has a narrow range of total selenium values (1.52 to 5.57 nmole/kg), with selenate being the primary selenium species.[19,27,28,47]

D. Rain and Snow

In order to complete the examination of freshwater selenium speciation, the deposition of water in the form of rain and snow to the surface of the earth should be reviewed. Speciation data have only been recently obtained, and the size of Table 4 reflects this. In the two precipitation samples from urban Antwerp,[28] the concentrations of total selenium (actually selenite + selenate) are similar, but the speciation patterns are nearly opposite. Jiang et al.[29] note that the atmospheric environment near Antwerp can be contaminated by selenium emissions from a nearby coal-fired power plant and a smelter.

The largest data sets for selenium speciation in rainwater have been obtained in Japan,[48] and from samples taken at the U.S. mid-Atlantic coast (Lewes, Delaware) and the island of Bermuda.[49] Suzuki and co-workers[48] observed a large range of selenium concentrations in Tokyo rainwater over a period of $2^1/_2$ years. They also note a decrease in selenium concentration with increasing rainfall amount. Selenite is the predominant chemical form in the Tokyo samples. For western Atlantic precipitation, Cutter and Church[49] showed that increasing total selenium in wet deposition strongly correlates with increasing acidity (total selenium vs. protons, $r = .9589$; $n = 15$). Based on these data, it appears that rainwater selenium is enriched by fossil fuel combustion in the eastern U.S.[49] In this respect, the elevated total selenium in rain from Lewes compares favorably with the values from Antwerp (Table 4). The rainwater selenite/selenate ratio at Lewes and Bermuda, 1.26 ± 0.95 ($n = 10$),[49] is half that found in Tokyo rain (2.98 ± 2.22, $n = 81$).[48] The study of selenium in precipitation is just beginning, but the available data indicate that precipitation can be an important source of selenium to surface waters.

V. SELENIUM IN SEDIMENTS

On a total mass basis, most of the selenium in lakes and reservoirs can be found in the sediments. This is due to abiotic and biotic scavenging of dissolved ions from the water column and burial in the underlying sediments. In this manner sediments reflect the net removal processes of dissolved selenium. Most studies of sedimentary selenium focus on the total concentration without regard to the sediment type (e.g., biogenic, authigenic) or the chemical constituents (phases) which make up sediment particles; Table 5 presents a compilation of selenium concentrations in several North American lake sediments. Unfortunately most studies of the aquatic environment are motivated by pollution problems rather

Table 3
DISSOLVED SELENIUM — DRINKING AND GROUND WATERS

Location	Selenium concentration (nmole/kg)[a]						
	Total selenium	Selenite	Selenate	Se(−II+0)	Method[b]	Remarks	Ref.
Belgium							
Miscellaneous ground-waters	<0.63—1.68	<0.51—2.66	—	—	XRF	d	27
Antwerp tap water	3.93—4.05	0.63—0.89	3.17—3.29	—	XRF	d	27
Brussels groundwater	1.58—2.23	0.32	—	—	EAAS	e	45
France							
Pyrenees groundwater	29.9	<0.51	29.3	—	XRF		28
Roche-Posay							
Du Geslin spring water	513	50.7	469	—	FS	e	44
St. Cyprien spring water	513	40.5	456	—	FS	e	44
St. Savin spring water	450	34.2	418	—	FS	e	44
Germany							
Darmstadt drinking water	1.52	<0.51	1.10	—	XRF	d	28
Israel, Jerusalem							
Drinking water	5.57	0.76	4.81	—	XRF	d	28
Groundwater	11.4—342	<0.51	<0.63—329	—	XRF	d	28
Netherlands							
Drinking water	2.03	1.27	0.76	—	NAA	d	19
U.S.							
North Dakota groundwater	6.08—8.87	0.51—2.53	5.07—5.91	—	HAAS	d	46
U.S.S.R.							
Kvasy spring water	15.2	—	—	—	FS	c	44
Moscow							
Drinking water	1.58	0.63	0.95	—	FS	d	47
Well water	1.20—12.0	1.14—10.8	<0.01—1.27	—	FS	d	47
Vraged spring water	38.0—101	—	—	—	FS	c	44

Overall ranges	0.63—513	0.51—50.7	<0.01—469

[a] Conversions: nmol/kg × 0.0790 → μg/kg; μmol/kg × 0.0790 → mg/kg

[b] Analytical techniques: HAAS — atomic absorption spectrometry with hydride generation, EAAS — atomic absorption spectrometry with electrothermal atomization, GC — gas chromatography following derivatization, FS — fluorescence spectroscopy following derivatization, NAA — neutron activation analysis, XRF — X-ray fluorescence spectroscopy.

[c] Total selenium actually Se IV.

[d] Total selenium actually Se (IV + VI).

[e] Speciation may be incorrect.

Table 4
DISSOLVED SELENIUM — RAIN AND SNOW

Location	Selenium concentration (nmole/kg)[a]				Method[b]	Remarks	Ref.
	Total Selenium	Selenite	Selenate	Se(−II+0)			
Belgium, Antwerp, rain, snow	3.16	0.63	2.53	—	XRF	c	28
	3.67	2.78	0.89	—	XRF	c	28
Bermuda, rain	0.22—1.15	0.11—0.51	0.06—0.27	0.02	HAAS		49
Japan, Tokyo, rain	0.06—1.49	0.06—1.32	—	—	FS	c	48
U.S.							
San Diego, CA, rain	—	0.66	<0.02	—	HAAS		13
Delaware, Lewes, rain	0.42—4.13	0.29—1.64	0.03—3.41	1.06	HAAS	d	49
Overall ranges	0.06—4.13	0.06—2.78	<0.02—3.41	1.06			

[a] Conversions: nmol/kg × 0.0790 → μg/kg; μmol/kg × 0.0790 → mg/kg
[b] Analytical techniques: HAAS — atomic absorption spectrometry with hydride generation, EAAS — atomic absorption spectrometry with electrothermal atomization, GC — gas chromatography following derivatization, FS — fluorescence spectroscopy following derivatization, NAA — neutron activation analysis, XRF — X-ray fluorescence spectroscopy.
[c] Total selenium actually Se (IV + VI).
[d] Only one sample had detectable Se (−II + 0)

Table 5
SELENIUM IN SEDIMENTS

Location	Selenium concentration (μmole/kg)[a]				Method[b]	Ref.
	Total selenium	Selenite	Selenate	Se(−II+0)		
Canada, Quebec						
Lake Dufault	17.7—184	—	—	—	HAAS	43
Lake Duparquet	2.53—10.1	—	—	—	HAAS	43
U.S.						
Michigan						
Lake Erie	1.27—9.49	—	—	—	NAA	42
New York						
Sagamore Lake	20.3	—	—	—	HAAS	51
Woods Lake	58.2	—	—	—	HAAS	51
North Carolina						
Catfish Lake	121 ± 7.0	1.23 ± 0.32	8.05 ± 0.37	113 ± 7.0	HAAS	14
Hyco Reservoir	366 ± 30	9.29 ± 0.91	39.2 ± 2.60	318 ± 3.0	HAAS	14
Belews Lake	83.8	—	—	—	NAA	52
South Carolina						
Robinson Impoundment	21.0 ± 1.40	<0.02	0.58 ± 0.07	20.4 ± 1.40	HAAS	14
Savannah River Project	27.8—77.2	—	—	—	NAA	53
South Dakota						
Belle Fourche Reservoir	19.0	—	—	—	FS	50
Wisconsin						
Mendota Lake	24.1	—	—	—	FS	50
Weber Lake	34.2	—	—	—	FS	50
Overall ranges	2.53—366	<0.02—9.29	0.58—3.92	20.4—318		

[a] Conversions: nmol/kg × 0.0790 → μg/kg; μmol/kg × 0.0790 → mg/kg.
[b] Analytical techniques: HAAS — atomic absorption spectrometry with hydride generation, EAAS — atomic absorption spectrometry with electrothermal atomization, GC — gas chromatography following derivatization, FS — fluorescence spectroscopy following derivatization, NAA — neutron activation analysis, XRF — X-ray fluorescence spectroscopy.

than by attempts to understand natural processes. The available sediment data reflect this trend, but information on native selenium levels can still be gained.

Most total sedimentary selenium concentrations are in the 20- to 50-μmole/kg range (i.e., <4 mg/kg). In terms of anthropogenic contamination, the most pristine lake is Weber in northern Wisconsin,[50] which has a total sedimentary selenium concentration of 34.2 μmole/kg or 2.7 mg/kg. However, certain "contaminated" lakes (e.g., Erie) have lower sedimentary selenium concentrations.[42,50,51] Sediments in reservoirs which receive fossil fuel combustion products (e.g., fly ash) display elevated selenium concentrations. Hyco Reservoir, which has high concentrations of dissolved selenium (Table 2), is typical, with the highest total sedimentary selenium in this compilation (366 μmole/kg).[14] Sediments in other reservoirs associated with power plants (e.g., Belews,[52] Savannah River Project[53]) also have increased selenium concentrations.

Using the methods described by Tessier et al.,[54] Cutter[14] has analyzed the phase distributions of selenium in sediments from three power plant-receiving waters; the data are presented in Figure 1. In addition, the same analysis was performed on coal fly ash, which appears to be the major source of selenium to these reservoirs. Within sediments, selenium is primarily in the organic phase (operational definition; may also be Se(0) or a sulfide phase in anoxic sediments). The inorganic exchangeable (surface) carbonate and manganese and iron oxides phases contain less than 10% of the total selenium in sediments. These results agree with those for sediments in natural (nonimpacted) sediments.[14,23] The quantity of

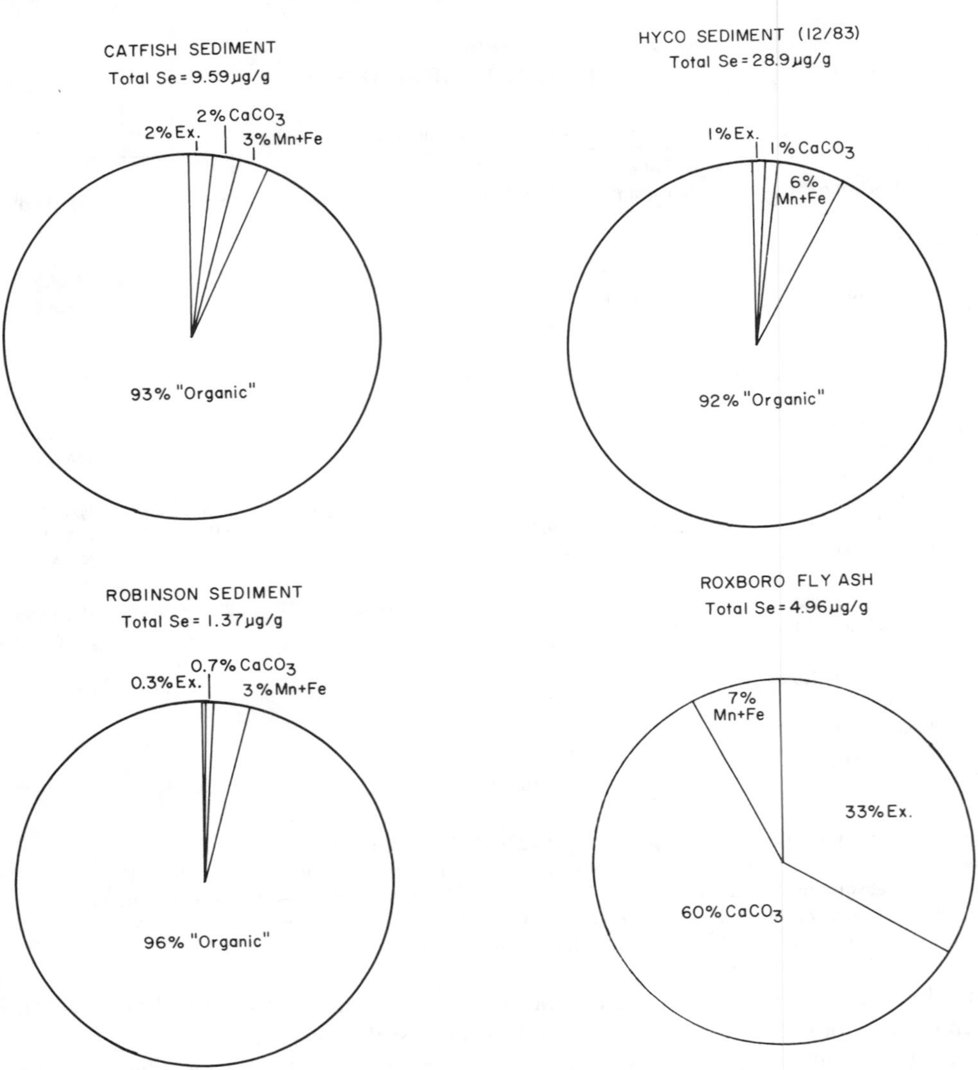

FIGURE 1. Phase distributions for selenium in sediments from three coal-fired power plant cooling reservoirs and in a coal fly ash sample. Four phases are analyzed: exchangeable (Ex.), carbonate (Carb.), manganese and iron oxides (Mn-Fe), and organic + sulfide (''organic''). Above each diagram is the total selenium concentration in the material. The amount in each phase is expressed as a percentage of the total selenium. Sample locations are Robinson sediment — Robinson Impoundment (North Carolina); Catfish sediment — Catfish Lake (North Carolina); Hyco sediment — Hyco Reservoir (North Carolina); Roxboro fly ash — Roxboro Steam Electric Plant (North Carolina). (From Cutter, G. A., Speciation of Selenium and Arsenic in Natural Waters and Sediments, Vol. 1, EA-4641, Electrical Power Research Institute, Palo Alto, CA, 1986. With permission.)

selenite and selenate associated with the Hyco, Robinson, and Sutton (Catfish Lake) sediments was also determined by Cutter.[14] Concentrations of sedimentary selenite + selenate are very low ($\leq$10% of the total selenium), with selenate being the predominant species. These chemical speciation results are also similar to those from natural sediments.[14,23]

In general, conclusions regarding selenium in sediments must be made with caution due to many complex environmental factors. Even in ''polluted'' lakes such as Robinson Impoundment, a power plant cooling reservoir, total sedimentary selenium concentrations are less than those in ''pristine'' systems such as Weber Lake. Some complicating factors include sedimentation rate (increased rate decreases concentration by dilution), biological produc-

Table 6
SELENIUM IN FISH AND PLANKTON

Location and sample type	Total selenium concentration (μmole/kg)[a]	Method[b]		Ref.
Canada				
Quebec				
Lake Dufault (fish muscle)	13.9—38.0	HAAS	W	43
Lake Duparquet (fish muscle)	2.53—7.85	HAAS	W	43
Ontario				
Moose Lake (whole fish)	2.15—3.04	NAA	W	56
Lake Ontario (whole fish)	4.81	NAA	W	56
Lake St. Pierre (whole fish)	4.68	NAA	W	56
Lake Erie (whole fish)	2.41—4.05	NAA	W	56
Finland				
Average freshwater fish	23.6—31.6	NAA	D	57
U.S.				
Kentucky				
Kentucky Lake (whole fish)	1.01—5.82	EAAS	W	41
North Carolina				
Hyco Reservoir				
Zooplankton	8.10—69.3	HAAS	W	55
Fish muscle	3.04—159	HAAS	W	55
Belews Lake (fish muscle)	122—354	NAA	D	58
Colorado				
Sweitzer Lake (fish muscle)	161—258	HAAS	D	58
Highline Reservoir (fish muscle)	92.4—197	HAAS	D	58
Michigan				
Lake Erie				
Whole fish	19.0—126	NAA	D	42
Zooplankton	9.88—49.1	NAA	D	42
South Carolina				
Savannah River Project				
Phytoplankton	16.5—152	NAA	D	53
Higher plants	7.95—72.1	NAA	D	53
Whole fish	82.2—119	NAA	D	53
Overall ranges				
Plankton	8.10—152			
Fish	1.01—354			

[a] Conversions: nmol/kg × 0.0790 → μg/kg; μmol/kg × 0.0790 → mg/kg.
[b] Analytical techniques: HAAS — atomic absorption spectrometry with hydride generation, EAAS — atomic absorption spectrometry with electrothermal atomization, GC — gas chromatography following derivatization, FS — fluorescence spectroscopy following derivatization, NAA — neutron activation analysis, XRF — X-ray fluorescence spectroscopy.
[d] Only one sample had detectable Se (–II + 0).

tivity (increased scavenging and removal), and sediment type (carbonate sediments contain low selenium concentrations; Figure 1).

VI. SELENIUM IN FISH AND PLANKTON

In addition to sediments and the water itself, biological organisms contain appreciable concentrations of selenium. Studies of the marine system have emphasized the importance of biological activity and its by-products on the selenium cycle.[5,6,9] In fresh water, biology should be just as important. From the dissolved state, selenium can be taken up by primary producers (phytoplankton) and moved through the "food web" to higher trophic level organisms (fish and perhaps humans). Table 6 contains data on total selenium in plankton and fish from some areas for which dissolved selenium values are reported (Table 2). For all of the biogenic matter, selenium is enriched over dissolved selenium from approximately 100 (Kentucky Lake[41]) to 20,000 times (Lake Dufault[43]).

The range of planktonic selenium is 8.1 to 152 μmoles/kg,[42,53,55] which is comparable to concentrations in marine plankton (Chapter 11). Selenium concentrations in fish range from 1.0 to 354 μmoles/kg (0.08 to 28 mg/kg).[41-43,53,55-58] There is a problem in comparing the fish values directly since some are whole-body analyses, while others are analyses of muscle tissue only. If selenium is fractionated within an organism (i.e., concentrated in certain organs), then only whole body comparisons are legitimate. In addition, many wet weight determinations are reported, whereas rigorously, only dry weight values should be compared directly since moisture content varies. If we assume an average 80% moisture content for fish,[59] the range of total selenium in fish can be reported as 5 to 1770 μmoles/kg on a dry basis.

VII. SUMMARY

State-of-the-art analytical procedures now allow environmental scientists to examine the chemical forms of elements as well as their concentrations in the aquatic system. This is particularly relevant for selenium, but the data for freshwater systems is only beginning to be acquired. This chapter attempts to summarize some of these initial works with a focus on the different geochemical compartments (''reservoirs'') in which selenium can be found. Within the dissolved state, selenium speciation varies depending on the location. Freshwater speciation is certainly not at thermodynamic equilibrium, an observation similar to marine findings (Chapter 11). For particulate and biogenic selenium, speciation studies are almost nonexistent. The few available biological and chemical data suggest that selenide is the predominant form of particulate selenium. It is hoped that this compilation will serve as the start of, or basis for, future research.

ACKNOWLEDGMENTS

The author thanks the Department of Oceanography, Old Dominion University, Norfolk, VA, for supporting the preparation of this chapter, C. Krahforst for literature review, and L. S. Abbott for editorial assistance.

REFERENCES

1. **Robberecht, H. and Van Grieken, R.,** Selenium in environmental waters: determination, speciation, and concentration levels, *Talanta,* 29, 823, 1982.
2. **Turner, D. R., Whitfield, M., and Dickson, A. G.,** The equilibrium speciation of dissolved components in freshwater and seawater at 25°C and 1 atm. pressure, *Geochim. Cosmochim. Acta,* 45, 855, 1981.
3. **Cutter, G. A.,** Selenium in reducing waters, *Science,* 217, 829, 1982.
4. **Sillen, L. G.,** The physical chemistry of seawater, in *Oceanography,* Sears, M., Ed., American Association for the Advancement of Science, Washington, D.C., 1961, 549.
5. **Measures, C. I., Grant, B. C., Mangum, B. J., and Edmond, J. M.,** The relationship of the distribution of dissolved selenium IV and VI in three oceans to physical and biological processes, in *Trace Metals in Seawater,* Wong, C. S., Boyle, E., Bruland, K. W., Burton, J. D., and Goldberg, E. D., Eds., Plenum Press, New York, 1983, 73.
6. **Cutter, G. A. and Bruland, K. W.,** The marine biogeochemistry of selenium: a re-evaluation, *Limnol. Oceanogr.,* 29, 1179, 1984.
7. **Sugimura, Y., Suzuki, Y., and Miyake, Y.,** The content of selenium and its chemical form in sea water, *J. Oceanogr. Soc. Jpn.,* 32, 235, 1976.

8. **Howard, J. H., III,** Geochemistry of selenium: formation of ferroselite and selenium behavior in the vicinity of oxidizing sulfide and uranium deposits, *Geochim. Cosmochim. Acta,* 41, 1665, 1977.
9. **Wrench, J. J. and Campbell, N. C.,** Protein bound selenium in some marine organisms, *Chemosphere,* 10, 1155, 1981.
10. **Measures, C. I. and Burton, J. D.,** Gas chromatographic method for the determination of selenite and total selenium in sea water, *Anal. Chim. Acta,* 120, 177, 1980.
11. **Robberecht, H. J. and Van Grieken, R. E.,** Sub-part-per-billion determination of total dissolved selenium and selenite in environmental waters by X-ray fluorescence spectrometry, *Anal. Chem.,* 52, 449, 1980.
12. **Cheam, V. and Agemian, H.,** Preservation and stability of inorganic selenium compounds at ppb levels in water samples, *Anal. Chim. Acta,* 113, 237, 1980.
13. **Cutter, G. A.,** Species determination of selenium in natural waters, *Anal. Chim. Acta,* 98, 59, 1978.
14. **Cutter, G. A.,** Speciation of Selenium and Arsenic in Natural Waters and Sediments, Vol. 1, EA-4641, Electric Power Research Institute, Palo Alto, CA, 1986.
15. **Fourie, H. O. and Peisach, M.,** Loss of trace elements during dehydration of marine zoological material, *Analyst,* 102, 193, 1977.
16. **Shimoishi, Y. and Tôei, K.,** The gas chromatographic determination of selenium (IV) and total selenium in natural waters with 1,2 diamino-3,5-dibromobenzene, *Anal. Chim. Acta,* 100, 65, 1978.
17. **Sugimura, Y. and Suzuki, Y.,** A new fluorometric method of analysis of selenium in sea water, *J. Oceanogr. Soc. Jpn.,* 33, 23, 1977.
18. **Takayanagi, K. and Wong, G. T. F.,** Fluorimetric determination of selenium(IV) and total selenium in natural waters, *Anal. Chim. Acta,* 148, 263, 1983.
19. **Massée, R., Van der Sloot, H. A., and Das, H. A.,** Total selenium and selenium IV determinations using neutron activation analysis, *J. Radioanal. Chem.,* 38, 157, 1977.
20. **Cutter, G. A.,** Elimination of nitrite interference in the determination of selenium by hydride generation, *Anal. Chim. Acta,* 149, 391, 1983.
21. **Welz, B., Schlemmer, G., and Voellkopf, U.,** Influence of the valency state on the determination of selenium in graphite furnace atomic absorption spectrometry, *Spectrochim. Acta Part B,* 39, 501, 1984.
22. **Campbell, A. D.,** Determination of selenium in biological materials and water, *Pure Appl. Chem.,* 56, 645, 1984.
23. **Cutter, G. A.,** Determination of selenium speciation in biogenic particles and sediments, *Anal. Chem.,* 57, 2951, 1985.
24. **Kharkar, D. P., Turekian, K. K., and Bertine, K. K.,** Stream supply of dissolved silver, molybdenum, antimony, selenium, chromium, cobalt, rubidium and cesium to the oceans, *Geochim. Cosmochim. Acta,* 32, 285, 1968.
25. **Measures, C. I. and Burton, J. D.,** Behavior and speciation of dissolved selenium in estuarine waters, *Nature (London),* 273, 293, 1978.
26. **Burton, J. D.,** Behavior of some trace chemical constituents in estuarine waters, *Pure Appl. Chem.,* 50, 385, 1978.
27. **Robberecht, H., Van Grieken, R., Van Sprundel, M., Vanden Berghe, D., and Deelstra, H.,** Selenium in environmental and drinking waters of Belgium, *Sci. Total Environ.,* 26, 163, 1983.
28. **Robberecht, H. and van Grieken, R.,** Selenium content and speciation in environmental waters determined by X-ray fluorescence spectroscopy, in *Trace Substances in Environmental Health — 14,* Hemphill, D. D., Ed., University of Missouri, Columbia, 1980, 362.
29. **Jiang, S., Robberecht, H., and Adams, F.,** Identification and determination of alkylselenide compounds in environmental air, *Atmos. Environ.,* 17, 111, 1983.
30. **Uchida, H., Shimoishi, Y., and Tôei, K.,** Gas chromatographic determination of selenium (−II,0), (IV), and (VI) in natural waters, *Environ. Sci. Tech.,* 14, 541, 1980.
31. **Suzuki, Y., Sugimura, Y., and Miyake, Y.,** Selenium content and its chemical form in rivers of Japan, *Jpn. J. Limnol.,* 42, 89, 1981.
32. **Takayanagi, K. and Wong, G. T. F.,** Total selenium and selenium (IV) in the James River estuary and southern Chesapeake Bay, *Estuarine Coastal Shelf Sci.,* 18, 113, 1984.
33. **Takayanagi, K. and Wong, G. T. F.,** Organic and colloidal selenium in southern Chesapeake Bay and adjacent waters, *Mar. Chem.,* 14, 141, 1984.
34. **Velinsky, D. J. and Cutter, G. A.,** Unpublished data, 1985.
35. **Takayanagi, K. and Cossa, D.,** Speciation of dissolved selenium in the upper St. Lawrence estuary, in *Marine and Estuarine Geochemistry,* Sigleo, A. C. and Hattori, A., Eds., Lewis, Chelsea, MI, 1985, 275.
36. **Cutter, G. A.,** The estuarine behavior of selenium in the San Francisco Bay, 1987, in preparation.
37. **Sinemus, H. W., Melcher, M., and Welz, B.,** Influence of valence state on the determination of antimony, arsenic, bismuth, selenium, and tellurium in lake water using the hydride AA technique, *At. Spectrosc.,* 2, 81, 1981.
38. **Lindstrom, K.,** *Peridinium cinctum* bioassays of selenium in Lake Erken, *Arch. Hydrobiol.,* 89, 110, 1980.

39. **Lindstrom, K.,** Selenium as a growth factor for plankton algae in laboratory experiments and in some Swedish lakes, *Hydrobiologia,* 101, 35, 1983.
40. **Cooke, T. D.,** Processes Affecting Selenium Speciation in Natural Waters: a Case Study of the Kesterson Reservoir, M.S. thesis, University of California, Santa Cruz, 1985.
41. **McClellan, B. E. and Frazer, K. J.,** An environmental study of the origin, distribution, and bioaccumulation of selenium in Kentucky and Barkely Lakes, Rep. No. 122, Water Resources Research Institute, University of Kentucky, Lexington, 1980.
42. **Adams, W. J. and Johnson, H. E.,** Survey of the selenium content in the aquatic biota of western Lake Erie, *J. Great Lakes Res.,* 3, 10, 1977.
43. **Speyer, M. R.,** Mercury and selenium concentrations in fish, sediments, and water of two northwestern Quebec lakes, *Bull. Environ. Contam. Toxicol.,* 24, 427, 1980.
44. **Goleva, G. A. and Lushnikov, V. V.,** Occurrence of selenium in the ground waters of ore deposits and some types of mineral waters, *Geokhimiya,* (4), 438, 1967.
45. **Nève, J., Hanocq, M., and Molle, L.,** Nouvelles propositions pour la détermination du sélénium dans les eaux naturelles, *Int. J. Environ. Anal. Chem.,* 8, 177, 1980.
46. **Roden, D. R. and Tallman, D. E.,** Determination of inorganic selenium species in ground waters containing organic interferences by ion chromatography and hydride generation/atomic absorption spectrometry, *Anal. Chem.,* 54, 307, 1982.
47. **Nazarenko, I. I. and Kislova, I. V.,** Determination of different forms of selenium in water, *Zh. Anal. Khim.,* 33, 157, 1978.
48. **Suzuki, Y., Sugimura, Y., and Miyake, Y.,** The content of selenium and its chemical form in rain water and aerosol in Tokyo, *J. Meteorol. Soc. Jpn.,* 59, 405, 1981.
49. **Cutter, G. A. and Church, T. M.,** Selenium in western Atlantic precipitation, *Nature (London),* 322, 720, 1986.
50. **Wiersma, J. H. and Lee, G. F.,** Selenium in lake sediments — analytical procedure and preliminary results, *Environ. Sci. Tech.,* 5, 1203, 1971.
51. **Heit, M., Tan, Y., Klusek, C., and Burke, J. C.,** Anthropogenic trace elements and polycyclic aromatic hydrocarbon levels in sediment cores from two lakes in the Adirondack acid lake region, *Water Air Soil Pollut.,* 15, 441, 1981.
52. **Cumbie, P. N.,** Deposition of selenium and arsenic in sediments of Belews Lake, North Carolina, in *Workshop Proc.: The Effects of Trace Elements on Aquatic Ecosystems,* EA-3329, Electric Power Research Institute, Palo Alto, CA, 1984.
53. **Guthrie, R. K. and Cherry, D. S.,** Trophic level accumulation of heavy metals in a coal ash basin drainage system, *Water Resour. Bull.,* 15, 244, 1979.
54. **Tessier, A., Campbell, P. G. C., and Bisson, M.,** Sequential extraction procedure for the speciation of particulate trace metals, *Anal. Chem.,* 51, 844, 1979.
55. **Woock, S. E. and Summers, P. B.,** Selenium monitoring in Hyco Reservoir (N.C.) waters (1977—1981) and biota (1977—1980), in *Workshop Proc.: The Effects of Trace Elements on Aquatic Ecosystems,* EA-3329, Electric Power Research Institute, Palo Alto, CA, 1984.
56. **Uthe, J. F. and Bligh, E. G.,** Preliminary survey of heavy metal contamination of Canadian freshwater fish, *J. Fish. Res. Board Can.,* 28, 786, 1971.
57. **Sandholm, M., Oksanen, H. E., and Pesonen, L.,** Uptake of selenium by aquatic organisms, *Limnol. Oceanogr.,* 18, 496, 1973.
58. **Baumann, P. C. and May, T. W.,** Selenium residues in fish from inland waters of the United States, in *Workshop Proc.: The Effect of Trace Elements on Aquatic Ecosystems,* EA-3329, Electric Power Research Institute, Palo Alto, CA, 1984.
59. **Giesy, J. P. and Wiener, J. G.,** Frequency distribution of trace metal concentrations in five freshwater fishes, *Trans. Am. Fish. Soc.,* 106, 393, 1977.
60. **Welz, B.,** Personal communication.

Chapter 11

THE MARINE ENVIRONMENT

K. W. Michael Siu and Shier S. Berman

TABLE OF CONTENTS

I. INTRODUCTION

Selenium occurs in geological formations primarily as selenide in sulfide deposits, with natural selenate and selenite minerals being rare. An estimate of the ratio of sulfur to selenium in igneous rocks has been given as 6000 to 1. The average selenium crustal abundance is estimated to be 0.05 to 0.09 mg/kg.[1,2] Concentrations in coal and fuel oil range from 0.47 to 8.1 and 2.4 to 7.5 mg/kg, respectively.[3] In soil, selenium concentration varies tremendously and can approach 80 mg/kg in seleniferous soil.[4,5]

The combustion of fuel oils introduces selenium into the atmosphere and subsequently into the ocean. The annual mobilization of selenium from fossil fuel burning is assessed as 4.5×10^8 g. In comparison, the annual flux from river discharges during normal weathering processes is 7.2×10^9 g.[6] Thus, it seems unlikely that anthropogenic mobilization would lead to any global elevations of selenium, although local effects may be significant.[7]

A survey on the riverborne contribution of selenium to the oceans from seven North American rivers, the Rhone, and the Amazon reported a nonweighted average of 0.20 μg/l.[8] A more recent study on several chalk streams in southern England yielded a nonweighted average of 0.27 μg/l.[9] The subject of whether selenium is conserved during estuarine mixing is controversial. Equilibration studies suggested desorption of selenium from riverborne particulates in seawater.[8] A study on the total selenium content of the coastal waters off Honshu Island, Japan, indicated that selenium is removed from the seawater during mixing.[10] Data on Southhampton water and the estuarine section of the River Test (southern England), however, show conservation of both selenite and total selenium.[9] Further, results of a survey on the selenium concentration in the James River and Chesapeake Bay (northeastern U.S.) imply that selenite and total selenium are removed from the river water and selenate is produced during estuarine mixing.[11]

II. DISSOLVED INORGANIC SELENIUM

Although selenium can exist in nature as Se(−II), (0), (IV), and (VI), only two oxidation states, (IV) and (VI), are thought to be important inorganic selenium species in seawater.

The presence of selenium in seawater was first recorded by Goldschmidt and Strock[12] in 1935. The reported value of 3.8 μg/l was later confirmed by Wattenberg[13] and by Ishibashi et al.[14] Much later, Schutz and Turekian[15] found a comparatively lower average of 0.09 μg/l in 23 samples of surface and deep seawaters during a survey of the Pacific and Atlantic Oceans and the Caribbean Sea. Chau and Riley[16] reported a Se(IV) concentration range of 0.4 to 0.5 μg/l, with no detectable Se(VI) in the seawater of the English Channel. They considered this predominance of tetravalent selenium to be in agreement with the redox potential of the selenite-selenate pair. This view was reiterated by Shimoshi[17] who reported a Se(IV) range of 0.04 to 0.08 μg/l and no detectable Se(VI) in the coastal and offshore seawater at Shibukawa, Japan. This reported absence of selenium(VI) in seawater is questionable in view of thermodynamics,[18] and is contrary to the results of later investigations.

The existence of Se(VI) was first reported by Sugimura et al.[10] in a study of North Pacific waters, in which the concentration of Se(VI) was operationally defined as the difference between the concentrations of total selenium and Se(IV). They showed that the total selenium concentrations ranged from 0.06 to 0.12 μg/l at the surface and increased to 0.20 μg/l in deeper waters, while the ratio of Se(IV) to total selenium ranged from 0.5 to 0.8 at the surface and 0.4 to 0.6 in deeper parts. Se(IV) exhibited quite a uniform distribution with depth. Se(VI), on the other hand, appeared to become more abundant with increasing depth. These authors were also the first to suggest that the coexistence of Se(IV) and (VI), which did not agree with earlier determinations and predictions that either Se(IV)[16,17] or Se(VI)[18] should predominate, might be due to biochemical processes which take place dynamically

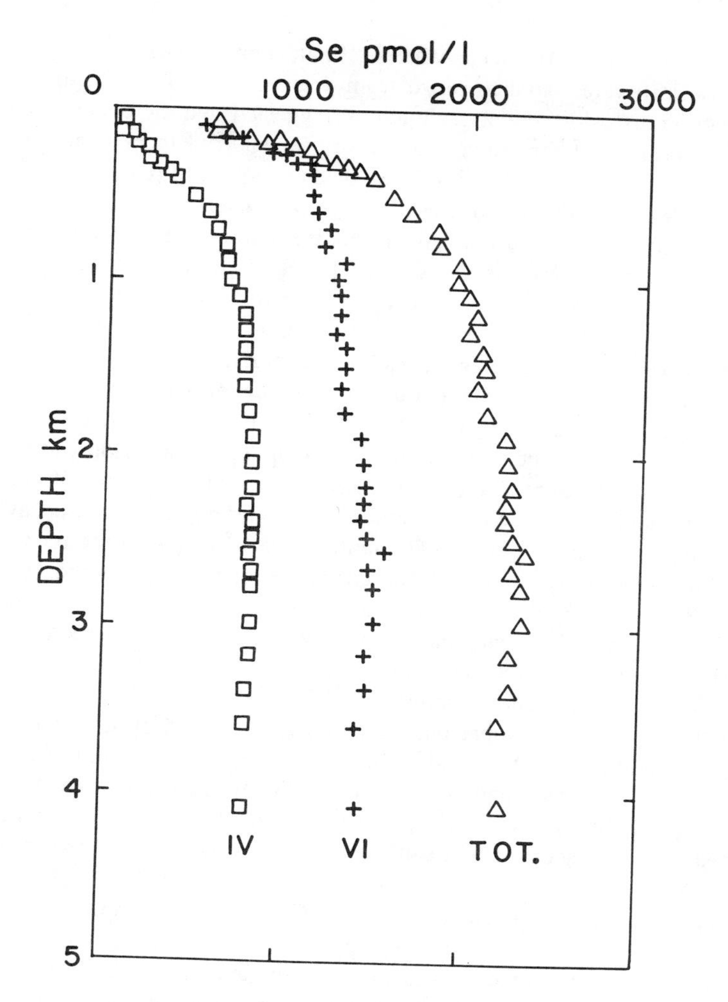

FIGURE 1. A typical vertical profile of selenium. (From Measures, C. I., McDuff, R. E., and Edmond, J. M., *Earth Planet. Sci. Lett.*, 49, 102, 1980. With permission.)

and constantly in the ocean. The existence of both tetra- and hexavalent selenium in seawater was later confirmed in a study of waters from the North Sea[19] and Japan.[20] For the latter, reported total Se [Se(IV) + Se(VI)] and Se(IV) concentrations ranged from 33 to 47 and 12 to 32 ng/l, respectively, with Se(IV)/(total Se) ranging from 0.4 to 0.7.[20]

In a study of California seawater, Cutter[21] observed an increase of Se(IV) concentration with depth from a low of <5 ng/l at the surface to a high of 70 ng/l at 1810 m. (The analytical methodology used was capable of detecting volatile methyl selenides; none, however, was found.) Measures and Burton[22] reported a detailed survey of the selenium(IV) and the total selenium concentrations in two areas of the northeastern Atlantic Ocean. The observed ranges were 2 to 40 and 25 to 125 ng/l for Se(IV) and total selenium, respectively. As in previous studies, the difference between total selenium and Se(IV) was assigned as Se(VI). Depth profiles of Se(IV) and (VI) bore a resemblance to and correlated well with those of the micronutrients, silicate, and phosphate (Figure 1). The coexistence of both Se(IV) and (VI) plus their correlation with micronutrients suggested that selenium is involved in a biogeochemical cycle which sees its removal at the surface, incorporation by organisms with or without reduction, and regeneration in deeper waters. These conclusions were supported in a later study of Pacific Ocean waters off San Diego.[23] The concentrations of Se(IV) and assigned Se(VI) were found to range from 4 to 63 and 40 to 120 ng/l, respectively. Box and advection-diffusion model calculations suggested that the coexistence of both oxidation states could result from either a downward transport of incorporated Se(IV) formed during biogeochemical uptake at the surface or an unique redox cycle involving surface oxidation of Se(IV) and reduction of Se(VI) in deeper waters.

On the analytical side, Uchida et al.[24] devised a speciation scheme that would differentiate Se(−II,0), Se(IV), and Se(VI). In analyses of Japanese coastal waters, they found <2 to 6 ng/l of Se(−II,0) <2 to 31 ng/l of Se(IV), and 18 to 51 ng/l of Se(VI), with a ratio of Se(−II,0) to total selenium ranging from <0.04 to 0.11.

Uptake and incorporation of selenium by marine organisms were observed in a coastal ecosystem.[25] Wrench and Measures reported the depletion of Se(IV) as well as the nutrients, silicate, phosphate, and nitrate during a phytoplankton bloom, and their regeneration afterwards. The drop of Se(VI) concentration during the rapid growth period was small. As well, the particulate selenium concentration was seen to vary with chlorophyll counts and concentrations of particulate organic carbon and nitrogen during the bloom-decay cycle.

A different type of selenium spatial distribution was observed in the reducing waters of the Saanich Inlet, Vancouver Island.[26] The concentrations of Se(IV) and (VI) were found to decrease with decreasing oxygen concentration. Very little (ca. 0.8 ng/l) Se(IV) and (VI) were detected in the anoxic water, and contrary to thermodynamic predictions, very little hydrogen selenide (ca. 0.8 ng/l) was found. The predominant selenium species in suboxic and anoxic waters was apparently organic selenide.

Based on the results of a global oceanic survey, Measures et al.[27] concluded that the Atlantic, Pacific, and Indian Oceans all exhibit similar selenium spatial distributions which resembled those of phosphate and silicate. In the Pacific and Indian Oceans, Se(IV) rose from a surface concentration of 4 ng/l to a concentration of 62 ng/l in deeper waters, whereas Se(VI) started with a concentration of 40 ng/l at the surface and increased to 110 ng/l in deeper parts. Atlantic Ocean concentrations were found to be 30 to 40% lower. The deep water distribution pattern has been confirmed by Cutter and Bruland.[28]

Thus, it appears that selenium concentrations in unpolluted open ocean waters are well below the 1-μg/l level. The higher values measured in the early studies were probably results of sample contamination during collection, laboratory handling, and/or analytical errors. Table 1 lists the selenium concentrations reported in the last 2 decades. Care should be exercised when comparing or using data, as sampling depths are not always reported and concentrations at a given site may exhibit temporal variations.

III. DISSOLVED ORGANIC SELENIUM

The presence of organic selenium* in seawater was first reported in 1978.[29] Sugimura et al. separated an organic selenium fraction (27 to 45% of the total) from the surface waters of the northeastern Pacific. The nature of this organic fraction is unknown. Cutter[26] reported the presence of organic selenide as the predominant selenium species in the suboxic and anoxic water of the Saanich Inlet. Takayanagi and Wong[30] studied waters of Chesapeake Bay and the Atlantic Ocean off Cape Henry. About 60% of the total selenium found was organically bound. Further, Suzuki et al.[31] observed a 23% organic selenium fraction in western Pacific surface waters corresponding to a concentration of 0.02 μg of Se per liter. Cutter and Bruland[28] studied the distribution of selenium in the northeastern tropical Pacific. Significant concentrations of organic selenide were found in a well-characterized site. Two distribution maxima, one at 45 to 60 m and the other at 350 m, were noted, and these

* A term used loosely here to include all organically bound selenium species of different oxidation states.

Table 1
SELENIUM IN SEAWATER

Date	Location	Selenium concentration in different oxidation states (µg Se/l)[a]					Ref.
		Total	−II, 0	IV	VI	Organic	
1965	Long Island Sound	0.10—0.12					15
	Caribbean	0.11					15
	NW Atlantic	0.096					15
	NE Atlantic	0.088					15
	SW Atlantic	0.075					15
	E Pacific	0.087					15
	Antarctic	0.052					15
	English Channel			0.4—0.5			16
1973	Off Shibukawa, Japan			0.04—0.08			17
	Off Misaki, Japan			0.01—0.07			34
1976	N Pacific, Surface	0.06—0.12		0.03—0.10	0.03—0.09		10
	N Pacific, 4000 m	0.20		0.07	0.12		10
1977	North Sea	0.13		0.10	0.03		19
	NW Pacific	0.06—0.12		0.04—0.06	0.02—0.06		35
	North Sea	0.045 ± 0.003					36
	Offshore, Japan			0.020—0.055	0.036—0.28		37
1978	Seashore, Japan	0.033—0.047		0.012—0.032			20
	Santa Catalina Basin, CA						
	Surface			<0.005			21
	120 m			0.021			21
	250 m			0.033			21
	960 m			0.051			21
	1310 m			0.070			21
	Near shore, Oahu, HI			0.40 ± 0.12			38
1979	North Sea	0.07 ± 0.01					39
	Wadden Sea	0.24 ± 0.02					39
	Seashore, Shibukawa, Japan		0.023				40
	North Sea	0.074—0.078					41

Table 1 (continued)
SELENIUM IN SEAWATER

Date	Location	Selenium concentration in different oxidation states (μg Se/l)[a] Total	−II, 0	IV	VI	Organic	Ref.
1979	W Pacific, surface water	0.036—0.17		0.007—0.061	0.006—0.094	0.008—0.031	31
	Example of vertical profile						
	0 m			0.038	0.032	0.014	31
	92 m			0.040	0.030	0.010	31
	235 m			0.048	0.030	0.004	31
	474 m			0.041	0.023	0.017	31
	710 m			0.073	0.048	0.007	31
	950 m			0.060	0.049	0.003	31
	1430 m			0.060	0.060	0	31
	1915 m			0.062	0.058	0	31
	2885 m			0.060	0.063	0	31
	3855 m			0.068	0.081	0.008	31
	4830 m			0.070	0.071	0.004	31
1980	N Atlantic	0.025—0.125		≤0.002—0.040			22
	Example of vertical profile						
	1 m	0.048		≤0.002	0.046		22
	60 m	0.028		≤0.002	0.026		22
	130 m	0.028		≤0.002	0.026		22
	468 m	0.050		0.006	0.044		22
	725 m	0.100		0.017	0.083		22
	1960 m	0.078		0.026	0.052		22
	2953 m	0.103		0.031	0.072		22
	3908 m	0.122		0.035	0.087		22
	4868 m	0.138		0.035	0.103		22
	Pacific, off San Diego, CA	0.045—0.18		0.004—0.063	0.040—0.12		23
	50 m	0.045		0.004	0.041		23
	130 m	0.054		0.005	0.049		23
	169 m	0.074		0.010	0.064		23
	199 m	0.079		0.013	0.066		23
	260 m	0.089		0.015	0.074		23

	326 m	0.104		0.020	0.084		23
	503 m	0.124		0.036	0.088		23
	703 m	0.143		0.046	0.097		23
	1001 m	0.152		0.053	0.099		23
	1400 m	0.164		0.060	0.104		23
	2062 m	0.178		0.064	0.114		23
	2994 m	0.184		0.065	0.119		23
	4089 m	0.175		0.062	0.113		23
	Seashore, Japan	0.048—0.067	<0.002—0.006	<0.002—0.031	0.018—0.051		24
	North Sea, surface	0.11—0.27		0.07—0.14			42
1982	Saanich Inlet, Vancouver Island	0.050—0.11	0.001	0.001—0.025	0.001—0.060	0.001—0.1	26
	10 m	0.061		0.001	0.060		26
	40 m	0.068		0.010	0.058	0.001	26
	80 m	0.080		0.025	0.040	0.015	26
	100 m	0.080		0.020	0.030	0.030	26
	120 m	0.11		0.015	0.025	0.065	26
	140 m	0.091		0.006	0.020	0.065	26
	160 m	0.075	0.001	0.003	0.002	0.070	26
	180 m	0.079	0.001	0.003	0.001	0.075	26
	196 m	0.098	0.001	0.001	0.001	0.095	26
1983	Pacific and Indian Oceans	0.044—0.17		0.004—0.062	0.040—0.11		27
1984	E tropical Pacific						
	15 m	0.083		0.006	0.011	0.065	28
	60 m	0.141		0.002	0.059	0.081	28
	125 m	0.144		0.024	0.101	0.019	28
	350 m	0.168		0.027	0.100	0.041	28
	500 m	0.156		0.037	0.111	0.008	28
	700 m	0.156		0.054	0.116	0.014	28
	1100 m	0.176		0.067	0.102	0.007	28
	1500 m	0.182		0.077	0.104	0.001	28
	2000 m	0.185		0.079	0.106		28
	3000 m	0.202		0.089	0.114		28
	3250 m	0.208		0.080	0.129		28

[a] Concentrations are reported as ranges, means, or mean ± standard deviation.

corresponded to the maximum of biological activity and the location of a suboxic zone. The concentrations of organic selenium measured in these studies are also listed in Table 1.

Compared to its inorganic counterpart, organic selenium is much less well characterized. It is an operationally defined fraction, the exact identity of which is unknown. Cutter[26] has produced some evidence that the organic selenide fraction in anoxic waters may be associated with dissolved amino acids. Thus, the implication is that dissolved organic selenium may exist in the form of selenoamino acid-containing peptides in certain environments.[26,28] Takayanagi and Wong[30] found that their organic fraction had a nominal molecular weight less than 10,000. It was also linearly correlated with the concentration of dissolved organic carbon.

Various procedures have been used to measure the organic fraction: adsorption on macroreticulate polystyrene-divinylbenzene copolymer (XAD®-2),[29,31] oxidation by potassium persulphate,[26,28] acid hydrolysis plus ligand-exchange chromatography,[26] and oxidation by hydrogen peroxide plus ultraviolet irradiation.[30] It is interesting to note that Measures et al.[9,22,23,25,27,32] used a photochemical technique not unlike this last procedure to reduce inorganic Se(VI) to Se(IV). The possibility thus exists that some of their selenate data may have had contributions from organic selenium.[28,30,33] Having said that, it should be stressed that both Cutter et al. (who measured organic selenium) and Measures et al. (who did not) obtained rather similar concentrations for selenite and selenate, at least in deep waters, at particular sites.[23,28,33] Evidently, more work is required with regard to this analytical aspect of selenium chemistry, which then has implications to the biogeochemical cycle of selenium.

IV. THE MARINE CYCLE

The general consensus at present is that the vertical distributions of selenite and selenate resemble those of micronutrients.[28,33] Surface depletion of the selenium species is due to removal and incorporation of selenium into particulate matter. Like other micronutrients, the organically associated selenium is transported downwards via sinking detritus. Inorganic selenite and selenate are regenerated in deep waters and delivered eventually to the surface by vertical mixing — thus completing a selenium cycle. Details of the cycle are still debatable; for example, whether Se(IV) is associated with silicate,[22,23,28] whether Se(VI) is associated with phosphate,[22,23,28] whether bioincorporation of selenium involves selenium reduction,[23,28] whether dissolved organic selenium exists in seawater,[22,26,28] and whether regeneration of inorganic selenium entails oxidation.[22,23,28] Figure 2 illustrates one interpretation of the selenium marine cycle.[28]

V. BIOLOGICAL TISSUES

Selenium is a ubiquitous element. As discussed previously, its concentration in seawater is well below 1 μg/l. Marine sediments usually contain 0.1 to 2 mg/kg, while selenium concentrations in biological tissues can range from 0.1 to 1000 mg/kg, depending on the type of material (Table 2).

Marine flora and fauna accumulate selenium via the food chain and from the water and sediments with which they come into contact.[43] In general, selenium concentrations in marine fish and shellfish are higher than those in terrestrial animals. For fish, there is strong evidence that the major selenium uptake route is not accumulation from water,[44] but rather via the foodchain.[5,45] As far as direct uptake from water is concerned, mussels (*Mytilus galloprovincialis*) appear to accumulate Se(IV) to a greater degree than Se(VI).[45] Raising the water temperature increases the rate of selenium uptake as well as excretion. This temperature dependence is not apparent in benthic shrimp (*Lysmata seticaudata*) which molt faster with increasing temperature.[45] Hence, for custaceans, surface adsorption appears to play a sig-

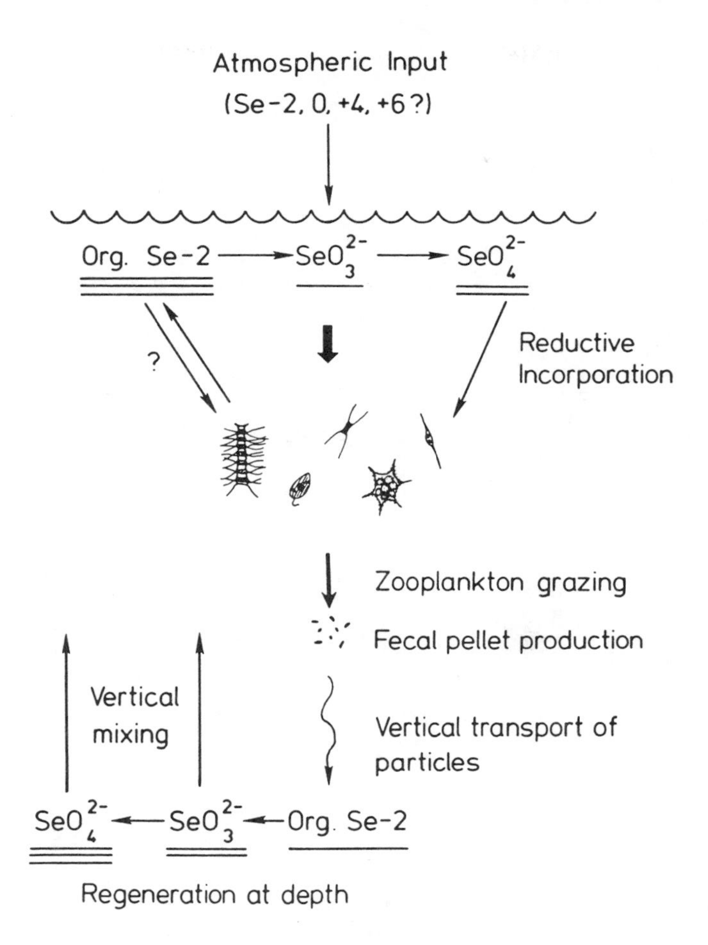

FIGURE 2. The marine cycle. (From Cutter, G. A. and Bruland, K. W., *Limnol. Oceanogr.*, 29, 1179, 1984. With permission.)

nificant role in the direct uptake of selenium from water. The species of selenium accumulated has no effect on the subsequent excretion rate of the element, pointing to the possibility of some kind of biochemical transformation of selenium within the animal. Of more importance, the excretion of selenium assimilated via the food chain is slower than that accumulated from the water because a significant fraction of selenium accumulated from the latter route is lost with molting.

Aside from water and food, particulate matter and sediments may also serve as selenium sources. This is especially true for bivalves, in particular oysters (*Crassostrea virginica*), which are efficient retainers of particulate matter.[46] No correlation between selenium concentration in the livers of a bottom dweller, dover sole (*Microstomus pacificus*), and the sediments of its habitat, however, has been observed.[5,47]

The distribution of selenium in a given tissue from a single specimen may not be uniform. Significant concentration variation in the muscle and liver of a single specimen of skipjack tuna has been reported.[48] Evidently, this stresses the importance of representative sampling and subsequent sample homogenization. More striking, however, is the variation in the selenium level among different tissues from a given specimen. As an example, the reported selenium contents in a single 129-lb dolphin are the following: muscle, 1.9; spleen, 10; kidney, 15; and liver, 29 mg/kg.[48] The organs of detoxification, such as the liver and kidney of marine fish and mammals and the hepatopancreas of shellfish, usually contain higher

Table 2
SELENIUM IN MARINE BIOLOGICAL TISSUES

Species	Location	Tissue[a]	Selenium conc. (mg/kg)[b]	Weight basis[c]	Ref.
Finfish, 159 species	Coastal waters, U.S.	Muscle	0.1—2.0	NS	75
Finfish, 79 species	Coastal waters, U.S.	Liver	0.6—30	NS	75
Finfish, 17 species	Coastal waters, U.S.	Whole	0.3—2.0	NS	75
Mollusca, 18 species	Coastal waters, U.S.	Soft parts	0.1—0.9	NS	75
Crustacea, 16 species	Coastal waters, U.S.	Soft parts	0.2—2.0	NS	75
Dolphin, 129-lb specimen		Muscle	1.9	NS	48
Dolphin, 129-lb specimen		Spleen	10	NS	48
Dolphin, 129-lb specimen		Kidney	15	NS	48
Dolphin, 129-lb specimen		Liver	29	NS	48
Dolphin, 3 small specimens		Liver	4.6 ± 1.2	NS	48
Dolphin, 2 large specimens		Liver	42 ± 19	NS	48
Killer whale, "Shamu"	San Diego, *Sea World*	Muscle	2.3 ± 0.2	NS	48
Killer whale, "Shamu"	San Diego, *Sea World*	Liver	82 ± 3	NS	48
Albacore tuna, a few specimens		Muscle	1.4 ± 0.3	NS	48
Albacore tuna, a few specimens		Liver	41 ± 14	NS	48
Plant, *Pelvetia canaliculata*			0.84	D	16
Mammals, 22 specimens incl. seals, dolphins, and porpoises		Liver	0.60—134	W	54
Guillemot, *Uria aalge*		Liver	2.4—4.6	W	54
Guillemot, *Uria aalge*		Brain	0.46—1.1	W	54
Razorbill, *Alca torda*		Liver	3.6	W	54
Razorbill, *Alca torda*		Brain	0.69	W	54
Eelgrass, *Zostera marina*	Prince William Sound, AK	Leaves	2.0	W	76
Mullusca, *Saxidomus gigantea*		Siphon	0.33—0.43	W	76
Mullusca, *Prototheca staminea*		Whole	0.2—0.4	W	76
Fish, *Sebastes roseceus*		Eye	22—25	W	76
Fish, *Sebastes roseceus*		Skin	100—110	W	76
Fish, *Sebastes pancispinus*		Eye	26	W	76
Fish, *Sebastes constellatus*		Eye	0.4—0.8	W	76
Fish, *Pleuronectes platessa*		Eye	18	W	76
Fish, *Pleuronectes platessa*		Skin	5.4	W	76

Black marlin, *Makaira indica Cuvier*	Off Cairns, N Queensland, Australia	Muscle	0.4—4.3	W	55
Black marlin, *Makaira indica Cuvier*	Off Cairns, N Queensland, Australia	Liver	1.4—13.5	W	55
Plankton	Tsukumo Bay, Japan	Whole	0.47	D	77
Dover sole, *Microstomus pacificus*	Point Vicente, Palos Verdes, CA	Muscle	0.27 ± 0.05	W	78
Dover sole, *Microstomus pacificus*	Santa Barbara, CA	Muscle	0.38 ± 0.006	W	78
Crab, *Cancer anthonyi*	Point Vicente, Palos Verdes, CA	Muscle	5.1 ± 1.0	W	78
Crab, *Cancer anthonyi*	Santa Barbara, CA	Muscle	16 ± 3	W	78
Halibut			2.7—2.9	NS	79
Swordfish			2.6—3.4	NS	79
Haddock			1.6	D	80
Perch			2.2—2.6	D	80
Flounder			0.9—1.2	D	80
Cod			1.2—1.4	D	80
Clam, *Rangia cuneata*	San Antonio Bay, TX		0.54	D	81
Windowpand flounder, *Scophthalmus aquosus*	Ocean dump sites off New York City, New Haven, CT, and Delaware Bay	Muscle	≤0.3—0.75	W	52
Windowpand flounder, *Scophthalmus aquosus*		Liver	1.1—5.1	W	52
Rock crab, *Cancer irroratus*		Muscle	1.0—5.5	W	52
Rock crab, *Cancer irroratus*		Hepatopancreas	1.3—3.7	W	52
Rock crab, *Cancer irroratus*		Gills	0.6—0.9	W	52
Fluke, *Paralichthys dentatus*		Muscle	0.7	W	52
Ling, *Urophycis tenuis*		Muscle	0.3	W	52
Lind, *Urophycis tenuis*		Liver	1.7	W	52
Red hake, *Urophycis chuss*		Muscle	0.42	W	52
Red hake, *Urophycis chuss*		Liver	1.9	W	52
Spiny dogfish, *Squalus acanthias*		Muscle	0.94	W	52
Yellowtail flounder, *Limanda ferruginea*		Muscle	0.50—0.70	W	52
Winter flounder, *Pseudopleuronectes americanus*		Muscle	0.40	W	52
Winter flounder, *Pseudopleuronectes americanus*		Liver	2.8	W	52
Surf clam, *Spisula solidissima*		Muscle	0.7—0.8	W	52
Surf clam, *Spisula solidissima*		Digestiv.	1.1	W	52
Lobster, *Homarus americanus*		Muscle	2.0	W	52
Lobster, *Homarus americanus*		Digestiv.	2.9	W	52
Channelled whelk, *Busycon canaliculatum*		Muscle	<0.2—<0.4	W	52
Zooplankton, *Meganyctiphanes norvegica*	Monaco	Whole	4.4	W	82
Zooplankton, *Meganyctiphanes norvegica*	Monaco	Moults	1.9	W	82
Zooplankton, *Meganyctiphanes norvegica*	Monaco	Fecal Pellets	6.6	W	82
Microplankton	Monaco	Whole	2.7	W	82
Halibut		Muscle	0.70 ± 0.03	W	83

Table 2 (continued)
SELENIUM IN MARINE BIOLOGICAL TISSUES

Species	Location	Tissue[a]	Selenium conc. (mg/kg)[b]	Weight basis[c]	Ref.
Swordfish		Muscle	0.73—0.78	W	83
Oyster		Muscle	0.30—0.37	W	83
Cod		Muscle	0.28 ± 0.02	W	83
Yellowfin bream, *Acanthopagrus australis*	Off New South Wales, Australia	Muscle	0.1—0.8	NS	84
Dusky flathead, *Platycephalus fuscus*	Off New South Wales, Australia	Muscle	0.2	NS	84
Sea mullet, *Mugil cephalus*	Off New South Wales, Australia	Muscle	0.1—0.3	NS	84
Snapper, *Chrysophris auratus*	Off New South Wales, Australia	Muscle	0.1—0.6	NS	84
Tailor, *Pomatomus saltatrix*	Off New South Wales, Australia	Muscle	0.1—0.6	NS	84
Mulloway, *Sciaena antarctica*	Off New South Wales, Australia	Muscle	0.1—0.4	NS	84
Yellowtail kingfish, *Seriola grandis*	Off New South Wales, Australia	Muscle	0.3	NS	84
Australian salmon, *Arripis tuna*	Off New South Wales, Australia	Muscle	0.3—0.5	NS	84
Yellowfin tuna, *Thunnus albacares*	Off New South Wales, Australia	Muscle	0.4—0.7	NS	84
Lobster, *Homarus americanus*	Off Halifax, NS, Canada	Digestiv.	0.58—1.9	W	50
Lobster, *Homarus americanus*	North Lake, PEI, Canada	Digestiv.	0.82—1.8	W	50
Lobster, *Homarus americanus*	Off Shediac, NB, Canada	Digestiv.	0.81—2.7	W	50
Lobster, *Homarus americanus*	Victoria Beach, NS, Canada	Digestiv.	0.78—1.9	W	50
Lobster, *Homarus americanus*	Off Petit Rocher, NB, Canada	Digestiv.	0.69—1.4	W	50
Lobster, *Homarus americanus*	Off Petit Rocher, NB, Canada	Claw muscle	0.34—0.56	W	50
Rock crab, *Cancer irroratus*	Victoria Beach, NS, Canada	Digestiv.	0.58—2.1	W	50
Albacore tuna			0.58	NS	85
Mako shark			0.24	NS	85
Northern pike			0.44	NS	85
Oyster	Off Séte, France	Soft parts	2.3 ± 0.1	D	86
Oyster			4.8—5.1	D	87
Herring meal			1.2—1.4	D	87
Shrimp			1.9	D	87
Crab			2.9	D	87
Cod liver product			3.7	D	87
Mangrove, *Rhizophora stylosa*	Queensland, Australia	Leaves	0.063	D	88
Oyster, *Saccostrea cucullata*	Queensland, Australia	Muscle	2.6	D	88
Yellowtail kingfish, *Seriola lalandi*	Queensland, Australia	Muscle	1.5	D	88

Yellowtail kingfish, *Seriola lalandi*	Queensland, Australia	Eye	3.2	D	88
Banana prawn, *Penaeus merguiensis*	Queensland, Australia	Muscle	2.2	D	88
Panda prawn, *Penaeus monodon*	Queensland, Australia	Muscle	1.9	D	88
Swordfish, *Xiphias gladius*	NW Atlantic	Muscle	0.29—1.3	NS	57
Ringed seal, *Phoca hispida*	Baltic Sea	Muscle	0.44—0.92	W	89
Ringed seal, *Phoca hispida*	Baltic Sea	Liver	6.1—110	W	89
Ringed seal, *Phoca hispida*	Baltic Sea	Kidney	2.5—3.3	W	89
Oyster			1.8—2.2	NS	90
Swordfish			2.4—3.3	NS	90
Flounder			1.4—1.6	NS	90
Dory, *Cyttoidopus* spp.	Off New Zealand	Muscle	0.31	NS	91
English hake, *Merluccius australis*	Off New Zealand	Muscle	0.19	NS	91
Kingklip, *Genypterus blacodes*	Off New Zealand	Muscle	0.08	NS	91
Hoki, *Macruronus novaezelandidae*	Off New Zealand	Muscle	0.19	NS	91
Silver fish, *Seriollela maculata*	Off New Zealand	Muscle	0.24	NS	91
Tarakihi, *Cheilodactylus macropterus*	Off New Zealand	Muscle	0.38	NS	91
Sea perch, *Helicolenus percoides*	Off New Zealand	Muscle	0.26	NS	91
Monkfish, *Kathetostoma giganteum*	Off New Zealand	Muscle	0.19	NS	91
Southern kingfish, *Rexea solandri*	Off New Zealand	Muscle	0.23	NS	91
Barracuda, *Thyrsites atum*	Off New Zealand	Muscle	0.26	NS	91
Jack mackeral, *Trachurus novaezelandiae*	Off New Zealand	Muscle	0.38	NS	91
Bream, *Hyperoglyphe antarctica*	Off New Zealand	Muscle	0.16	NS	91
Pampanito, *Stromateus brasiliensis*	Patagonia	Muscle	0.46	NS	91
Taragisu, *Mugiloides somnambura*	Patagonia	Muscle	0.34	NS	91
Patagonian silver trevalla, *Seriolella porosa*	Patagonia	Muscle	0.43	NS	91
Lenguado, *Paralichthys patagonicus*	Patagonia	Muscle	0.33	NS	91
Oyster, *Crassostrea virginica*	Texas Gulf	Soft parts	0.14	W	92
Horse mackerel	Off Shikoku Island, Japan	Muscle	0.75—1.7	D	93
Sillago	Off Shikoku Island, Japan	Muscle	1.1—1.6	D	93
Sardine	Off Shikoku Island, Japan	Muscle	1.6—1.8	D	93
Shrimp	Off Shikoku Island, Japan	Muscle	2.5	D	93
Seaweed, *Undaria pinnatifida*	Off Shikoku Island, Japan		0.39	D	93
Seaweed, purple laver	Off Shikoku Island, Japan		0.16	D	93
Blue marlin, *Makaira migricans Lacépède*	Off Hawaiian Islands	Muscle	0.63—5.3	NS	53
Blue marlin, *Makaira migricans Lacépède*	Off Hawaiian Islands	Liver	2.5—61	NS	53
Blue marlin, *Makaira migricans Lacépède*	Off Hawaiian Islands	Kidney	2.6—56	NS	53
Blue marlin, *Makaira migricans Lacépède*	Off Hawaiian Islands	Spleen	0.63—24	NS	53
Blue marlin, *Makaira migricans Lacépède*	Off Hawaiian Islands	Stomach	1.4—4.0	NS	53

Table 2 (continued)
SELENIUM IN MARINE BIOLOGICAL TISSUES

Species	Location	Tissue[a]	Selenium conc. (mg/kg)[b]	Weight basis[c]	Ref.
Blue marlin, *Makaira migricans Lacépède*	Off Hawaiian Islands	Pyloric cecum	2.3—10	NS	53
Blue marlin, *Makaira migricans Lacépède*	Off Hawaiian Islands	Gill	0.71—2.2	NS	53
Blue marlin, *Makaira migricans Lacépède*	Off Hawaiian Islands	Gonad	1.2—3.8	NS	53
Blue marlin, *Makaira migricans Lacépède*	Off Hawaiian Islands	Blood	0.72—2.3	NS	53
School shark, *Galeorhinus australis*	Off SE Australia	Muscle	0.2—0.8	W	94
Gummy shark, *Mustelus antarcticus*	Off SE Australia	Muscle	0.2—0.5	W	94
Ling, *Molva molva*	Northern North Sea	Muscle	0.31—0.36	W	51
Ling, *Molva molva*	Northern North Sea	Liver	2.8	W	51
Redfish, *Sebastes* spp.	South of Greenland	Muscle	0.26—0.53	W	51
Redfish, *Sebastes* spp.	South of Greenland	Liver	1.8	W	51
Plaice, *Pleuronectus platessa*	Central North Sea	Muscle	0.26—1.0	W	51
Plaice, *Pleuronectus platessa*	Central North Sea	Liver	1.8	W	51
Sole, *Solea solea*	Central North Sea	Muscle	0.15—0.29	W	51
Sole, *Solea solea*	Central North Sea	Liver	2.0—2.1	W	51
Mackerel, *Scomber scombrus*	Dutch coast	Muscle	0.22—0.44	W	51
Mackerel, *Scomber scombrus*	Dutch coast	Liver	2.8—4.6	W	51
Blue Whiting, *Gadus poutasson*	South of Fär Öer	Muscle	0.25—0.43	W	51
Blue Whiting, *Gadus poutasson*	South of Fär Öer	Liver	0.45—1.1	W	51
Cod, *Gradus morhua*	North Sea	Muscle	0.17—0.43	W	51
Cod, *Gradus morhua*	North Sea	Liver	0.86—2.0	W	51
Shrimp, *Crangon crangon*	Texel	Soft parts	0.28—0.36	W	51
Mussel, *Mytilus edulis*	Hammen	Soft parts	0.46—0.51	W	51
Mackerel					
Scomber japonicus colias	Aegean Sea	Otoliths	0.07—0.28	D	95
Epinephelus fario	Off Takarajima Island, Japan	Muscle	0.40 ± 0.36	NS	96
Cirrhitus pinnulatus	Off Takarajima Island, Japan	Muscle	0.26 ± 0.09	NS	96
Hapalogenys nigripinnis	Off Takarajima Island, Japan	Muscle	0.27 ± 0.17	NS	96
Thalassoma spp.	Off Takarajima Island, Japan	Muscle	0.27 ± 0.06	NS	96
Xanthichthys spp.	Off Takarajima Island, Japan	Muscle	0.40 ± 0.18	NS	96
Holocentrus spp.	Off Takarajima Island, Japan	Muscle	0.81	NS	96
Girella punctata	Off Takarajima Island, Japan	Muscle	0.17	NS	96

Parathunnus sibi	Off Amami Island, Japan	Muscle	0.62	NS	96
Coryphaena hippurus	Off Amami Island, Japan	Muscle	0.95	NS	96
Acanthocybium solandri	Off Amami Island, Japan	Muscle	0.10	NS	96
Scombrops spp.	Off Amami Island, Japan	Muscle	0.23	NS	96
Scaridae	Off Amami Island, Japan	Muscle	0.17	NS	96
Panulirus lonipes	Off Takarajima Island, Japan	Muscle	0.69 ± 0.39	D	96
Purpura armigera	Off Takarajima Island, Japan	Muscle	0.70 ± 0.12	D	96
Thunnus thynnus	Ligurian-Sardinian Sea		0.66—1.3	W	97
Mullus barbatus	Tyrrenian Sea		0.32—1.2	W	97
Nephrops norvegicus	Tyrrenian Sea		0.70—2.6	W	97
Mackerel			0.093	W	98
Flatfish			0.14	W	98
Mussel					
Mytilus galloprovincialis	Kastella Bay, Adriatic Sea	Soft parts	0.19—1.6	W	99
Mullus barbatus	Kissamos Gulf, Greece		0.19	W	100
Mullus surmuletus	Kissamos Gulf, Greece		0.18	W	100
Pagellus acarne	Kissamos Gulf, Greece		0.45	W	100
Boops boops	Kissamos Gulf, Greece		0.43	W	100
Trachurus mediterraneus	Kissamos Gulf, Greece		0.37	W	100
Epinephelus guaza	Kissamos Gulf, Greece		0.62	W	100
Mullus barbatus	Gera Gulf, Greece		0.35	W	100
Pagellus acarne	Gera Gulf, Greece		0.34	W	100
Diplodus annularis	Gera Gulf, Greece		0.53	W	100
Diplodus sargus	Gera Gulf, Greece		0.47	W	100
Serranus scriba	Gera Gulf, Greece		0.17	W	100
Mugil labeo	Gera Gulf, Greece		0.12	W	100
Maena smaris	Gera Gulf, Greece		0.57	W	100
Pagellus acarne	Antikyra Gulf, Greece		0.77	W	100
Boops boops	Antikyra Gulf, Greece		1.0	W	100
Sarranus cabrilla	Antikyra Gulf, Greece		0.55	W	100
Scorpaena scrofa	Antikyra Gulf, Greece		0.75	W	100
Conger conger	Antikyra Gulf, Greece		0.88	W	100
Mullus barbatus	Saronikos Gulf, Greece		0.47	W	100
Oyster tissue, NBS SRM 1566		Soft parts	2.1 ± 0.5	D	101
Clam, *Mya arenia*	Nova Scotia, Canada	Hepatopancreas	0.82	W	49
Lobster, *Homarus americanus*	Charlo, NB, Canada	Hepatopancreas	0.61	W	49
Lobster, *Homarus americanus*	Charlo, NB, Canada	Muscle	0.18	W	49
Lobster, *Homarus americanus*	Dalhousie, NB, Canada	Hepatopancreas	0.91	W	49

Table 2 (continued)
SELENIUM IN MARINE BIOLOGICAL TISSUES

Species	Location	Tissue[a]	Selenium conc. (mg/kg)[b]	Weight basis[c]	Ref.
Lobster, *Homarus americanus*	Dalhousie, NB, Canada	Muscle	0.39	W	49
Pteropod, *Clione limacina*	Arctic waters, Jones Sound	Whole	0.53	W	49
Sculpin, *Myxocephalus scorpius*	Arctic waters, Jones Sound	Liver	0.56	W	49
Albacore tuna, NBS RM 50	Off San Diego, CA	Muscle	3.6 ± 0.4	D	102
Tuna, wahoo		Muscle	0.33[d]	W	58
Tuna, wahoo		Muscle	0.10[e]	W	58
Tuna, yellowfin		Muscle	0.53[d]	W	58
Tuna, yellowfin		Muscle	0.098[e]	W	58
Bluefish		Muscle	0.42[d]	W	58
Bluefish		Muscle	0.082[e]	W	58
Dolphin		Muscle	0.16[d]	W	58
Dolphin		Muscle	0.058[e]	W	58
Flounder		Muscle	0.14[d]	W	58
Flounder		Muscle	0.028[e]	W	58
Red snapper		Muscle	0.32[d]	W	58
Red snapper		Muscle	0.067[e]	W	58
Mako shark		Muscle	0.24[d]	W	58
Mako shark		Muscle	0.12[e]	W	58
Swordfish		Muscle	0.42[d]	W	58
Swordfish		Muscle	0.068[e]	W	58
Macroalga, *Lobospira bicuspida*			0.14	D	103
Fish, *Sillaginodes punctatulus*			0.38	D	103
Scallop, *Pecten alba*			0.76	D	103
Bluefish, *Pomatomus saltatrix*	Off Exmore, VA	Muscle	0.30—0.57	D	104
Oyster, *Crassostrea virginica*	St. Louis Bay, MS	Soft parts	0.28	W	46
Clam, *Rangia cuneata*	St. Louis Bay, MS	Soft parts	0.49	W	46
Cod			1.1—1.4	D	105
Haddock			1.1—1.6	D	105
Perch			2.1—2.6	D	105
Flounder			0.8—1.2	D	105
Menhaden fish	Atlantic Ocean		0.03	NS	106

Mackerel		Muscle	0.23[d]	W	59
Mackerel		Muscle	0.11[e]	W	59
Salmon, pink		Muscle	0.30[d]	W	59
Salmon, pink		Muscle	0.15[e]	W	59
Mako shark		Muscle	0.16[d]	W	59
Mako shark		Muscle	0.072[e]	W	59
Swordfish		Muscle	0.15[d]	W	59
Swordfish		Muscle	0.052[e]	W	59
Tuna	Off N Peru	Muscle	0.15[d]	W	59
Tuna	Off N Peru	Muscle	0.54[d]	W	59
Tuna	Off N Peru	Muscle	0.085[e]	W	59
Tuna	Off N Peru	Muscle	0.11[e]	W	59
Octopus		Muscle	0.18[d]	W	59
Octopus		Muscle	0.093[e]	W	59
Squid		Muscle	0.37[d]	W	59
Squid		Muscle	0.58[e]	W	59
Lobster, *Homarus americanus*, NRC SRM TORT-1	Off Prince Edward Island	Hepatopancreas[f]	6.88	D	107
Algae					
Laminaria saccharina	Kimmeridge Bay, Dorset, U.K.		0.41	D	108
Rhodomela confervoides	Kimmeridge Bay, Dorset, U.K.		0.26	D	108
Codium tomentosum	Kimmeridge Bay, Dorset, U.K.		0.13	D	108
Mollusca					
Merceneria merceneria	Southampton water		0.73—0.87	W	108
Mytilus edulis	Solent		0.63—0.87	W	108
Patella vulgata	Solent		0.22—0.38	W	108
Fish					
Blennius pholis	English Channel	Muscle	0.36	W	108
Blennius pholis	English Channel	Digestiv.	0.92	W	108
Pleuronectes platessa	English Channel	Muscle	0.42	W	108
Pleuronectes platessa	English Channel	Digestiv.	0.74	W	108
Lithothamnium japonicum	Japan		0.35—1.7	NS	109
Corallina pilulifera	Japan		0.28	NS	109
Hemirhamphus australis	St. Vincent's Gulf, S Australia	Muscle	0.56—0.81	D	110
Hemirhamphus australis	St. Vincent's Gulf, S Australia	Digestiv.	1.3—2.0	D	110
Sillaginodes punctatus	St. Vincent's Gulf, S Australia	Muscle	1.1—1.7	D	110
Sillaginodes punctatus	St. Vincent's Gulf, S Australia	Digestiv.	2.0—2.6	D	110
Arripis georgianus	St. Vincent's Gulf, S Australia	Muscle	0.72—0.98	D	110

Table 2 (continued)
SELENIUM IN MARINE BIOLOGICAL TISSUES

Species	Location	Tissue[a]	Selenium conc. (mg/kg)[b]	Weight basis[c]	Ref.
Arripis georgianus	St. Vincent's Gulf, S Australia	Digestiv.	1.1—1.8	D	110
Callogobius mucosus	St. Vincent's Gulf, S Australia	Muscle	0.40—0.63	D	110
Callogobius mucosus	St. Vincent's Gulf, S Australia	Digestiv.	0.79—1.2	D	110
Crustacea					
Penaeus latisulcatus	St. Vincent's Gulf, S Australia	Muscle	3.7—5.6	D	110
Jasus novae hollandiae	St. Vincent's Gulf, S Australia	Muscle	2.5—2.9	D	110
Jasus novae hollandiae	St. Vincent's Gulf, S Australia	Digestiv.	3.0—3.5	D	110
Crangon novae zelandiae	St. Vincent's Gulf, S Australia	Muscle	3.4—3.9	D	110
Helograpsus spp.	St. Vincent's Gulf, S Australia	Muscle	1.8—3.3	D	110
Helograpsus spp.	St. Vincent's Gulf, S Australia	Soft parts	1.6—2.7	D	110
Calogobins mucosus	St. Vincent's Gulf, S Australia	Muscle	3.0—3.6	D	110
Calogobins mucosus	St. Vincent's Gulf, S Australia	Soft parts	2.6—3.2	D	110
Mollusca					
Mytilus edulis planulatus	St. Vincent's Gulf, S Australia	Muscle	0.7—1.5	D	110
Mytilus edulis planulatus	St. Vincent's Gulf, S Australia	Digestiv.	1.1—2.3	D	110
Pecten alba	St. Vincent's Gulf, S Australia	Muscle	1.6—2.5	D	110
Pecten alba	St. Vincent's Gulf, S Australia	Digestiv.	1.4—2.7	D	110
Cephalopoda, *Sepioteuthis australis*	St. Vincent's Gulf, S Australia	Muscle	0.9—2.6	D	110
Macroalgae, Ecklonia radiata	St. Vincent's Gulf, S Australia	Whole	0.25 ± 0.01	W	111
Fish, *Hemir hamphus australis*	St. Vincent's Gulf, S Australia	Muscle	0.18 ± 0.001	W	111
Scallop, *Pecten alba*	St. Vincent's Gulf, S Australia	Muscle	0.77 ± 0.05	W	111
Prawn, *Penaeus latisulcatus*	St. Vincent's Gulf, S Australia	Muscle	1.1 ± 0.03	W	111
Fish, *Sillaginodes punctatus*	St. Vincent's Gulf, S Australia	Muscle	0.39 ± 0.01	W	111
Crayfish, *Jasus novae hollandiae*	St. Vincent's Gulf, S Australia	Muscle	0.66 ± 0.03	W	111
Fish, bolty			0.40 ± 0.07	NS	112
Tuna, *Thunnus alalunga*	Atlantic Ocean	Liver	18—61	D	113
Tuna, *Thunnus thynnus*	Mediterranean	Pancreas	5.3	D	113
Tuna, *Thunnus thynnus*	Mediterranean	Liver	20	D	113
Tuna, *Thunnus thynnus*	Mediterranean	Heart	12	D	113
Tuna, *Thunnus thynnus*	Mediterranean	Stomach	7.1	D	113
Tuna, *Thunnus thynnus*	Mediterranean	Intestine	8.8	D	113

Tuna, *Thunnus thynnus*	Mediterranean	White muscle	0.4—4.6	D	113
Tuna, *Thunnus obesus*	Indian Ocean	Liver	11—31	W	113
Tuna, *Thunnus obesus*	Indian Ocean	Spleen	13—44	W	113
Tuna, *Thunnus alalunga*	Indian Ocean	Liver	15—31	W	113
Tuna, *Thunnus alalunga*	Indian Ocean	Spleen	14—26	W	113
Tuna, *Thunnus albacares*	Indian Ocean	Liver	12—36	W	113
Tuna, *Thunnus albacares*	Indian Ocean	Spleen	21—43	W	113
Tuna, *Thunnus albacares*	Indian Ocean	Dark muscle	5.2	W	113
Tuna, *Thunnus albacres*	Indian Ocean	White muscle	0.54	W	113
Tetrapturus andax	Indian Ocean	Liver	5.0—18	W	113
Tetrapturus andax	Indian Ocean	Spleen	4.0—14	W	113
Dolphin, *Stenella coeruleoalba*	France	Liver	61—780	W	113
Dolphin, *Stenella coeruleoalba*	France	Kidney	20—80	W	113
Dolphin, *Stenella coeruleoalba*	France	Spleen	10—220	W	113
Shark					
Black tip, *Carcharhinus limbatus*	Off Charleston, SC	Liver	4.5 ± 1.4	W	114
Great hammerhead, *Sphyrna mokarran*	Off Charleston, SC	Liver	3.0 ± 1.3	W	114
Tiger, *Galeocerdo cuvieri*	Off Charleston, SC	Liver	0.70	W	114
Sandtiger, *Odotaspis taurus*	Off Charleston, SC	Liver	0.50	W	114
Sandbar, *Carcharhinus plumbeus*	Off Charleston, SC	Liver	7.9	W	114
Shortfin mako, *Isurus oxyrinchus*	Southeastern coast, U.S.	Liver	0.45—0.78	W	114
Blue, *Prionace glauca*	Southeastern coast, U.S.	Liver	0.55—0.96	W	114
Blue, *Prionace glauca*	Southeastern coast, U.S.	Muscle	0.12 ± 0.05	W	114
Yellowfin tuna, *Thunnus albacares*	Off Charleston, SC	Liver	0.48 ± 0.03	W	114
Yellowfin tuna, *Thunnus albacares*	Off North Carolina	Liver	9.7 ± 2.4	W	114
Yellowfin tuna, *Thunnus albacares*	Off North Carolina	Muscle	0.28 ± 0.06	W	114
Dogfish		Liver	2.5 ± 0.1	D	115
Dogfish		Muscle	1.0 ± 0.1	D	115
Copepod, IAEA MA-A-1			3.00 ± 0.20[g]	D	154
Copepod, IAEA MA-A-1			0.02 ± 0.01[h]	D	154
Copepod, IAEA MA-A-1			0.02 ± 0.01[e]	D	154
Oyster tissue, NBS SRM 1566		Soft parts	2.60 ± 0.30[g]	D	154
Oyster tissue, NBS SRM 1566		Soft parts	<0.01[h]	D	154
Oyster tissue, NBS SRM 1566		Soft parts	<0.01[e]	D	154
Euphausiids	Northern Pacific Ocean	Whole	3.60[g]	D	154
Euphausiids	Northern Pacific Ocean	Whole	0.11[h]	D	154

Table 2 (continued)
SELENIUM IN MARINE BIOLOGICAL TISSUES

Species	Location	Tissue[a]	Selenium conc. (mg/kg)[b]	Weight basis[c]	Ref.
Euphausiids	Northern Pacific Ocean	Whole	<0.01[e]	D	154

[a] Digestiv. — digestive gland or tissue; NS — not specified.
[b] Concentrations are given as ranges, means, or mean ± standard deviation.
[c] W — wet weight or fresh weight basis; D — dry weight basis; NS — not specified.
[d] Se(−II) and Se(IV).
[e] Se(VI).
[f] Lipid removed.
[g] Total selenium; difference between total selenium and Se(IV) + Se(VI) regarded as organically bound selenium.
[h] Se(IV).

concentrations of selenium than other types of tissue, such as flesh.[48-53] As well, there is some evidence that the selenium level, in some tissues at least, is positively correlated to the age or size of the specimen. For example, the following correlations have been observed: selenium value of dolphin liver vs. size,[48] seal brain and liver vs. age,[54] black marlin muscle and liver vs. weight, length, and girth,[55] striped mullet muscle vs. age,[56] swordfish muscle vs. fork length,[57] and blue marlin muscle, liver, kidney, as well as spleen vs. weight.[53]

There is evidence that in biological tissues only a small fraction of Se(IV) and (VI) is adsorbed, and that most selenium is organically bound.[162] Assimilated selenium appears to reside primarily in protein. This is especially true for the clam and lobster hepatopancreas.[49] As well, the highest concentrations of selenium are also found in these tissues of the specimens. It is noteworthy that the lobster muscle contains three to four times as much protein as the hepatopancreas. Since selenium concentration in the muscle is comparatively low, it is evident that protein itself does not determine the quantity of the element in a given tissue.[49] Both Se(IV) and (VI) coexist in fish, mollusks, and crustaceans.[58,59] Over half of the selenium in protein, averaging about 60%, is water extractable. This fraction includes ionic and neutral selenium, polar proteins, peptides, and amino acids. On the average, about 10% of the total selenium is associated with polar proteins having molecular weights higher than 100,000. A lipid-soluble, lipoprotein-bound selenium compound has been isolated in marine fish.[60] The presence of an acid-labile selenium species has been demonstrated in the proteins of rat liver.[61] Further, there is some evidence that this selenium may be part of a nonheme iron protein.[62] A selenoprotein with a molecular weight of approximately 10,000 has been found in the muscle of lamb.[63] This protein may have a structure similar to cytochromes.[64] Selenoproteins having catalytic activities have been discovered in the last decade,[65] with one of the most discussed being glutathione peroxidase.[66] The form of selenium in this enzyme, however, is still not identified. Glycine reductase, another selenium-containing enzyme, is also noteworthy.[67] Again, details of its molecular composition are lacking.

Selenium may replace sulfur in iron-sulfur proteins. This has been demonstrated in putidaredoxin, an iron-sulfur protein from *Pseudomonas putida*.[68] Evidence of selenotrisulfide linkages in proteins has been established.[69,70] In a study on seleniferous wheat, selenium has been found to be an integral part of the protein and could be released by hydrolysis.[71] Selenium was discovered to be associated with fractions of the hydrolysate that contained cystine and methionine, indicating the probable presence of selenium-containing analogues. Enzyme hydrolysis of the protein of the marine phytoplanktons *Tetraselmis tetrathele* and *Dunaliella minuta* revealed the presence of seleno-analogues of the sulfur amino acids, selenocystine, selenomethionine, selenocysteic acid, selenomethionine selenoxide, and Se-methylselenocysteine.[72] A small quantity of this protein-bound selenium could be volatilized by acids.[61,72] Selenium amino acids were also found in nonprotein extracts.

Despite being an essential trace element, selenium is highly toxic at elevated concentrations. In fact, its toxic and essential levels are estimated to be only about tenfold apart.[73] Acute toxicity of selenium (as sodium selenite) to three invertebrates, *Cyclaspis usitata*, a cumacean, juveniles of *Notocallista* spp., a bivalve, and two stocks of *Allorchestes compressa*, an amphipod, has been studied.[74] The 96-h LC50 values for these four groups were 6.12, 2.88, 6.17, and 4.77 mg of Se per liter, respectively. These results suggest that juveniles may be more susceptible to selenium poisoning than adults.

VI. INTERACTION OF SELENIUM WITH OTHER ELEMENTS

There is strong evidence that the accumulation of trace elements and the physiological effects exerted on an organism may be modified by the presence of selenium.[116] Antagonistic behavior has been observed between selenium and the following elements: arsenic,[117,118] cadmium,[119,120] mercury,[121-123] and thallium.[124] The most noteworthy interaction is that between selenium and mercury. The number of studies devoted to this particular interdisciplinary subject is voluminous. Many of these cover areas in food science and biochemistry and are beyond the scope of this chapter. For these, interested readers are referred to the original literature, for example, the work of Ganther et al.[122,123,125,126,128,138]

The ability of selenium compounds to decrease the toxic action of both organic and inorganic mercury in experimental animals has been established beyond doubt. The first known report was made by Parizek and Ostadalova.[121] Later, Japanese quails fed a diet of tuna containing a natural concentration of 0.5 mg of Hg per kilogram were reported to show no signs of mercury poisoning. Further, when methylmercury was added to the same diet, its toxicity was much lessened. The agent responsible for suppressing mercury toxicity in this tuna diet was identified as selenium.[122,123] These results were later confirmed with other animals and different experimental parameters.[127-129]

It is not clear how selenium protects against mercury toxicity. Conclusions drawn from different studies are often conflicting. For example, selenium has been reported by some to enhance demethylation of methylmercury and facilitate its excretion,[130,131] and by others to enhance methylmercury accumulation in some organs, like the brain.[129,132-136] This increased mercury retention, and yet decreased toxicity in the presence of selenium, leads some to suggest the possible formation of a low-toxic and/or low-bioavailable selenium-mercury compound[137] which as yet has to be identified. Besides, several studies reported the enhanced accumulation of the species methylmercury, rather than mercury or a nontoxic compound.[128,136,138]

Information pertaining to the marine environment is equally confusing. The two main questions here are whether there is any correlation between the concentrations of selenium and mercury in marine fauna, and whether the uptake of these two elements is in a one-to-one molar ratio. Koeman et al.[54] observed a linear relationship between the concentrations of selenium and mercury in the livers of marine mammals. As well, a molar one-to-one increment was reported. This equal molar accumulation was also claimed in dolphins,[139] tuna,[122] and other marine fish.[140] Significant correlation between the concentrations of the two elements was also noted in the muscle tissue of black marlin,[55] the dorsal muscle of yellowtail,[141] the muscle, liver, kidney, and spleen of blue marlin,[53] the liver and kidney of dolphin,[140] and the liver of ringed seal.[89] Observations to the contrary were reported in the dorsal muscle of swordfish,[57] the muscle tissue of tuna,[58,97] the muscle of horse mackerel, sillago, and sardine,[93] and the liver and brain of the marine birds guillemot and razorbill.[54]

Despite all this conflicting information, certain conclusions are apparent. For one, the equal molar accumulation of selenium and mercury appears to be a rather limited phenomenon, restricted to marine mammals. In marine fish, concentrations of selenium are usually several times higher than those of mercury,[54,57-59,75,91,96,113] and the early observation of equal molar accumulation has been recognized as an exception rather than the rule.[126]

Selenium has been reported to affect the accumulation, distribution, and toxicity of cadmium.[119,120] In an investigation on the uptake of these two elements and their mutual interaction in shore crab, *Carcinus maenas*, it was found that selenium enhanced cadmium accumulation in the gill. No augmentation, however, was observed for the hepatopancreas, carapace, and muscle, although a positive correlation in the concentrations of selenium and cadmium was apparent in the hepatopancreas and carapace.[142]

VII. SEDIMENTS

Compared to seawater and biological tissues, data on sediments are relatively scarce (Table 3). Most selenium concentrations lie within the range of 0.1 to 2 mg/kg, despite very different geographical locations. Significant concentration variation with depth, 0.07 to 0.43 mg/kg,

Table 3
SELENIUM IN SEDIMENTS

Location	Selenium concentration (mg/kg)[a]	Weight basis[b]	Ref.
Atlantic Ocean	1.6	D	16
North Pacific Ocean	0.1—1.7	NS	143
Pacific Ocean	0.11—0.13	NS	144
Tyrrenian Sea	0.1—1.6	NS	145
Pacific Ocean (34°52′N 151°55′E)	0.053—0.43	D	146
Off Japan	0.02—2.23	D	147
Chesapeake Bay, NBS SRM 1646	0.6	D	148
Tyrrenian Sea	0.1—0.3	NS	149
Suruga Bay, Japan	0.47 ± 0.02	D	150
Pacific Ocean and Japan Sea	4.5—9.6	D	77
Baie des Chaleurs, NB, NRC SRM BCSS-1	0.45 ± 0.03	D	151, 152
Miramichi River, NB, NRC SRM MESS-1	0.36 ± 0.02	D	151, 152
Spencer's Gulf, South Australia			153
Seagrass flat	0.46	D	153
Sand flat	0.52	D	153
Mangrove	0.60	D	153
Estuarine	0.82	D	153
Supratidal	1.12	D	153

[a] Concentrations are given as ranges, means or mean ± standard deviation.
[b] D — dry weight basis; NS — not specified.

Table 4
SELENIUM SPECIATION IN SEDIMENTS[a]

Location	Selenium concentration in different oxidation states (mg/kg, dry weight)		
	Total	IV	VI
Indiana Harbour Canal (NBS SRM 1645)	1.70 ± 0.30	0.02 ± 0.01	0.08 ± 0.03
Chesapeake Bay (NBS SRM 1646)	0.43 ± 0.02	0.001 ± 0.0006	0.04 ± 0.02
North Pacific Ocean[b]	3.24	0.04	0.21
	2.48	<0.01	0.16

[a] Difference between total and Se(IV) + (VI) is "organically bound" Selenium, Se(−II + 0).
[b] Sediments from 105- and 232-m depths, respectively.

From Cutter, G. A., *Anal. Chem.*, 57, 2951, 1985. With permission.

has been reported.[146] Storage of sediment samples may be a problematic area. Significant loss of selenium has been reported after a 1-d storage.[76] Most selenium in sediments appears to reside in the "organic" fraction,[154] operationally defined using a sequential extraction procedure designed to separate the various sedimentary phases.[155] Table 4 lists the few data available on selenium speciation.

The effect of sediment selenium on the surrounding flora and fauna is not clear. One study indicates that selenium concentrations in sediments are apparently not reflected by those in the livers of the dover sole, a bottom dweller.[5,47]

VIII. SELECTED ANALYTICAL CHEMISTRY

An extensive survey on the analytical methodology pertinent to marine materials is beyond the scope of this chapter. It is also unnecessary, as this is covered by two recent reviews on this area.[156,157]

Marine biological tissues and sediments require dissolution prior to analysis. Mixed-acid decomposition in either open or closed vessels is the most commonly used procedure,[158-161] although other techniques like fusion,[152] ashing,[105,160,162] and various combustion methods[157] have also been tried. For biological tissues, nitric acid or a mixture of nitric and sulfuric acids, sometimes with the addition of perchloric acid,[158,163] hydrogen peroxide,[159] phosphoric acid,[160] or vanadium pentoxide,[161] are frequently used. Sediments usually require hydrofluoric acid, plus a combination of nitric, perchloric, and sulphuric acids, for complete dissolution.[151] Certain digestion procedures call for evaporating the samples to dryness in order to completely volatilize hydrogen fluoride and silicon tetrafluoride.[151] It is noteworthy that selenium is not lost under such conditions, whereas in a similarly set up blank, i.e., a selenium-spiked acid mixture, the loss of selenium may approach 100%. The mechanism by which the sample retains selenium has yet to be elucidated. Less vigorous digestion procedures may be used to yield more specific data. A procedure involving several sequential extractions may be used to separate the various sedimentary phases with progressively more aggressive reagents.[154,155] This type of operation gives information on the locality of an element in the sediment.

Nearly all, if not all, modern instrumental methods have been used for selenium determination.[157] Instrumental neutron activation analysis is distinguished for requiring potentially the least degree of sample pretreatment. Often, some kind of sample work-up, for instance, formation of a selenium complex for chromatography[17] or extraction,[98] is required. Other methods, for example, fluorometry,[163] require masking agents to alleviate interferences. To date, the most sensitive techniques for selenium are, arguably, atomic absorption spectroscopy with hydride generation[21] and gas chromatography with electron capture detection.[32] The choice is admittedly partially personal. Both methods are capable of determining selenium at nanogram-per-liter concentrations in seawater without the need of a separate preconcentration step. Moreover, with judicious sample treatment, both techniques are amenable to speciation studies. This is achieved by selectively converting the selenium species of interest to Se(IV), as this is the lone species seen in both hydride generation and gas chromatography. In the former, only tetravalent selenium reacts with sodium borohydride to form hydrogen selenide, and in the latter technique, only Se(IV) couples with *o*-phenylenediamines to form the volatile piazselenols.

For seawater, reduction of Se(VI) to (IV) is most conveniently achieved by heating with hydrochloric[21] or hydrobromic[24] acid. Bromine may be added to oxidize any Se(−II) and Se(0) to Se(IV).[24] UV radiolysis[30] and persulfate oxidation[26] have been used to decompose organoselenium compounds.

Compared to seawater, speciation work on biological tissues and sediments is more difficult, as vigorous dissolution conditions are normally required[151,155,158-161] with the danger that selenium speciation may be altered during sample decomposition.

IX. CONCLUSIONS

Much work is required to better our understanding of selenium in the marine environment. In seawater, the importance of organoselenium is slowly being recognized. The determination of these species is difficult and has led to much debate. In sediments and biological tissues, the majority of selenium appears to be organically associated or bound. Little, however, is known about these selenium species.

It is apparent that selenium plays many roles in an organism. It is incorporated into amino acids, enzymes, and other proteins. The structures of most of these selenocompounds, however, are unknown. Advances here would hopefully shed new light on the biochemistry of selenium. Further, it may clarify the antagonistic effect of the element on mercury and other elements. A note of caution appears to be in order here in that selenium is both toxic and essential to an organism. Thus, it is likely that the element is metabolized via different pathways depending on its rate of intake. Hence, biochemical conclusions drawn from elevated selenium intake, for example, in a tracer experiment, may not be applicable to normal circumstances where selenium is present at lower concentrations. Further complications may arise from the mode of intake, i.e., whether by ingestion or by absorption through the skin or gills, which may affect the metabolic pathway of selenium.

REFERENCES

1. **Turekian, K. K. and Wedepohl, K. H.,** Distribution of the elements in some major units of the earth's crust, *Bull. Geol. Soc. Am.*, 72, 175, 1961.
2. **Frost, D. V.,** The two faces of selenium — can selenophobia be cured? *CRC Crit. Rev. Toxicol.*, p. 467, 1972.
3. **Kut, D. and Sarikaya, Y.,** Determination of selenium in atmospheric particulate material of Ankara and its possible sources, *J. Radioanal. Chem.*, 62, 161, 1981.
4. **Duchaigne, A. and Arvy, M. P.,** Le sélénium en biologie, *Ann. Biol.*, 17, 529, 1978.
5. **Wilber, C. G.,** Toxicology of selenium: a review, *Clin. Toxicol.*, 17, 171, 1980.
6. **Bertine, K. K. and Goldberg, E. D.,** Fossil fuel combustion and the major sedimentary cycle, *Science*, 173, 233, 1971.
7. **Weiss, H. V., Koide, M., and Goldberg, E. D.,** Selenium and sulfur in a Greenland ice sheet: relation to fossil fuel combustion, *Science*, 12, 261, 1971.
8. **Kharkar, D. P., Turekian, K. K., and Bertine, K. K.,** Stream supply of dissolved silver, molybdenum, antimony, selenium, chromium, cobalt, rubidium and cesium to the oceans, *Geochim. Cosmochim. Acta*, 32, 285, 1968.
9. **Measures, C. I. and Burton, J. D.,** Behaviour and speciation of dissolved selenium in estuarine waters, *Nature (London)*, 273, 293, 1978.
10. **Sugimura, Y., Suzuki, Y., and Miyake, Y.,** The content of selenium and its chemical form in sea water, *J. Oceanogr. Soc. Jpn.*, 32, 235, 1976.
11. **Takayanagi, K. and Wong, G. T. F.,** The distribution of total selenium and Se(IV) in the James River and southern Chesapeake Bay, *EOS*, 63, 352, 1982.
12. **Goldschmidt, V. M. and Strock, L. W.,** Zur geochemie des selens. II, *Nachr. Akad. Wiss. Goettingen Math. Physik. Kl. Fachgruppe, 1*, 123, 1935.
13. **Wattenberg, H.,** Selen in Meerwasser, *Z. Anorg. Chem.*, 236, 339, 1938.
14. **Ishibashi, M., Shigematsu, T., and Nagasawa, Y.,** Determination of selenium in sea water, *Rec. Oceanogr. Works Jpn. New Ser.*, 1, 44, 1935.
15. **Schutz, D. F. and Turekian, K. K.,** The investigation of the geographical and vertical distribution of several trace elements in sea water using neutron activation analysis, *Geochim. Cosmochim. Acta*, 29, 259, 1965.
16. **Chau, Y. K. and Riley, J. P.,** The determination of selenium in sea water, silicates and marine organisms, *Anal. Chim. Acta*, 33, 36, 1965.
17. **Shimoishi, Y.,** The determination of selenium in sea water by gas chromatography with electron-capture detection, *Anal. Chim. Acta*, 64, 465, 1973.
18. **Sillen, L. G.,** The physical chemistry of sea water, in *Oceanography*, Sears, M., Ed., American Association for the Advancement of Science, Washington, D.C., 1961, 549.
19. **Massée, R., Van der Sloot, H. A., and Das, H. A.,** The determination of selenium in water samples from the environment, *J. Radioanal. Chem.*, 35, 157, 1977.
20. **Shimoishi, Y. and Tôei, K.,** The gas chromatographic determination of selenium(IV) and total selenium in natural waters with 1,2-diamino-3,5-dibromobenzene, *Anal. Chim. Acta*, 100, 65, 1978.
21. **Cutter, G. A.,** Species determination of selenium in natural waters, *Anal. Chim. Acta*, 98, 59, 1978.
22. **Measures, C. I. and Burton, J. D.,** The vertical distribution and oxidation states of dissolved selenium in the northeast Atlantic Ocean and their relationship to biological processes, *Earth Planet. Sci. Lett.*, 46, 385, 1980.
23. **Measures, C. I., McDuff, R. E., and Edmond, J. M.,** Selenium redox chemistry at GEOSECS I-reoccupation, *Earth Planet. Sci. Lett.*, 49, 102, 1980.

24. **Uchida, H., Shimoishi, Y., and Tôei, K.,** Gas chromatographic determination of selenium(− II, 0), − (IV), and − (VI) in natural waters, *Environ. Sci. Technol.,* 14, 541, 1980.
25. **Wrench, J. J. and Measures, C. I.,** Temporal variations in dissolved selenium in a coastal ecosystem, *Nature (London),* 299, 431, 1982.
26. **Cutter, G. A.,** Selenium in reducing waters, *Science,* 217, 829, 1982.
27. **Measures, C. I., Grant, B. C., Mangum, B. J., and Edmond, J. M.,** The relationship of the distribution of dissolved selenium IV and VI in three oceans to physical and biological processes, in *Trace Metals in Sea Water,* Wong, C. S., Boyle, E., Bruland, K. W., Burton, J. D., and Goldberg, E. D., Eds., Plenum Press, New York, 1983, 73.
28. **Cutter, G. A. and Bruland, K. W.,** The marine biogeochemistry of selenium: a re-evaluation, *Limnol. Oceanogr.,* 29, 1179, 1984.
29. **Sugimura, Y., Suzuki, Y., and Miyake, Y.,** Chemical forms of minor metallic elements in the ocean, *J. Oceanogr. Soc. Jpn.,* 34, 93, 1978.
30. **Takayanagi, K. and Wong, G. T. F.,** Organic and colloidal selenium in southern Chesapeake Bay and adjacent waters, *Mar. Chem.,* 14, 141, 1984.
31. **Suzuki, Y., Sugimura, Y., and Miyake, Y.,** The content of selenium in sea water in the western north Pacific and its marginal seas, in *Proc. 4th Symp. for the Cooperative Study of the Kuroshio and Adjacent Regions,* Saikon Publishing, Tokyo, 1980, 396.
32. **Measures, C. I. and Burton, J. D.,** Gas chromatographic method for the determination of selenite and total selenium in sea water, *Anal. Chim. Acta,* 120, 177, 1980.
33. **Measures, C. I. and Wrench, J. J.,** Selenium in the Marine Environment, Rep. MCWG 1983/7.7, International Council for the Exploration of the Sea, Kobenhaunk, Denmark, 1983.
34. **Hiraki, K., Yoshii, O., Hirayama, H., Nishikawa, Y., and Shigematsu, T.,** Fluorometric determination of selenium in sea-water, *Bunseki Kagaku,* 22, 712, 1973.
35. **Sugimura, Y. and Suzuki, Y.,** A new fluorometric method of analysis of selenium in sea water, *J. Oceanogr. Soc. Jpn.,* 33, 23, 1977.
36. **Lieser, K. H., Calmano, W., Heuss, E., and Neitzert, V.,** Neutron activation as a routine method for the determination of trace elements in water, *J. Radioanal. Chem.,* 37, 717, 1977.
37. **Yoshii, O., Hiraki, K., Nishikawa, Y., and Shigematsu, T.,** Fluorometric determination of selenium(IV) and selenium(VI) in sea water and river water, *Bunseki Kagaku,* 26, 91, 1977.
38. **Tzeng, J.-H. and Zeitlin, H.,** The separation of selenium from sea water by adsorption colloid flotation, *Anal. Chim. Acta,* 101, 71, 1978.
39. **Van der Sloot, H. A.,** Actieve kool als sporenvanger, *Chem. Weekbl.,* May, 297, 1979.
40. **Nakashima, S.,** Flotation separation and atomic absorption spectrometric determination of selenium(IV) in water, *Anal. Chem.,* 51, 654, 1979.
41. **Heuss, E. and Lieser, K. H.,** Abtrennung von spurenelementen aus meerwasser durch adsorption und ihre bestimmung durch neutronenaktivierungsanalyse, *J. Radiolanal. Chem.,* 50, 289, 1979.
42. **Robberecht, H. and Van Grieken, R.,** Selenium content and speciation in environmental waters determined by X-ray fluorescence spectroscopy, in *Trace Substances in Environmental Health — 14,* Hemphill, D. D., Ed., University of Missouri, Columbia, 1980, 362.
43. **Hodson, P. V. and Hilton, J. W.,** The nutritional requirements and toxicity to fish of dietary and waterborne selenium, in *Environmental Biogeochemistry (Ecol. Bull.),* Vol. 35, Hallberg, R., Ed., Ecological Bulletins, Stockholm, 1983, 335.
44. **Klaverkamp, J. F., Turner, M. A., Harrison, S. E., and Hesslein, R. H.,** Fates of metal radiotracers added to a whole lake: accumulation in slimy sculpin (*Cottus cognatus*) and white sucker *(Catostomus commersoni), Sci. Total Environ.,* 28, 119, 1983.
45. **Fowler, S. W. and Benayoun, G.,** Influence of environmental facors on selenium flux in two marine invertebrates, *Mar. Biol.,* 37, 59, 1976.
46. **Lytle, T. F. and Lytle, J. S.,** Heavy metals in oysters and clams of St. Louis Bay, Mississippi, *Bull. Environ. Contam. Toxicol.,* 29, 50, 1982.
47. **de Goeij, J. J. M., Guinn, V. P., Young, D. R., and Mearns, A. J.,** Neutron activation analysis trace-element studies of dover sole liver and marine sediments, in *Comparative Studies of Food and Environmental Contamination,* International Atomic Energy Agency, Vienna, 1974, 189.
48. **Guinn, V. P. and Kishore, R.,** Results from multi-trace-element neutron activation analyses of marine biological specimens, *J. Radioanal. Chem.,* 19, 367, 1974.
49. **Wrench, J. J. and Campbell, N. C.,** Protein bound selenium in some marine organisms, *Chemosphere,* 10, 1155, 1981.
50. **Chou, C. L. and Uthe, J. F.,** Heavy metal relationships in lobster (*Homarus americanus*) and rock crab *(Cancer irroratus)* digestive glands, Rep. E:15/C.M., International Council for the Exploration of the Sea, Kobenhaunk, Denmark, 1978.
51. **Luten, J. B., Ruiter, A., Ritskes, T. M., Rauchbaar, A. B., and Riekwel-Booy, G.,** Mercury and selenium in marine and freshwater fish, *J. Food Sci.,* 45, 416, 1980.

52. **Grieg, R. A. and Jones, J.,** Nondestructive neutron activation analysis of marine organisms collected from ocean dump sites of the middle eastern United States, *Arch. Environ. Contam. Toxicol.*, 4, 420, 1976.
53. **Shultz, C. D. and Ito, B. M.,** Mercury and selenium in blue marlin, *Makaira nigricans,* from the Hawaiian Islands, *Fish. Bull.*, 76, 872, 1979.
54. **Koeman, J. H., van de Ven, W. S. M., de Goeij, J. J. M., Tjioe, P. S., and van Haaften, J. L.,** Mercury and selenium in marine mammals and birds, *Sci. Total Environ.*, 3, 279, 1975.
55. **MacKay, N. J., Kazacos, M. N., Williams, R. J., and Leedow, M. I.,** Selenium and heavy metals in black marlin, *Mar. Pollut. Bull.*, 6, 57, 1975.
56. **Leonzio, C., Focardi, S., and Bacci, E.,** Complementary accumulation of selenium and mercury in fish muscle, *Sci. Total Environ.*, 24, 249, 1982.
57. **Freeman, H. C., Shum, G., and Uthe, J. F.,** The selenium content in swordfish (*Xiphias gladius*) in relation to total mercury content, *J. Environ. Sci. Health,* 13A, 235, 1978.
58. **Cappon, C. J. and Smith, J. C.,** Mercury and selenium content and chemical form in fish muscle, *Arch. Environ. Contam. Toxicol.*, 10, 305, 1981.
59. **Cappon, C. J. and Smith, J. C.,** Chemical form and distribution of mercury and selenium in edible seafood, *J. Anal. Toxicol.*, 6, 10, 1982.
60. **Lunde, G.,** Location of lipid-soluble selenium in marine fish to the lipoproteins, *J. Sci. Food Agric.*, 23, 987, 1972.
61. **Diplock, A. T., Caygill, C. P. J., Jeffrey, E. H., and Thomas, C.,** The nature of the acid-volatile selenium in the liver of the male rat, *Biochem. J.*, 134, 283, 1973.
62. **Caygill, C. P. J., Diplock, A. T., and Jeffrey, E. H.,** Studies on selenium incorporation into, and electron-transfer function of, liver microsomal fractions from normal and vitamin E-deficient rats given phenobarbitone, *Biochem. J.*, 136, 851, 1973.
63. **Pedersen, N. D., Whanger, P. D., Weswig, P. H., and Muth, O. H.,** Selenium binding proteins in tissues of normal and selenium responsive myopathic lambs, *Bioinorg. Chem.*, 2, 33, 1972.
64. **Whanger, P. D., Pedersen, N. D., and Weswig, P. H.,** Selenium proteins in ovine tissue. II. Special properties of a 10,000 molecular weight selenium protein, *Biochem. Biophys. Res. Commun.*, 53, 1031, 1973.
65. **Williams, R. J. P.,** A short note on selenium biochemistry, in *New Trends in Bio-inorganic Chemistry,* Williams, R. J. P. and de Silva, J. R. R. F., Eds., Academic Press, London, 1978, 253.
66. **Flohé, L.,** Die Glutathionperoxidase: Enzymologie und biologische Aspekte, *Klin. Wochenschr.*, 49, 669, 1971.
67. **Turner, D. C. and Stadtman, T. C.,** Purification of protein components of the clostridial glycine reductase system and characterization of protein A as a selenoprotein, *Arch. Biochem. Biophys.*, 154, 366, 1973.
68. **Tsibris, J. C. M., Namtvedt, M. J., and Gunsalus, I. C.,** Selenium as an acid labile sulfur replacement in putidaredoxin, *Biochem. Biophys. Res. Commun.*, 30, 323, 1968.
69. **Jenkins, K. J.,** Evidence for the absence of selenocystine and selenomethionine in the serum proteins of chicks administered selenite, *Can. J. Biochem.*, 46, 1417, 1968.
70. **Ganther, H. E. and Corcoran, C.,** Selenotrisulfides. II. Cross-linking of reduced pancreatic ribonuclease with selenium, *Biochemistry,* 8, 2557, 1969.
71. **Rosenfeld, I. and Beath, O. A.,** *Selenium,* Academic Press, New York, 1964.
72. **Wrench, J. J.,** Selenium metabolism in the marine phytoplanktens *Tetraselmis tetrathele* and *Dunaliella minuta, Mar. Biol.*, 49, 231, 1978.
73. **Flinn, C. G. and Aue, W. A.,** Photometric detection of selenium compounds for gas chromatography, *J. Chromatogr.*, 153, 49, 1978.
74. **Ahsanullah, M. and Palmer, D. H.,** Acute toxicity of selenium to three species of marine invertebrates, with notes on a continuous-flow test system, *Aust. J. Mar. Freshwater Res.*, 31, 795, 1980.
75. **Hall, R. A., Zook, E. G., and Meaburn, G. M.,** National Marine Fisheries Service Survey of Trace Elements in the Fishery Resource, National Oceanic and Atmospheric Administration Tech. Rep. NMFS SSRF-721, 1978.
76. **Gosink, T. A. and Reynolds, D. J.,** Selenium analysis of the marine environment, gas chromatography and some results, *Mar. Sci. Commun.*, 1, 101, 1975.
77. **Terada, K., Ooba, T., and Kiba, T.,** Separation and determination of selenium in rocks, marine sediments and plankton by direct evolution with the bromide-condensed phosphoric acid reagent, *Talanta,* 22, 41, 1975.
78. **Fowler, B. A., Fay, R. C., Walter, R. L., Willis, R. D., and Gutknecht, W. F.,** Levels of toxic metals in marine organisms collected from southern California coastal waters, *Environ. Health Perspect.*, 12, 71, 1975.
79. **Ihnat, M.,** Selenium in foods: evaluation of atomic absorption spectrometric techniques involving hydrogen selenide generation and carbon furance atomization, *J. Assoc. Off. Anal. Chem.*, 59, 911, 1976.

80. **Fiorino, J. A., Jones, J. W., and Capar, S. G.,** Sequential determination of arsenic, selenium, antimony and tellurium in foods via rapid hydride evolution and atomic absorption spectrometry, *Anal. Chem.*, 48, 120, 1976.
81. **Sims, R. R., Jr. and Presley, B. J.,** Heavy metal concentrations in organisms from an actively dredged Texas bay, *Bull. Environ. Contam. Toxicol.*, 16, 520, 1976.
82. **Fowler, S. W.,** Trace elements in zooplankton particulate products, *Nature (London)*, 269, 51, 1977.
83. **Shum, G. T. C., Freeman, H. C., and Uthe, J. F.,** Flameless atomic absorption spectrophotometry of selenium in fish and food products, *J. Assoc. Off. Anal. Chem.*, 60, 1010, 1977.
84. **Bebbington, G. N., MacKay, N. J., Chvojka, R., Williams, R. J., Dunn, A., and Auty, E. H.,** Heavy metals, selenium and arsenic in nine species of Australian commercial fish, *Aust. J. Mar. Freshwater Res.*, 28, 277, 1977.
85. **Cappon, C. J. and Smith, J. C.,** Determination of selenium in biological materials by gas chromatography, *J. Anal. Toxicol.*, 2, 114, 1978.
86. **Fukai, R., Oregioni, B., and Vas, D.,** Interlaboratory comparability of measurements of trace elements in marine organisms: results of intercalibration exercise on oyster homogenate, *Oceanol. Acta*, 1, 391, 1978.
87. **Egaas, E. and Julshamn, K.,** A method for the determination of selenium and mercury in fish products using the same digestion procedure, *At. Absorpt. Newsl.*, 17, 135, 1978.
88. **Bycroft, B. M. and Clegg, D. E.,** Gas-liquid chromatographic determination of selenium in biological materials, using 4-bromo- and 4-chloro-1,2-diaminobenzene as derivatizing reagents, *J. Assoc. Off. Anal. Chem.*, 61, 923, 1978.
89. **Kari, T. and Kauranen, P.,** Mercury and selenium contents of seals from fresh and brackish waters in Finland, *Bull. Environ. Contam. Toxicol.*, 19, 273, 1978.
90. **Watkinson, J. H.,** Semi-automatic fluorometric determination of nanogram quantities of selenium in biological material, *Anal. Chim. Acta*, 105, 319, 1979.
91. **Kobayashi, R., Hirata, E., Shiomi, K., Yamanaka, H., and Kikuchi, T.,** Heavy metal contents in deep-sea fishes, *Nippon Suisan Gakkaishi*, 45, 493, 1979.
92. **Guthrie, R. K., Davis, E. M., Cherry, D. S., and Murray, H. E.,** Biomagnification of heavy metals by organisms in a marine microcosm, *Bull. Environ. Contam. Toxicol.*, 21, 53, 1979.
93. **Noda, K., Hirai, S., Sunayashiki, K., and Danbara, H.,** Neutron activation analyses of selenium and mercury in marine products from along the coast of Shikoku Island, *Agric. Biol. Chem.*, 43, 1381, 1979.
94. **Glover, J. W.,** Concentrations of arsenic, selenium and ten heavy metals in school shark, *Galeorhinus australis* (Macleay), and gummy shark, *Mustelus antarcticus* Günther, from south-eastern Australian waters, *Aust. J. Mar. Freshwater Res.*, 30, 505, 1979.
95. **Papadopoulou, C., Kanias, G. D., and Moraitopoulou-Kassimati, E.,** Trace element content in fish otoliths in relation to age and size, *Mar. Pollut. Bull.*, 11, 68, 1980.
96. **Suzuki, T., Satoh, H., Yamamoto, R., and Kashiwazaki, H.,** Selenium and mercury in foodstuff from a locality with elevated intake of methylmercury, *Bull. Environ. Contam. Toxicol.*, 24, 805, 1980.
97. **Orvini, E., Caramella-Crespi, V., and Genova, N.,** Activation analysis of As, Hg and Se in some marine organisms, in *Analytical Techniques in Environmental Chemistry*, Albaiges, J., Ed., Pergamon Press, New York, 1980, 441.
98. **Kamada, T. and Yamamoto, Y.,** Use of transition elements to enhance sensitivity for selenium determination by graphite-furnace atomic-absorption spectrophotometry combined with solvent extraction with the APDC-MIBK system, *Talanta*, 27, 473, 1980.
99. **Stegnar, P., Vukadin, I., Smodis, B., Vakselj, A., and Prosenc, A.,** Trace elements in sediments and organisms from Kastela Bay, *Journées d'Études sur les Pollutions Marines en Méditerranéee, 5th*, Commission Internationale pour l'Exploration Scientifique de la Mer Méditerranée, Monaco, 1981, 595.
100. **Grimanis, A. P., Zafiropoulos, D., Papadopoulou, C., and Vassilaki-Grimani, M.,** Trace elements in the flesh of different fish species from three gulfs of Greece, *Journées d'Études sur les Pollutions Marines en Méditerranée, 5th*, Commission Internationale pour l'Exploration Scientifique de la Mer Méditerranée, Monaco, 1981, 407.
101. Certificate of Analysis, Standard Reference Material 1566, Oyster Tissue, National Bureau of Standards, Washington, D.C., 1979.
102. **LaFleur, P. D. and Reed, W. P.,** Report of Investigation, Research Material 50, Albacore Tuna, National Bureau of Standards, Washington, D.C., 1977.
103. **Maher, W. A.,** Fluorometric determination of selenium in some marine materials after digestion with nitric and perchloric acids and co-precipitation of selenium with lanthanum hydroxide, *Talanta*, 29, 1117, 1982.
104. **Heit, M. and Klusek, C. S.,** The effects of dissecting tools on the trace element concentrations of fish and mussel tissues, *Sci. Total Environ.*, 24, 129, 1982.
105. **May, T. W.,** Recovery of endogenous selenium from fish tissues by open system dry ashing, *J. Assoc. Off. Anal. Chem.*, 65, 1140, 1982.

106. **Whitacre, M. and Latshaw, J. D.,** Selenium utilization from menhaden fish meal as affected by processing, *Poult. Sci.,* 61, 2520, 1982.
107. Certificate of Analysis, TORT-1, Lobster Hepatopancreas Marine Reference Material, National Research Council of Canada, Ottawa, 1983.
108. **Maher, W. A.,** Determination of selenium in marine organisms using hydride generation and electrothermal atomic absorption spectroscopy, *Anal. Lett.,* 16, 801, 1983.
109. **Harada, T., Oishi, K., and Koyama, M.,** Radioactivation analysis of calcareous algae and laminariales, and the regularity of distribution of elements in the organisms, from bacteria to mammals, *Nippon Suisan Gakkaishi,* 49, 1135, 1983.
110. **Maher, W. A.,** Selenium in marine organisms from St. Vincent's Gulf, South Australia, *Mar. Pollut. Bull.,* 14, 35, 1983.
111. **Maher, W. A.,** An investigation of trace element losses during lyophilization of marine biological samples, *Sci. Total Environ.,* 26, 173, 1983.
112. **Askar, A. and Bielig, H. J.,** Selenium content of food consumed by Egyptians, *Food Chem.,* 10, 231, 1983.
113. **Thibaud, Y.,** *Interaction Entre Élements Metalliques lors de Processus d'Assimilation chez les Organismes Marins,* Rapport Definitif, Convention 78-36 (8064) avec le Ministere de l'Environnement, Institut Scientifique et Technique des Pêches Maritimes, Nantes, France, 1983.
114. **Braddon, S. A. and Sumpter, C. R.,** *Selenium Levels in Yellowfin Tuna (Thunnus albacares) and Sharks from the Carolinas,* National Oceanic and Atmospheric Administration Tech. Memo. NMFS-SEFC-83, National Oceanic and Atmospheric Administration, Washington, DC, 1981.
115. **Landsberger, S. and Hoffman, E.,** Rapid determination of selenium in various marine species by instrumental neutron activation analysis, *J. Radioanal. Nucl. Chem.,* 87, 41, 1984.
116. **Ganther, H. E.,** Biochemistry of selenium, in *Selenium,* Zingaro, R. A. and Cooper, W. C., Eds., Van Nostrand Reinhold, New York, 1974, 546.
117. **Moxon, A. L.,** The effect of arsenic on the toxicity of seleniferous grains, *Science,* 88, 81, 1938.
118. **Palmer, I. S. and Bonhorst, C. W.,** Modification of selenite metabolism by arsenite, *J. Agric. Food Chem.,* 5, 928, 1957.
119. **Parizek, J., Ostadalova, I., Benes, I., and Babicky, A.,** Pregnancy and trace elements: the protective effect of compounds of an essential trace element — selenium — against the peculiar toxic effects of cadmium during pregnancy, *J. Reprod. Fertil.,* 16, 507, 1968.
120. **Chen, R. W., Whanger, P. D., and Weswig, P. H.,** Selenium-induced redistribution of cadmium binding to tissue proteins: a possible mechanism of protection against cadmium toxicity, *Bioinorg. Chem.,* 4, 125, 1975.
121. **Parizek, J. and Ostadalova, I.,** The protective effect of small amounts of selenite in sublimate intoxication, *Experientia,* 23, 142, 1967.
122. **Ganther, H. E., Goudie, C., Sunde, M. L., Kopecky, M. J., Wagner, P., Oh, S.-H., and Hoekstra, W. G.,** Selenium: relation to decreased toxicity of methylmercury added to diets containing tuna, *Science,* 175, 1122, 1972.
123. **Ganther, H. E. and Sunde, M. L.,** Effect of tuna fish and selenium on the toxicity of methylmercury: a progress report, *J. Food Sci.,* 39, 1, 1974.
124. **Hollo, Z. M. and Zlatarov, S.,** The prevention of thallium death by sodium selenate, *Naturwissenschaften,* 4, 87, 1960.
125. **Ganther, H. E.,** Letter to the Hearing Clerk, U.S. Food and Drug Administration, February 28, 1975 (and references therein).
126. **Ganther, H. E.,** Interactions of vitamin E and selenium with mercury and silver, *Ann. N.Y. Acad. Sci.,* 355, 212, 1980.
127. **Potter, S. and Matrone, G.,** Effect of selenite on the toxicity of dietary methylmercury and mercuric chloride in the rat, *J. Nutr.,* 104, 638, 1974.
128. **Ganther, H. E., Wagner, P. A., Sunde, M. L., and Hoekstra, W. G.,** Protective effects of selenium against heavy metal toxicities, in *Trace Substances in Environmental Health — 6,* Hemphill, D. D., Ed., University of Missouri, Columbia, 1973, 247.
129. **Stillings, B. R., Lagally, H., Bauersfeld, P., and Soares, J.,** Effect of cystine, selenium, and fish protein on the toxicity and metabolism of methylmercury in rats, *Toxicol. Appl. Pharmacol.,* 30, 243, 1974.
130. **Iwata, H., Okamoto, H., and Ohsawa, Y.,** Effect of selenium on methylmercury poisoning, *Res. Commun. Chem. Pathol. Pharmacol.,* 5, 673, 1973.
131. **Ohi, G., Nishigaki, S., Seki, H., Tamura, Y., Maki, T., Konno, H., Ochiai, S., Yamada, H., Shimamura, Y., Mizoguchi, I., and Yagyu, H.,** Efficacy of selenium in tuna and selenite in modifying methylmercury intoxication, *Environ. Res.,* 12, 49, 1976.
132. **Alexander, J. and Norseth, T.,** The effect of selenium on the biliary excretion and organ distribution of mercury in the rat, *Acta Pharmacol. Toxicol.,* 44, 168, 1979.

133. **Chen, R. W., Lacy, V. L., and Whanger, P. D.,** Effect of selenium on methylmercury binding to subcellular and soluble proteins in rat tissues, *Res. Commun. Chem. Pathol. Pharmacol.,* 12, 297, 1975.
134. **Magos, L. and Webb, M.,** The effect of selenium on the brain uptake of methylmercury, *Arch. Toxicol.,* 38, 201, 1977.
135. **Ohi, G., Nishigaki, S., Seki, H., Tamura, Y., Maki, T., Maeda, H., Ochiai, S., Yamada, H., Shimamura, Y., and Yagyu, H.,** Interaction of dietary methylmercury and selenium on accumulation and retention of these substances in rat organs, *Toxicol. Appl. Pharmacol.,* 32, 527, 1975.
136. **Stoewsand, G. S., Bache, C. A., and Lisk, D. J.,** Dietary selenium protection of methylmercury intoxication of Japanese quail, *Bull. Environ. Contam. Toxicol.,* 11, 152, 1974.
137. **Beijer, K. and Jernelöv, A.,** Ecological aspects of mercury-selenium interactions in the marine environment, *Environ. Health Perspect.,* 25, 43, 1978.
138. **Chen, R. W., Ganther, H. E., and Hoekstra, W. G.,** Studies on the binding of methylmercury by thionein, *Biochem. Biophys. Res. Commun.,* 51, 383, 1973.
139. **Thibaud, Y.,** Présence simultanée de mercure et de sélénium chez le dauphin *Stenella coeruleoalba* et le thon rouge *Thunnus thynnus* de Méditerranée, *Journées d'Etudes sur les Pollutions Marines en Méditerranée, 4th,* Commission Internationale pour l'Exploration Scientifique de la Mer Méditerranée, Monaco, 1979, 193.
140. **Koeman, J. H., Peeters, W. H. M., Koudstaal-Hol, C. H. M., Tjioe, P. S., and de Goeij, J. J. M.,** Mercury-selenium correlations in marine mammals, *Nature (London),* 245, 385, 1973.
141. **Takeda, M. and Ueda, T.,** Accumulation of mercury and selenium in cultured yellowtail, *Nippon Suisan Gakkaishi,* 45, 901, 1979.
142. **Bjerregaard, P.,** Accumulation of cadmium and selenium and their mutual interaction in the shore crab *Carcinus maenas* (L.), *Aquat. Toxicol.,* 2, 113, 1982.
143. **Sokolova, E. G. and Pilipchuk, M. F.,** Geochemistry of selenium in deposits in the north-western part of the Pacific Ocean, *Geokhimiya,* (10), 1537, 1973.
144. **Tamari, Y., Hiraki, K., and Nishikawa, Y.,** Fluorometric determination of selenium in sediments with 2,3-diaminonaphthalene, *Bunseki Kagaku,* 28, 164, 1979.
145. **Pilipchuk, M. F. and Sokolova, E. G.,** Selenium in upper quaternary formations of the Tyrrhenian Sea, *Geokhimiya,* p. 1374, 1979.
146. **Tamari, Y.,** Neutron activation analysis of selenium and 17 elements in sediment, *Radioisotopes,* 28, 1, 1979.
147. **Sakai, T., Yamamoto, H., and Arakawa, Y.,** Concentration of selenium in sea and river water and in the sediments, *Mizu Shori Gijutsu,* 22, 503, 1981.
148. Certificate of Analysis, Standard Reference Material 1646, Estuarine Sediment, National Bureau of Standards, Washington, D.C., 1982.
149. **Pilipchuk, M. F.,** Molybdenum, tungsten and selenium in holocene sediments of the Tyrrhenian Sea, *Dokl. Acad. Sci. U.S.S.R., Earth Sci. Sect.,* 239, 166, 1978.
150. **Hiraki, K., Tamari, Y., Nishikawa, Y., and Shigematsu, T.,** Homogeneities of trace amounts of selenium in powdered samples and the variation of oxidation-state of the element during storage, *Bull. Inst. Chem. Res. Kyoto Univ.,* 58, 228, 1980.
151. **Siu, K. W. M. and Berman, S. S.,** Determination of selenium in marine sediments by gas chromatography with electron capture detection, *Anal. Chem.,* 55, 1603, 1983.
152. **de Oliveira, E., McLaren, J. W., and Berman, S. S.,** Simultaneous determination of arsenic, antimony, and selenium in marine samples by inductively coupled plasma atomic emission spectrometry, *Anal. Chem.,* 55, 2047, 1983.
153. **Maher, W. A.,** Fluorimetric determination of selenium in marine geological materials, *Anal. Lett.,* 16, 491, 1983.
154. **Cutter, G. A.,** Determination of selenium speciation in biogenic particles and sediments, *Anal. Chem.,* 57, 2951, 1985.
155. **Tessier, A., Campbell, P. G. C., and Bisson, M.,** Sequential extraction procedure for the speciation of particulate trace metals, *Anal. Chem.,* 51, 844, 1979.
156. **Robberecht, H. and Van Grieken, R.,** Selenium in environmental waters: determination, speciation and concentration levels, *Talanta,* 29, 823, 1982.
157. **Raptis, S. E., Kaiser, G., and Tölg, G.,** A survey of selenium in the environment and a critical review of its determination at trace elevels, *Fresenius Z. Anal. Chem.,* 316, 105, 1983.
158. **Agemian, H. and Thomson, R.,** Simple semi-automated atomic-absorption spectrometric method for the determination of arsenic and selenium in fish tissue, *Analyst,* 105, 902, 1980.
159. **Schreiber, B. and Linder, H. R.,** Programmgesteuertes gerät zum nassaufschluss organischer substanzen, *Fresenius Z. Anal. Chem.,* 298, 404, 1979.
160. **Reamer, D. C. and Veillon, C.,** Preparation of biological materials for determination of selenium by hydride generation-atomic absorption spectrometry, *Anal. Chem.,* 53, 1192, 1981.

161. **Masson, M. R.,** Microdetermination of selenium, tellurium, and arsenic in organic compounds, *Mikrochim. Acta,* I, 399, 1976.
162. **Poole, C. F., Evans, N. J., and Wibberley, D. G.,** Determination of selenium in biological samples by gas-liquid chromatography with electron-capture detection, *J. Chromatogr.*, 136, 73, 1977.
163. **Ihnat, M.,** Fluorometric determination of selenium in foods, *J. Assoc. Off. Anal. Chem.*, 57, 368, 1974.

Chapter 12

THE ATMOSPHERE

Byard W. Mosher and Robert A. Duce

TABLE OF CONTENTS

I. INTRODUCTION

The atmosphere is an important reservoir and transport medium for the biogeochemical cycling of many trace elements, including selenium. Selenium is known to exist in the atmosphere in the aerosol phase, as a gas, and in precipitation. In this chapter information will be discussed regarding the concentration of particulate and vapor-phase selenium, as well as selenium in precipitation. This will be followed by a discussion of the anthropogenic and natural sources of atmospheric selenium. Both natural and anthropogenic processes are important in the atmospheric cycling of selenium.

II. PARTICULATE SELENIUM IN THE ATMOSPHERE

A. Selenium Concentrations

The atmospheric aerosol, defined as a dispersed system of liquid and solid particles suspended in a gas, exhibits highly complex composition and physical properties resulting from the myriad of sources and chemical transformations that can occur. The sources of atmospheric selenium in aerosols, such as sea spray, windblown mineral dust, biogenic materials, volcanic effluvia, direct anthropogenic emissions, and the products of gas to particle conversion, are complex and diverse. The reader is referred to several review articles for an overview of the aerosol system.[1-3]

In Table 1 atmospheric particulate selenium concentrations from urban, semiurban, and rural continental regions are reported. In general, the concentration of urban particulate selenium is in the range of 1 to 10 ng/m^3 (total selenium, unless otherwise stated) at standard temperature and pressure (STP). These elevated concentrations observed in urban areas reflect input from anthropogenic industrial sources, and as one moves to semiurban and rural continental locations, such as Vermont and Montana in the U.S., concentrations of 0.3 to 1 ng/m^3 are typical. Exceptionally high concentrations of over 100 ng/m^3 have been reported in the urban Ankara, Turkey atmosphere by Ölmez and Aras,[32] and a single value of 120 ng/m^3 was found near a Belgian smelter.[26] As we shall see, fossil fuels such as coal and peat as well as metal refining can be major sources of atmospheric selenium. The Ankara samples were collected during the fall and winter of the year in an area where peat is the major residential fuel. Emission controls on residential fuel consumption are probably minimal in Ankara, and thus it is not surprising to find highly elevated selenium concentrations in such a location.

Most of the studies cited in Table 1 are of very short duration (weeks to months), with few lasting as long as a year or more at a particular location. Thus, it is difficult, if not impossible, to draw any conclusions as to the long-term temporal trends of particulate selenium concentrations in the U.S. or most other areas. However, atmospheric sampling was conducted for 18 years at a site in the U.K., and this record has been examined for long-term distributions of numerous constituents.[23] In Figure 1 the selenium concentration in the atmospheric aerosol at Chilton, U.K., is presented for the period 1957 to 1974. The data are presented as four quarterly moving means, and a linear regression line for the period 1957 to 1974 is also plotted. It is evident from the data and the linear regression analysis that the selenium concentration at this semirural site decreased significantly during the first few years following the enactment in the U.K. of the Clean Air Act of 1956. After this initial drop, the decline in concentration was more gradual, with the 1957 to 1974 period showing a 2.8% annual decrease. It is interesting to note that petrol consumption in the U.K. doubled in the period 1960 to 70,[37] with no obvious effect on atmospheric particulate selenium concentrations.

If we now move from the continents to marine locations (Table 2), the selenium concentrations decrease by a factor of 2 to 100 compared with continental locations. Areas such

Table 1
ATMOSPHERIC PARTICULATE SELENIUM AT URBAN AND SEMIRURAL LOCATIONS

Location	Mean selenium concentration (ng/m^3)	N[a]	Date and remarks	Ref.
U.S. Locations				
U.S. cities	2.3	79	From data for 16 cities of the 25 cities EPA Inhalable Particulate Network, 24-h sampling every 6th day, 1980	4
Boston, MA	1.2	3	~100 m^3 samples, May and June 1965	5
Boston, MA	1.23	90	8 sites, metropolitan Boston, March 25, 1970—April 27, 1970, May 12, 1970—June 29, 1970	6
Watertown, MA	0.60	332	Dichotomous sampler, 24-h samples at least every other day, June 4, 1979—June 5, 1981	7
Narragansett, RI	0.98	171	24-h samples, summer 1982, winter 1982—83	8
Underhill, VT	0.48	135	24-h samples, summer 1982, winter 1982—83	8
New York, NY	1.9	6	Semiweekly samples, midtown Manhattan, winter 1977—68	9
Buffalo, NY	6.5	18	4 sampling stations, periodic sampling 1968—69	10
Washington, D.C.	2.5	130	10 urban sites, 24-h samples, summer 1976	11
Charlestown, WV	7	NR[b]	Dichotomous sampler, 24-h samples, August 25—September 14, 1976	12
Cleveland, OH	6.9	31	24-h samples, August—November 1971	13
Ohio River Valley	2	1211	3 sites, Ohio River Valley, 12- and 24-h samples, May 1980—August 1981	14
NW Indiana	2.2	25	25 urban and semiurban sites, 24-h samples, June 11—12, 1969	15
Great Smoky Mtns. National Park, TN	1.6	14	Dichotomous sampler, 12-h samples, September 20—26, 1978	16
St. Louis, MO	2.7	4	4 sites, downtown St. Louis, 20, 40, and 60 km downwind, July 17, 1973	17
Chicago, IL	7.7	22	22 sites, metropolitan Chicago, 24-h samples, April 4, 1968	18
Chicago, IL	2.01	35	24-h samples, July 1981—January 1982	19
Colstrip, MT	0.27	48	3 sites, 1—3-d samples, May—September 1975	20
San Francisco, CA	1.1	9	9 bay area sites, 24-h samples, July 23, 1970	21
Non-U.S. Locations				
Birkenes, Norway	0.54	177	24-h samples, August 1978—June 1979	22
Chilton, England	1.28	~5000	24-h samples, collected each working day, 1954—1974; portion of filters combined in 3-month intervals for analysis	23
Rural England	1.3	~48	8 sites, monthly samples, January and December 1972, June 1972—May 1973	24
Glascow, Scotland	3.3	156	48-h samples	25
Petten, Holland	0.96	12	West coast of Holland, monthly sampling, June 1972—May 1973	24
Antwerp, Belgium	120	1	Near smelter, September 1981	26
Ghent, Belgium	11	~49	7 sites, ≤24-h sampling	27

Table 1 (continued)
ATMOSPHERIC PARTICULATE SELENIUM AT URBAN AND SEMIRURAL LOCATIONS

Location	Mean selenium concentration (ng/m^3)	N[a]	Date and remarks	Ref.
Semirural Belgium	3.7	44	Weekly sampling, 1981, 30 km SE Brussels	28
Rural Belgium	0.62	18	Sampling only when $[SO_2] \leq 18$ $\mu g/m^3$	29
Karlsruhe, W. Germany	2.1	24	Monthly sampling, 5 min/h	30
Munich, W. Germany	5.1	72	6 sites, monthly sampling, January—December 1971	31
Ankara, Turkey	116	NR	4 sites, fall and winter	32
Tokyo, Japan	3.0	30	~200 m^3 samples, January—December 1977	33
Tokyo, Japan	1.57	20	14-d samples, 7 sites, 5 periods, May 1979—January 1980	34
Mizushima, Japan	2.8	2	Fine & coarse particles	35
Yokohama, Japan	8.3	6	Fine & coarse particles	35
Xinglong, China	1.2	3	12-h samples, March 7, 8, 10, 1980	36

[a] N — number of samples analyzed.
[b] NR — not reported.

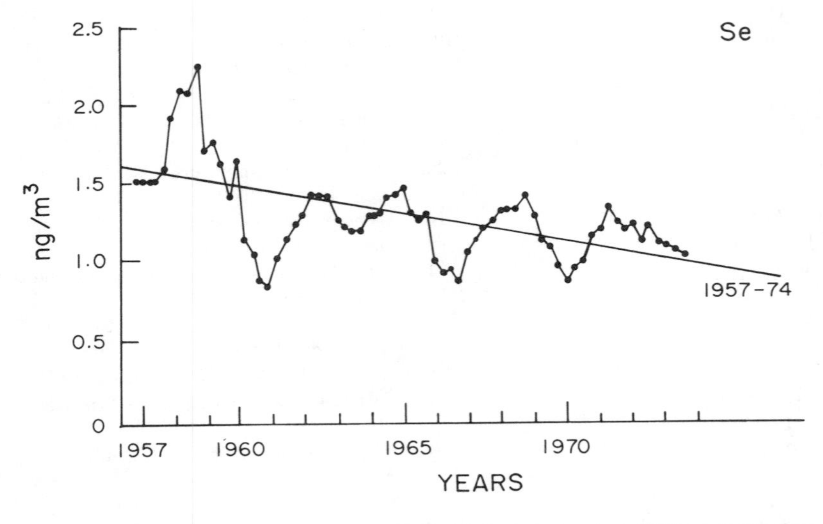

FIGURE 1. Selenium concentration in the atmospheric aerosol at Chilton, U.K.; four quarterly moving means and a linear regression line for the period 1957 to 1974. (From Salmon, L., Atkins, D. H. F., Fisher, E. M. R., Healy, C., and Law, D. V., *Sci. Total Environ.*, 9, 161, 1978. With permission.)

as the Shetland Islands off the coast of Norway are obviously affected by the transport of anthropogenic aerosol from the European continent and the U.K. Influence of North American pollution sources is also evident in some of the data collected over the North Atlantic. Particulate selenium concentrations observed in these locations can be similar to levels found in semiurban continental locations. The relatively high concentrations found along the coast of Peru can also be ascribed to the transport of pollution aerosol, in this case, smelter aerosol from several copper refineries 400 to 500 km upwind.[41]

Table 2
ATMOSPHERIC PARTICULATE SELENIUM AT MARINE LOCATIONS

Location	Mean selenium concentration (ng/m^3)	N[a]	Date and remarks	Ref.
Bermuda	0.13	29	Sector-controlled samples, April—December 1973	38
Shetland Islands	0.38	22	2 sites, monthly samples, January—December 1972, January 1972—May 1973	24
E tropical Atlantic	0.30	8	24-h shipboard samples, Chesselet et al., 1975, from Prospero[39]	39
Tropical N Atlantic	0.43	58	500—2000 m^3, shipboard samples, wind direction controlled, 1974 and 1975	40
Peru Coast	0.58	2	Shipboard samples, March—April 1981	41
Spitsbergen Island Area	0.12	4	Shipboard samples, July and September 1980	41
Enewetak Atoll	0.13	55	April—August 1979	42
Shemya, AK	0.25	52	Weekly samples, January 1981—May 1983	43
Midway Island	0.13	119	Weekly samples, January 1981—June 1983	43
Oahu, HI	0.11	114	Weekly samples, April 1981—July 1984	43
Enewetak Atoll	0.10	80	Weekly samples, January 1981—June 1983	43
Fanning Island	0.38	146	Weekly samples, April 1981—July 1984	43
Guam	0.11	62	Weekly samples, January 1981—September 1982	43
Belau	0.10	28	Weekly samples, February 1981—April 1983	43
American Samoa	0.062	4	Sector-controlled samples, January—February 1981	41
Cape Reinga, NZ	0.11	11	Sector-controlled samples, June—August 1983	43
Mauna Loa, HI	0.016	2	Mountain top samples (3400 m), September—October 1981	41
Mauna Loa, HI	0.011	180	3400 m, "Clean season", downslope winds only	44
Mauna Loa, HI	0.021	50	3400 m, "Asian dust season"	44

[a] N — number of samples analyzed.

At remote Northern Hemisphere marine locations as diverse as Spitsbergen Island, Bermuda, Enewetak Atoll, and the Hawaiian Islands, particulate selenium concentrations are remarkably similar and fall in the range of 0.1 to 0.2 ng/m^3. These levels may represent a Northern Hemisphere "background level". It is interesting to note that Priest et al.[29] used sulfur/selenium ratios observed in the rural Belgian atmosphere, extrapolated to very low sulfur levels, to predict a "background concentration" of 0.17 ng/m^3 for particulate selenium, very similar to concentrations found in remote marine locations. The only measurements from marine locations in the southern hemisphere are those of Mosher and Duce[41,43] at the island sites of American Samoa and New Zealand. At Samoa, the mean particulate selenium concentration during a 2-month period in early 1981 was 0.062 ng/m^3 (range 0.043 to 0.080 ng/m^3), or roughly half typical concentrations at the Northern Hemisphere locations cited. Somewhat higher particulate concentrations were observed at Cape Reinga, New Zealand, where a mean of 0.108 ng/m^3 and a range of 0.069 to 0.191 ng/m^3 were found for 11 samples. Air mass trajectories indicated that in some cases the air sampled at Cape Reinga had recently crossed the Australian continent. While it is tempting to speculate on interhemispheric differences, the Southern Hemisphere data base at present is far too small to draw any definite conclusions in this regard.

Also presented in Table 2 are data from approximately 700 samples from the SEAREX (sea/air exchange) Program Asian Dust Study (SADS) network.[43] The data provide some

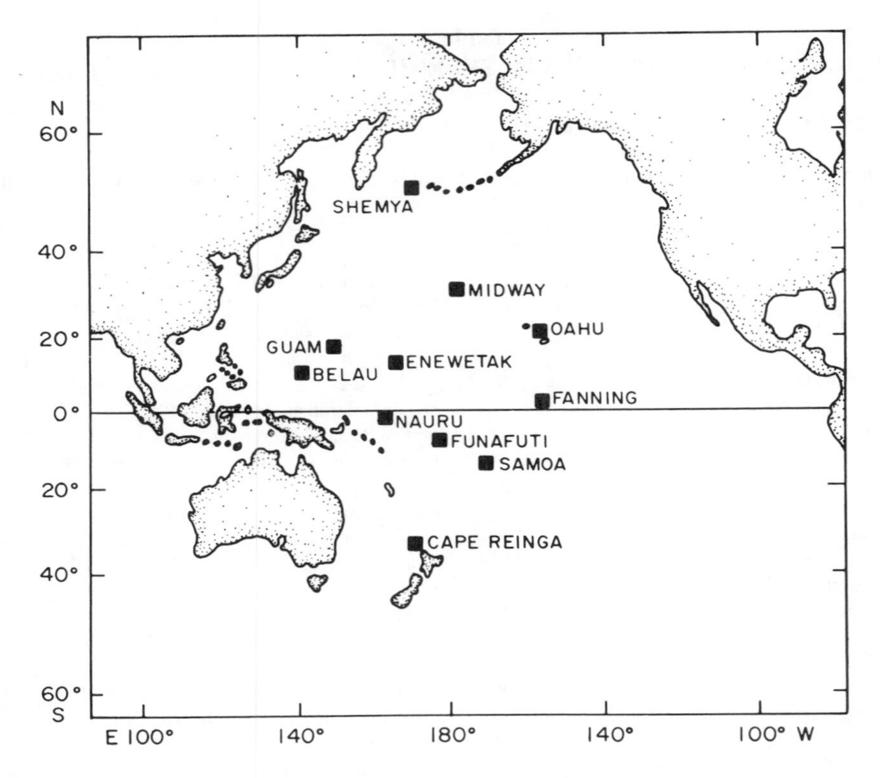

FIGURE 2. Locations of SEAREX island aerosol sampling sites in the North and South Pacific Ocean.

information concerning the sources of selenium in the remote North Pacific atmosphere. The SADS network consist of seven island stations: Shemya, Midway, Oahu, Guam, Enewetak, Belau, and Fanning[45] (Figure 2). Continuous aerosol samples were collected for 1-week periods starting in early 1981 at these locations. In Figure 3 the mean particulate selenium concentrations at these sites have been plotted as a function of latitude. Also plotted are the excess (non-sea salt) sulfate concentrations observed by Prospero et al.[46] at the SADS sites along with additional excess $SO_4^=$ data from a similar network of island sampling sites in the South Pacific[47] (Figure 2). There are interesting similarities in the behavior of particulate selenium and excess (non-sea salt) sulfate over the Pacific. Both species exhibit relatively high concentrations at the North Pacific island site of Shemya at approximately 50° N, where vigorous meterological circulation could certainly result in transport of pollutants from the U.S.S.R., China, and Japan. As one moves southward to the central Pacific basin islands of Midway, Oahu, and Enewetak, both particulate selenium and excess sulfate concentrations decrease by approximately a factor of two. In the vicinity of the equator both species exhibit relative maxima in concentration. Prospero et al.[46] have suggested an equatorial oceanic source for the excess sulfate. This marine source is probably sulfate precursors in the form of reduced organosulfur species, such as dimethyl suflide (DMS), which are subsequently oxidized in the atmosphere to SO_2 and $SO_4^=$.

The elevated concentrations found near the equator could result from the upwelling of cool relatively nutrient-rich waters and the attendant increased biological production that occurs in the equatorial Pacific. In fact, Cline and Bates[48] have observed a pronounced maxima in surface water DMS concentration between 2° and 4° N along a transect 10° to the east of Fanning. New data concerning particulate concentrations of methane sulfonates (oxidation products of DMS) collected at Shemya[47] suggests that the high excess sulfate

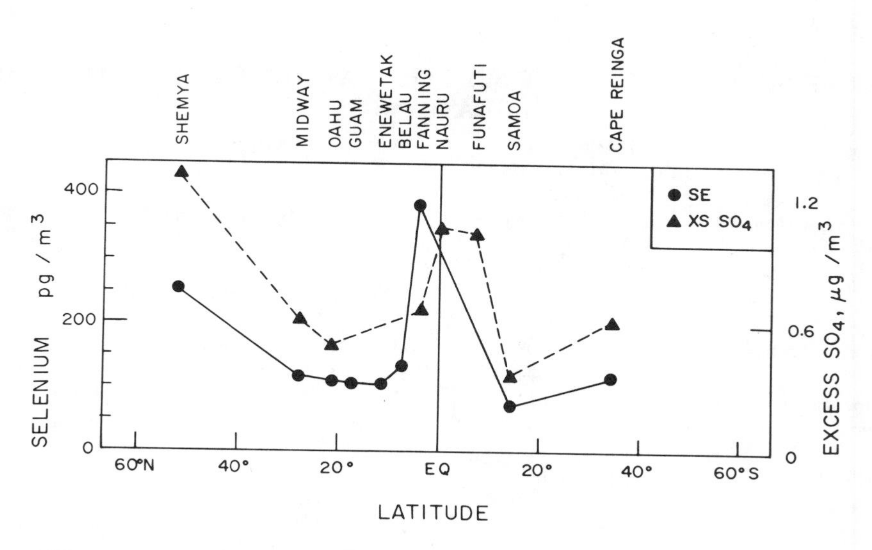

FIGURE 3. Selenium and non-sea salt sulfate as a function of latitude in the North and South Pacific Ocean.

observed in the Aleutians could also be largely natural due to the high seasonal biological productivity in the surrounding waters. Andreae and Raemdonck[49] present DMS data for 628 surface seawater samples collected in the major world oceans. They observed that the DMS vertical distribution, patchiness, and distribution in oceanic ecozones was similar to the pattern exhibited by primary biological production. The similarities in the excess sulfate and particulate selenium distribution suggest that this may also be the case for selenium. Fitzgerald et al.[50] found significant increases in total gaseous mercury near the ocean surface in the equatorial region (4° N to 10° S) of the Pacific at 160° N which they also attributed to increased biological activity.

Simultaneous sampling at two island sites in Hawaii, Oahu, and Mauna Loa, revealed that particulate selenium concentrations were a factor of ten lower in the free troposphere (3400 m elevation) than in the marine boundary layer[41] (Table 2). This distribution also suggests an oceanic source if one assumes that a vapor-phase species produced in the ocean surface has a longer residence time in the atmosphere than the particulate phase. The authors suggested that release of volatile oceanic organoselenium compounds followed by gas-to-particle conversion could support the particulate selenium concentrations observed in remote marine regions.

In Table 3 we see that particulate concentrations typical of remote continental areas are extremely low, and this fact again supports the suggestion that the ocean can be a major source of atmospheric particulate selenium. The extremely low concentrations observed at the South Pole may be more typical of lower stratospheric levels due to the elevation (2800 m) and meteorology of the polar plateau. Concentrations at Barrow, Alaska, during the winter months can reach levels found in semirural regions, and this has been attributed to long-range transport of pollutant aerosols from Eurasia.[60,61] Selenium concentrations can be much lower (0.03 to 0.05 ng/m^3) at Barrow at times when pollution aerosol is not transported into the arctic.[58]

Thus, in moving from continental urban areas to remote marine and continental locations, the particulate selenium appears to decrease by roughly a factor of 100 (5 down to 0.05 ng/m^3). These differences in observed selenium concentration, while significant, are actually quite small. This becomes especially evident when one compares the atmospheric distribution of selenium with an element such as lead, which in both urban and remote locations is

Table 3
ATMOSPHERIC PARTICULATE SELENIUM AT REMOTE CONTINENTAL LOCATIONS

Location	Mean selenium concentration (ng/m³)	N[a]	Date and remarks	Ref.
South Pole, Antarctica	0.0066	NR[b]	2800 m, Austral summer 1971, 1975, 1976, 1978 and winter periods in 1975—76	51—53
South Pole, Antarctica	0.0084	30	2800 m, Austral summer	
	0.0048	159	Austral winter	54
North America, Gulf of Mexico	0.024	20	Aircraft samples, free troposphere	55
Jungfraujoch, Switzerland	0.042	18	3752 m, monthly samples, August 1973—August 1975	56
Chacaltaya, Bolivia	0.05	NR	5220 m, 14-d—3-week samples, July 1976—August 1977	57
Barrow, AK	0.11	21	Sea level, 3—4 samples, December 1977—March 1978	58
Northern Greenland	0.026	3	Sea level, several-day samples, June—August 1974	59

[a] N — number of samples analyzed.
[b] NR — not reported.

primarily derived from anthropogenic processes. Settle et al.[62] have reported that the atmospheric particulate lead concentration is five orders of magnitude lower in Samoa than in urban areas. The relatively uniform distribution of particulate selenium in the atmosphere is another indication that an areally uniform source (i.e., the ocean) may be important in the cycling of selenium through the atmosphere.

Before discussing the distribution of the gas phase and selenium in precipitation, it will perhaps be useful to discuss two general properties of atmospheric particulate selenium: particle size distribution and selenium enrichment.

B. Selenium Particle Size Distribution

One feature of atmospheric particulate selenium which provides information about the sources and cycling of the element is its particle size distribution. Particle size distribution is also an important factor in assessing the physiological importance of airborne particulate matter for several reasons. It is known that submicrometer particles have a relatively long residence time in the atmosphere, are least efficiently removed by pollution control devices, may possess high surface concentrations of toxic trace elements such as selenium,[63-65] and are most efficiently trapped in the lungs.[66] In addition, elements such as selenium which can be volatilized during combustion will preferentially adsorb on and/or form very small particles ($\leq$1 μm radius).

In Figure 4 the results of the analysis of a size-separated cascade impactor aerosol sample collected in Bermuda by Duce et al.[38] are plotted. The major mass of the crustal elements, such as aluminium, is found on particles of radii about 1.5 μm. While selenium and several other rather volatile elements, such as lead, do show secondary concentration maxima on stage two (cutoff radius of 1.5 μm), their major mass is on particles of radii less than 0.75 μm. The relationships observed in this Bermuda sample agree well with the compilation of particle size distribution data reported by Rahn.[3] Rahn[3] found a median mass-median diameter (MMD)* for selenium of 0.8 μm in 14 separate investigations of Se size distributions, while

* MMD — the particle size for which 50% of the mass is on larger particles and 50% on smaller.

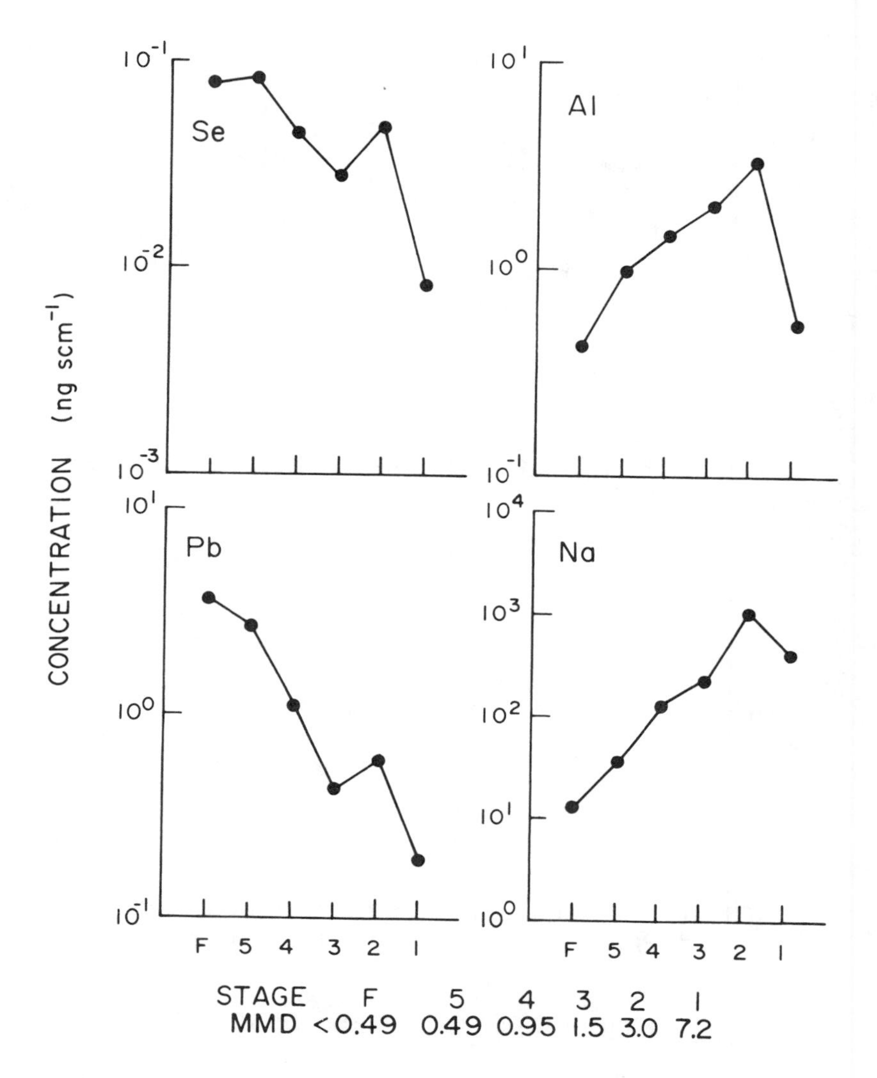

FIGURE 4. Mass size distributions of atmospheric aerosol particles for selenium (Se), aluminum (Al), lead (Pb), and sodium (Na) at Bermuda. SCM: standard cubic meter; 1, 2, 3, 4, 5, F are stages of the cascade impactor sampler with median mass median diameter (MMD) particle sizes express in micrometers. (From Duce, R. A., Ray, B. J., Hoffman, G. L., and Walsh, P. R., *Geophys. Res. Lett.*, 3, 339, 1976. With permission.)

the median MMDs for elements derived from crustal weathering processes, such as aluminium (MMD = 5 μm), and elements derived from the ocean, such as sodium (MMD = 3 μm), were substantially larger. Numerous studies of size-fractionated aerosol in urban areas,[67] power plant plumes,[68] fly ash,[69] remote continental areas, such as Barrow, Alaska,[70] and remote marine locations,[42] also indicate that selenium is concentrated in the submicrometer size range.

Particle size distribution data for aluminium, sodium, lead, and selenium at the remote Pacific island of Enewetak determined by Duce et al.[42] are presented in Figure 5. Due to the much greater distance from continental source areas, the absolute concentrations of lead and aluminium at Enewetak are much lower than those found at Bermuda. The particle size distributions for these two elements are very similar at these two sites, however. This is also true for the sea salt element sodium, although the absolute sodium concentration at

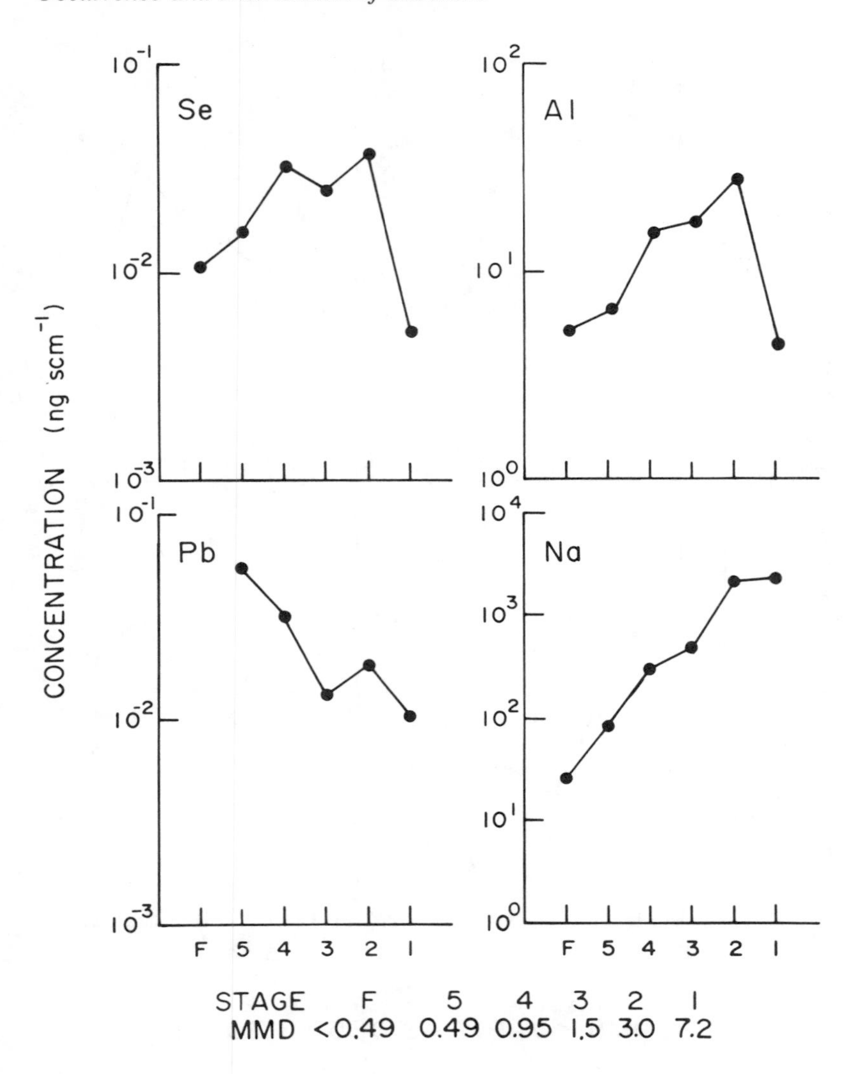

FIGURE 5. Mass size distributions of atmospheric aerosol particles for selenium (Se), aluminium (Al), lead (Pb), and sodium (Na) at Enewatak. SCM: standard cubic meter; 1, 2, 3, 4, 5, F are stages of the cascade impactor sampler with median mass median diameter (MMD) particle sizes expressed in micrometers. (From Duce, R. A., Arimoto, R., Ray, B. J., Unni, C. K., and Harder, P. J., *J. Geophys. Res.*, 88, 5321, 1983. With permission.)

Enewetak is slightly larger. At this remote marine location, the particulate selenium size distribution is different from that at Bermuda, with the major difference in the distribution at Enewetak being the absence of a maximum in concentration in the smallest (final and fifth stages) particle size range, as observed in Bermuda.

In order to gain some insight into the processes responsible for the selenium particle size distribution found at this remote marine region, it is useful to examine the particle size distribution of two additional particulate species which are both believed to have sources similar to those of selenium. While both marine non-sea salt sulfate and aerosol nitrate are thought to be produced from gaseous precursors, there are major differences in their marine particle size distributions. At marine locations such as Barbados, West Indies, and Sal Island, Cape Verde Islands, Savoie and Prospero[71] have investigated the nitrate, non-sea salt sulfate, and sea salt particle size distributions. They found that the non-sea salt sulfate, being derived primarily from gaseous precursors, is contained on submicrometer particles. Although particulate nitrate is also presumed to be secondary in nature (resulting from gas to particle conversion), Savoie and Prospero[71] found that the nitrate size distribution was distinctly different from that excess sulfate and similar to that of sea salt (i.e., sodium). In fact, these authors found nitrate concentrated on particles larger than 1 μm with its *mass* distribution very similar to that of the sea salt *surface area*. These facts have prompted several authors[71-73] to suggest that gaseous NO_X/HNO_3 may produce aerosol nitrate by the adsorption and subsequent reaction of NO_X or dissolution of HNO_3 on alkaline sea salt particles rather than reacting with the submicrometer acidic particles containing sulfate and ammonium ions. It has been suggested[71] that the extremely low concentration of nitrate on submicrometer particles could be due to the relative volatility of HNO_3 and NH_4NO_3 when compared to the sulfur species H_2SO_4, NH_4HSO_4, and $(NH_4)_2SO_4$ which would form in these submicrometer particles. If this is the case, then obviously sea salt particles play an important role in the air-to-sea flux of odd nitrogen (NO_X, HNO_3) species, and the deposition or removal rates of nitrate and sulfate should be quite different.

The similarities in remote marine nitrate and selenium particle size distributions and the observed nitrate particle size/sea salt surface area distribution relationship suggest that reaction of vapor-phase organoselenium compounds with sea salt may represent a major mechanism for gaseous selenium as well as gaseous NO_X/HNO_3. Removal of gaseous selenium on the surface of sea salt particles may also mean that biogenic selenium compounds would exhibit a shorter residence time in the marine atmosphere than their sulfur analogues since sea salt particles are in general much larger than sulfate-containing particles, which are generally less than 1 μm.

C. Selenium Enrichment in Atmospheric Particulate Matter

A second informative feature of atmospheric particulate selenium concentration is what has been termed the "anomalous enrichment" of selenium on aerosol particles. Most trace elements found in aerosol samples have a natural geochemical origin (i.e., windblown crustal material, sea spray, etc.). It has been estimated that 80 to 95% of the global atmospheric particulate burden results from natural sources.[1] Of these natural sources sea salt is estimated to account for 15 to 65% of the total mass, and windblown dust several to 25% of the total. For atmospheric aerosols collected near the surface of the earth it is useful to compare the aerosol composition with that of these two major sources. This can be done conveniently by calculating enrichment factors[3,52,74,75] relative to either crustal or seawater composition. Such an enrichment factor can be defined as

$$EF\,(X)_{source} = \frac{(X/Ref)_{aerosol}}{(X/Ref)_{source}} \tag{1}$$

where $EF(X)_{source}$ is the enrichment factor with respect to concentration for element X in the aerosol relative to the source of interest, $(X/Ref)_{aerosol}$ is the ratio of the concentration of the element of interest to that of a reference element in the aerosol, and $(X/Ref)_{source}$ is this same ratio in the source material, i.e., crustal material or bulk seawater.

The most commonly used crustal reference elements are aluminium, silicon, iron, or scandium, while sodium is generally used for sea salt. Elements with enrichment factors near unity are essentially present in crustal (seawater) proportions, are termed "nonenriched", and probably have major crustal (sea salt) sources. Those elements with EF significantly greater than unity relative to both crustal and oceanic sources are termed "anomalously enriched elements" (AEE) and may have other major sources in addition to the crust or sea.

One must be aware of the limitations involved in the use of enrichment factors. In polluted

Table 4
SELENIUM ENRICHMENT FACTORS FOR ATMOSPHERIC AEROSOL PARTICULATES

Location	Enrichment factor relative to Crust	Enrichment factor relative to Sea	Ref.
Enewetak Atoll	23,000	8.2×10^3	42
Buffalo, NY	4,500	1.5×10^6	76
Bermuda	2,600	2.4×10^4	38
South Pole, Antarctica	12,300	5.2×10^5	51
St. Louis, MO	2,900—7,000	$0.6—1.4 \times 10^6$	17
San Francisco, CA	1,900	8×10^4	21
Chacalytaya, Bolivia	1,000	1.4×10^5	57
NW Territories, Canada	1,060	6.6×10^5	74

areas, high enrichment factors may be attributed almost exclusively to anthropogenic sources. Also, high enrichment factors observed in remote areas may be due to long-range transport of urban pollutants, other natural sources such as forest fires, biological or vegetative production, volcanoes, and chemical enrichment at the sea surface during atmospheric sea salt production. Also, EF near unity can result from anthropogenic activities, since processes such as coal combustion and activities at construction sites can emit material of essentially crustal composition. With these factors in mind, let us examine some enrichment factors for selenium derived from aerosol measurements around the globe.

Table 4 presents some representative EF_{crust} and EF_{sea} values for selenium. It is evident from the table that one must consider other sources in addition to crustal weathering and aerosol formation from bulk seawater in order to account for observed aerosol selenium concentrations. Crustal enrichment factors of 1000 or greater are common in both urban and remote areas, while EF_{sea} values are several orders of magnitude larger. In urban areas such as Boston[6] and Washington, D.C.,[11] multivariate analysis and chemical element balance (CEB) model techniques have been applied to the results of extensive particulate analysis in an effort to identify sources of aerosol components. In the case of selenium, these approaches have met with only limited success. Kowalcyzk et al.[11] considered coal, oil, and refuse combustion, soil dust, marine aerosols, motor vehicle emissions, and limestone source functions in a study of the aerosol trace element distribution in Washington, D.C. While the authors identified coal combustion as a major source of atmospheric selenium, arsenic, chromium, and copper, the predicted selenium levels generated by their CEB model were only approximately 35% of the observed concentrations. The authors suggested that vapor-phase selenium emissions produced during combustion processes might explain this discrepancy.

The similarity in enrichment factors for selenium observed in both Northern and Southern Hemisphere sites has led Duce et al.[75] to speculate that this anomalous enrichment may originate from some natural vapor-phase source such as volcanism, biological methylation, or low-temperature volatilization (recall earlier discussion of a possible vapor-phase selenium release from the ocean). In part, this is because roughly 90% of anthropogenic emissions are estimated to occur in the Northern Hemisphere.[77] Atmospheric residence times for aerosols are generally on the order of a week or so,[1] and tropospheric interhemispheric mixing times are approximately 1 year.[78] It is thus difficult to explain these similarities as a result of transport of pollutant aerosols from major source areas in the Northern Hemisphere to the relatively clean atmosphere of the Southern Hemisphere. In addition to selenium, a number of other elements such as zinc, copper, antimony, lead, cadmium, silver, arsenic, and mercury are also commonly observed to be anomalously enriched on aerosol particles, although generally not to the extent of selenium. Lest this discussion and the term "anomalous enrichment" leave the reader with the impression that selenium comprises a large fraction of the total aerosol burden, we should point out that total suspended particulate matter in the atmosphere is typically on the order of 1 to 200 $\mu g/m^3$, while selenium concentrations vary from several picograms to roughly 20 ng/m^3. Thus, atmospheric selenium levels range from several to approximately 100 mg/kg of the total aerosol by weight.

Both enrichment factor calculations and particle size distribution data suggest that the major mass of AEE, and of selenium in particular, is associated with particles having an important source(s) other than the crust or sea salt produced directly by the ocean.

III. VAPOR-PHASE SELENIUM IN THE ATMOSPHERE

In the preceding section, several lines of evidence have been suggested that vapor-phase species are important in the atmospheric cycling of selenium: (1) similarities in particulate non-sea salt sulfate and selenium distributions over the remote Pacific coupled with a likely natural vapor-phase source for excess $SO_4^=$ near the productive equatorial upwelling regions; (2) significant natural vapor-phase mercury emissions along the equator apparently related to increased biological productivity; (3) much lower particulate selenium concentrations above the remote marine boundary layer; (4) very low remote continental particulate selenium levels; (5) relatively uniform global distribution of particulate selenium; (6) preferential concentration of selenium on small particles; (7) inability of urban source strength models to account for observed selenium concentrations; and (8) similarities in Northern and Southern hemisphere selenium enrichment factors. These observations all point to the potential importance of vapor-phase selenium.

Before reviewing the literature concerning vapor-phase selenium species in the atmosphere, the term "vapor phase" should be defined. In practice, "vapor phase" is operationally defined as that portion of atmospheric selenium which passes through particulate collection filters. An obvious disequilibrium will exist between particulate matter on the filter and the air passing through it. Calculations have suggested that significant quantities of compounds such as SeO_2 can be lost during and after filter collection.[79,80] This is due to the relatively high equilibrium vapor pressure of SeO_2 and tends to complicate the definition of "vapor phase". Loss of selenium from filters has been reported by Shendrikar and West.[81] Smoke from trash burning was sampled with a filter/H_2O impinger system for a fixed sampling period; the impinger was then refilled with fresh water, and clean laboratory air was pulled through the filters for 4 h. Desorbed selenium was trapped in the impinger with observed losses averaging 40%.

Several sampling methods have been used to characterize the particulate/vapor partitioning of selenium in the atmosphere. In some of the first published work, Pillay and Thomas[76] and Pillay et al.[10] employed impingers filled with a solution of lead acetate behind particulate filters. The authors sampled intermittently for 1 year in Buffalo, New York, and found that an average of 56% of the selenium passed through filter which were nearly 100% efficient for particles of diameter equal to or greater than 0.1 μm.

Activated charcoal has also been used to collect total gaseous selenium. Laboratory studies[41,82] have shown that activated charcoal is an efficient collection substrate for vapor-phase SeO_2 and methylated selenium compounds such as dimethyselenide, $(CH_3)_2Se$, and dimethyldiselenide, $(CH_3)_2Se_2$. In the urban Maryland atmosphere, Reamer[82] used activated charcoal filter disks and found 10 to 40% of selenium in the vapor phase, with total concentrations of 0.69 to 1.1 ng/m^3. A similar partitioning (15 to 36%) was observed by Mosher and Duce[41] at urban and semirural locations in Rhode Island. Total vapor-phase selenium concentrations were found to be in the range of 0.1 to 0.6 ng/m^3. Vapor-phase selenium concentrations and available speciation data are reported in Table 5.

Table 5
ATMOSPHERIC VAPOR-PHASE SELENIUM CONCENTRATIONS AND SPECIATION

Location	Mean selenium concentration in vapor phase (ng Se/m³)	N[a]	Speciation[b]	Percent of total atmospheric selenium in vapor phase	Date and remarks	Ref.
Buffalo, NY	3.51	14	Total	44—63	4 sites, lead acetate traps, 1968—69	10
Narragansett, RI	0.25	17	Total	15—36	Activated charcoal[c], December 1979—September 1980	41
Providence, RI	0.41	3	Total	18—28	Activated charcoal, February 1980—September 1980	41
Coastal Peru	0.065	2	Total	10	Ship samples, activated charcoal, March—April 1981	41
Oahu, HI	0.035	4	Total	13—29	Sea level, activated charcoal, September—October 1982	41
Mauna Loa, HI	0.0012	2	Total	39—48	3400 m, activated charcoal September—October 1982	41
American Samoa	0.022	2	Total	17—35	Activated charcoal, January—February 1981	41
Piscataway, MD	2.02	5	$(CH_3)_2Se$	13—58	Collected next to sewage digestion tank, chrom./GC[d]	82
Antwerp, Belgium	1.04	7	$(CH_3)_2Se$	NR[e]	4 sites, cryo/AAS[f], April—October 1981	26
Antwerp, Belgium	0.50	5	$(CH_3)_2Se_2$	NR	2 sites, cryo/AAS[f], April—October 1981	26
Antwerp, Belgium	0.38	3	$(CH_3)_2SeO_2$	NR	3 sites, cryo/AAS[f], April—October 1981	26

[a] N — number of samples analyzed.
[b] Selenium compound, the concentration of which was determined (expressed as ng Se/m³ in Column 2).
[c] Activated charcoal collection.
[d] Chromatographic column trapping/gas chromatographic detection.
[e] NR — not reported.
[f] Cryogenic trapping/atomic absorption detection.

Sampling for total vapor-phase selenium has also been conducted at four remote locations in the Pacific. While vapor selenium concentrations at the Southern Hemisphere site of Samoa may be slightly lower than those found at sea level in the northern hemisphere (Oahu), the vapor/particle partitioning appears to be quite similar at both sites (13 to 35%). Both the particulate and vapor-phase concentrations above the marine boundary layer (3400 m) at Mauna Loa, Hawaii, are extremely low, and it is interesting to note that the vapor-phase fraction is substantially higher (39 to 48%) than that observed at sea level locations such as Oahu. The Mauna Loa samples were collected during the same time period as the Oahu samples at locations approximately 300 km apart. The differences observed in absolute and relative selenium concentrations of both species provide insights into the sources and residence times of particulate and gaseous selenium. The overall decrease in absolute concentration of both species and the increase in vapor fraction as one moves away from the ocean surface are both consistent with the suggestion of an oceanic source of vapor-phase selenium which has a longer residence time in the atmosphere than the particulate phase.

Data concerning the actual speciation of vapor-phase selenium in the atmosphere are very limited. One study of coal-fired power plants indicated that vapor-phase selenium was in the elemental form.[83] As the melting point of elemental selenium is 685°C, this form would not remain in the vapor phase for an appreciable length of time. The most important vapor-phase species of selenium found in the atmosphere appear to be the methylated forms such as $(CH_3)_2Se$ and $(CH_3)_2Se_2$. Sampling and analysis strategies such as cryogenic trapping/thermal desorption/atomic absorption detection[84] and chromatographic column collection/solvent elution/gas chromatographic detection[82] have been utilized to study these species.

Laboratory studies using elevated selenium levels have shown that these methylated compounds can be formed in soil, sediments, and sewage sludge,[85-91] plants,[92,93] and animals.[94] In ambient sampling, Reamer[82] detected dimethylselenide (0.2 to 5.4 ng/m^3) in the immediate vicinity of sewage digestion tanks. Jiang et al.[26] also identified $(CH_3)_2Se$ and $(CH_3)_2Se_2$ at concentrations of up to 2.4 ng/m^3 in the urban Belgian atmosphere. In the latter study, several sewage treatment works, a coal-fired power plant, and a selenium-producing smelter were in the immediate vicinity of the sampling site, and a very high particulate selenium concentration (120 ng/m^3) was detected in one sample.

A third organoselenium compound, tentatively identified as dimethylselenone, $(CH_3)_2SeO_2$, has been reported[26,86] in several ambient samples at levels around 0.30 ng/m^3. At this time, the practical limitations involved with cryogenic or chromatographic column sampling of organoselenium compounds have only allowed measurements to be made in urban areas.

IV. SELENIUM IN PRECIPITATION

We have seen thus far that the atmosphere is an important component in the biogeochemical cycling of selenium. Perhaps the most important removal mechanism for selenium from this reservoir is scavenging by precipitation. As is the case with data on the distribution and speciation of vapor-phase selenium, there is at present a real paucity of data concerning the removal of atmospheric selenium by precipitation. Precipitation concentration data available for selenium are presented in Table 6.

The bulk samples, where a bucket or similar collection device is left continuously open to the atmosphere, exhibit much higher selenium concentrations than the "rain only" samples. This is due to appreciable quantities of dry deposition, particularly resuspended local material, collected in bulk samplers. Arimoto et al.,[99] after correcting for the estimated 15% of the selenium mass attributable to recycled sea salt, have calculated that in a remote marine location such as Enewetak Atoll, net dry deposition may account for roughly 30% of total selenium removal from the atmosphere. Reported rain concentrations in urban areas are in the range of 0.1 to 0.4 μg/l with values in remote marine regions approximately a factor of three to ten less. An inverse relationship between total rain volume and selenium concentration has also been observed in 31 rain samples collected in Tokyo.[33]

Weiss et al.[100] examined the selenium and sulfur content of 14 dated glacial ice samples from Camp Century, Greenland. These samples spanned the time period from 800 B.C. to 1965. The authors observed that while sulfur concentrations increased substantially after the period of the industrial revolution in the Northern Hemisphere, a similar increase in selenium levels was not apparent. They concluded that fossil fuel combustion had not yet contributed to the long-range transport of selenium as it had for sulfur. In Antarctica, Boutron et al.[101] also observed similar selenium concentrations in preindustrial ice and in contemporary snow samples. They likewise concluded that the remote polar regions of the Southern Hemisphere have been little affected by global pollution in the case of selenium.

Table 6
SELENIUM CONCENTRATIONS IN PRECIPITATION

Location	Sample type	Mean selenium concentration	N[a]	Ref.
	Rain (μg/l)			
Swansea, U.K.	Bulk[b]	1.1	NR[c]	95[d]
N Norway	Bulk	1	NR	95[e]
U.K., 6 sites	Bulk	0.56	NR	24
Wraymires, U.K.	Bulk	0.34	NR	96
Gas platform, North Sea	Bulk	1.75	NR	24
Boston, MA	Rain	0.4	NR	5
Quillayute, WA	Rain	0.07	2	97
Tokyo, Japan	Rain	0.069	92	33
Ghent, Belgium	Rain	0.14	2	98
Enewetak Atoll	Rain	0.036	12	99
	Ice (μg/kg)			
Greenland Ice Sheet	Ice	0.012	14	100
Coastal Antarctica	Ice (12,000 years BP)	0.010	NR	101
Dome C, Antarctica	Ice (present-day surface snow)	0.012—0.030	NR	101
	Snow (μg/kg)			
Boston, MA	Snow (falling)	0.20	12	5
Boston, MA	Snow (ground)	0.08	7	5

[a] N — number of samples analyzed.
[b] Bulk — wet and dry deposition.
[c] NR — not reported.
[d] Work of N. J. Pattenden (A. E. R. E. Rep. R-7729, Her Majesty's Stationery Office, London) as reported in Galloway et al.[95]
[e] Work of Hanssen, J. E., Ramback, J. P., Semb, A., and Steinnes, E., (Proc. Int. Conf. Ecological Impact of Acid Precipitation, Sandefjord, Norway, SNSF Project, As-NHL, Norway, 1980) as reported in Galloway et al.[95]

V. ANTHROPOGENIC SOURCES OF ATMOSPHERIC SELENIUM

A. Coal Combustion

The major anthropogenic source of atmospheric selenium appears to be the combustion of coal. A number of studies have indicated that a substantial portion of the selenium contained in coal may escape even very efficient pollution control devices and be released to the atmosphere in the form of a gas and/or very fine particles. Andren et al.[83] sampled coal, slag, fly ash, and precipitator inlet and outlet flue gases at a coal-fired power plant in Memphis, Tennessee, and were able to construct a selenium mass balance for the facility. They found that only 0.3% of the selenium contained in the coal was incorporated into the slag, while roughly 68% of the selenium was found on fly ash particles. The remainder (32%) was discharged in the vapor phase. This particular plant was equipped with electrostatic precipitators, which are very efficient (99.6%) for the removal of fly ash. Consequently, more than 90% of the selenium emitted to the atmosphere was in the vapor state. These authors also found that all selenium in the fly ash and slag existed in the elemental form and speculated that excess SO_2 produced in the coal combustion process acts as a reducing agent for any selenium dioxide formed (Equation 2):

$$SeO_2 + 2SO_2 \rightleftharpoons Se^0 + 2SO_3 \quad (2)$$

In addition they suggested that vapor-phase selenium would also be in the form of Se^0.

From this mass balance, Andren et al.[83] were able to estimate that combined vapor phase and fly ash atmospheric discharges and fly ash removal to land fills and slag ponds are 2.5 and 1.5 times larger, respectively, than river mobilization of selenium in the U.S. and the world. These findings have been confirmed by other studies of coal-fired power plants, such as those of Ondov et al.,[102] Conzemius et al.,[103] and Block and Dams,[104] where between 7 and 60% of the selenium contained in coal was emitted in the vapor phase from plants equipped with modern pollution control devices. Kaakinen et al.[105] also found more than 50% of the selenium unaccounted for at a coal-fired power plant equipped with a high-efficiency mechanical dust collector followed by an electrostatic precipitator and a wet scrubber in parallel. Similar results were reported by Shendrikar et al.[106] for fabric baghouse control devices, where selenium showed penetrations an order of magnitude greater than that of the total particulate mass and of most other elements. Finally, numerous studies of coal-fired power plant aerosols indicate that selenium is preferentially concentrated on small fly ash particles.[67-69,107-111] As was mentioned previously, the larger surface-to-volume ratio of these small particles results in a volatile trace element concentration which is inversely related to particle diameter.

Submicrometer particles are not efficiently collected by pollution control devices,[112] and these particles are enriched in potentially toxic and mutagenic inorganic species.[113] While fugitive fly ash or gaseous selenium emissions do not appear to pose a significant environmental problem in the vicinity of coal-fired power plants,[114] the potential for the contamination of water resources through the disposal of fly ash does exist. Dreesen et al.[115] found that approximately 5% of the selenium content of fly ash was soluble in redistilled water (pH 7.4). As the pH was decreased, solubility increased, with as much as 78% solubilization at a pH of 0.5 (1.0 *M* HNO_3). Crecelius[116] investigated the solubility of coal fly ash and marine aerosol in seawater and reported a 16 to 69% release of selenium from fly ash and a 25 to 39% release from marine aerosols collected on the Washington coast. Hardy and Crecelius[117] collected aerosols in St. Louis, Missouri, and Seattle, Washington, and found that between 28 and 36% of the selenium content was soluble in seawater. Similar results were reported for 31 aerosol samples collected in Tokyo,[33] where 54% of the selenium retained by filters was water soluble. Coal combustion in the U.S. alone generates at least 40 million tons of fly ash annually,[118] and thus it is evident that a substantial quantity of selenium can find its way into natural waters. This fly ash-derived selenium may then be taken up by plants, foraging animals, and aquatic species.[119-121]

Literature estimates of the magnitude of global atmospheric emissions of selenium to the atmosphere as a result of coal combustion are presented in Table 7. The assumed fraction of selenium released during coal combustion varies widely from author to author (5 to 90%), with flux estimates, reflecting this variation, exhibiting roughly a factor of difference of 21 between the extremes.

B. Fuel Oil Combustion

While similar quantities of coal and fuel oil are combusted globally, emissions from oil-fired power plants and other fuel oil emission sources introduce significantly less selenium into the atmosphere. This is due to the lower selenium content of fuel oils, typically 0.5 to 0.8 mg/kg, compared to 2 to 4 mg/kg in coal, and the fact that emission of particulate matter from an oil-fired plant with no emissions controls is comparable to a coal-fired unit with better than 99% collection efficiency.[127] Estimates of selenium emissions as a result of fuel oil combustion are also presented in Table 7 and are typically a factor of ten less than those due to coal combustion.

Table 7
GLOBAL ESTIMATES OF ANTHROPOGENIC ATMOSPHERIC SELENIUM EMISSIONS

Mass of Se emitted (10^9 g Se/year)[a]	Emission factor (g Se/g source)	% Se released[b]	Ref.
	Coal Combustion		
0.42	1.5×10^{-7}	5	122
4.3—5.7	1.4×10^{-6}	55	123
0.82	4.1×10^{-7}	32	83
9.5	2.7×10^{-6}	90	124
0.68	1.5×10^{-6}	52	125
	Fuel Oil Combustion		
0.03	1.7×10^{-8}	10	122
0.5	1.7×10^{-7}	90	124
0.13—0.42	1.0×10^{-7}	55	126
0.06	6.7×10^{-7}	100	125
	Refuse Incineration		
<0.0005	1×10^{-8}	<2	125
	Metal Mining and Refining		
2.0	NR[c]	NR	124
2.1	—	NR	123, 125, 126
0.29	$0.14—7.5 \times 10^{-9}$	0.2—18	125
	Industrial		
0.07	$0.5—1 \times 10^{-3}$	0.2—50	125
0.17	—	NR	123, 125, 126

[a] Estimated mass of selenium emitted from all sources = 0.8×10^9 to 12×10^9 g Se per year.
[b] Percent of selenium present in the material released to the atmosphere upon combustion or processing.
[c] NR — not reported.

C. Other Combustion Sources

We must also consider the incineration of refuse and sewage sludge as a source of atmospheric selenium. The use of refuse or refuse-derived fuel as a supplement to oil and coal in power plants has increased in recent years. Johnson[128] conducted a study of the selenium content of municipal solid waste and emissions from an incinerator. He found a range of 0.3 to 4.4 mg/kg in the waste, with emission factors for particulate selenium of 1×10^{-6} to 2×10^{-9} g/g refuse. Roughly 60% of the refuse at this plant consisted of paper products, which have been shown to contain 0.01 to 2 mg/kg.[129-131] Studies of several other municipal refuse incinerators[132,133] report in-stack particulate selenium concentrations of 7 to 122 mg/kg. However, at the high temperature of the stack gases, much of the selenium is probably in the vapor phase, as is the case with coal combustion.

Many communities have also built sewage sludge incinerators, as land burial sites are closed and concerns mount regarding the toxic components of waste water sludge. Furr et al.[134] have studied the composition of municipal sludge from 16 U.S. cities and found selenium concentrations of 1.7 to 8.7 mg/kg. Although there appears to be some enrichment of the more volatile elements, such as selenium, in released particles (0.1 to 45 $\mu g/m^3$),[135,136] particulate emissions of selenium are quite small when compared with other sources (Table 7).

D. Metal Mining and Refining

Another important source of atmospheric selenium is the roasting and refining of sulfide ores. As is the case with coal combustion, many trace elements emitted from copper smelters are found on submicrometer particles which can be transported long distances.[137] Studies of five copper smelters in Arizona[138,139] indicated that approximately 10% of the selenium emitted was in the vapor phase, with in-stack concentrations of total selenium as high as 4.7 mg/m^3. While typical plume concentrations, measured from aircraft, of the five smelters were quite variable (42 to 920 ng/m^3) due to the batch nature of the copper smelting process, it is evident that emissions of selenium are important. In the vicinity of a lead smelting complex in Kellogg, Idaho, Ragaini et al.[140] considered soil profiles, grass analysis, and aerosol measurements and concluded that the smelter had significantly contaminated the Kellogg environment with toxic elements such as selenium, cadmium, arsenic, nickel, lead, and mercury. In addition, Mosher and Duce[41] suggested that the anomalously high particulate selenium concentrations that they observed along the Peruvian coast could result from transport of smelter aerosol from distant (400 to 500 km) copper refineries. Global atmospheric emissions of selenium due to mining, smelting, and refining of ores appear to be comparable to those resulting from fossil fuel combustion (Table 7).

E. Miscellaneous Anthropogenic Sources

Finally, the miscellaneous industrial activities which consume selenium, such as the production of glass, ceramics, pigments, electronic components, and iron and steel alloys, should be considered. Atmospheric emission factors are estimated to vary from 1×10^{-3} to 0.5 g of Se per gram Se consumed for the electronics and iron and steel alloy industries, respectively. Estimates of global selenium emissions range from 7×10^7 to 17×10^7 g/year from these sources.

Xerography is a minor source of selenium globally, but concentrations have been found to be significantly elevated in the workplace. Because of its unique photoelectric properties, selenium plates or drums are used in the photosensitizing step that can lead to the emission of vapor-phase selenium. Harkin et al.[141] reported total selenium concentrations of 20 to 60 ng/m^3 in an unventilated room housing a xerographic machine which produced about 450 copies daily. This early work has since been confirmed by Mosher and Duce[43] who found between 3.5 and 45 ng/m^3 in an air-conditioned xerography room. Typical outdoor levels at this location were ≤ 1 ng/m^3, with roughly 20% of the total selenium in the vapor phase. In the xerography room 65 to 85% of the atmospheric selenium was gaseous. The only vapor-phase selenium exposure limits established by the U.S. Occupational Safety and Health Administration (OSHA) are for the compound selenium hexafluoride (SeF_6). These regulations state that an employee's time-weighted average exposure for SeF_6 in any 8-h workshift of a 40-h work week shall not exceed 0.05 mg/kg. Expressed as selenium, this OSHA limit is 850 ng/m^3. Thus, the levels observed in this study were well below this recommended limit. However, the toxicity of an element is very much a function of its chemical form, and thus more information regarding the form of selenium emissions during the xerographic process is necessary before the true physiological effects, if any, can be determined.

VI. NATURAL SOURCES OF ATMOSPHERIC SELENIUM

As we have discussed, approximately 80 to 95% of the total atmospheric particulate matter is estimated to be natural in origin. Roughly half of this natural aerosol is thought to result from direct production of particles via processes such as wave breaking and bubble bursting at the ocean surface, injection and transport of mineral dust by winds, volcanic emissions, and forest fires. The remaining natural aerosol is thought to be secondary in nature, i.e., particles resulting from the gas-to-particle conversion of substances such as sulfur and nitrogen species and hydrocarbons in the atmosphere. In the case of selenium, because so few measurements of gas phase species are available, flux estimates for this natural secondary aerosol are much more difficult to quantify than is the case with anthropogenic emissions. In the previous section we discussed the major anthropogenic sources of both particulate and vapor-phase selenium, and we shall now examine the natural sources of both species in the atmosphere. By comparing the relative magnitude of natural and anthropogenic emissions we shall gain some insight into the degree to which human activity has perturbed the atmospheric selenium cycle.

One should keep in mind that while the term ''natural'' refers to nonanthropogenic sources, some of the fluxes of natural species must be considered a direct result of human activities. This is perhaps best illustrated by the work of Jiang et al.[26] in Antwerp, Belgium. These authors detected ''natural'' biologically produced species such as $(CH_3)_2Se$ in concentrations as high as 1.4 ng/m^3. This was in an area where a particulate concentration of 120 ng/m^3 had also been measured. While selenium is an essential micronutrient, it has been suggested that at toxic selenium levels, bacteria detoxify and methylate selenium salts to volatile selenium compounds.[85] The relatively high concentrations of methylated selenium species in the Antwerp atmosphere are in all probability a biological response to the potentially toxic selenium levels that result from anthropogenic activities in the area. With this caveat in mind, the major nonanthropogenic sources of particulate selenium in the atmosphere are examined below.

A. Windblown Mineral Dust

An important source of atmospheric aerosols is continentally derived mineral aerosol. Studies of aerosol composition in the North Pacific have shown that dust storm injection of crustal material into the atmosphere in arid regions of Asia followed by long-range transport by westerly winds aloft can represent a significant source of sedimentary material for the North Pacific.[45] Sediments in many areas of the world ocean reflect the mineralogy and meteorological characteristics of adjoining continental regimes (see Windom[142] for a review of aeolian contributions to recent marine sediments). If we know the annual continental dust flux and the selenium content of this material, we can make a reasonable estimate of the selenium flux. Estimates of global aeolian particulate production as windblown dust vary from 0.6×10^{14} to 5×10^{14} g $year^{-1}$.[1,2,143] There are a number of questions concerning the strata most representative of wind blown dust: (1) are soil or rock values most representative of aeolian material; (2) should one use a local or globally averaged reference composition; (3) is bulk or a certain soil size fraction more applicable? The reader is referred to Rahn[3] for a discussion of these factors. An approximate range of 0.05 to 0.9 mg/kg[3,144,145] is observed for the selenium content of soils, rocks, and sedimentary deposits (see also Chapter 9). Multiplying the continental dust flux estimates by this compositional range, we see that 0.03 to 0.45×10^9 g of Se per year are injected into the atmosphere in the form of continentally derived mineral material. Literature estimates are presented in Table 8.

B. Sea Salt

Physical processes at the sea surface such as breaking waves and bubble formation and bursting can result in the formation of spray droplets (for a review of these processes, see Blanchard[147]) which can to a large extent control aerosol chemical processes in the marine boundary layer. As has been mentioned, sea salt particles are relatively large (0.5 to 20

Table 8
GLOBAL ESTIMATES OF NATURAL ATMOSPHERIC SELENIUM EMISSIONS

Nature (phase) of the element	Source	Mass of Se emitted (10^9 g Se/year)[a]	Ref.
Particulate	Mineral dust	0.25	126
		0.3	22, 124
		0.03—0.045	Text
	Sea salt	0.021—0.55	126
		0.004—0.04	Text
	Volcanic	0.04—1.2	Derived from 126
		0.1	22, 124, 146
	Wood/vegetation burning	0.5	124
Gaseous	Marine	4.8	124
		5.2	Derived from 126
	Continental	3.0	124
	Soils[b]	0.24	126
	Plants	4.2	Derived from 126
	Volcanic[c]	0.013	124

[a] Estimated mass of selenium emitted from all sources = 8×10^9 to 12×10^9 g Se per year.
[b] For the region 30° to 90° N; not included in total.
[c] Not included in total, see text.

μm), and thus a large fraction of this sea salt will be recycled rather quickly back to the sea surface. Estimates of the magnitude of the sea salt flux through the atmosphere annually range from 10^{12} to 10^{13} kg.[148-150] Selenium concentrations in near surface seawater of approximately 40 ng/kg have been reported by Measures and Burton.[151] There is some experimental evidence to suggest that in the sea salt aerosol production process selenium is enriched in the resulting aerosol relative to its concentration in bulk surface seawater. Heaton[152] has conducted experiments at sea where sea salt aerosols were produced artificailly using a device known as the "bubble interfacial microlayer sampler" (BIMS). BIMS is essentially a catamaran-mounted device designed for *in situ* bubble generation and aerosol collection. In samples collected along the Peru coast, Heaton[152] has observed enrichments or fractionation of selenium in the aerosol of roughly a factor of 10 to 100 over bulk seawater concentrations. If we multiply this range of enrichments by average surface seawater selenium concentrations and the sea salt flux estimates, we find that sea salt production may introduce 0.0004×10^9 to 0.04×10^9 g Se per year into the atmosphere. It is evident from Table 8 that this process is a relatively unimportant source of particulate selenium. Ross[126] has estimated the global flux of selenium from sea salt to be 0.021×10^9 to 0.55×10^9 g Se per year. The lower value was calculated using Ryaboshapko's[153] sea salt sulfur flux estimate of 140×10^{12} g S per year and a seawater selenium-to-sulfur ratio of 1.5×10^{-7} from Broecker and Peng.[154] The author has produced a second estimate of the selenium flux from the ocean associated with sea salt by normalizing selenium concentrations in seawater to those of sodium. Ross[126] used the estimate of Duce et al.[42] of the annual sodium flux from the oceans of 1.5×10^{15} g, a seawater sodium concentration of 1.1×10^{-1} g Na per kilogram of seawater,[155] and a selenium surface seawater concentration of 4×10^{-8} g Se per kilogram seawater[151] to arrive at the larger flux estimate of 0.55×10^9 g Se per year. The upper limit of 0.55×10^9 g Se per year is a factor of 100 too high. This is due to the fact that an incorrect sea salt sodium concentration value was used. The correct range should be 0.0055×10^9 to 0.021×10^9 g Se per year. It should be pointed out that a number of the original selenium flux estimates made by Ross[123,126] were for the area 30° to 90° N. The estimates in Table 8 have been scaled up by an appropriate factor in order to reflect global source strengths.

C. Gaseous Marine Emissions

Although at the present time there are not available measurements of vapor phase/methylated selenium compounds in the marine atmosphere, several lines of evidence suggest that the ocean is a significant source of vapor-phase selenium (Section III). The biological production of organoselenium compounds in surface seawater, evasion of these compounds into the atmosphere, and subsequent gas-to-particle conversion could support the anomalously high particulate selenium concentrations observed in remote marine regions.[41,156] We have also seen that methylated gaseous selenium compounds are produced in urban areas and, thus, must consider both continental and oceanic sources of vapor-phase emissions.

In order to arrive at an estimate of the ocean to atmosphere flux, Lantzy and Mackenzie[124] assumed that the atmosphere is in steady state with respect to selenium and that the flux from the sea equals the rainout flux. They then calculated the flux of selenium from the atmosphere to the ocean. This reverse flux estimate was made using the selenium concentration data in the Greenland ice sheet[100] of 16.2 ng/kg as an estimate of the selenium concentration in preindustrial revolution rain, an annual global rainfall of 4.2×10^{20} g per year and the assumption that $^2/_3$ of this rainfall occurs over the oceans. They thus calculated an atmosphere-to-ocean flux of 4.8×10^9 g Se per year, and from their steady state assumption stated that this also equals the ocean-to-atmosphere flux. One problem with this approach is that it is not always appropriate to assume that aerosol composition can be derived directly from precipitation data. Junge[157] expressed reservations concerning this approach, and the more recent work of Rahn and McCaffrey[158] suggests that there may be large fractionation of "biophile" elements such as selenium between aerosols and snows in the Alaskan Arctic (higher enrichment in aerosols than snows).

In order to calculate this flux of biogenic selenium from the ocean to atmosphere, Ross[126] has assumed that in clean marine air, the only sources of particulate selenium are the gas-to-particle conversion and scavenging onto aerosols of volatile organoselenium compounds. He chose a value of 0.12 ng/m^3 as representative of particulate values in clean marine air and assumed that this aerosol would have a residence time of 3 d (Kritz and Rancher[159]). Ross[126] used a marine boundary layer height of 1000 m and calculated an annual production of 1.46×10^{-5} g Se per square meter over the ocean surface (0.120×10^{-9} g Se per cubic meter $\times$ 1000 m $\times$ 365 d per year $\div$ 3 d). Multiplying this value by the surface area of the ocean (3.55×10^{14} m^2), we arrive at a global ocean to atmosphere flux of 5.2×10^9 g Se per year.

D. Gaseous Continental Emissions

Laboratory studies using elevated selenium levels have shown that bacteria, fungi, plants, and animals are capable of producing volatile selenium compounds. Flux estimates of organoselenium compounds from continental areas are difficult to make, as no direct measurements are available. Lantzy and Mackenzie[124] have computed an estimated continental selenium vapor flux of 2.26×10^{-9} g Se per square centimeter from the work of Chau et al.[85] This estimate was then multiplied by the world continental area less ice (1.33×10^{18} cm^2) to arrive at a flux of 3.0×10^9 g Se per year. This number should be considered an upper limit, as the laboratory studies of Chau et al.[85] were conducted with elevated selenium concentrations and lake sediments collected near a large smelter complex at Sudbury, Ontario.

Ross[126] has assumed that the metabolic pathways which volatilize selenium and sulfur are similar. Using the work of Adams et al.,[160] a study of sulfur emissions from soils in the southeastern U.S., and a Se/S ratio in organisms of 1×10^{-4}, Ross[126] has calculated a flux for the region 30° to 90° N of 0.24×10^9 g Se per year. The author states that the land area of the region considered in his study is roughly 43% of the global total, and thus we could correct this estimate to reflect global emissions to arrive at a flux of 0.55×10^9 g Se per year. This linear scaling factor probably results in an underestimation of the true global gaseous continental selenium flux because the work of Adams et al.[160] predicts an exponential north-to-south increase in the biological total sulfur gas flux, with an additional increase of roughly 25-fold between 25° N and the equator. For this reason the 30° to 90° N flux estimate of Ross[126] has been included in Table 8, but this value has not been considered in the summation of global natural selenium sources given at the bottom of the table.

Ross[126] has also considered the release of volatile selenium by plants. The author used a leaf-mediated volatilization rate of 0.05% of leaf selenium content per hour from the work of Lewis et al.[161] This number was then multiplied by an average dry weight leaf concentration of 0.076 mg/kg to obtain a per gram leaf flux of 1.1×10^{-14} g Se per second. He further assumed an average leaf lifetime of 4 months. In a manner analogous to that used by Ross[126] for the area 30° to 90° N, we have calculated a global litter flux of 496×10^{14} g per year (75% leaves, from Reiner[162]) to arrive at a global leaf-mediated selenium flux to the atmosphere of 4.2×10^9 g Se per year. Once again this laboratory study of Lewis et al.[161] employed elevated selenium concentrations, and thus this flux estimate must be considered as an upper limit.

E. Volcanic Emissions

Emissions of both particulate and gas-phase species from volcanoes is another high-temperature process, in this case natural, which must be considered in evaluating the atmospheric cycling of selenium. Here again the number of actual measurements of particulate, and especially gaseous, selenium emissions is quite limited. There are also several important questions that remain to be answered before more precise estimates of volcanic selenium emissions can be made: (1) are emissions during quiescent or eruptive periods more representative of global source strength over an annual cycle; (2) what are the relative contributions of andesitic and basaltic volcanoes; (3) what are the chemical forms and particulate/vapor partitioning of selenium?

1. Particulate Volcanic Emissions

The majority of volcanic data (Table 9) have been collected from aircraft sampling in plume or ash clouds after major eruptive periods. The three data sets collected at ground level in the vicinity of volcanic vents and lava exhibit higher selenium concentrations than those found in the more dilute aircraft samples. As Table 9 indicates, the range of particulate selenium concentrations is ca. 1 ng/m^3 to 4.5 μg/m^3 or roughly a factor of 4500. In the Mediterranean region, Buat-Ménard and Arnold[163] have estimated that particulate selenium emissions from Mount Etna, Sicily, of 6.3×10^5 g Se per day are much larger than industrial emissions (6×10^4 g Se per day) from the countries bordering the Mediterranean. Additional estimates of particulate selenium fluxes from individual volcanoes vary from 4×10^3 g per day for El Chichón[168] to 1×10^5 g Se per day at Mount St. Helens.[146] Estimates of global volcanic particulate selenium emissions are found in Table 8. Ross[126] has used the method of Phelan et al.,[146] where the Se/S concentration ratio in emissions is multiplied by an estimate of the total sulfur flux in order to derive a volcanic selenium flux. Following this approach on a global basis, we calculate that 0.04 to 1.2×10^9 g Se per year are emitted to the atmosphere from volcanoes.

2. Gaseous Volcanic Emissions

The only available data concerning gaseous emissions of volcanic selenium come from

Table 9
SELENIUM CONCENTRATIONS IN VOLCANIC PARTICULATES AND GASES

Volcano	Selenium concentration	Date and remarks	Ref.
	Particulate Selenium (ng/m³)		
Mt. Etna, Sicily	2100—4500	Main plume and hot vents, ground based	163
Mt. Etna, Sicily	250—2400	Plume and vent samples, ground based	43
Kilauea, HI	3.2—10.1	50 km upwind, ground based	164
Augustine, AK	0.95—7.7	Aircraft plume samples	165
Heimaey, Iceland	7.6—25	Ground-based sampling in lava fields	166
Mt. St. Helens, WA	1.1	Aircraft plume samples	146
Mt. St. Helens, WA	0.72—0.77	13—18 km stratospheric aircraft plume samples	167
El Chichón, Mexico	1.3	Aircraft plume samples, November 1982	168
	Total Selenium (mg/kg)		
Nasudake, Japan	0.01—0.31[a]	Fumarole condensates	169

[a] Particulate and gaseous selenium condensable at 0°C.

the work of Suzuoki[169] at Nasudake Volcano on Honshu, Japan. In this study, hot volcanic gases (120 to 420°C) and particulate matter emitted from fumaroles were collected in ice-cooled traps and analyzed for their selenium and sulfur contents. Suzuoki[169] found that total selenium concentrations ranged from 0.01 to 0.31 mg/kg (author's units are ppm) in the condensed liquid and reported that the selenium concentration increased with increasing temperature. Lantzy and Mackenzie[124] have used the work of Suzuoki[169] to derive what they term a volcanic gas flux. They multiplied the arithmetic mean condensate concentration (0.17 mg/kg) found by Suzuoki[169] by an annual total gas flux of 7.5×10^{13} g per year (Cadle[170]) to arrive at a flux estimate of 0.013×10^{9} g Se per year. There seem to be problems with this approach in that in the field work of Suzuoki,[169] only particles and gases condensable at the temperature of the ice traps were collected. On the other hand, the gas flux estimate of Cadle[170] includes gases such as CO_2 which were certainly not collected in Suzuoki's[169] ice-water traps. This estimate of Lantzy and Mackenzie[124] has been listed in Table 8, but not included in the summation.

F. Miscellaneous Natural Sources

Selenium emissions from the combustion of vegetation and wood may result both from human activities and natural processes such as agriculturally related burning, cooking and home heating with wood, and forest fires. Lantzy and Mackenzie[124] have made a rough estimate of the atmospheric selenium flux from the burning of wood and vegetation. They multiplied an average selenium concentration in land plants of 0.2 mg/kg[171] by an estimate of the total carbon released to the atmosphere by burning of wood and forest fires of 2500×10^{12} g C per year[172] to arrive at a global flux estimate of 0.5×10^{9} g Se per year.

Significant quantities of some trace elements may be released in the particulate form from plant surfaces. Beauford et al.[173,174] have documented the release of particulates containing zinc and lead from the surface of pea plants and pine tree seedlings. Lantzy and Mackenzie[124] have calculated that this flux of particulate matter could be globally significant in the case of zinc. At the present time it is impossible to make an estimate of this flux strength for selenium due to the lack of data.

Finally, it has been suggested[175,176] that direct sublimation or low-temperature volatilization of inorganic constituents from the surface of rocks could be responsible for a portion of the aerosol burden of the earth in the case of trace elements such as selenium, mercury, and arsenic. However, a consideration of the volatility of the oxides of these elements has led Brimblecombe and Hunter[177] to conclude that at ambient temperatures this process cannot account for observed atmospheric concentrations.

VII. CONCLUSIONS

Although it is apparent that substantial work remains to be done before the atmospheric cycling of selenium is fully understood, there are a number of conclusions that can be drawn concerning selenium in the atmosphere. We have seen that particulate selenium is the dominant phase in the atmosphere, with usual concentrations ranging from 5 ng/m^3 in urban areas to roughly 0.05 to 0.1 ng/m^3 in remote marine and continental areas. The global distribution of atmospheric selenium is relatively uniform, emphasizing the important role of natural sources, particularly the biota, in its cycling. Selenium is generally found concentrated on submicrometer aerosol particles. In urban areas this is the result of the high temperature nature of the combustion sources of selenium and gas-to-particle conversion of both natural and anthropogenic gaseous species. Remote marine particle size distribution measurements suggest that reaction of gaseous organoselenium compounds with sea salt particles may represent a major removal mechanism in these locations. We have also seen that selenium is highly enriched on the aerosol relative to both crustal and seawater source functions and that a substantial portion of particulate selenium in remote marine regions may result from the gas-to-particle conversion of biologically produced gaseous species.

A second important phase of selenium in the atmosphere is the gaseous component. Roughly 25% of the total atmospheric selenium appears to be in the vapor phase. Pollution sources rich in SO_2 emit selenium in an elemental vapor-phase form. The major anthropogenic sources of atmospheric selenium appear to be coal combustion and sulfide mineral refining, with various industrial activities and fuel oil combustion of secondary importance. Substantial portions of selenium contained in coal and ores are also known to escape even highly efficient pollution control devices.

Organoselenium compounds may be the major vapor-phase selenium species in the atmosphere, and while actual measurements are very limited, the major sources of natural selenium appear to be marine and continental biologically mediated vapor-phase emissions. Within the accuracy of present estimates, it would appear that anthropogenic and natural source strengths are of the same order of magnitude, very roughly 1×10^9 to 10×10^9 g Se per year.

VIII. RECOMMENDATIONS FOR FUTURE WORK

While data concerning the distribution of particulate selenium in the Northern Hemisphere are substantial, data from the Southern Hemisphere are very sparse. Additional measurements of particulate and vapor-phase selenium in the Southern Hemisphere should help us estimate the true extent of human influence on the atmospheric cycling of selenium.

With the few measurements of total and vapor phase speciation data available we are just now beginning to appreciate the importance of vapor-phase emissions in the biogeochemical cycling of selenium. Much more vapor-phase/speciation data are needed if we are to accurately quantify the sources and sinks of atmospheric selenium and fully understand the importance of the atmosphere as a reservoir and pathway for the biogeochemical cycling of selenium.

REFERENCES

1. **Prospero, J. M., Charlson, R. J., Mohnen, V., Jaenicke, R., Delany, A. C., Moyers, J., Zoller, W., and Rahn, K.,** The atmospheric aerosol system: an overview, *Rev. Geophys. Space Phys.,* 21, 1607, 1983.
2. **Jaenicke, R.,** Atmospheric aerosols and global climate, *J. Aerosol Sci.,* 11, 577, 1980.
3. **Rahn, K. A.,** *The Chemical Composition of the Atmospheric Aerosol,* Technical Report, University of Rhode Island, Kingston, 1976.
4. **Davis, B. L., Johnson, L. R., Stevens, R. K., Courtney, W. J., and Safriet, D. W.,** The quartz content and elemental composition of aerosols from selected sites of the EPA inhalable particulate network, *Atmos. Environ.,* 18, 771, 1984.
5. **Hashimoto, Y. and Winchester, J. W.,** Selenium in the atmosphere, *Environ. Sci. Technol.,* 1, 338, 1967.
6. **Hopke, P. K., Gladney, E. S., Gordon, G. E., Zoller, W. H., and Jones, A. G.,** The use of multivariate analysis to identify sources of selected elements in the Boston urban aerosol, *Atmos. Environ.,* 10, 1015, 1976.
7. **Thurston, G. D. and Spengler, J. D.,** A quantitative assessment of source contributions to inhalable particulate matter pollution in metropolitan Boston *Atmos. Environ.,* 19, 9, 1985.
8. **Rahn, K. A. and Lowenthal, D. H.,** Pollution aerosol in the northeast: northeastern-midwestern contributions, *Science,* 228, 275, 1985.
9. **Rahn, K. A. and Lowenthal, D. H.,** The promise of elemental tracers as indicators of source areas of pollution aerosol in the eastern United States, in *Trace Substances in Environmental Health — 17,* Hemphill, D. D., Ed., University of Missouri, Columbia, 1983, 189.
10. **Pillay, K. K. S., Thomas, C. C., Jr., and Sondel, J. A.,** Activation analysis of airborne selenium as a possible indicator of atmospheric sulfur pollutants, *Environ. Sci. Technol.,* 5, 74, 1971.
11. **Kowalczyk, G. S., Gordon, G. E., and Rheingrover, S. W.,** Identification of atmospheric particulate sources in Washington, D.C., using chemical element balances, *Environ. Sci. Technol.,* 16, 79, 1982.
12. **Lewis, C. W. and Macias, E. S.,** Composition of size-fractionated aerosol in Charleston, West Virginia, *Atmos. Environ.,* 14, 185, 1980.
13. **Kuykendall, W. E., Jr., Fite, L. E., and Wainerdi, R. E.,** Instrumental neutron activation analysis of air filter samples, *J. Radioanal. Chem.,* 19, 351, 1974.
14. **Shaw, R. W., Jr. and Paur, R. J.,** Composition of aerosol particles collected at rural sites in the Ohio River Valley, *Atmos. Environ.,* 17, 2031, 1983.
15. **Dams, R., Robbins, J. A., Rahn, K. A., and Winchester, J. W.,** Quantitative relationships among trace elements over industrialized N.W. Indiana, in *Nuclear Techniques in Environmental Pollution,* International Atomic Energy Agency, Vienna, 1971, 139.
16. **Stevens, R. K., Dzubay, T. G., Shaw, R. W., Jr., McClenny, W. A., Lewis, C. W., and Wilson, W. E.,** Characterization of the aerosol in the Great Smoky Mountains, *Environ. Sci. Technol.,* 14, 1491, 1980.
17. **Tanner, T. M., Young, J. A., and Cooper, J. A.,** Multielement analysis of St. Louis aerosols by nondestructive techniques, *Chemosphere,* 3, 211, 1974.
18. **Brar, S. S., Nelson, D. M., Kline, J. R., Gustafson, P. F., Kanabrocki, E. L., Moore, C. E., and Hattori, D. M.,** Instrumental analysis for trace elements present in Chicago area surface air, *J. Geophys. Res.,* 75, 2939, 1970.
19. **Scheff, P. A., Wadden, R. A., and Allen, R. J.,** Development and validation of a chemical element mass balance for Chicago, *Environ. Sci. Technol.,* 18, 923, 1984.
20. **Crecelius, E. A., Lepel, E. A., Laul, J. C., Rancitelli, L. A., and McKeever, R. L.,** Background air particulate chemistry near Colstrip, Montana, *Environ. Sci. Technol.,* 14, 422, 1980.
21. **John, W., Kaifer, R., Rahn, K., and Wesolowski, J. J.,** Trace element concentrations in aerosols from the San Francisco Bay area, *Atmos. Environ.,* 7, 107, 1973.
22. **Pacyna, J. M., Semb, A., and Hanssen, J. E.,** Emission and long-range transport of trace elements in Europe, *Tellus,* 36B, 163, 1984.
23. **Salmon, L., Atkins, D. H. F., Fisher, E. M. R., Healy, C., and Law, D. V.,** Retrospective trend analysis of the content of U.K. air particulate material 1957—1974, *Sci. Total Environ.,* 9, 161, 1978.
24. **Peirson, D. H., Cawse, P. A., and Cambray, R. S.,** Chemical uniformity of airborne particulate material, and a maritime effect, *Nature (London),* 251, 675, 1974.

25. **McDonald, C. and Duncan, H. J.,** Atmospheric levels of trace elements in Glascow, *Atmos. Environ.*, 13, 413, 1979.
26. **Jiang, S., Robberecht, H., and Adams, F.,** Identification and determination of alkylselenide compounds in environmental air, *Atmos. Environ.*, 17, 111, 1983.
27. **Heindryckx, R. and Dams, R.,** Continental, marine and anthropogenic contributions to the inorganic composition of the aerosol of an industrial zone, *J. Radioanal. Chem.*, 19, 339, 1974.
28. **Hallet, J. Ph., Ronneau, C., and Cara, J.,** Sulfur and iron as indicators of pollution status in a rural atmosphere, *Atmos. Environ.*, 18, 2191, 1984.
29. **Priest, P., Navarre, J.-L., and Ronneau, C.,** Elemental background concentration in the atmosphere of an industrialized country, *Atmos. Environ.*, 15, 1325, 1981.
30. **Vogg, H. and Hartel, R.,** Experience in the analysis of atmospheric aerosols at the Karlsruhe Nuclear Research Center, *J. Radioanal. Chem.*, 37, 857, 1977.
31. **Schramel, P., Samsahl, K., and Pavlu, J.,** Determination of 12 selected microelements in air particles by neutron activation analysis, *J. Radioanal. Chem.*, 19, 329, 1974.
32. **Olmez, I. and Aras, N. K.,** Trace elements in the atmosphere determined by nuclear activation analysis and their interpretation, *J. Radioanal. Chem.*, 37, 671, 1977.
33. **Suzuki, Y., Sugimura, Y., and Miyake, Y.,** The content of selenium and its chemical form in rain water and aerosol in Tokyo, *J. Meteorol. Soc. Jpn.*, 59, 405, 1981.
34. **Fukino, H., Mimura, S., Inoue, K., and Yamane, Y.,** Correlations among atmospheric elements, airborne particulate matter, benzene extracts, benzo(a)pyrene, NO, NO_2 and SO_2 concentrations in Japan, *Atmos. Environ.*, 18, 983, 1984.
35. **Kobayashi, R. and Hashimoto, Y.,** A study on emission sources of selenium in the atmosphere, *J. Jpn. Soc. Air Pollut.*, 17, 96, 1982.
36. **Winchester, J. W., Wang, M.-X., Ren, L.-X., Lu, W.-X., Hansson, H.-C., Lannefors, H., Darzi, M., and Leslie, A. C. D.,** Nonurban aerosol composition near Beijing, China, *Nucl. Instrum. Methods*, 181, 391, 1981.
37. **Parkinson, G. S.,** Lead in gasoline. 3-UK levels, *Pet. Rev.*, 25, 289, 1971.
38. **Duce, R. A., Ray, B. J., Hoffman, G. L., and Walsh, P. R.,** Trace metal concentration as a function of particle size in marine aerosols from Bermuda, *Geophys. Res. Lett.*, 3, 339, 1976.
39. **Prospero, J. M., Ed.,** *The Tropospheric Transport of Pollutants and Other Substances to the Oceans*, National Academy of Sciences, Washington, D.C., 1978.
40. **Buat-Ménard, P. and Chesselet, R.,** Variable influence of the atmospheric flux on the trace metal chemistry of oceanic suspended matter, *Earth Planet. Sci. Lett.*, 42, 399, 1979.
41. **Mosher, B. W. and Duce, R. A.,** Vapor phase and particulate selenium in the marine atmosphere, *J. Geophys. Res.*, 88, 6761, 1983.
42. **Duce, R. A., Arimoto, R., Ray, B. J., Unni, C. K., and Harder, P. J.,** Atmospheric trace elements at Enewetak Atoll. I. Concentrations, sources, and temporal variability, *J. Geophys. Res.*, 88, 5321, 1983.
43. **Mosher, B. W. and Duce, R. A.,** Unpublished data, 1985.
44. **Parrington, J. R., Zoller, W. H., and Gordon, G. E.,** Personal communication, 1985.
45. **Uematsu, M., Duce, R. A., Prospero, J. M., Chen, L., Merrill, J. T., and McDonald, R. L.,** Transport of mineral aerosol from Asia over the north Pacific ocean, *J. Geophys. Res.*, 88, 5343, 1983.
46. **Prospero, J. M., Savoie, D. L., Nees, R. T., Duce, R. A., and Merrill, J.,** Particulate sulfate and nitrate in the boundary layer over the North Pacific Ocean, *J. Geophys. Res.*, 90, 10586, 1985.
47. **Prospero, J. M.,** Unpublished data, 1985.
48. **Cline, J. D. and Bates, T. S.,** Dimethyl sulfide in the equatorial Pacific Ocean: a natural source of sulfur to the atmosphere, *Geophys. Res. Lett.*, 10, 949, 1983.
49. **Andreae, M. O. and Raemdonck, H.,** Dimethyl sulfide in the surface ocean and the marine atmosphere: a global view, *Science*, 221, 744, 1983.
50. **Fitzgerald, W. F., Gill, G. A., and Kim, J. P.,** An equatorial Pacific Ocean source of atmospheric mercury, *Science*, 224, 597, 1984.
51. **Maenhaut, W., Zoller, W. H., Duce, R. A., and Hoffman, G. L.,** Concentration and size distribution of particulate trace elements in the south polar atmosphere, *J. Geophys. Res.*, 84, 2421, 1979.
52. **Zoller, W. H., Gladney, E. S., and Duce, R. A.,** Atmospheric concentrations and sources of trace metals at the South Pole, *Science*, 183, 198, 1974.
53. **Cunningham, W. C. and Zoller, W. H.,** The chemical composition of remote area aerosols, *J. Aerosol Sci.*, 12, 367, 1981.
54. **Tuncel, G. and Zoller, W. H.,** Personal Communication, 1985.
55. **Phelan-Kotra, J., Ölmez, I., and Zoller, W. H.,** Personal communication, 1985.
56. **Dams, R. and De Jonge, J.,** Chemical composition of Swiss aerosols from the Jungfraujoch, *Atmos. Environ.*, 10, 1079, 1976.
57. **Adams, F., Van Craen, M., Van Espen, P., and Andreuzzi, D.,** The elemental composition of atmospheric aerosol particles at Chacaltaya, Bolivia, *Atmos. Environ.*, 14, 879, 1980.

58. **Rahn, K. A., Lewis, N. F., Lowenthal, D. H., and Smith, D. L.,** Noril'sk only a minor contributor to Arctic haze, *Nature (London),* 306, 459, 1983.
59. **Flyger, H. and Heidam, N. Z.,** Ground level measurements of the summer tropospheric aerosol in northern Greenland, *J. Aerosol Sci.,* 9, 157, 1978.
60. **Rahn, K. A.,** Relative importances of North America and Eurasia as sources of Arctic aerosol, *Atmos. Environ.,* 15, 1447, 1981.
61. **Rahn, K. A. and McCaffrey, R. J.,** On the origin and transport of the winter Arctic aerosol, *Ann. N.Y. Acad. Sci.,* 338, 486, 1980.
62. **Settle, D. M., Patterson, C. C., Turekian, K. K., and Cochran, J. K.,** Lead precipitation fluxes at tropical oceanic sites determined from ^{210}Pb measurements, *J. Geophys. Res.,* 87, 1239, 1982.
63. **Natusch, D. F. S., Wallace, J. R., and Evans, C. A., Jr.,** Toxic trace elements: preferential concentration in respirable particles, *Science,* 183, 202, 1974.
64. **Hansen, L. D. and Fisher, G. L.,** Elemental distribution in coal fly ash particles, *Environ. Sci. Technol.,* 14, 1111, 1980.
65. **Biermann, A. H. and Ondov, J. M.,** Application of surface-deposition models to size-fractionated coal fly ash, *Atmos. Environ.,* 14, 289, 1980.
66. **Natusch, D. F. S. and Wallace, J. R.,** Urban aerosol toxicity: the influence of particle size, *Science,* 186, 695, 1974.
67. **Gladney, E. S., Small, J. A., Gordon, G. E., and Zoller, W. H.,** Composition and size distribution of in-stack particulate material at a coal-fired power plant, *Atmos. Environ.,* 10, 1071, 1976.
68. **Wangen, L. E.,** Elemental composition of size-fractionated aerosols associated with a coal-fired power plant plume and background, *Environ. Sci. Technol.,* 15, 1080, 1981.
69. **Smith, R. D., Campbell, J. A., and Nielson, K. K.,** Concentration dependence upon particle size of volatilized elements in fly ash, *Environ. Sci. Technol.,* 13, 553, 1979.
70. **Rahn, K. A.,** Unpublished data, 1985.
71. **Savoie, D. L. and Prospero, J. M.,** Particle size distribution of nitrate and sulfate in the marine atmosphere, *Geophys. Res. Lett.,* 9, 1207, 1982.
72. **Gravenhorst, G., Muller, K. P., and Franken, H.,** Inorganic nitrogen in marine aerosols, *Ges. Aerosolforsch., Conf., (Aerosols Sci., Med. Technol.-Biomed. Influence Aerosol-Conf., 7th),* 7, 182, 1980.
73. **Duce, R. A.,** Biogeochemical cycles and the air-sea exchange of aerosols, in *The Major Biogeochemical Cycles and Their Interactions,* Scope 21, Bolin, B. and Cook, R. B., Eds., Scientific Committee on Problems of the Environment, John Wiley & Sons, New York, 1983, chap. 16.
74. **Rahn, K. A.,** Sources of Trace Elements in Aerosols — an Approach to Clean Air, Ph.D. thesis, University of Michigan, Ann Arbor, 1971.
75. **Duce, R. A., Hoffman, G. L., and Zoller, W. H.,** Atmospheric trace metals at remote northern and southern hemisphere sites: pollution or natural?, *Science,* 187, 59, 1975.
76. **Pillay, K. K. S. and Thomas, C. C., Jr.,** Determination of the trace element levels in atmospheric pollutants by neutron activation analysis, *J. Radioanal. Chem.,* 7, 107, 1971.
77. **Robinson, E. and Robbins, R. C.,** Emission, Concentrations and Fate of Particulate Atmospheric Pollutants, American Petroleum Institute, Menlo Park, CA, 1971.
78. **Newell, R. E., Boer, G. J., Jr., and Kidson, J. W.,** An estimate of the interhemispheric transfer of carbon monoxide from tropical general circulation data, *Tellus,* 26, 103, 1974.
79. **Pupp, C., Lao, R. C., Murray, J. J., and Pottie, R. F.,** Equilibrium vapour concentrations of some polycyclic aromatic hydrocarbons, As_4O_6 and SeO_2 and the collection efficiencies of these air pollutants, *Atmos. Environ.,* 8, 915, 1974.
80. **Pupp, C., Lao, R. C., Murray, J. J., and Pottie, R. F.,** Equilibrium vapour concentrations of some polycyclic aromatic hydrocarbons, As_4O_6, and SeO_2, *Atmos. Environ.,* 9, 367, 1975.
81. **Shendrikar, A. D. and West, P. W.,** Air sampling methods for the determination of selenium, *Anal. Chim. Acta,* 89, 403, 1977.
82. **Reamer, D. C.,** Methods for the Determination of Atmospheric Tetra-alkyllead and Alkyl Selenide Species using a GC-microwave Plasma Detector, Ph.D. thesis, University of Maryland, College Park, 1978.
83. **Andren, A. W., Klein, D. H., and Talmi, Y.,** Selenium in coal-fired steam plant emissions, *Environ. Sci. Technol.,* 9, 856, 1975.
84. **Jiang, S., De Jonghe, W., and Adams, F.,** Determination of alkylselenide compounds in air by gas chromatography-atomic absorption spectrometry, *Anal. Chim. Acta,* 136, 183, 1982.
85. **Chau, Y. K., Wong, P. T. S., Silverberg, B. A., Luxon, P. L., and Bengert, G. A.,** Methylation of selenium in the aquatic environment, *Science,* 192, 1130, 1976.
86. **Reamer, D. C. and Zoller, W. H.,** Selenium biomethylation products from soil and sewage sludge, *Science,* 208, 500, 1980.
87. **Fleming, R. W. and Alexander, M.,** Dimethylselenide and dimethyltelluride formation by a strain of *Penicillium, Appl. Microbiol.,* 24, 424, 1972.

88. **Francis, A. J., Duxbury, J. M., and Alexander, M.,** Evolution of dimethylselenide from soils, *Appl. Microbiol.*, 28, 248, 1974.
89. **Doran, J. W. and Alexander, M.,** Microbial transformations of selenium, *Appl. Environ. Microbiol.*, 33, 31, 1977.
90. **Abu-erreish, G. M., Whitehead, E. I., and Olson, O. E.,** Evolution of volatile selenium from soils, *Soil Sci.*, 106, 415, 1968.
91. **Barkes, L. and Fleming, R. W.,** Production of dimethylselenide gas from inorganic selenium by eleven soil fungi, *Bull. Environ. Contam. Toxicol.*, 12, 308, 1974.
92. **Asher, C. J., Evans, C. S., and Johnson, C. M.,** Collection and partial characterization of volatile selenium compounds from *Medicago sativa L.*, *Aust. J. Biol. Sci.*, 20, 737, 1967.
93. **Evans, C. S., Asher, C. J., and Johnson, C. M.,** Isolation of dimethyl diselenide and other volatile selenium compounds from *Astragalus racemosus (Pursh.)*, *Aust. J. Biol. Sci.*, 21, 13, 1968.
94. **McConnell, K. P. and Portman, O. W.,** Excretion of dimethyl selenide by the rat, *J. Biol. Chem.*, 195, 277, 1952.
95. **Galloway, J. N., Thornton, J. D., Norton, S. A., Volchok, H. L., and McLean, R. A. N.,** Trace metals in atmospheric deposition: a review and assessment, *Atmos. Environ.*, 16, 1677, 1982.
96. **Peirson, D. H., Cawse, P. A., Salmon, L., and Cambray, R. S.,** Trace elements in the atmospheric environment, *Nature (London)*, 241, 252, 1973.
97. **Rancitelli, L. A. and Perkins, R. W.,** Trace element concentrations in the troposphere and lower stratosphere, *J. Geophys. Res.*, 75, 3055, 1970.
98. **Schutyser, P., Maenhaut, W., and Dams, R.,** Instrumental neutron activation analysis of dry atmospheric fall-out and rain-water, *Anal. Chim. Acta*, 100, 75, 1978.
99. **Arimoto, R., Duce, R. A., Ray, B. J., and Unni, C. K.,** Atmospheric trace elements at Enewetak Atoll. II. Transport to the ocean by wet and dry deposition, *J. Geophys. Res.*, 90, 2391, 1985.
100. **Weiss, H. V., Koide, M., and Goldberg, E. D.,** Selenium and sulfur in a Greenland ice sheet: relation to fossil fuel combustion, *Science*, 172, 261, 1971.
101. **Boutron, C., Leclerc, M., and Risler, N.,** Atmospheric trace elements in Antarctic prehistoric ice collected at a coastal ablation area, *Atmos. Environ.*, 18, 1947, 1984.
102. **Ondov, J. M., Ragaini, R. C., and Biermann, A. H.,** Emissions and particle-size distributions of minor and trace elements at two western coal-fired power plants equipped with cold-side electrostatic precipitators, *Environ. Sci. Technol.*, 13, 946, 1979.
103. **Conzemius, R. J., Welcomer, T. D., and Svec, H. J.,** Elemental partitioning in ash depositories and material balances for a coal burning facility by spark source mass spectrometry, *Environ. Sci. Technol.*, 18, 12, 1984.
104. **Block, C. and Dams, R.,** Inorganic composition of Belgian coals and coal ashes, *Environ. Sci. Technol.*, 9, 146, 1975.
105. **Kaakinen, J. W., Jorden, R. M., Lawasani, M. H., and West, R. E.,** Trace element behavior in coal-fired power plant, *Environ. Sci. Technol.*, 9, 862, 1975.
106. **Shendrikar, A. D., Ensor, D. S., Cowen, S. J., Woffinden, G. J., and McElroy, M. W.,** Size-dependent penetration of trace elements through a utility baghouse, *Atmos. Environ.*, 17, 1411, 1983.
107. **Coles, D. G., Ragaini, R. C., Ondov, J. M., Fisher, G. L., Silberman, D., and Prentice, B. A.,** Chemical studies of stack fly ash from a coal-fired power plant, *Environ. Sci. Technol.*, 13, 455, 1979.
108. **Lee, R. E., Jr., Crist, H. L., Riley, A. E., and MacLeod, K. E.,** Concentration and size of trace metal emissions from a power plant, a steel plant, and a cotton gin, *Environ. Sci. Technol.*, 9, 643, 1975.
109. **Block, C. and Dams, R.,** Study of fly ash emission during combustion of coal, *Environ. Sci. Technol.*, 10, 1011, 1976.
110. **Smith, R. D., Campbell, J. A., and Nielson, K. K.,** Characterization and formation of submicron particles in coal-fired plants, *Atmos. Environ.*, 13, 607, 1979.
111. **Davison, R. L., Natusch, D. F. S., Wallace, J. R., and Evans, C. A., Jr.,** Trace elements in fly ash: dependence of concentration on particle size, *Environ. Sci. Technol.*, 8, 1107, 1974.
112. **Drehmel, D. C.,** Fine particle control technology: conventional and novel devices, *J. Air Pollut. Control Assoc.*, 27, 138, 1977.
113. **Chrisp, C. E., Fisher, G. L., and Lammert, J. E.,** Mutagenicity of filtrates from respirable coal fly ash, *Science*, 199, 73, 1978.
114. **Wangen, L. E. and Williams, M. D.,** Elemental deposition downwind of a coal-fired power plant, *Water Air Soil Pollut.*, 10, 33, 1978.
115. **Dreesen, D. R., Gladney, E. S., Owens, J. W., Perkins, B. L., Wienke, C. L., and Wangen, L. E.,** Comparison of levels of trace elements extracted from fly ash and levels found in effluent waters from a coal-fired power plant, *Environ. Sci. Technol.*, 11, 1017, 1977.
116. **Crecelius, E. A.,** The solubility of coal fly ash and marine aerosols in seawater, *Mar. Chem.*, 8, 245, 1980.

117. **Hardy, J. T. and Crecelius, E. A.,** Is atmospheric particulate matter inhibiting marine primary productivity?, *Environ. Sci. Technol.,* 15, 1103, 1981.
118. **Brackett, C. E.,** Information Circ. 8488, U.S. Bureau of Mines, Washington, D.C., 1970, 11.
119. **Furr, A. K., Parkinson, T. F., Hinrichs, R. A., Van Campen, D. R., Bache, C. A., Gutenmann, W. H., St. John, L. E., Jr., Pakkala, I. S., and Lisk, D. J.,** National survey of elements and radioactivity in fly ashes. Absorption of elements by cabbage grown in fly ash-soil mixtures, *Environ. Sci. Technol.,* 11, 1194, 1977.
120. **Gutenmann, W. H., Bache, C. A., Youngs, W. D., and Lisk, D. J.,** Selenium in fly ash, *Science,* 191, 966, 1976.
121. **Scanlon, D. H. and Duggan, J. C.,** Growth and element uptake of woody plants on fly ash, *Environ. Sci. Technol.,* 13, 311, 1979.
122. **Bertine, K. K. and Goldberg, E. D.,** Fossil fuel combustion and the major sedimentary cycle, *Science,* 173, 233, 1971.
123. **Ross, H. B.,** Atmospheric Selenium, Rep. CM-66, Department of Meteorology, University of Stockholm, Stockholm, 1984, 68.
124. **Lantzy, R. J. and Mackenzie, F. T.,** Atmospheric trace metals: global cycles and assessment of man's impact, *Geochim. Cosmochim. Acta,* 43, 511, 1979.
125. **National Academy of Sciences,** *Selenium,* National Academy of Sciences, Washington, D.C., 1976, 203.
126. **Ross, H. B.,** An atmospheric selenium budget for the region 30°N to 90°N, *Tellus,* 37B, 78, 1985.
127. **Pacyna, J. M.,** Estimation of Emission Factors of Trace Metals from Oil-Fired Power Plants, Tech. Rep. 2/82, Norwegian Institute for Air Research, Lillestrom, Norway, 1982, 22.
128. **Johnson, H.,** Determination of selenium in solid waste, *Environ. Sci. Technol.,* 4, 850, 1970.
129. **West, P. W., Sachdev, S. L., and Shendrikar, A. D.,** Selenium in papers, cigarette papers and tobaccos, *Environ. Lett.,* 2, 225, 1972.
130. **Olson, O. E. and Frost, D. V.,** Selenium in papers and tobaccos, *Environ. Sci. Technol.,* 4, 686, 1970.
131. **Anderson, L. W. and Acs, L.,** Selenium in North American paper pulps, *Environ. Sci. Technol.,* 8, 462, 1974.
132. **Greenberg, R. R., Gordon, G. E., Zoller, W. H., Jacko, R. B., Neuendorf, D. W., and Yost, K. J.,** Composition of particles emitted from the Nicosia municipala incinerator, *Environ. Sci. Technol.,* 12, 1329, 1978.
133. **Greenberg, R. R., Zoller, W. H., and Gordon, G. E.,** Composition and size distributions of particles released in refuse incineration, *Environ. Sci. Technol.,* 12, 566, 1978.
134. **Furr, A. K., Lawrence, A. W., Tong, S. S. C., Grandolfo, M. C., Hofstader, R. A., Bache, C. A., Gutenmann, W. H., and Lisk, D. J.,** Multielement and chlorinated hydrocarbon analysis of municipal sewage sludges of American cities, *Environ. Sci. Technol.,* 10, 683, 1976.
135. **Bennett, R. L. and Knapp, K. T.,** Characterization of particulate emissions from municipal wastewater sludge incinerators, *Environ. Sci. Technol.,* 16, 831, 1982.
136. **Greenberg, R. R., Zoller, W. H., and Gordon, G. E.,** Atmospheric emissions of elements on particles from the Parkway sewage-sludge incinerator, *Environ. Sci. Technol.,* 15, 64, 1981.
137. **Parungo, F. and Pueschel, R.,** Nucleation properties of plume aerosol from a copper smelter, *Proc. Int. Conf. Sens. Environ. Pollut.,* 89, 156, 1977.
138. **Germani, M. S., Small, M., Zoller, W. H., and Moyers, J. L.,** Fractionation of elements during copper smelting, *Environ. Sci. Technol.,* 15, 299, 1981.
139. **Small, M., Germani, M. S., Small, A. M., Zoller, W. H., and Moyers, J. L.,** Airborne plume study of emissions from the processing of copper ores in southeastern Arizona, *Environ. Sci. Technol.,* 15, 293, 1981.
140. **Ragaini, R. C., Ralston, H. R., and Roberts, N.,** Environmental trace metal contamination in Kellogg, Idaho, near a lead smelting complex, *Environ. Sci. Technol.,* 11, 773, 1977.
141. **Harkin, J. M., Dong, A., and Chesters, G.,** Elevation of selenium levels in air by xerography, *Nature (London),* 259, 204, 1976.
142. **Windom, H. L.,** Eolian contributions of marine sediments, *J. Sediment. Petrol.,* 45, 520, 1975.
143. **Goldberg, E. D.,** Atmospheric dust, the sedimentary cycle and man, private communication, 1971.
144. **Taylor, S. R.,** Abundance of chemical elements in the continental crust: a new table, *Geochim. Cosmochim. Acta,* 28, 1273, 1964.
145. **Mason, B. H.,** *Principles of Geochemistry,* 3rd ed., John Wiley & Sons, New York, 1966, 329.
146. **Phelan, J. M., Finnegan, D. L., Ballantine, D. S., Zoller, W. H., Hart, M. A., and Moyers, J. L.,** Airborne aerosol measurements in the quiescent plume of Mount St. Helens: September, 1980, *Geophys. Res. Lett.,* 9, 1093, 1982.
147. **Blanchard, D. C.,** The production, distribution and bacterial enrichment of the sea-salt aerosol, in *Air-Sea Exchange of Gases and Particles,* Liss, P. S. and Slinn, W. G. N., Eds., Reidel, Boston, 1983, 407.
148. **Eriksson, E.,** The yearly circulation of chloride and sulfur in nature; meteorological, geochemical and pedological implications. II, *Tellus,* 12, 63, 1960.

149. **Blanchard, D. C.,** The electricification of the atmosphere by particles from bubbles in the sea, *Prog. Oceanogr.,* 1, 71, 1963.
150. **Petrenchuk, O. P.,** On the budget of sea salts and sulfur in the atmosphere, *J. Geophys. Res.,* 85, 7439, 1980.
151. **Measures, C. I. and Burton, J. D.,** The vertical distribution and oxidation states of dissolved selenium in the northeast Atlantic Ocean and their relationship to biological processes, *Earth Planet. Sci. Lett.,* 46, 385, 1980.
152. **Heaton, R. W.,** Enrichment of Arsenic, Antimony, Selenium, Copper, Zinc, and Iron on Sea-Salt Aerosol, Ph.D. thesis, University of Rhode Island, Kingston, 1985.
153. **Ryaboshapko, A. G.,** The atmospheric sulphur cycle, in *The Global Biogeochemical Sulphur Cycle,* Scope 19, Ivanov, M. V. and Freney, J. R., Eds., Scientific Committee on Problems of the Environment, John Wiley & Sons, New York, 1983, chap. 4.
154. **Broecker, W. S. and Peng, T.-H.,** *Tracers in the Sea,* Lamont-Doherty Geological Observatory, Columbia University, Eldigo Press, Palisades, NY, 1982, 26.
155. **Goldberg, E. D.,** The oceans as a chemical system, in *The Sea,* Vol. 2, Hill, M. N., Ed., Wiley-Interscience, New York, 1963, chap. 1.
156. **Zoller, W. H., Parrington, J. R., Vollmerhausen, J., Kelly, W. R., Koch, W. F., and Paulsen, P. J.,** The source of sulphate and selenium in the marine atmosphere, abstract, *CACGP Symp. on Tropospheric Chemistry,* Pergamon Press, Oxford, 1983, 102.
157. **Junge, C. E.,** Processes responsible for the trace content in precipitation, in *Proc. Symp. Isotopes and Impurities in Snow and Ice,* IAHS Publ. No. 118, International Association of Hydrological Sciences, Grenoble, France, 1977, 63.
158. **Rahn, K. A. and McCaffrey, R. J.,** Compositional differences between Arctic aerosol and snow, *Nature (London),* 280, 479, 1979.
159. **Kritz, M. A. and Rancher, J.,** Circulation of Na, Cl, and Br in the tropical marine atmosphere, *J. Geophys. Res.,* 85, 1633, 1980.
160. **Adams, D. F., Farwell, S. O., Robinson, E., Pack, M. R., and Bamesberger, W. L.,** Biogenic sulfur source strengths, *Environ. Sci. Technol.,* 15, 1493, 1981.
161. **Lewis, B. G., Johnson, C. M., and Delwiche, C. C.,** Release of volatile selenium compounds by plants. Collection procedures and preliminary observations, *J. Agric. Food Chem.,* 14, 638, 1966.
162. **Reiners, W. A.,** Terrestrial detritus and the carbon cycle, in *Carbon and the Biosphere,* Vol. 30, Proc. 24th Brookhaven Symp. in Biology, Woodwell, G. M. and Pecan, E. V., Eds., U.S. Atomic Energy Commission, Washington, D.C., 1973, 303.
163. **Buat-Ménard, P. and Arnold, M.,** The heavy metal chemistry of atmospheric particulate matter emitted by Mount Etna Volcano, *Geophys. Res. Lett.,* 5, 245, 1979.
164. **Zoller, W. H., Parrington, J. R., and Phelan-Kotra, J. M.,** Iridium enrichment in airborne particles from Kilauea Volcano: January 1983, *Science,* 222, 1118, 1983.
165. **Lepel, E. A., Stefansson, K. M., and Zoller, W. H.,** The enrichment of volatile elements in the atmosphere by volcanic activity: Augustine Volcano 1976, *J. Geophys. Res.,* 83, 6213, 1978.
166. **Mroz, E. J. and Zoller, W. H.,** Composition of atmospheric particulate matter from the eruption of Heimaey, Iceland, *Science,* 190, 461, 1975.
167. **Vossler, T., Anderson, D. L., Aras, N. K., Phelan, J. M., and Zoller, W. H.,** Trace element composition of the Mount St. Helens plume: stratospheric samples from the 18 May eruption, *Science,* 211, 827, 1981.
168. **Phelan-Kotra, J., Finnegan, D. L., Zoller, W. H., Hart, M. A., and Moyers, J. L.,** El Chichón: composition of plume gases and particles, *Science,* 222, 1018, 1983.
169. **Suzuoki, T.,** A geochemical study of selenium in volcanic exhalation and sulfur deposits, *Bull. Chem. Soc. Jpn.,* 37, 1200, 1964.
170. **Cadle, R. D.,** Volcanic emissions of halides and sulfur compounds to the troposphere and stratosphere, *J. Geophys. Res.,* 80, 1650, 1975.
171. **Bowen, H. J. M.,** *Trace Elements in Biochemistry,* Academic Press, London, 1966, 202.
172. **Bolin, B.,** Changes of land biota and their importance for the carbon cycle, *Science,* 196, 613, 1977.
173. **Beauford, W., Barber, J., and Barringer, A. R.,** Heavy metal release from plants into the atmosphere, *Nature (London),* 256, 35, 1975.
174. **Beauford, W., Barber, J., and Barringer, A. R.,** Release of particles containing metals from vegetation into the atmosphere, *Science,* 195, 571, 1977.
175. **Goldberg, E. D.,** Rock volatility and aerosol composition, *Nature (London),* 260, 128, 1976.
176. **Desaedeleer, G. and Goldberg, E. D.,** Rock volatility — some initial experiments, *Geochem. J.,* 12, 75, 1978.
177. **Brimblecombe, P. and Hunter, K. A.,** Rock volatility and aerosol composition, *Nature (London),* 265, 761, 1977.

Chapter 13

GLOBAL CYCLING OF SELENIUM

Jerome O. Nriagu

TABLE OF CONTENTS

I. INTRODUCTION

The biogeochemistry of selenium has been a matter of long-standing interest because selenium is an essential trace element in human and animal nutrition at low concentration, but turns into an environmental toxin at elevated levels.[1] Historically, research on selenium has emphasized its distribution in various compartments of the biosphere, especially the atmosphere, soils, fresh waters, the oceans, plants, and sediments. Although such data are invaluable in calculating the steady-state inventories (or reservoirs) of selenium in the different compartments of the selenium cycle, they tell us little about its transfer rates from one compartment to the other. The ultimate fate and effects of the rapidly increasing quantities of selenium being supplied to the environment is determined primarily by the dispersion (fluxes) along key biogeochemical pathways.

Different phenomenological models of the selenium cycle have been described in the literature. The first of these was published in 1939 and emphasized the role of plants and animals on the cycling of this element.[2] A geochemical cycle of selenium was described by Lakin and Davidson,[3] and the flow of selenium along the soil-plant-animal chain has been presented in several reports.[4,5] These early models focused primarily on the processes and reaction mechanisms that control the distribution of selenium in particular ecological niches (Figure 1).

Models which depict the human influence on the regional cycling of selenium have been presented by Ross[6] and Pacyna et al.[7] Full-scale global models of the selenium cycle were first described by Mackenzie and colleagues.[8,9] These models, however, were based mostly on old chemical measurements and emission factors that have become outdated. The present report uses the most recent information available to derive an inventory of selenium emissions from natural and anthropogenic sources. It presents improved estimates of the selenium flux within a seven-compartment global system made up of the atmosphere, soil, terrestrial and marine biota, the ocean (surface/mixed and deep layers), and marine sediments. The important role of the biota in the internal cycling of selenium within each compartment is highlighted. In fact, the model serves to illustrate the fact that human activities are now having major impacts on the selenium cycle on regional (especially) and global scales.

II. SELENIUM INVENTORIES IN GLOBAL RESERVOIRS

Selenium is the 13th element in terms of cosmic abundance (about 0.68 atom per 10,000 atoms of silicon), but ranks 70th according to its crustal abundance of only 0.05 mg/kg.[10] Selenium forms stable compounds with 16 other elements, but because of its crystallochemical properties, its geochemistry is closely linked to that of the more abundant sulfur. Selenium, though rarely found in the native state, is widely distributed in nature, usually as an isomorphic substituent in heavy metal sulfides formed by magmatic, pegmatitic, and hydrothermal processes (see Chapter 9). Although selenium is the major component of about 40 mineral species and a minor constituent of 37 others,[11] most of these seleniferous minerals occur as small crystals in small quantities which are never large enough to constitute an exploitable selenium ore deposit. This basic geochemistry of selenium governs its natural distribution in the various environmental compartments.

Practically all of the selenium used commercially is won as a by-product from the beneficiation of sulfide deposits of copper, nickel, zinc, silver, and, to a lesser extent, lead and mercury. The known selenium reserves, in association with nonferrous metal sulfide deposits, amount to about 112,000 t, whereas the selenium resources have been estimated to be about 291,000 t.[11] The worldwide consumption in 1984 was about 2000 t.[12]

A compartmental model of the global selenium cycle consists of a defined number of reservoirs or pools between which the transfer of selenium can occur. The capacity of a

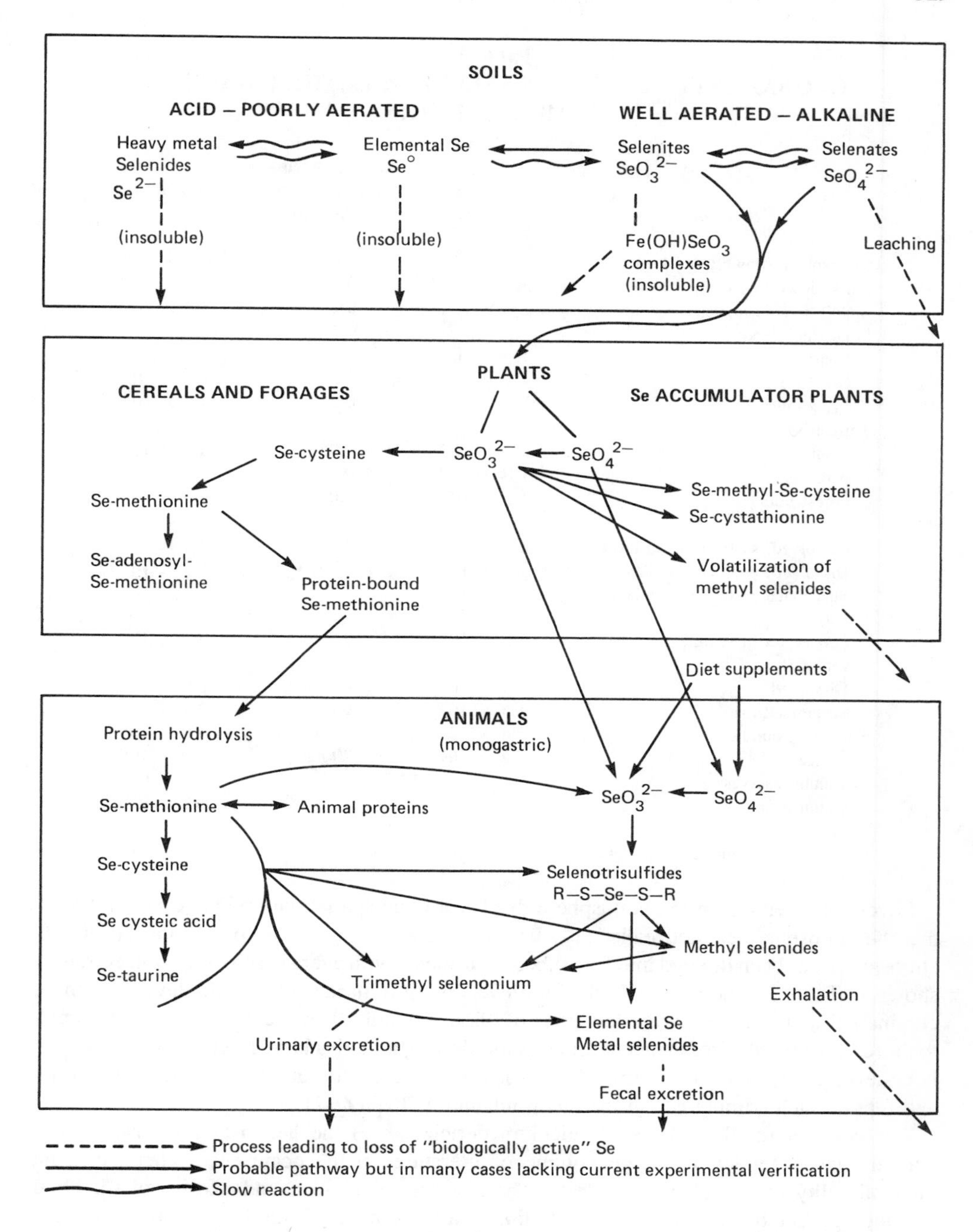

FIGURE 1. Biogeochemical cycling of selenium in the soil-plant-animal systems.[4,5]

given reservoir (or its burden) is derived as the product of the average selenium concentration therein and the total mass of the medium in the reservoir. Usually, the number of reservoirs considered in a model is determined by the availability of reliable data and the understanding of the basic biogeochemistry of the particular element. This report describes a seven-reservoir geochemical model for selenium. Table 1 shows the reservoir masses, the average selenium concentration, and selenium burden in each reservoir. It should be immediately obvious that the available data bases are grossly inadequate for a detailed discussion of the global cycling of selenium.

Table 1
GLOBAL INVENTORIES OF SELENIUM IN THE PRINCIPAL COMPARTMENTS[a]

Reservoir	Reservoir mass (g)	Average selenium concentration (mg/kg)	Selenium pool (g)
Lithosphere (down to 45 km)	57×10^{24}	0.05	2.8×10^{18}
Soils (down to 1.0 m)	3.3×10^{20}	0.4	1.3×10^{14}
Soil organic matter	1.6×10^{18}	0.2	3.2×10^{11}
Fossil fuel deposits			
Coal	1.0×10^{19}	3.4	3.4×10^{13}
Oil shale	4.6×10^{19}	2.3	1.1×10^{14}
Crude oil	2.3×10^{17}	0.2	4.6×10^{10}
Terrestrial biomass			
Plants	1.2×10^{18}	0.05	6.0×10^{10}
Animals	1.0×10^{15}	0.15	1.5×10^{8}
Forest litter	1.2×10^{17}	0.08	9.6×10^{9}
Oceans			
Dissolved, surface mixed layer	2.8×10^{22}	30 ng/kg	8.4×10^{11}
Dissolved, deep ocean	1.4×10^{24}	95 ng/kg	1.3×10^{14}
Suspended particulates, mixed layer	7×10^{16}	3	2.1×10^{11}
Pore water in sediments	3.3×10^{23}	0.3 μg/kg	9.9×10^{13}
Rivers			
Dissolved	3.4×10^{19}	60 ng/kg	2.0×10^{9}
Suspended load	1.5×10^{16}	0.8	2.4×10^{12}
Shallow groundwater	4×10^{8}	0.2 μg/kg	8.0×10^{8}
Polar ice	2×10^{22}	20 ng/kg	4.0×10^{11}
Selenium reserves	—	—	112×10^{9}
Selenium resources	—	—	291×10^{9}

[a] Compiled from literature data.

Levels of selenium in the atmosphere display a large spatial variability (Chapter 12), so that the atmosphere cannot realistically be regarded as a single reservoir. In this report, the atmosphere has been divided into five subcompartments with average selenium concentrations shown in Table 2. The ratios of selenium levels in air at urban areas to the levels at remote continental and remote marine locations of about 10 and 20, respectively, are comparable to those of many other toxic metals and metalloids.[13] Although a large fraction of the selenium is believed to be in volatile form (Table 2), little is currently known about the distribution and fate of such compounds in the environmental (Chapter 12).

The concern for the effects of selenium deficiency on the health and growth of farm animals has engendered a large volume of literature on the occurrence, chemistry, and bioavailability of this element in soils. The average selenium abundance in soils is about 0.4 mg/kg, but the observed concentrations vary from 0.005 mg/kg in some selenium-impoverished regions of Finland to over 8000 mg/kg in the Tuva area of the U.S.S.R. (Chapter 9).

Accurate data on the concentration of selenium in continental waters are very sparse. The mean concentrations of 0.2 μg/l for groundwater and 60 ng/l for rivers are tenuous, and the distribution between the dissolved and particulate phases is even less certain. The low selenium concentrations and the small sizes of the selenium burdens imply that fresh waters are expected to be very sensitive to selenium pollution. Lack of research on the fate and effects of pollutant selenium in fresh waters should be considered a major oversight in the field of environmental studies.

Table 2
REGIONALIZED AVERAGE DISTRIBUTION OF SELENIUM IN THE GLOBAL ATMOSPHERE

Region	Area[a] (m^2)	Selenium concentration: Particulate[b] (ng/m^3)	Selenium concentration: Total (ng/m^3)	Volatile (% of total)[c]	Selenium burden (10^8 g)[d]
Urban	1.0×10^{12}	3.7	4.9	25	0.49
Rural	10×10^{12}	1.2	1.6	25	1.6
Remote (clean) continental	133×10^{12}	0.3	0.43	30	5.6
Oceans, productive	136×10^{12}	0.21	0.35	40	4.8
Oceans, nonproductive	231×10^{12}	0.15	0.25	40	5.8

[a] Data from Andreae and Raemdonck[21] and Ryaboshapko.[33]
[b] Based on the compilations by Ross;[6] Mosher and Duce (Chapter 12).
[c] These estimates presume large outflux of volatile organoselenium compounds from the oceans as well as from soils and wetlands at remote regions.
[d] Assuming uniform distribution of selenium up to 10 km elevation.

The most recent data on the vertical profiles of selenium in the oceans are closely similar to those of the nutrients, hence, suggesting that selenium is involved in biological cycling (Chapter 11). Reported concentrations can be represented by simple multivariate functions of phosphate and silicate concentrations (Table 3). The mean concentrations of dissolved selenium in surface and deep waters of 30 and 95 ng/kg, respectively, are based on several reports on selenium, phosphate, and silica distribution in the oceans. The mean selenium content of oceanic particulates are based on the data of Fowler,[14] Buat-Ménard and Chesselet,[15] and Cutter and Bruland.[16] It should be noted that organic selenium usually is the principle dissolved species in the surface (250 m) layers of the ocean where its concentration can reach three times that of the inorganic components. With increasing depth, however, inorganic Se(IV) and Se(VI) become the dominant species, suggesting that selenium is regenerated by the oxidation of organic matter in the deeper waters.[16,17]

The average concentrations of selenium in the biota listed in Table 1 are based on several literature sources. The quality of the available data bases are generally poor, and the average abundances given should be used in the circumspect manner.

III. SOURCES OF SELENIUM IN THE ENVIRONMENT

A. Natural Sources

The atmosphere is a critical medium for the long-range dispersion of selenium released from both natural and anthropogenic sources. Table 4 shows the worldwide contributions of the various natural sources to the atmospheric selenium burden. It is clear that, on a global basis, forest fires and sea salt sprays represent only minor sources of the airbone element.

From the total annual dust emission of 330 million t and the average selenium concentration in soils of 0.4 mg/kg, a selenium emission of 132 t/year is calculated. The other value (i.e., 99 t/year) given in the table is derived by assuming that only 15% of the resuspended soil particles is fine grained ($\leq$5 μm or less),[18] and that there is a fivefold enrichment of selenium in the very fine aerolsols.

There is little quantitative information on biogenic selenium emissions from terrestrial sources.[6] The release of selenium from lands (J_{Se}) was therefore estimated using the empirical relation:

$$J_{Se} = 0.1 \times J_S \times (C_{Se}/C_S) \tag{1}$$

Table 3
LINKAGES BETWEEN THE SELENIUM AND NUTRIENT CYCLES IN THE MARINE ENVIRONMENT[a]

Pacific Ocean

GEOSECS I — off San Diego[34]

Se(IV) = −21.5 + 60.7 PO_4 + 3.98 Si

Se(VI) = 554.3 + 113.7 PO_4 + 3.31 Si

Σ Se = 532.8 + 174.4 PO_4 + 7.28 Si

Atlantic Ocean

Cruise 79 — eastern Atlantic around 20°—25° N[35]

Se(IV) = −2.2 + 59.5 PO_4 + 7.08 Si

Se(VI) = 352 + 286 PO_4 + 3.92 Si

Σ Se = 354 + 363 PO_4 + 9.87 Si

Cruise 88 — eastern Atlantic around 35° N[35]

Se(IV) = 38.5 + 13.4 Si

Se(VI) = 547 + 301 PO_4

Σ Se = 546 + 429 PO_4 + 7.08 Si

Atlantic, Pacific, and Indian Oceans Combined[36]

Se(IV) = 63 + 50 PO_4 + 4.21 Si	r^2 = 0.94
Se(VI) = 448 + 181 PO_4 + 3.32 Si	r^2 = 0.83

Gulf of Mexico[17]

Se(IV) = 107.9 + 101.1 PO_4 + 4.24 Si	r^2 = 0.96
Se(VI) = 192.8 − 192.0 PO_4 + 21.2 Si	r^2 = 0.89
Σ Se = 300.7 − 90.9 PO_4 + 25.5 Si	r^2 = 0.96

Note: All selenium species are in picomoles per kilogram except in the Atlantic, where they are in picomoles per liter; phosphate and silicate are in micromoles per liter.

[a] Σ Se = Se(IV) + Se(VI).

From Takayanagi, K. and Wong, G. T. F., *Geochim. Cosmochim. Acta,* 49, 539, 1985. With permission.

where C is the mean crustal abundance of sulfur or selenium and J_S is the global biogenic emission of sulfur assigned the value of 64 million t/year.[19] The factor of 0.1 was applied to account for the lower diffusivity of organoselenium compounds in soils. The present estimate of 1230 t/year lies between the 560[6] and 3000 t/year[9] that have been reported in the literature. Ross[6] also estimated the annual emission from leaf-mediated selenium volatilization to be 4200 t. In the present report, leaf exudates are believed to be a minor source of selenium aerosols, accounting for only about 11 t/year.

Biogenic emission from the oceans was estimated assuming that only 0.08 to 0.5% of the organoselenium compounds, with average concentration in mixed ocean layer of about 80 ng/l,[16,17] occurs in volatile forms. Furthermore, the exchange (pipe) velocity of the volatile selenium compounds was assigned a value of 0.5 m/d, which is about $^1/_{10}$ that of dimethylsulfide.[20] According to these data, the mean biogenic emission rate is estimated to be 15 to 73 μg m^{-2} $year^{-1}$. Using a total surface area of the ocean of 367 × 10^6 km^2,[21] the annual global biogenic flux is 5510 to 26,800 t. The ocean is clearly implicated as a major

Table 4
WORLDWIDE EMISSIONS OF SELENIUM FROM NATURAL SOURCES

Source category	Global production (10^6 t/year)	Emission factor	Global flux (t/year)
Dust resuspension	330[a]	0.4 mg/kg[a]	132
	49.5[b]	2 mg/kg[c]	99
Volcanoes and hot springs	40	8 mg/kg	320
Forest wild fires	250	0.15 mg/kg	38
Sea salt spray	500	1.5 μg/kg[d]	0.8
Vegetative exudates	75	0.15 mg/kg	11
Biogenic emissions			
Land (soils, marshes, & wetlands)	See text	See text	1230
Ocean	See text	See text	6700 (5510—26,800)
Total			8399

[a] Emission factor assumed to be the same as the average selenium content of soils.
[b] Assumes that 15% of the resuspended soil particles are less than 5 μm in diameter (see Walsh et al.[18]).
[c] Assumes a fivefold enrichment of selenium in the finer grained airborne particulates compared to the average abundance in soils.
[d] From mean selenium concentration in mixed ocean layer (30 ng/kg) and a 50-fold enrichment of selenium in the sea spray compared to ocean water.

natural source of selenium in the atmosphere. In order to obtain a reasonable balance in the global cycle, the mean concentration of volatile selenium compounds in seawater has been assigned a value of 0.1 ng/kg, which corresponds to a global biogenic flux of 6700 t/year. This mean value is higher than the rates (4800 to 5200 t/year) reported by Lantzy and Mackenzie[9] and Ross.[6]

The total selenium emission from natural sources is thus estimated to be about 8400 t/year (Table 4). Nearly 80% of total natural emission can be attributed to biogenic processes in the ocean. Land-based biogenic processes account for about 15% of the global flux so that about 95% of the selenium released from natural sources is in volatile form. Clearly, the distribution and fate of vapor-phase selenium compounds in the atmosphere remains a critical gap in our understanding of the natural biogeochemical cycle of this element. Several lines of evidence which suggest that vapor-phase compounds represent a large fraction of selenium in air are discussed by Mosher and Duce (Chapter 12). Ross[6] in particular drew attention to the very high enrichment (up to 23,000-fold) of selenium in airborne particulates at remote continental and marine locations and attributed the unusual enrichment to the high flux of selenium of biogenic origin.

B. Anthropogenic Sources

It is estimated that about 3781 t of particulate selenium and 2521 t of gaseous selenium are currently (1984 base year) being released to the atmosphere as a result of various industrial activities (Table 5). About 30% of the total particulate emission comes from the mining and smelting of base metals, with the copper/nickel industry being the leading selenium emitter. The combustion of fossil fuels, especially hard coals, accounts for about 36% of the pollutant selenium released. Slash and agricultural burning contribute about 23% of the pollutant selenium aerosols, with much smaller quantities derived from wood fuel, refuse incineration, and fertilizer production (Table 5).

Mackenzie et al.[8] estimated the annual global emission of selenium from industrial operations and the burning of fossil fuels to be 12,000 t. They made an unrealistic assumption that about 90% of the selenium content of fossil fuels was released to the atmosphere and derived the global selenium flux from this source alone to be about 10,000 t/year. More

Table 5
WORLDWIDE ANTHROPOGENIC EMISSIONS OF PARTICULATE SELENIUM DURING 1984

Source category	Global production/ consumption (10^6 t/year)	Emission factor[a] (g/t)	Selenium emission (t)
Copper/nickel production	8.8	120	1056
Lead smelting	3.6	25	90
Zinc smelting	1.5	20	30
Selenium extraction and use	2550 t	0.5%	13
Fossil fuel combustion			
Hard coal	3450	0.28	966
Lignite	1250	0.2	250
Fuel oil	3500	0.05	175
Refuse incineration	150	0.32	48
Wood fuel	850	0.15	128
Slash burning	3800	0.15	570
Burning, pasture lands	2000	0.15	300
Iron and steel production	650	0.04	26
Phosphate fertilizers	143	0.8	114
Fugitive dust sources	5	3	15
Total (particulates)			3781
Volatile selenium emissions (40% of total emissions)			2521
Total			6302

[a] Compiled from the literature; see in particular Ross[6] and Pacyna.[37]

recently, Ross[6] estimated the total anthropogenic emission of selenium for the 1973 base year to be 7200 to 8300 t, made up of 2300 t from metal smelting and industrial processes and about 4900 to 6000 t from fossil fuel combustion. The estimates of Ross are in reasonable agreement with the data in Table 5.

It follows from the data in Tables 4 and 5 that the global flux of selenium to the atmosphere is about 14,700 t/year and that about 43% of the total emission comes from anthropogenic sources. By contrast, the worldwide mine production of selenium is about 2000 t/year. The low ratio of selenium production to emission reflects the fact that selenium is an energy-related pollutant and also underscores its by-product relationships to the exploitation of base metal sulfides. Attempts to reduce the release of selenium to the environment must, therefore, pay special attention to these particular industrial sectors.

IV. SELENIUM FLUXES AMONG GLOBAL COMPARTMENTS

The following discussions are based on a global model which assumed that all the fluxes are donor controlled and obey first-order kinetics. Thus, F_{ij}, the flux from the i-th to the j-th compartment is related to the inventory Q_i in the i-th compartment by

$$F_{ij} = Q_i \cdot k_{ij} \qquad (2)$$

where k_{ij} is the first-order rate constant for the fractional transfer from one compartment to the other. The reciprocal of the fractional transfer rate ($1/k_{ij}$) is the partial mean residence time in the i-th compartment due to the particular transport pathway. The mean residence time, t_{ij}, in the i-th compartment is obtained by summing the inverse of all the coupled

Table 6
FLUXES ALONG THE MAJOR PATHWAYS OF THE GLOBAL SELENIUM CYCLE

Pathway	Material flux (g/year)	Selenium concentration (mg/kg)	Annual selenium flux (t)
Atmosphere — Inputs			
Natural sources	See Table 4	See Table 4	8,400
Anthropogenic sources	See Table 5	See Table 5	6,300
Atmosphere — Outputs			
Deposition on land	See text	See text	6,600
Deposition into oceans	See text	See text	8,700
Oceans — Inputs			
Atmospheric fallout	See text	See text	8,700
Waste discharges and storm runoff	1.7×10^{18}	0.8 μg/kg	1,400
Rivers, dissolved	3.4×10^{19}	60 ng/kg	2,000
Rivers, suspended load	1.5×10^{16}	0.8	12,000
Hydrothermal emanations	—	—	—
Oceans — Outputs			
Sea salt spray	See Table 4	See Table 4	8
Sedimentation	See text	See text	500
Soils — Inputs			
Atmospheric fallout	See text	See text	6,600
Fertilizer and soil additives	9.4×10^{13}	125	12,000
Waste disposal	2×10^{12}	3.4	7
Litter fall	6.8×10^{16}	0.08	5,400
Soils — Outputs			
Windblown dusts	See Table 13.4	See Table 13.4	132
Erosion (river transport)	—	—	14,000
Biological assimilation			
Oceans, primary production	6×10^{16}	0.27	16,000
Oceans, zooplankton	2×10^{16}	0.15	3,000
Continents, primary production	1.2×10^{17}	0.05	6,000
Continents, animal	7.8×10^{15}	0.15	1,200
Industrial production and use	—	—	2,550

transfer rates; i.e., $t_{ij} = {}_{j}^{c}(1/k_{ij})$. More detailed descriptions of the mathematical formulations applicable to the global cycles of the elements are given by Mackenzie et al.,[8] Kocher,[22] and Bennett.[23]

Transfer rates of selenium along the key pathways between the seven global reservoirs are shown in Table 6, and the compartmental inventories and fluxes are shown schematically in Figure 2. As to be expected, most of the flux determinants pertain to atmospheric and water transports, the principal vehicles in the global redistribution of pollutants.

A. The Atmosphere

The regionalized average distribution of selenium in the different atmospheric subcompartments is shown in Table 2. Global depositions of selenium from the atmosphere to land and oceans were calculated from the following mean data:

Region	Area (10^{12} m^2)	Average Se flux (ng cm^{-2} $year^{-1}$)	Total Se deposition (t)
Urban	1.0	60	600
Rural	10	20	2000
Continental, clean	133	3.0	3990
Ocean, productive	136	3.0	4080
Ocean, "sterile"	231	2.0	4620

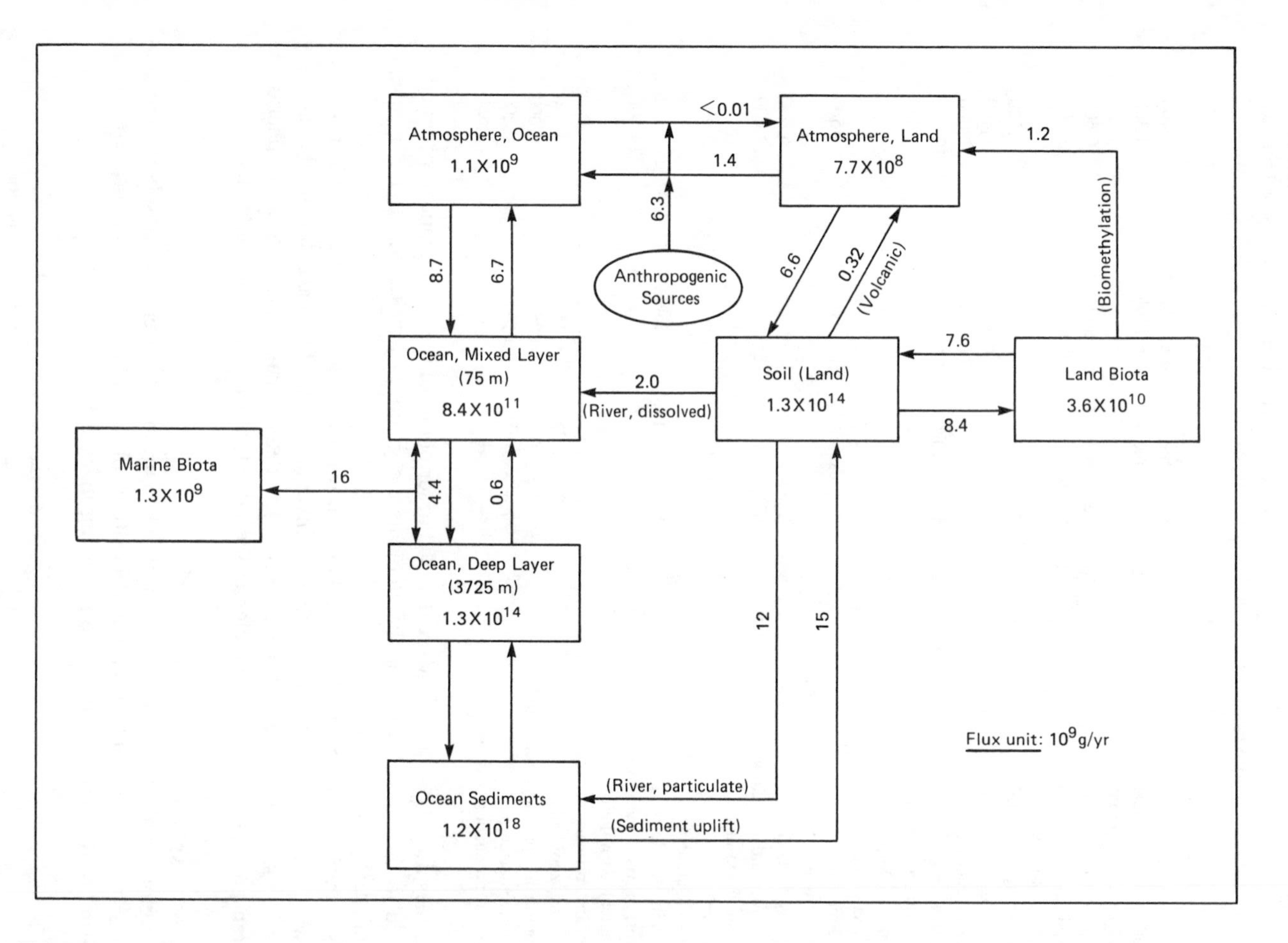

FIGURE 2. Global biogeochemical cycle of selenium showing the compartmental inventories and steady-state fluxes. Units for fluxes are 10^9 g/year.

The calculated total deposition rate of 15,290 t/year is in good agreement with the annual total emissions of 14,700 t (Tables 4 and 5). The export from land to the oceans of about 1410 t/year was obtained as the difference between continental emissions (8000 t from natural and anthropogenic sources) and the estimated fallout on land (6590 t). There is an imbalance of 590 t between the deposition rate into the ocean (8110 t/year) and the sum of marine emission and terrestrial import (8700 t/year). This difference can be attributed to the recycling of biogenic selenium in the marine atmosphere; it amounts to about 7% of the selenium deposition and is less than the selenium recycling rate of 17% reported by Arimoto et al.[24]

The mean residence time for selenium in the atmosphere, as defined above, is

$$Q/F = (1829\ t)/(14{,}700\ t/year) = 0.124\ year = 45\ d$$

where Q is the total atmospheric selenium pool (Table 2) and F is the total annual flux selenium to the atmosphere. This long residence time supports the suggestion that Se is emitted mostly in volatile form. By comparison, mean residence times for pollutant metal particulates typically fall between 3 and 15 d.[25]

B. The Hydrosphere

The dissolved selenium reservoir in fresh waters is small compared to the other compartmental inventories (Table 1) and, therefore, is the one most liable to be drastically impacted by anthropogenic selenium inputs. The quality of the available data base on the distribution of selenium in lakes and rivers is too poor and fragmented to permit a realistic evaluation of the pollution component in these ecosystems. Nevertheless, the analyses of dated lake sediment layers clearly show that the present-day input of selenium into lakes in Europe and North America exceeds the rate in prehistoric times by several fold.[26,27] A discussion of the freshwater component of the selenium cycle is given by Cutter (Chapter 10). It should be noted that only 15% of the worldwide selenium load of rivers is in dissolved form (Table 6).

In the present model, the ocean water is divided into two subcompartments, namely, a surface mixed to a depth of 75 m[16] and deep waters from 75 to 3700 m. There is also the complementary biological subcompartment. The annual exchange of selenium between the subcompartments is depicted in Figure 2. The fluxes indicated correspond approximately to the following average production/fallout rates in micrograms per square meter per year:

Downward particulate selenium flux from mixed layer	15
Atmospheric fallout	24
Dissolved river input	5.4
Biogenic release to the atmosphere	15—73
Upwelling flux from deep waters	1.6
Particulate selenium deposition from deep waters	14
Biological assimilation	52

The transfer of particulate selenium from the mixed layer to the deep ocean is based on particulate selenium concentration of 3 mg/kg; use of the higher concentrations reported in the literature would have resulted in a significant mass imbalance.

The selenium flux to ocean sediments was also estimated from an annual accumulation of 3.2×10^{15} g,[8] and the average selenium content of clays of 0.17 mg/kg.[28] This method yields the flux to sediments of 5300 t/year, in good agreement with the estimate based on ocean-wide particulate selenium deposition rate of 14 $\mu g\ m^{-2}\ year^{-1}$. As expected, most of the 12,000 t of the bedload of selenium delivered by the rivers is deposited in the coastal and nearshore marine zones.

The total input of selenium into the mixed (75 m) ocean layer from atmospheric fallout, dissolved river load, and upwelling from deep ocean is about 12,000 t. The fact that the rate of biological assimilation (19,000 t/year) exceeds this total input suggests that the turnover of selenium particularly in the productive zones must be very rapid. From the rate of input and the selenium burden (840,000 t), the mean residence time for selenium in the mixed layer is estimated to be about 70 years. This is short compared to, for example, the mean life of selenite in the deep ocean of about 1100 years.[16] The biogeochemistry of selenium in the marine environment is described in detail by Siu and Berman (Chapter 11).

V. HUMAN INFLUENCE ON THE SELENIUM CYCLE

There is no doubt that human activities are interfering with some segments of the global selenium cycle. Selenium is an energy-related pollutant and has been used as a probe in identifying atmospheric emissions associated with fossil fuel combustion.[29] The increase in the quantity of selenium released to the environment must, therefore, parallel the sharp rise in rate of energy consumption since the turn of this century.[30] Selenium also occurs together with sulfur in heavy metal ores whence the two elements are released together to the environment. Ecosystems around base metal smelters and power plants are thus most susceptible to selenium contamination. On a worldwide basis, about 40% of the selenium released annually to the atmosphere comes from anthropogenic sources. This particular gain of selenium by the biosphere is at the expense of the lithospheric selenium reservoir.

The intensity of selenium emission in the urban areas is about 20-fold higher than that at remote continental regions (see above). This high *human interference factor* for the atmospheric compartment is evidence that urban areas and the subadjacent areas are being heavily contaminated with selenium. Most of the pollutant selenium is released either in volatile form or is bound to fine-grained (≤ 5 μm diameter) particles,[31] and can be transported to far-removed ecosystems. In many selenium-starved ecosystems the atmosphere has thus become the major source of selenium, an essential nutrient.

Anthropogenic inputs should also dominate the selenium cycle in freshwater ecosystems. All it would take to double the small selenium reservoir of fresh waters is the addition of just 10% of the anthropogenic selenium released to the atmosphere. This quantity most likely is less than the selenium additions to fresh waters from waste effluents and other industrial discharges; the available data are inadequate for quantifying the extent of selenium pollution due to waste waters and urban runoff. Increased selenium loading has been associated with increased nuisance algal growth and general degredation of water quality.[32] The need for proper assessment of the impacts of selenium contamination of fresh waters (a critical resource and life-support system for mankind) cannot be overemphasized.

Soils are currently gaining about 5200 t/year of selenium from atmospheric fallout, plus an unknown but potentially large amount associated with industrial and domestic waste disposal on land. Since soils are generally selenium deficient, the gain of selenium is unlikely to engender undesirable environmental changes except on a local scale.

The ocean also gains about 1400 t of selenium per year from man-made emissions to the atmosphere. This added flux represents only about 0.2% of the selenium burden in the mixed layer. In the short term, the effect of such a gain on the marine chemistry of selenium is likely to be insignificant. At the same rate of gain, however, the mixed layer burden will be increased by well over 20% in 100 years or so, and long-term consequences can become quite important.

REFERENCES

1. **Shamberger, R. J.,** *Biochemistry of Selenium,* Plenum Press, New York, 1983.
2. **Moxon, A. L., Olson, O. E., and Searight, W. V.,** Selenium in rocks, soils and plants, *S.D. Agric. Exp. Stn. Tech. Bull.,* 2, 1939.
3. **Lakin, H. W. and Davidson, D. F.,** The relation of the geochemistry of selenium to its occurrence in soils, in *Selenium in Biomedicine,* Muth, O. H., Oldfield, J. E., and Weswig, P. H., Eds., AVI Publishing, Westport, CT, 1967, 27.
4. **Allaway, W. H., Cary, E. E., and Ehlig, C. F.,** The cycling of low levels of selenium in soils, plants and animals, in *Selenium in Biomedicine,* Muth, O. H., Oldfield, J. E., and Weswig, P. H., Eds., AVI Publishing, Westport, CT, 1967, chap 17.
5. **National Academy of Sciences,** *Selenium,* National Academy of Sciences, Washington, D.C., 1976.
6. **Ross, H. B.,** An atmospheric selenium budget for the region 30°N to 90°N, *Tellus,* 37B, 78, 1985.
7. **Pacyna, J. M., Semb, A., and Hanssen, J. E.,** Emission and long-range transport of trace elements in Europe, *Tellus,* 36B, 163, 1984.
8. **Mackenzie, F. T., Lantzy, R. J., and Paterson, V.,** Global trace metal cycles and predictions, *J. Int. Assoc. Math. Geol.,* 11, 99, 1979.
9. **Lantzy, R. J. and Mackenzie, F. T.,** Atmospheric trace metals: global cycles and assessment of man's impact, *Geochim. Cosmochim. Acta,* 43, 511, 1979.
10. **Mason, B. and Moore, C. B.,** *Principles of Geochemistry,* 4th ed., John Wiley & Sons, New York, 1982.
11. **Elkin, E. M.,** Selenium and selenium compounds, in, *Encyclopedia of Chemical Technology,* Vol. 20, John Wiley & Sons, New York, 1982, 576.
12. **McCutcheon, W. J.,** Selenium and Tellurium, in Canadian Minerals Yearbook, 1983—1984, Mineral Rep. 33, Energy, Mines and Resources Canada, Ottawa, 1984, 51.1.
13. **Nriagu, J. O. and Davidson, C. I., Eds.,** *Toxic Metals in the Atmosphere,* John Wiley & Sons, New York, 1986.
14. **Fowler, S. W.,** Trace elements in zooplankton particulate products, *Nature (London),* 269, 51, 1977.
15. **Buat-Ménard, P. and Chesselet, R.,** Variable influence of the atmospheric flux on the trace metal chemistry of oceanic suspended matter, *Earth Planet. Sci. Lett.,* 42, 399, 1979.
16. **Cutter, G. A. and Bruland, K. W.,** The marine biogeochemistry of selenium: a re-evaluation, *Limonl. Oceanogr.,* 29, 1179, 1984.
17. **Takayanagi, K. and Wong, G. T. F.,** Dissolved inorganic and organic selenium in the Orca Basin, *Geochim. Cosmochim. Acta,* 49, 539, 1985.
18. **Walsh, P. R., Duce, R. A., and Fasching, J. L.,** Considerations of the enrichment, sources and flux of arsenic in the troposphere, *J. Geophys. Res.,* 84, 1719, 1979.
19. **Adams, D. F., Farwell, S. O., Robinson, E., Pack, M. R., and Bamesberger, W. L.,** Biogenic sulfur source strengths, *Environ. Sci. Technol.,* 15, 1493, 1981.
20. **Liss, P. S. and Slater, P. G.,** Flux of gases across the air-sea interface, *Nature (London),* 247, 181, 1974.
21. **Andreae, M. O. and Raemdonck, H.,** Dimethyl sulfide in the surface ocean and the marine atmosphere: a global view, *Science,* 221, 744, 1983.
22. **Kocher, D. C.,** A dynamic model of the global iodine cycle and estimation of dose to the world population from releases of iodine-129 to the environment, *Environ. Int.,* 5, 15, 1981.
23. **Bennett, B. G.,** *Exposure Commitment Assessments of Environmental Pollutants,* Vol. 2, MARC Rep. No. 28, Monitoring and Assessment Research Centre, Chelsea College, University of London, London, 1982.
24. **Arimoto, R., Duce, R. A., Ray, B. J., and Unni, C. K.,** Atmosphere trace elements at Enewetak Atoll. II. Transport to the ocean by wet and dry deposition, *J. Geophys. Res. D,* 90(D1), 2391, 1985.
25. **Muller, J.,** Residence time and deposition of particle-bound atmospheric substances. in, *Deposition of Atmospheric Pollutants,* Georgii, H. W. and Pankrath, J., Eds., D., Reidel, Dordrecht, Netherlands, 1982, 43.
26. MARC, *Historical Monitoring,* Rep. No. 31, Monitoring and Assessment Research Centre, Chelsea College, University of London, London, 1985.
27. **Nriagu, J. O. and Wong, H. K.,** Selenium pollution of lakes near the smelters at Sudbury, Ontario, *Nature (London),* 301, 55, 1983.
28. **Bowen, H. J. M.,** *Environmental Chemistry of the Elements,* Academic Press, London, 1979.
29. **Rahn, K. A. and Lowenthal, D. H.,** Pollution aerosols in the northeast: northeastern-midwestern contributions, *Science,* 228, 275, 1985.
30. **Nriagu, J. O.,** Global inventory of natural and anthropogenic emissions of trace metals to the atmosphere, *Nature (London),* 279, 409, 1979.
31. **Chiou, K. Y. and Manuel, O. K.,** Tellurium and selenium in aerosols, *Environ. Sci. Technol.,* 20, 987, 1986.

32. **Wehr, J. D. and Brown, L. M.,** Selenium requirement of a bloom-forming planktonic alga from softwater and acidified lakes, *Can. J. Fish. Aquat. Sci.*, 42, 1783, 1985.
33. **Ryaboshapko, A. G.,** The atmospheric sulfur cycle, in, *The Global Biogeochemical Sulfur Cycle*, SCOPE 19, Scientific Committee on Problems of the Environment, Ivanov, M. V. and Freney, J. R., Eds., John Wiley & Sons, New York, 1983, 203.
34. **Measures, C. I., McDuff, R. E., and Edmond, J. M.,** Selenium redox chemistry at GEOSECS I re-occupation, *Earth Planet. Sci. Lett.*, 49, 102, 1980.
35. **Measures, C. I. and Burton, J. D.,** The vertical distribution and oxidation states of dissolved selenium in the northeast Atlantic Ocean and their relationship to biological processes, *Earth Planet. Sci. Lett.*, 46, 385, 1980.
36. **Measures, C. I., Grant, B. C., Mangum, B. J., and Edmond, J. M.,** The relationship of the distribution of dissolved selenium IV and VI in three oceans to physical and biological processes, in, *Trace Metals in Sea Water*, Wong, C. S., Boyle, E., Bruland, K. W., Burton, J. D., and Goldberg, E. D., Eds., Plenum Press, New York, 1983, 73.
37. **Pacyna, J. M.,** Atmospheric trace elements from natural and anthropogenic sources, in *Toxic Metals in the Atmosphere*, Nriagu, J. O. and Davidson, C. I., Eds., John Wiley & Sons, New York, 1986, 33.

APPENDIX

LIST OF UNITS, SYMBOLS AND ABBREVIATIONS

Units

cm	centimeter (10^{-2} meter)
°C	temperature in degrees celcius
d	day(s)
g	gram
kcal	kilocalorie (10^3 calorie)
kg	kilogram (10^3 g)
km	kilometer (10^3 meter)
l	liter
m	meter
mg	milligram (10^{-3} g)
min.	minute
mm	millimeter (10^{-3} meter)
ng	nanogram (10^{-9} gram)
nm	nanometer (10^{-9} meter)
nmol	nanomole (10^{-9} mole)
pmol	picomole (10^{-12} mole)
s	second
t	metric tonne (10^3 kilogram)
μm	micrometer (10^{-6} meter)
μmol	micromole (10^{-6} mole)
μg	microgram (10^{-6} gram)
μl	microlitre (10^{-6} litre)

Symbols and Abbreviations

A	ash weight basis
AAS	atomic absorption spectrometry
ADDT	acid digestion-distillation-titrimetry
ADF	acid decomposition-fluorometry
ADPLAS	acid digestion-precipitation-light absorption spectrometry
AEE	anomalously enriched elements
Agric.	Agricultural
AK	Alaska
anal.	analyzed
AOAC	Association of Official Analytical Chemists
AR	Arkansas
av.	average
AZ	Arizona
br.	brown
blk.	black
B.C.; BC	before Christ; British Columbia
BIMS	bubble interfacial microlayer sampler
BP	before present

BRM	biological reference material
ca.	circa (about)
calc.	calculated
China	People's Republic of China
chrom./GC	chromatographic column trapping/gas chromatographic detection
C	carbon; creatinine; mean crustal abundance
CA	California
CEB	chemical element balance
cf.	Compare also
ckd.	cooked
cnd.	canned
Co.	county
CO	Colorado
conc.	concentration
contg.	containing
comml.	commercial
cryo./AAS	cryogenic trapping/atomic absorption detection
CT	Connecticut
Ctr.	Center
D	dry weight basis
D.C.	District of Columbia
decaf.	decaffeinated
del.	delivery
digestiv.	digestive gland or tissue
DMS	dimethyl sulfide
E	east
EAAS	electrothermal atomization-atomic absorption spectrometry
ECD	electron capture detector (detection)
EF	enrichment factor
e.g.	(*exempli gratia*) for example
enr.	enriched
et al.	*et alii* (and others)
exp.	experiment
F	fluorometry; flux
FABMS	fast atom bombardment mass spectrometry
Fed. Rep. Germany	Federal Republic of Germany
F_{ij}	flux from the i-th to the j-th compartment
FL	Florida
frzn.	frozen
FS	fluorescence spectrometry following derivatization
GA	Georgia
GC	gas chromatography
German Dem. Rep.	German Democratic Republic
GCMS	gas chromatography-mass spectrometry
hmd.	homemade
HAAS	hydride generation-atomic absorption spectrometry
Hb	hemoglobin

HDL	high-density lipoprotein
HI	Hawaii
HPLC	high-pressure (performance) liquid chromatography (chromatographic)
i.e.	(*id est*) that is
incl.	including
IA	Iowa
ICPAES	inductively coupled plasma atomic emission spectrometry
ID	Idaho
IL	Illinois
IN	Indiana
INAA	instrumental neutron activation analysis
J	rate of release of element (t/year)
K_{ij}	first-order rate constant for material transfer from i-th to j-th compartments
KY	Kentucky
LA	Louisiana
LDL	low-density lipoprotein
M	molar
MA	Massachusetts
mac.	macroni
Man.	Manitoba
mat.	matter
MD	Maryland
MI	Michigan
MMD	median mass-median diameter
MO	Missouri
MS	Mississippi; mass spectrometry
MT	Montana
N	north; number of samples analyzed
n	number of samples analyzed
NAA	neutron activation analysis
NB	New Brunswick
NBS	National Bureau of Standards (U.S.)
NC	North Carolina
ND	North Dakota
NE	northeast; Nebraska
NFCS	Nationwide Food Consumption Survey (U.S.)
NM	New Mexico
NMR	nuclear magnetic resonance
NR	not reported
NS	Nova Scotia; not specified
NSW	New South Wales (Australia)
NTIS	National Technical Information Service (U.S.)
NV	Nevada
NW	northwest
NY	New York
NZ	New Zealand
Ont.	Ontario
Org.	Organic

OH	Ohio
OR	Oregon
OSHA	Occupational Safety and Health Administration (U.S.)
PA	Pennsylvania
PEI	Prince Edward Island
Peop. Rep. China	People's Republic of China
PIXE	proton-induced X-ray emission
ppb	part per billion (1 in 10^9)
ppm	part per million (1 in 10^6)
P.R. China	People's Republic of China
proc.	processed
Q	inventory (quantity, pool) of component in compartment
quad.	quadruplicate
Que.	Quebec
r	correlation coefficient
Ref.	Reference
reg.	regular
refrig.	refrigerated
Res.	Research
RI	Rhode Island
RNAA	neutron activation analysis with radiochemical separation
rough.	roughage
RSD	relative standard deviation
RTE	ready-to-eat
S	south
SADS	Asian Dust Study network
Sask.	Saskatchewan
SCM	standard cubic meter
SD	South Dakota; standard deviation
SE	southeast
SEAREX	Sea/Air Exchange (Program)
spag.	spaghetti
SSMS	spark source mass spectrometry
SRM	standard reference material
Stn	Station
STP	standard temperature and pressure (0°C and 1 atm)
SW	southwest
TIMS	Thermal ionization mass spectrometry
TLC	thin layer chromatography (chromatographic)
TN	Tennessee
Tr.	trace
TX	Texas
U.K.	United Kingdom
unckd.	uncooked
unpub.	unpublished
U.S.	United States of America
U.S.S.R.	Union of Soviet Socialist Republics
UT	Utah
UV	ultraviolet

VA	Virginia
VLDL	very low-density lipoprotein
w	with
W	west; western; wet weight basis
WA	Washington
W. Germany	Federal Republic of Germany (West Germany)
w/o	without
WV	West Virginia
WY	Wyoming
XRF	X-ray fluorescence (emission) spectrometry
yel.	yellow
λ_{max}	wavelength for maximum absorption
ξ	extinction coefficient
Σ	sum
$\sim$	about

INDEX

A

B

C

D

E

F

G

H

I

K

L

M

N

O

P

Q

R

S

T

U

V

W

X

Y

Z